Self-Assembled Quantum Dots

LECTURE NOTES IN NANOSCALE SCIENCE AND TECHNOLOGY

Volumes Published in this Series:
Volume 1: Self-Assembled Quantum Dots, Wang, Z.M., 2008

Volume 2: Nanoscale Phenomena: Basic Science to Device Applications, Tang, Z., and Sheng, P., 2008

Forthcoming Titles:

Volume 3: One-Dimensional Nanostructures, Wang, Z.M., 2008

Volume 4: Epitaxial Semiconductor Nanostructures, Wang, Z.M., and Salamo, G., 2008

Volume 5: B-C-N Nanotubes and Related Nanostructures, Yap, Y.K., 2008

Volume 6: Towards Functional Nanomaterials, Wang, Z.M., 2008

Self-Assembled Quantum Dots

Zhiming M. Wang
Editor

University of Arkansas,
Fayetteville, AR, USA

Series Editors:

Zhiming M. Wang
Department of Physics
University of Arkansas
Fayetteville, AR
USA

Andreas Waag
Institut für Halbleitertechnik
TU Braunsweig
Braunschweig
Germany

Gregory Salamo
Department of Physics
University of Arkansas
Fayetteville, AR
USA

Naoki Kishimoto
Quantum Beam Center
National Institute for Materials Science
Tsukuba, Ibaraki
Japan

ISBN-13: 978-0-387-74190-1 e-ISBN-13: 978-0-387-74191-8

Library of Congress Control Number: 2007939826

Printed on acid-free paper.

9 8 7 6 5 4 3 2 1

springer.com

Preface

Self-Assembled Quantum Dots, commonly referred to as self-organized quantum dots, form spontaneously under certain growth conditions during molecular-beam epitaxy or metal organic chemical vapor deposition, as a consequence of lattice-mismatch between the deposited material (generally semiconductors) and underlying substrate. The resulting semiconductor nanostructures consist of three-dimensional islands standing on a two-dimensional wetting layer. Such islands can be subsequently buried to realize quantum confinement. In the past 15 years, self-assembled quantum dots have provided vast opportunities for physical research and technological applications, including quantum cryptography, quantum computing, optics and optoelectronics. The present book is devoted to some of these fascinating aspects, including growth, properties and applications of quantum dots, both in theory and experiment.

The main body of the chapter comprises of contributions that focuses on InGaAs quantum dots, since this material has been intensively investigated as a model system. Specifically, chapters 1–3 offer a comprehensive perspective from a growth point of view. Chapter 1 reviews the recent advances on understanding the basic microscopic mechanisms driving the nucleation and evolution of quantum dots under the role of composition, coverage, and intermixing. On the other hand, the authors of Chapter 2 and Chapter 3 summarize their recent efforts to address two challenging issues of the quantum dot growth; these include random distribution on surface and non-uniformity in size and shape. Chapter 2 demonstrates the capability to control the position of InGaAs quantum dots arrays at the nanometer scale on (110) crystal surface. Chapter 3 reports the achievement of a particular kind of quantum dot ensemble, equally-shaped with a well-defined multimodal size distribution.

In Chapter 4–7, great emphasis is placed on systematic studies of optical properties of InGaAs quantum dots. Chapter 4 deals with the carrier transfer among quantum dots. Chapter 5 reviews the dynamics of carrier transfer into InGaAs quantum dots. Chapter 6 presents spin phenomena in quantum dots, showing a great potential for the field of spintronics. Chapter 7 is a theoretical contribution in nature and addresses the optical properties of quantum dots coupled with a continuum of extended states, which will challenge experimental physicists to study quantum dots more as a complete system including the wetting layer, barrier and contact.

The progress made towards quantum information processing is discussed in chapters 8 and 9. Chapter 8 offers an extensive investigation on coupling in quantum dot molecules for their potential application to scale up in quantum computation. Chapter 9 describes the site-control and electronic access of quantum dots as well as the achievement of single photon emitters in telecommunication bands.

Chapter 10 reports on experimental and theoretical investigations of mechanical stress relaxation in heterostructures with buried quantum dots. Chapter 11 presents capacitance-voltage spectroscopy as a powerful method to gain information on the energy level structure and the Coulomb interaction in InGaAs quantum dots.

Both Chapter 12 and 13 give overviews of epitaxial growth, device design and fabrication for particular applications. Chapter 12 starts with a description of metal organic chemical vapor deposition, followed by a detailed discussion of various growth parameters and their impacts on formation of quantum dots. Finally, the authors characterize the associated quantum dot lasers and present their efforts towards quantum dot based photonic integrated circuits. Chapter 13 reports on new site-control techniques for photonic crystal based all-optical devices, including selective area growth by a metal-mask method, and lateral position control using lithography patterned substrates and Nano-Jet probes.

While the above chapters focus on InGaAs quantum dots, two different material systems are the subjects of the last two chapters. Chapter 14 is devoted to detailed analysis of the shape-dependent deformation field in Ge islands on Si (001). Investigations on such a relative simple heteroepitaxial system with respect to InGaAs quantum dots provide more fundamental perspective that determines the behavior and growth mode of compound nanostructures. On the other hand, the authors of Chapter 15 deal with nitride semiconductor quantum dots, which is a more complicated material system. This chapter covers the growth and characterization of III-Nitride quantum dots and their application to light emitting diodes emitting in the visible part of the electromagnetic spectrum.

Last but not least, the editor is honored and greatly indebted to Gregory J. Salamo for his support in many ways, without which this book would be impossible. With his enthusiasm, diligence, and intelligence, Salamo is always able to create a congenial atmosphere for friends and colleagues to explore the seemingly impossible. May he be an example followed by many!

Fayetteville, September 2007 Zhiming M. Wang

Contents

Contributors

G. Abstreiter, Walter Schottky Institute, Technische Universität München, Am Coulombwall 3, 85748 Garching, Germany

Aleksey Andreev, Hitachi Cambridge Laboratory, Hitachi Europe Ltd. Cavendish Laboratory, JJ Thomson Avenue, Cambridge CB3 0HE, United Kingdom
and
Department of Physics and Advanced Technology Institute, University of Surrey, Guildford, Surrey, GU2 7XH, United Kingdom

F. Arciprete, Dipartimento di Fisica, Universit di Roma "Tor Vergata", Via della Ricerca Scientifica 1, I-00133 Roma, Italy

Kiyoshi Asakawa, Center for Tsukuba Advanced Research Alliance (TARA), University of Tsukuba, 1-1-1 Tennoudai, Tsukuba, Ibaraki 305-8577, Japan

A. Balzarotti, Dipartimento di Fisica, Universit di Roma "Tor Vergata", Via della Ricerca Scientifica 1, I-00133 Roma, Italy

J. Bauer, Walter Schottky Institute, Technische Universität München, Am Coulombwall 3, 85748 Garching, Germany

N. A. Bert, Ioffe Physico-Technical Institute, Russian Academy of Sciences, Polytechnicheskaya 26, St.Petersburg 194021, Russia

M. Bichler, Walter Schottky Institute, Technische Universität München, Am Coulombwall 3, 85748 Garching, Germany

V. V. Chaldyshev, Ioffe Physico-Technical Institute, Russian Academy of Sciences, Polytechnicheskaya 26, St.Petersburg 194021, Russia

M. Fanfoni, Dipartimento di Fisica, Universit di Roma "Tor Vergata", Via della Ricerca Scientifica 1, I-00133 Roma, Italy

J. J. Finley, Walter Schottky Institute, Technische Universität München, Am Coulombwall 3, 85748 Garching, Germany

R. Gatti, L-NESS and Department of Materials Science, University of Milano-Bicocca, Via R. Cozzi 53, I-20125 Milano, Italy

Alexander O. Govorov, Department of Physics and Astronomy, Ohio University, Athens, Ohio 45701, USA

Naoki Ikeda, National Institute for Materials Science (NIMS), 1-2-1 Sengen, Tsukuba, Ibaraki 305-0047, Japan
and
Ultrafast Photonic Devices Laboratory, National Institute of Advanced Industrial Science and Technology (AIST), 1-1-1 Umezono, Tsukuba, Ibaraki 305-8568, Japan
and
Center for Tsukuba Advanced Research Alliance (TARA), University of Tsukuba, 1-1-1 Tennoudai, Tsukuba, Ibaraki 305-8577, Japan

C. Jagadish, Department of Electronic Materials Engineering, Research School of Physical Sciences and Engineering, The Australian National University, Canberra, ACT 0200, Australia

A.L. Kolesnikova, Institute of Problems of Mechanical Engineering, Russian Academy of Sciences, Bolshoj 61, Vas.Ostrov, St.Petersburg 199178, Russia

Saulius Marcinkevičius, Department of Microelectronics and Applied Physics, Royal Institute of Technology, 16440 Kista, Sweden

A. Marzegalli, L-NESS and Department of Materials Science, University of Milano-Bicocca, Via R. Cozzi 53, I-20125 Milano, Italy

Yuriy I. Mazur, Department of Physics, University of Arkansas, Fayetteville, Arkansas 72701, USA

Leo Miglio, L-NESS and Department of Materials Science, University of Milano-Bicocca, Via R. Cozzi 53, I-20125 Milano, Italy

S. Mokkapati, Department of Electronic Materials Engineering, Research School of Physical Sciences and Engineering, The Australian National University, Canberra, ACT 0200, Australia

F. Montalenti, L-NESS and Department of Materials Science, University of Milano-Bicocca, Via R. Cozzi 53, I-20125 Milano, Italy

A. Fontcuberta i Morral, Walter Schottky Institute, Technische Universität München, Am Coulombwall 3, 85748 Garching, Germany

Theodore D. Moustakas, Department of Electrical and Computer Engineering, Center of Photonics Research, Boston University, 8 Saint Mary's Street, Boston, MA 02215, USA

Shunsuke Ohkouchi, Fundamental and Environmental Research Laboratories, NEC Corporation, 34 Miyukigaoka, Tsukuba, Ibaraki 305-8501, Japan
and
Ultrafast Photonic Devices Laboratory, National Institute of Advanced Industrial Science and Technology (AIST), 1-1-1 Umezono, Tsukuba, Ibaraki 305-8568, Japan

Nobuhiko Ozaki, Center for Tsukuba Advanced Research Alliance (TARA), University of Tsukuba, 1-1-1 Tennoudai, Tsukuba, Ibaraki 305-8577, Japan

F. Patella, Dipartimento di Fisica, Universit di Roma "Tor Vergata", Via della Ricerca Scientifica 1, I-00133 Roma, Italy

E. Placidi, CNR-INFM, Via della Ricerca Scientifica 1, I-00133 Roma, Italy
and
Dipartimento di Fisica, Universit di Roma "Tor Vergata", Via della Ricerca, Scientifica 1, I-00133 Roma, Italy

Udo W. Pohl, Institut für Festkörperphysik Technische Universität Berlin, Hardenbergstr. 36, 10623 Berlin, Germany

D. Reuter, Lehrstuhl für Angewandte Festkörperphysik, Ruhr-Universität Bochum, Universitätsstraße 150 D-44799 Bochum, Germany

A.E. Romanov, Ioffe Physico-Technical Institute, Russian Academy of Sciences, Polytechnicheskaya 26, St.Petersburg 194021, Russia

Gregory J. Salamo, Department of Physics, University of Arkansas Fayetteville, Arkansas 72701, USA

D. Schuh, Institut für Angewandte und Experimentelle Physik Universität Regensburg 93040 Regensburg, Germany

K. Sears, Department of Electronic Materials Engineering, Research School of Physical Sciences and Engineering, The Australian National University, Canberra, ACT 0200, Australia

H. Z. Song, Nanotechnology Research Center, Fujitsu Lab. Ltd., Morinosato-Wakamiya 10-1, Atsugi, Kanagawa 243-0197, Japan

Yoshimasa Sugimoto, National Institute for Materials Science (NIMS), 1-2-1 Sengen, Tsukuba, Ibaraki 305-0047, Japan
and
Center for Tsukuba Advanced Research Alliance (TARA), University of Tsukuba, 1-1-1 Tennoudai, Tsukuba, Ibaraki 305-8577, Japan

H. H. Tan, Department of Electronic Materials Engineering, Research School of Physical Sciences and Engineering, The Australian National University, Canberra, ACT 0200, Australia

Georgiy G. Tarasov, Department of Physics, University of Arkansas, Fayetteville, Arkansas 72701, USA

Alexander Tartakovskii, Department of Physics and Astronomy, University of Sheffield, Sheffield, S3 7RH, United Kingdom

E. Uccelli, Walter Schottky Institute, Technische Universität München, Am Coulombwall 3, 85748 Garching, Germany

T. Usuki, Nanotechnology Research Center, Fujitsu Lab. Ltd., Morinosato-Wakamiya 10-1 Atsugi, Kanagawa 243-0197, Japan

G. Vastola, L-NESS and Department of Materials Science, University of Milano-Bicocca, Via R. Cozzi 53, I-20125 Milano, Italy

David A. Williams, Hitachi Cambridge Laboratory, Hitachi Europe Ltd. Cavendish Laboratory, JJ Thomson Avenue, Cambridge CB3 0HE, United Kingdom

Tao Xu, Department of Electrical and Computer Engineering, Center of Photonics Research, Boston University, 8 Saint Mary's Street, Boston, MA 02215, USA

Xiulai Xu, Hitachi Cambridge Laboratory, Hitachi Europe Ltd. Cavendish Laboratory, JJ Thomson Avenue, Cambridge CB3 0HE, United Kingdom

Chapter 1
The InAs/GaAs(001) Quantum Dots Transition: Advances on Understanding

E. Placidi, F. Arciprete, M. Fanfoni, F. Patella and A. Balzarotti

1.1 Introduction

In the heteroepitaxy of InAs/GaAs(001), the growing InAs layer remains planar up to a characteristic coverage (critical thickness) above which three-dimensional (3D) islands form. Such growth mode transition, conventionally called of the Stranski-Krastanov (S-K) type, is the most distinctive aspect of the InAs/GaAs(001) system and is at the basis of the formation of self-assembled quantum dots, which are very attractive for optoelectronic applications. Although of the S-K type, such transition has a more complex evolution and a variety of fundamental process mediated by surface diffusion leads to interesting properties for this system, including scale invariance, which are worth to be studied from a basic point of view. Specific issues that will be addressed in this review are:

i) the role of the composition of the InGaAs-alloy wetting layer (WL) at the transition [1, 2, 3];

ii) the sudden nucleation of $10^{10} - 10^{11}\ cm^{-2}$ coherent and partially relaxed QDs occurring within 0.2 ML of InAs deposition at the critical thickness for the 2D- to 3D- transition;

iii) the In-Ga intermixing in QDs;

iv) the total nucleated 3D volume, far larger than that being deposited in the narrow coverage range where nucleation is completed.

The interesting perspectives of QDs for optoelectronic devices, quantum computing and quantum cryptography have driven much attention on other important issues affecting the optical performance of self-assembled QDs. These are the homogeneity of their size and shape and the correlation of their lateral position on the surface. The fabrication of ordered arrays of dots by epitaxy, minimizing lithographic processes requires the capability to control at the nanoscale, microscopic processes involved in the 2D–3D transition. The issues listed above are among those lacking, at present, sufficient understanding with reference,in particular, to the dependence on substrate morphology, surface stress and kinetics of growth [4, 5, 6, 7].

Having in mind this research perspective, we report here the main results of our studies on the InAs/GaAs(001) heterostructure aiming at investigating the basic microscopic mechanisms driving the nucleation of QDs at each stage of their

Z. M. Wang, *Self-Assembled Quantum Dots.*

evolution. The experimental approach adopted is to growth the InAs/GaAs interface using different Molecular Beam Epitaxy (MBE) procedures, and to examine its surface morphology by atomic force microscopy (AFM) and reflection high energy electron diffraction (RHEED). Thus, experimental size distributions, volume and number density of dots, kinetics and scaling properties for different morphological conditions were comparatively analyzed for very small increments of InAs coverage throughout the 2D–3D transition. The strategy for tracing the evolution of 2D–3D transition in great experimental detail was to obtain the entire transition, from the WL to QDs, on the same sample. This is obtained is explained in the next section, where are shortly described the growth procedures and characterization techniques utilized in these studies. The main results are discussed in relation to other experimental data and well-established theoretical approaches given in literature [8, 9].

1.2 Heteroepitaxial Growth of InAs on GaAs(001)

1.2.1 Sample Preparation and Characterization Techniques

The most important aspect of MBE is the slow deposition rate which allows the films to grow epitaxially. The same film thickness can be achieved in two independent MBE growths, typically within ±0.05 ML, because of the unavoidable small fluctuations of the beam fluxes, sample temperature, etc. between two distintive growths.

The angular dependence of the Knudsen-cell flux at the substrate produces, in principle, a in standard growth thickness gradient on large samples. Nonetheless, thickness uniformity can be obtained on substrates up to 3 inch by rotating the substrate. On the contrary, in the case of a non-rotating substrate, the growth velocity increases linearly along a substrate direction and can be accurately measured by RHEED oscillations. One can take advantage from the angular dependence of the flux to achieve thickness differences as small as 0.01 ML, because the range of thickness is realized simultaneously. For InAs on GaAs(001), such a procedure was adopted, on a 2-inch long sample, to obtain the full 2D–3D transition on the same surface. Thus, the structural evolution of InAs QDs, occurring within 0.2 ML above the critical thickness [10, 11, 12, 13, 14], was studied for coverage increments of 0.01 ML in the range 0.9 to 2.2 ML.

Samples were prepared using different growth procedures. In the sample labeled "GI", the In delivery was cycled in 5 s of evaporation followed by 25 s of growth interruption (GI) (with the As shutter always left open) until the onset of the 2D–3D transition was observed by RHEED at the center of the sample [14]. Sample labeled as "GI-flat", was grown with the same procedure on a GaAs(001) surface flattened by a 80-min post-growth annealing at 660 °C under As_4 flux [15]. Sample labeled "CG", was grown in the continuous growth (CG) mode, i.e. by delivering In for 80 s on a standard GaAs(001) buffer. Prior to InAs deposition, a GaAs buffer layer of approximate thickness of 500 nm was grown at 590 °C on the (001)-oriented substrate, in As_4 overflow, at a rate of 1 $\mu m/h$. After 10 min post-growth annealing, prior to the InAs deposition, the temperature was lowered to 500 °C, when a GaAs $c(4 \times 4)$ surface was observed. The growth was carried out at this temperature without rotating the substrate, so as to obtain the aforementioned thickness profile along the sample. As illustrated in Fig. 1.1, the sample was placed in the region (b) where the impinging flux increased linearly along the [110] direction of the substrate, from 0.011 ML/s to 0.030 ML/s, resulting in InAs coverages ranging from 0.87 ML to 2.40 ML for 80 s of growth.

The growth velocity of InAs was measured by RHEED oscillations at various locations of the GaAs substrate along the [110] direction. The spotty RHEED pattern signaled the formation of QDs (Fig. 1.2). The RHEED intensity was monitored by a CCD camera. The steep rise in the intensity marks the 2D–3D onset at the coverage θ_c, the critical thickness for the 2D–3D transition. The growth velocity of InAs, V_{InAs}, is a function of the substrate temperature and was calibrated by means of RHEED intensity oscillations. The morphology of the InAs QDs was examined at several sites along the [110] direction points by *ex-situ* Atomic Force Microscopy (VEECO Multiprobe), operating in the tapping mode and using non-conductive Si tips.

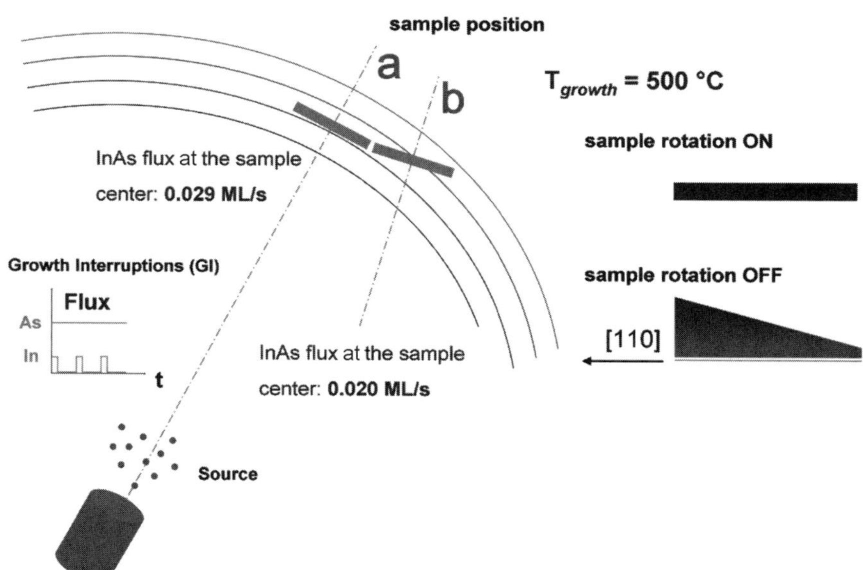

Fig. 1.1 Schematics of the growth geometry in the MBE reactor. The sample position (a) the flux wave front is nearly parallel to the surface while at oblique incidence (b) a coverage gradient is present

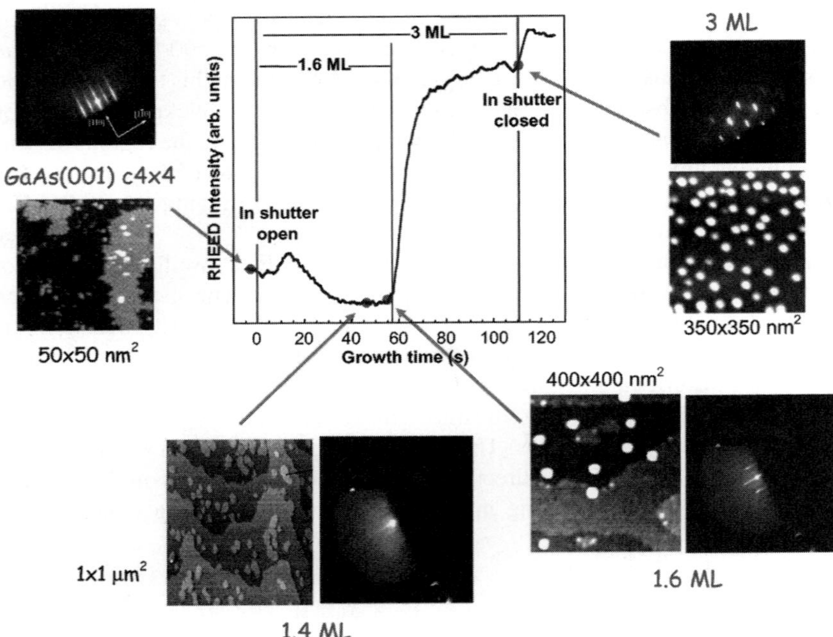

Fig. 1.2 The 2D–3D phase transition monitored by the RHEED intensity. RHEED patterns, STM $c(4 \times 4)$ and AFM images are shown at significant times (coverages) during the growth

1.2.2 The 2D Phase: Wetting Layer and Critical Thickness

Several experimental works in literature discuss In segregation and In-Ga intermixing in epitaxial ternary III-V alloys. Among group III elements, In has the highest segregation coefficient, leading to the formation of a quasi-binary surface. Moison et al. [16] estimated, by x-ray photoemission and Auger measurements on the $In_{0.2}Ga_{0.8}As$ bulk alloy grown at $480\,^{\circ}C$, an average surface composition of In of 0.7. Additional experimental evidence comes from the observation, by scanning tunneling microscopy (STM), of the surface reconstructions typical of In-rich InGaAs alloys [17, 18]. Dehaese, Wallart and Mollot (DWM) [19] proposed a simple kinetic model of segregation far from the equilibrium. In fact, thermodynamic equilibrium models fail to predict the concentration profiles at high fluxes and low growth-temperatures. Because of the very low bulk diffusion coefficient at the typical MBE growth temperatures, the DWM model rests on the hypothesis that the exchange between atoms A and B of a pseudobinary $A_{1-x}B_xC$ alloy takes place solely between the two topmost layers.

A recent significant advance is due to Cullis et al. [1, 2, 3]. They propose a model for the S-K growth of strained $In_xGa_{1-x}As/GaAs$ which focuses on the composition of WL, in particular on the strain dependence of the critical thickness as a consequence of In segregation. This model (called WCNH) is supported by transmission electron microscopy (TEM) measurements of the chemical composition of the WL and 3D islands. The data indicate that the 2D–3D transition is triggered by a critical value of the strain as soon as the In concentration of the top layer reaches a precise value. Around islands the In concentration is lower than the nominal value of the evaporated alloy, meanwhile it is considerably higher inside the islands. This finding is attributed to the vertical segregation of In, whose concentration in 3D islands is higher at the apex where the lattice parameter of InAs is closer to that of bulk. The WCNH mechanism was simulated using the Fukatsu-Dehase model [20] for In segregation. In this model, segregation continuously raises the In concentration in the top layer up to a saturation value (left panel of Fig. 1.3) which increases with increasing the nominal In concentration, x, of the alloy. Since $x = 0.25$ is the minimum value for islanding to occur (right panel of Fig. 1.3), the saturation concentration of the curve (c) has been identified as the critical one. Thus the 2D–3D transition occurs as soon as an In concentration of 80–85% is reached.

In the heteroepitaxy of Ge/Si and InAs/GaAs, Y. Tu and J. Tersoff [21] found a kinetic crossover to the instability regime of an initially small height-perturbation of the surface, due to the increase of surface In (Ge) content (hence of stress), as shown in Fig. 1.4. In their model, while the initial planar growth occurs due to intermixing of deposited material with the substrate, the transition is strictly kinetic in nature and the finite stiffness of the steps and surface mass diffusion are essential ingredients for the instability to set up.

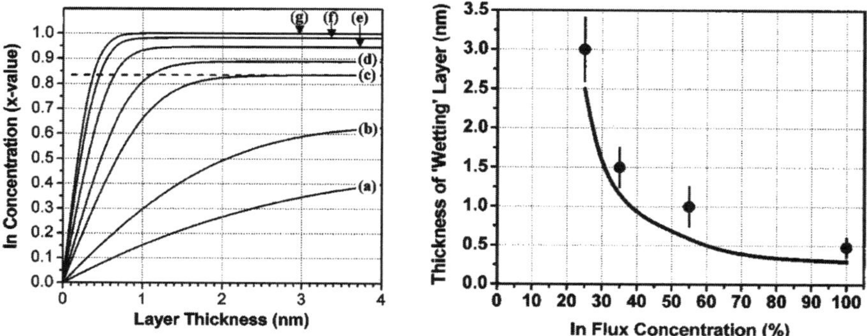

Fig. 1.3 Left panel: Composition variations in the surface monolayer, driven by In segregation to the surface, for deposition fluxes with (a) 5% In, (b) 10% In, (c) 25% In, (d) 35% In, (e) 55% In, (f) 80% In, and (g) 100% In. Right panel: Variation in the flat-layer critical thickness for the islanding transition as a function of In concentration in the deposition flux: measured data points and theoretically predicted values.Reprinted with permission from [2]. Copyright (2002) by the American Physical Society

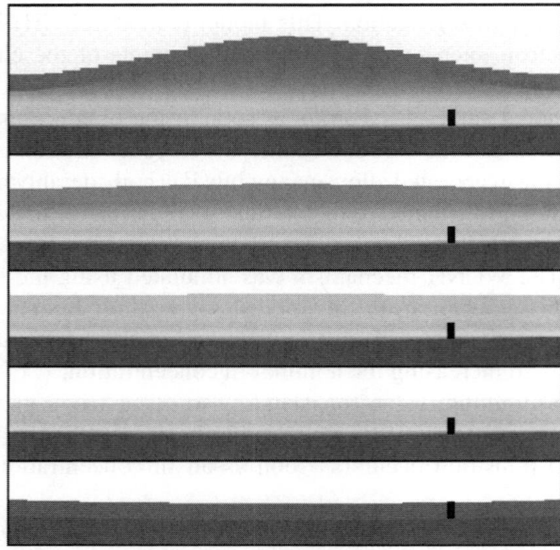

Fig. 1.4 Evolution of structure and composition during heteroepitaxy, for nominal $Si_{0.60}Ge_{0:40}$ on Si(001), at deposition rate of 10^4 (arbitrary units). The onset of nonplanar morphology is more abrupt at lower growth rates. The grey tones indicate composition, from pure Si substrate (bottom) to $Si_{0.62}Ge_{0.38}$ (top). The bottom panel is the initial surface (slightly nonplanar), and subsequent panels are at equal time intervals. The figure shows one unit cell of periodic system; the lateral size is $640w_s$. The surface-layer thickness w_s is indicated by a black rectangle in bottom panel; the vertical scale is greatly expanded to evidence the small perturbation. The rectangle is repeated in the same position in subsequent panels for reference. Reprinted with permission from [21]. Copyright (2004) by the American Physical Society

1.2.3 The 3D Phase: Evolution of the Surface Morphology During the 2D–3D Transition

As reported in several papers [10, 12, 13, 14, 17, 22, 23], the morphology of the InAs/GaAs(001) interface close to the critical thickness is quite rich. Figure 1.5 shows significant AFM topographies ($1\ \mu m \times 1\ \mu m$) for coverages between 1.42 and 1.60 ML for the samples CG, GI and GI-flat. The images testify a complex morphology of the WL comprising 2D islands 1-ML high, and large terraces one-step high. The samples display different initial morphologies of the WL. At coverage of 1.42 ML, while in the GI and CG samples the surface is considerably stepped, in the GI-flat sample only a few steps per $1\ \mu m^2$ area are present after the flattening process [15]. The morphological instabilities (mounds) observed on the GI sample are typical of the GaAs(001) growth [30]. As expected, QDs nucleation is strongly influenced by the underlying surface morphology. Infact, kinetic instabilities of the growth, due to the anisotropy of diffusion and the extra-barrier for down hopping at step edges [24, 25] cause the dots to nucleate mainly at step edges [10, 17, 22, 26, 27]. The first small QDs (present in all images) are recognizable for coverage of 1.45 ML and nucleate preferentially at the upper-step edges of 2D islands and

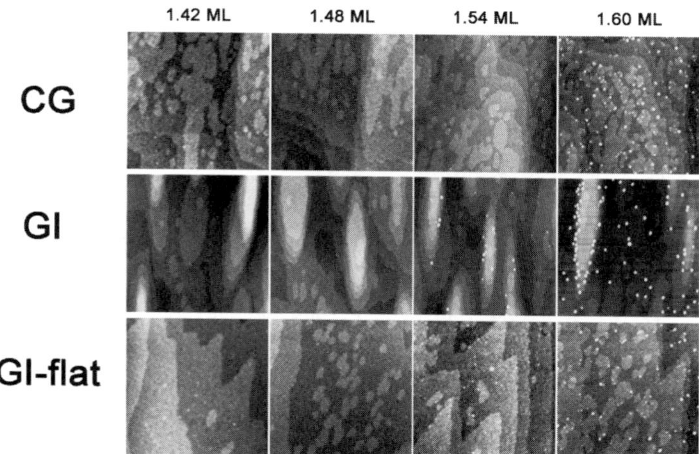

Fig. 1.5 $1 \times 1\ \mu m^2$ AFM topographies for the CG, GI and GI-flat samples. The images show the WL at 1.42 ML, free from $3D$ features, the first small QDs at 1.48 ML and the occurrence of large QDs at 1.54 ML for the GI sample and at 1.60 ML for the others

terraces by reason of a favorable strain condition at those sites. They have been reported several times [12, 22, 13, 17, 23] and often indicated as simple precursors of large QDs. However, this simplified picture seems unrealistic and we will show in the following sections that the process involves a more complex kinetic mechanism. At coverages higher than 1.54 ML the nucleation of large QDs is observed.

The number density evolution of both small and large QDs for GI sample is summarized in Fig. 1.6 as a function of InAs deposition. The number of the small QDs begins to increase around 1.45 ML of InAs deposit and maximizes at 1.57 ML to a value of $1.1 \times 10^{10}\ cm^{-2}$. Starting from 1.52 ML, the number of the large QDs increases steadily and then, between 1.57 and 1.61 ML, it undergoes a sudden rise,

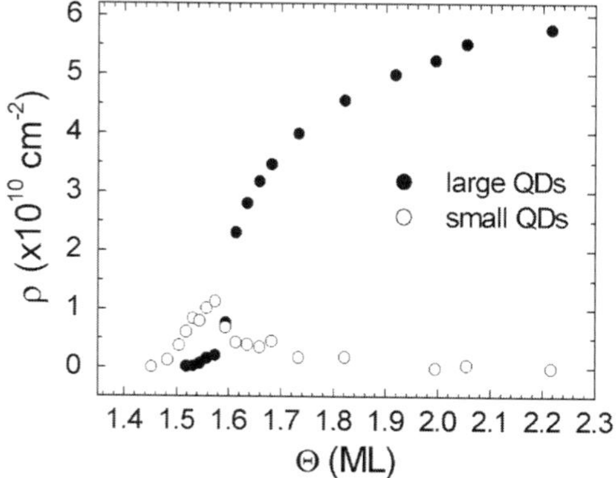

Fig. 1.6 Number density dependence of small and large QDs on InAs coverage

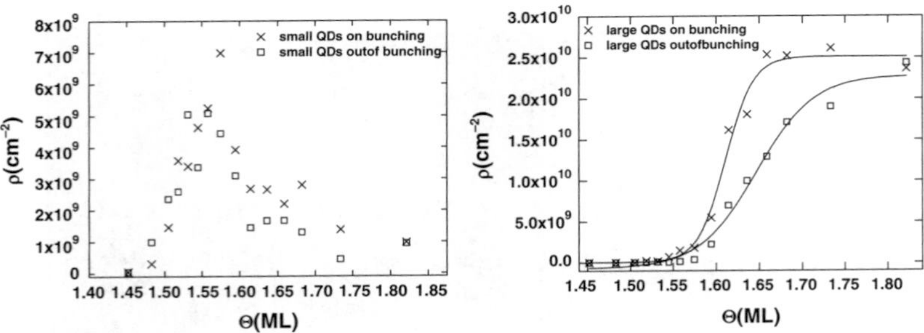

Fig. 1.7 Number density dependence on InAs coverage of small and large QDs

changing by an order of magnitude. At higher coverages ($> 1.8\,ML$) the density rise is reduced. In this region the average density is $6 \times 10^{10}\,cm^{-2}$ for all samples.

In Fig. 1.7 the density of small (left panel) and large QDs (right panel) on step bunching and elsewhere is shown. While for the small QDs the two distributions match, the nucleation of large QDs takes place early within 0.03 ML (comparing the middle height of the two distributions). Between 1.55 and 1.75 ML, the nucleation on step bunch is enhanced compared to that on other regions of the surface. The same is true for the volume increase of QDs. This result is supported by the morphologies shown in Fig. 1.5. Formation of large QDs takes place before (at 1.54 ML) in the GI sample, where step bunching is massively present. On the other hand the density of small QDs is not affected by the different strain due to step bunching, confirming that small QDs only require a step edge (more generally a defect) to nucleate [27].

1.3 Surface Mass Transport in the InAs/GaAs 2D–3D Transition

The diffusion of adparticles is the key controlling rate in most dynamical process taking place at surfaces. Its understanding is indeed a fundamental step in surface nanostructuring which acquired increasing technological importance in recent years. Surface diffusive motion may occur by many different mechanisms having different time scales. The simplest one is that of an isolated adatom performing thermally activated hopping between adjacent adsorption sites. At low temperatures (typically when the diffusion barrier $E_D \geq 4k_BT$) this is the dominant diffusion process [28] mainly invoked to explain nucleation and growth whenever the substrate is not active in the role of mass transport. Many other mechanisms are possible, however, when substrate atoms participate [29]. In heteroepitaxy, for instance, surface mediated mass transport can be due to intermixing which causes the increases of the substrate elastic strain, afterwards relieved by atoms ejected several nearest neighbor distances apart from site where exchange occurred. The major role in surface diffusion, however, is played by steps which are always present on real surfaces. Diffusion along and across steps is crucial for understanding surface growth morphologies as, for instance, the formation of mounds and the shape of growing 2D islands. Such morphologies result from the interplay among processes of attachment of flux atoms coming from the surrounding terrace, detachment and smoothing around the island edges due to diffusion along the lower step edge and corner crossing between adjacent steps. In most cases, particularly in homoepitaxy, the crossing of steps by diffusing adatoms is hindered by the additional barrier for down stepping due to the low coordination of the corner edge atom. That being the case, an upward adatom current causes kinetic instabilities and mounds formation [30]. Similar arguments can be applied to the morphology of the step edges [28, 31]. The diffusion along the lower step edge and the additional barrier for kink crossing control the instability of the ledge. The presence of steps can affect the adatom diffusion not only at their lower edge but also at larger distances, as revealed by the depletion of adatoms around 2D islands caused by the lowering of the diffusion barrier with respect to that in the middle of a terrace. Diffusion on the upper step edge, instead, is usually not very different from that on terraces unless strongly altered by local relaxation, which indeed occurs in the strained heteroepitaxy of InAs/GaAs. In such a case, a lowering of the Schwoebel barrier and even its inversion due to the combined effect of intermixing and local strain conditions, particularly around dots nucleated on steps, can strongly enhance dynamical step-edge processes.

Substantial surface mass transport, then, may come into play due to such "step erosion" [32]. In the growth of InAs QDs on GaAs, the fingerprint of wetting layer atoms participating to the dot growth is given by the behavior of total volume of dots as a function of coverage for different growth conditions. This is shown in Fig. 1.8 where is reported the total volume of dots (plotting separately the contribution due to coalescence) for samples grown with the standard continuous growth (CG samples) and samples where growth interruptions were applied (GI samples) and therefore surface mediated mass transport is enhanced.

1.3.1 Evidence by RHEED Intensity Analysis

Besides to volume behavior, the most striking experimental evidence of surface mass transport comes from RHEED intensity data. In Fig. 1.9 the RHEED spot intensity evolution as a function of the growing time is shown for $\sim 2\,ML$ of InAs deposited in the GI mode. On approaching θ_c (see the first three cycles of Fig. 1.9), the RHEED intensity is constant during GI; this fact indicates no substantial changes either in As stoichiometry and morphology of the alloyed $(2 \times 3) - (4 \times 3)$ WL [17, 18] resulting from the equilibrium of the dynamical processes of As incorporation-desorption and In-Ga detachment, migration and incorporation into different lattice sites. However, for a total delivery of 1.59 ML (330 s to 360 s in the time scale of Fig. 1.9) the large morphological change of the surface on formation of QDs begins; at the same time the RHEED intensity changes suddenly. The largest intensity variation, in Fig. 1.9, occurs in the two cycles above θ_c for a total coverage of 1.75 ML; two equal jumps follow in the next two cycles, when coverage reaches

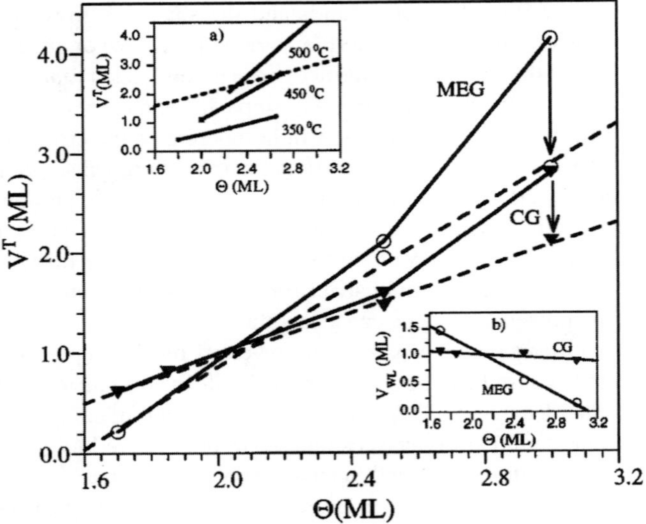

Fig. 1.8 Total volume of the dots, including ripened islands, as a function of the deposited InAs volume per unit surface, Θ. Both CG and MEG samples are grown at $500\,^{\circ}C$ at a rate of $0.028\,ML/s$ and at In/As flux ratio $1/15$. The full lines join the experimental points for an eye guide. The dashed lines are fit to the experimental points obtained by subtracting the contribution of the volume of ripened islands. The inset (a) shows data of Ref.[33] relative to samples grown at $500\,^{\circ}C$, $450\,^{\circ}C$, and $350\,^{\circ}C$, with a rate of $0.3\,ML/s$ and flux ratio $1/6$. The dashed line is the deposited volume of InAs. Notice that the lines relative to $450\,^{\circ}C$ and $350\,^{\circ}C$ samples are parallel to the dashed lines corresponding to the volume of non-ripened dots of GI and CG samples, respectively. In the inset (b) is reported the difference V_{WL} of the deposited InAs volume and that of the non-ripened dots

1.88 ML. Finally the intensity saturates in the last two ones, corresponding to a total coverage of 2.03 ML.

The side patterns, shown in Fig. 1.9(a–d), refer to the time interval after the last In shutter opening in the cycle corresponding to 1.59 ML total deposition. Remarkably, 3D spots start being observed after the interruption of the In flux. In fact, when the In shutter is closed, 5 s after the beginning of the cycle, the diffraction pattern is still like that observed in the preceding cycles, i.e. typical of a $2D$ surface with some degree of disorder; $3D$ spots appear about $2\,s$ after the GI and are fully developed after 24 s. During the GI the In (Ga) content of the sample is constant, since any In (Ga) desorption, at $500\,^{\circ}C$, is prevented by the high As flux, as also confirmed by the constant RHEED intensity during GI in the cycles preceding the $2D$–$3D$ transition [34]. Therefore we can conclude that the RHEED intensity variations observed during the GI, at InAs thickness larger than 1.59 ML, are solely due to the increase of the diffraction volume because of the large nucleation of QDs occurring on the time scale set by the processes of As desorption and cation (mainly In) surface diffusion.

To support this point, we report in Fig. 1.10 the QD number density versus InAs coverage for a sample grown, by MBE-GI, compared with the RHEED pattern

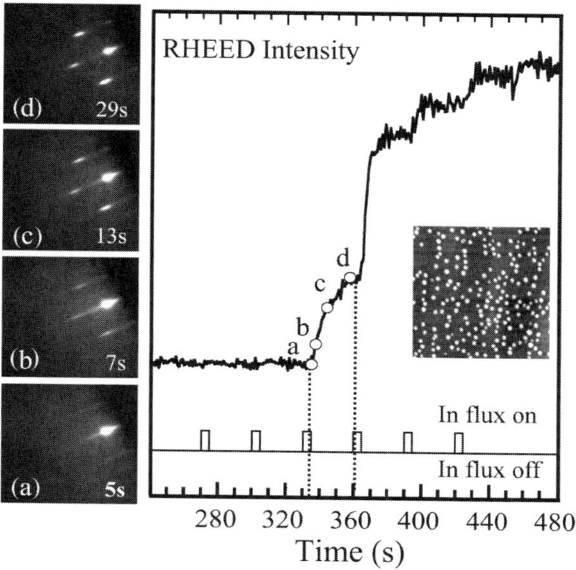

Fig. 1.9 RHEED intensity of the (01) diffraction reflex as a function of time during the MBE-GI growth of InAs on GaAs(001). A total InAs coverage of 2.03 ML is delivered in consecutive cycles, each consisting of 5 s of In deposition followed by 25 s of growth interruption, while the As_4 cell shutter if kept always open. The left panels, (a) to (d), show the sequence of RHEED patterns observed at the specified times of the cycle (marked in the figure) where the total InAs coverage reaches 1.59±0.01 ML. The $2D-3D$ transition occurs, meantime no further In has being deposited on the surface. The inset shows a $(600 \times 600)\ nm^2$ AFM image of the $3D$ QDs covering the surface, at the end of the growth

intensity of spotty features. The RHEED intensity data points reported in Fig. 1.10, are derived from Fig. 1.9 by assigning to the final coverage reached in each cycle (after 5 s of In evaporation) the intensity measured at the end of that cycle (after a growth interruption of 25 s). It is apparent that the RHEED changes closely follow the QD number density changes with coverage. Both start at about 1.59 ML, rapidly increase within 0.2 ML, and then saturate at approximately 2 ML, as expected, since each QD, independently on its size, contributes to diffraction only with a small fraction of its volume at the top. Hence, one can conclude that 3D nuclei are mostly formed within 0.2 ML from θ_c, their growth being triggered by the large surface diffusion of cations which sets up at this thickness. It is interesting to compare the transition evolution for GI and CG samples (see Fig. 5). In the two sets of samples there is a substantial difference in the absolute values of the intensity at equal coverage; the final morphologies also differ greatly. However, comparing GI and CG RHEED intensities, we found [60] that the onset values were identical, irrespective of the very different kinetics involved in the two growth modes. The 3D volume increase, in this narrow coverage range, is far larger than the deposited volume. Note, for instance, the steps of the 450 $^\circ C$ GI-sample curve that originate from

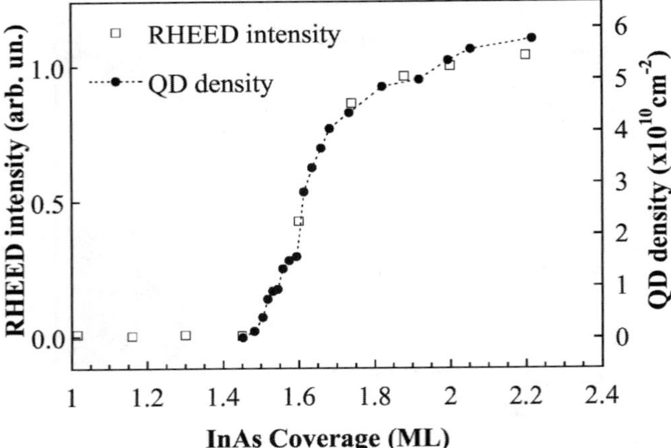

Fig. 1.10 Open squares: RHEED intensity of the (01) diffraction reflex as a function of InAs MLs deposited on GaAs(001) by MBE-GI, i.e. by consecutive cycles consisting in 5 s of InAs growth followed by 25 s of growth interruption, leaving the As_4 cell shutter open. The data points are obtained from Fig. 1.9, by assigning the RHEED intensity measured at the end of each 30 s cycle to the total InAs coverage reached in that cycle. Filled dots: measured QDs number density as function of the InAs coverage. The comparison is made by normalizing at 2 ML where both QD density and RHEED intensity saturate

the cut off of the RHEED intensity points relative to the growth interruption time of each cycle, when QD nucleation continues by means of surface mass transport (missing points are shown in the full curves of left panel). We speculate about the origin of such mass transport which is enabled by the As desorption process. As a result of the initial In-Ga intermixing the WL progressively enriches of In up to a critical concentration of about 83% after which the Stranski-Krastanov transition is expected [1]. This is confirmed by the STM topographies of the WL at 1.3 ML and of the strained $In_{0.2}Ga_{0.8}As/GaAs(001)$ alloy, which show the same (4×3) and $c(4 \times 6)$ domains characteristic of an In enriched surface by segregation, and, for the former, the additional presence of an incommensurate In-As (2×4) phase (zig-zag chains) above the intermixed layer [17]. According to calculations of the In tracer diffusion on ideal GaAs(001)$c(4 \times 4)$ and $In_{2/3}Ga_{1/3}As(001)(2 \times 3)$, that predict considerably lower binding energies and diffusion barriers for the latter surface [35], the morphology of the WL is suited to favor substantial mass transport. In addition to that, the compressive strain induced by the InAs island on the substrate is such that a few percent increase of strain, like that in the substrate around 3D dots, significantly weakens the binding energy of the In atoms which are moved away by the repulsive potential existing in that region [35]. It is plausible that the observed mass transport at θ_c (see Fig. 11) likely takes place around dots at step edges, due to step erosion, triggered by the lowering of the In binding energy due to the high compressive strain. At these sites, in fact, the As desorption [17, 33, 12] will allow for an anomalously high amount of In monomers to be released contributing to dot nucleation.

1.3.2 Evidence by AFM Volume Analysis

One of the puzzling aspects of the self-assembled QDs, is that the nucleated 3D volume is far larger than that being deposited in the narrow coverage range where the entire nucleation process is completed. We discuss here a peculiar feature, recently discovered [32], at the origin of this phenomenon, which is associated with mature 3D QDs, i.e. the erosion of the step edge surrounding dots. This is clearly shown in Fig. 1.11 which compares two areas of the surface of InAs, with coverage 1.54 ML (panel (a)) and 1.61 ML (panel (b)), where the average volume of the 3D-QDs amounts to $\sim 220\,nm^3$ and $\sim 340\,nm^3$, respectively. QDs are observed nucleating preferentially at the upper side of the step edges, as shown in panel (a). Comparing images for successive coverages one observes that QDs close to the step edge (A in panel (a)) progressively erode the step edge (B) until they appear detached (C in panel (b)), as shown in the schematic pictures A-D below the topographies. Such erosion is evident for dots over the step edges but is unclear for dots nucleated on flat areas of the surface. Nevertheless, signs of erosion of the surface plane surrounding dots (QDs labeled "D" in panel (c)) were observed for growths at higher temperature ($522\,^{\circ}C$) and coverages (2.4 ML). In the last case the larger size of the dot (average volume $1500\,nm^2$) and, consequently, the larger strain energy at its base, could account for the erosion. However, even at $500\,^{\circ}C$ a reduced plane-erosion cannot be excluded because of the finite tip resolution.

As for other systems [36, 37, 38, 39], substrate erosion may be favored in this case by the In-Ga intermixing which draws material from the substrate to create the alloyed island. This interpretation is supported by the work of Cullis et al. [2] on

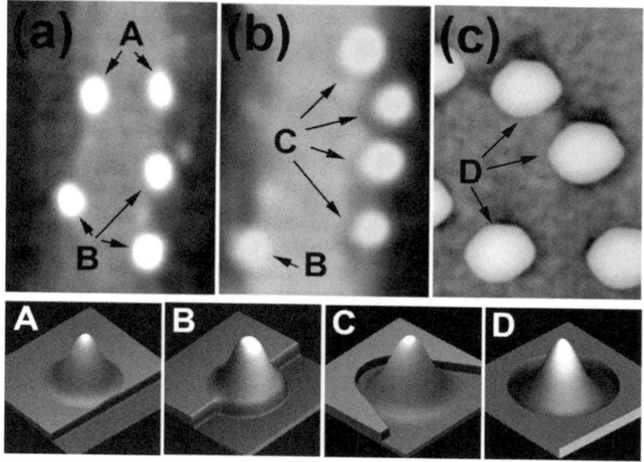

Fig. 1.11 $120 \times 160\,nm^2$ AFM images for coverages of 1.54 ML (a) and 1.61 ML (b). "A" are QDs nucleated over the step edge, "B" are QDs that have partially eroded the step edge, and "C" are QDs that have completely eroded it. (c) $80\,nm^2$ AFM image showing plane erosion occurring for 2.4 ML InAs growth at $522\,^{\circ}C$ ("D" are QDs after erosion). The bottom panels show schematic $3D$ views of QDs before and after the step or plane erosion [32]

$In_xGa_{1-x}As/GaAs$, which predicts, for $x = 1$, the largest elastic energy-per-atom and the highest In concentration in the upper layers of the QDs[1, 2, 40].

Around the QD, a compressive strained area forms where the elastic energy differs (see Fig. 1.12), from that of the WL far from the island by an amount that is large and positive; inside the dot, instead, it is negative [41, 42]. The minimum value is due to the partial strain relaxation inside the dot; the maximum, at the periphery, is due to the strain propagation along the interface which increases the misfit between the underlying substrate and the WL encircling the dot. We speculate that, because of this strain profile, the detachment rate of adatom from steps is higher than the attachment rate, favoring the consumption of the step. In Fig. 1.13 the total dot volume is plotted as a function of the evaporated In amount. Alike in other studies [14, 12, 33, 43], Fig. 1.13 highlights the increase of the dot volume in the 2D–3D transition region for CG, GI and GI-flat samples. This increase is well beyond the volume of the material being deposited (straight line in the figure). Two regions with different slopes can be distinguished: in the range 1.6–1.8 ML, the dependence of the dot volume on coverage is linear with slope $F_t \approx F_0$ for all the samples, where F_0 is the slope of straight line representing the volume being deposited by the impinging InAs flux.

Above 1.8 ML, the volume increases at the same rate F_0 of the coverage. It is worth underlining that this volume behavior is a clear evidence of the large surface mass transport from the 2D to the 3D phase at the very initial stage of the transition. This can be quantified $\Delta\theta = 0.9\,ML$ for the CG and GI samples, and $\Delta\theta = 0.5\,ML$ for the GI-flat. Such mass transfer ends at 1.8 ML, above which volume increases solely at the expense of the incoming flux. The ensuing discussion concerns the microscopic origin of the excess volume. AFM topographies allow for a lower esti-

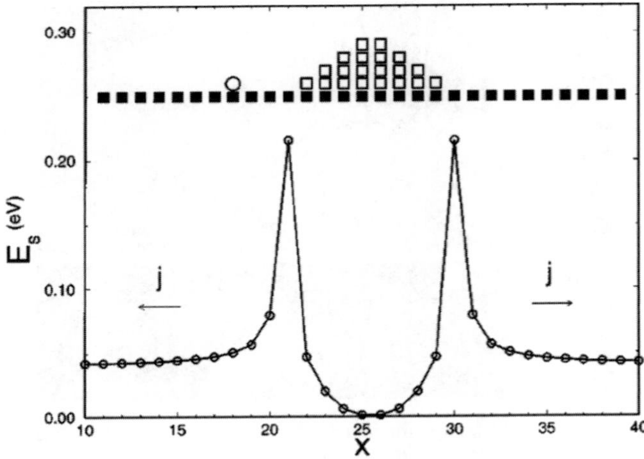

Fig. 1.12 Strain energy around a typical island. The substrate (filled squares) is shown on the upper part of the figure. E_s is the strain energy of an atom placed on the top of the substrate or on the island. Reused with permission from [41]. Copyright 1997, American Institute of Physics

mate of the amount of eroded material around islands, $\Delta V = 0.3\,ML$ [32] sets a lower limit to the surface mass current due to erosion. Therefore, in addition to surface segregation and substrate intermixing, a relevant fraction of the dot volume comes from step erosion; not enough, however, to account for all the 3D volume in excess. A further contribution can be accounted to floating In adatoms observed at the very beginning of nucleation [17, 44]. In fact STM images of a recent paper [45] indicate that an higher eroded volume is worth to be possible. A good input comes from the comparison of the total QDs volume in the CG, GI and GI-flat samples [46], shown in Fig. 1.13. While for the CG and GI samples the excess volume $\Delta\theta_{CG}$, $\Delta\theta_{GI} \approx 0.9\,ML$, for the GI-flat $\Delta\theta_{GI-flat} \approx 0.5\,ML$, thus suggesting an erosion dependence by the initial wetting layer morphology. This finding is still more interesting if compared with the step density per unit area for all the samples analyzed. Just before the transition starts, the CG and GI surfaces present a value $\rho^{CG}_{steps} = \rho^{GI}_{steps} = 23\,\mu m^{-1}$, while for the GI-flat is $\rho^{GI-flat}_{steps} = 15\,\mu m^{-1}$. The relation between the step density and the excess volume is thus apparent: in particular, if we notice that the ratio $\rho^{GI-flat}_{steps}/\rho^{CG}_{steps}$ is similar to $\Delta\theta_{GI-flat}/\Delta\theta_{CG}$ and that the step erosion occurs with the same intensity for each sample, we infer that large part of the excess volume has likely to be accounted by step erosion (and, in general, by wetting layer erosion). Such interpretation is strengthened by the disappearance of 2D islands during the large QDs density explosion, i.e. when step erosion is operating. Further support to the step (and WL) erosion contribution to the excess volume comes from STM images (see Fig. 1.14) of InAs islands deposited onto nominal and vicinal (1.0° off) GaAs(2511)A surfaces[47]. The authors find a massive WL erosion which well justify the InAs material QDs excess volume.

The origin of the transition between the two families of QDs can be understood if we consider the evolution, as a function of coverage, of the total volume contained in large QDs only. Above 1.8 ML, the QDs excess volume increase reverts to

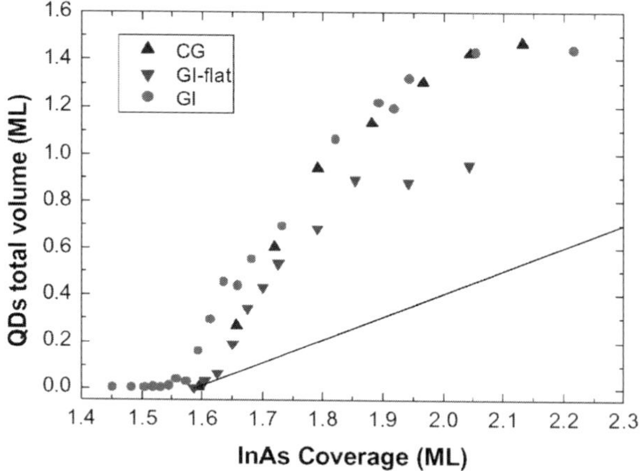

Fig. 1.13 QDs volume plotted as a function of InAs coverage for the CG, GI and GI-flat samples. The lowest straight curve indicates the InAs flux above the 2D–3D transition

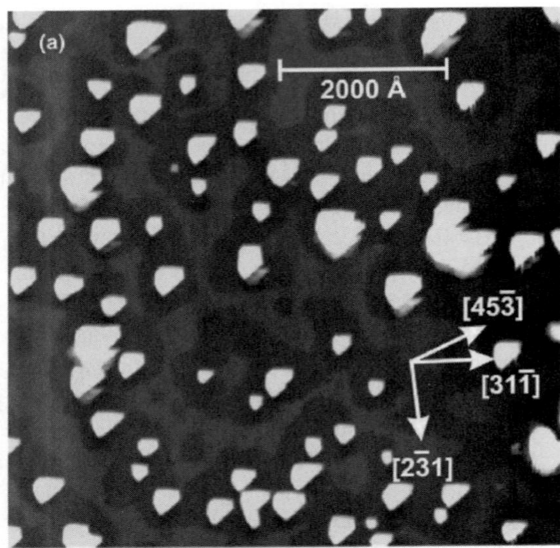

Fig. 1.14 STM image of InAs islands grown on nominal GaAs(2,5,11)A (6700 × 6700 ^{2}), $U = 2.5\,V$, $I = 0.15\,nA$, InAs thickness 5.25 grown at a sample temperature T. Reproduced with permission from Elsevier [47]

being compatible with F_0. The extra quantity of matter amounts to roughly 0.9 ML. The evidence of step erosion from QDs nucleated at step edges is confirmed by a recent work [45] confirms our finding: by looking at those data, an amount greater than 0.3 ML of eroded steps might be estimated. Even though the erosion could be responsible for the whole supplementary 0.9 ML, a further contribution could arise from substrate intermixing and In segregation.[14, 17] The total volume contained in large QDs is determined by the equation: $V^T_{large} = \rho_{large} < V_{large} >$, where ρ_{large} is the density of the large islands and $< V_{large} >$ is the mean volume of the single large island. To specify how the variation of the large QD density and mean volume contribute to the volume increase of V^T_{large}, we plot separately in Fig. 1.15 the two terms $((d\rho_{large})/(d\theta)) < V_{large} >$ and $\rho_{large}((d < V_{large} >)/(d\theta))$ as a function of θ. The derivatives of ρ_{large} and $< V_{large} >$ are calculated numerically by interpolating the experimental data. During the first stage of transition the volume increase is mainly due to the sudden nucleation of large islands and it is only subsequently that single island growth prevails. The conclusions are thus apparent: at the transition, the QD density explosion is triggered by the step erosion process [32, 45]; a great amount of monomers becomes available which allow for the explosive increase of the QD density at 1.6 ML (Figs. 1.5 and 1.15). At this stage the nucleation and growth is ruled by pure adatom diffusion. The growth instability leading to the 2D–3D transition is thermodynamic in character, this being caused, as pointed out by Cullis et al., [1] by the strain energy relaxation. Our data clearly show that the nucleation process begins at 1.45 ML of InAs deposition with the formation of small nuclei at the step edges. In accordance with the model proposed

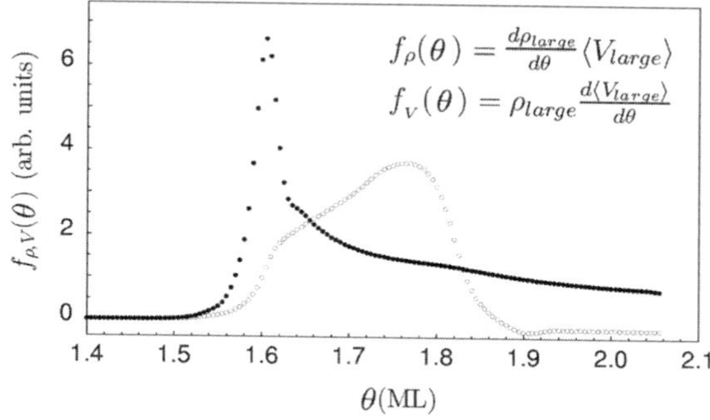

Fig. 1.15 Derivative terms function f_ρ (filled dots) and f_V (empty dots) versus InAs coverage for the GI sample

by Dehaese et al., [19] at 1.45 ML the average concentration of In in the uppermost layer is larger than 82% [14]. Although the system, from a thermodynamic point of view, prefers growing by forming 3D islands, the conditions for this to occur are first met at step edges.

1.4 Scaling Behavior of Self-assembled QDs

It has been shown, by numerical simulations that a growing 2D film characterized by nucleation and growth displays the interesting and intriguing property of being scale invariant as far as its morphology is concerned. In references [48, 49, 50, 51, 52] particular attention has been devoted to the scaling behaviour of the size distribution of the islands. The research is twofold: on one hand one would wish to determine the universal functional form of the distribution, on the other to find out what is the origins of scale invariance. In addition, it is quite rightful to wonder if the scale invariance holds for 3D islands as well. From the experimental side the corroboration of the numerical evidences have been reported by Stroscio and Pierce [53] who studied the homoepitaxial growth of Fe on Fe(001). Another fundamental experimental result was achieved by Ebiko et al.[54]. Not only do they substantiated the existence of scale invariance but, above all, they demonstrated that it holds also for 3D islands. This objective, which was reached studying the growth of InAs quantum dots on GaAs(001) substrate, has assumed a certain significance since it demonstrates that what counts is the number of atoms which make up island. Indeed they used the island volume as island size. In other words, one can regard a nucleus as a kind of attractive centre which exerts its influence on the monomers that cavort on the surface within a certain area around it. This area is called capture zone. [55]

The capture zones of a distribution of nuclei can be identified in quite a natural way by the associated Voronoi cells. By using the words of Okabe et al.[56],

"Given a finite set of distinct, isolated points in a continuous space, we associate all locations in that space with the closest member of the point set. The result is a partition of the space into a set of regions". In 2D these regions are polygon and are referred to as Voronoi cells, altogether as Voronoi diagram (VD). The atoms that land in the i-th cell are captured, on average, by the i-th nucleation centre, since, by definition, it is the closest one; in other words the nucleus grows in proportion to the area of its own Voronoi cell. This last statements have had numerical [52] and experimental [57] confirmations.

Naturally, the previous statement implies that if the film discloses a size scale invariance the VD will have to be, in fact, invariant. This last claim is evidently false because, even in the extreme case of exhausted nucleation process, that is when no more nuclei come out, due to the growth, the VD is bound to change. At most one can speak of "quasi" scale invariance (QSI) and, at any rate, the question remains: what are the conditions under which QSI emerges? To begin with, all the published experimental data [58, 59] clearly show a certain degree of spread and the QSI can be invoked, as obvious, within an experimental error. The latter also includes all the systematic effects which require (justify) the Q in the achronim. In order for the growth of nuclei to perturb slightly the VD, at least in an appropriate range of coverage, two conditions have to be fulfilled, at the limit, together:

i) the 2D projection of the nuclei on the substrate has to be negligible with respect to the correspondent Voronoi cells;

ii) the distribution of the areas of the Voronoi cells has to be as narrow as possible.

The former requires that nucleation is to be as fast as possible in order to make invariant, or almost so, the number of nuclei when the island sizes are still small. The latter is all the more satisfied the more correlated islands are. In that case the islands grow, on average, with a very close rate and, in turn, the tessellation does not change significantly.

In Fig. 1.16 the kinetics of the number of the InAs on GaAs(001) QDs and its derivative has been reported as a function of coverage. As it appears, the nucleation

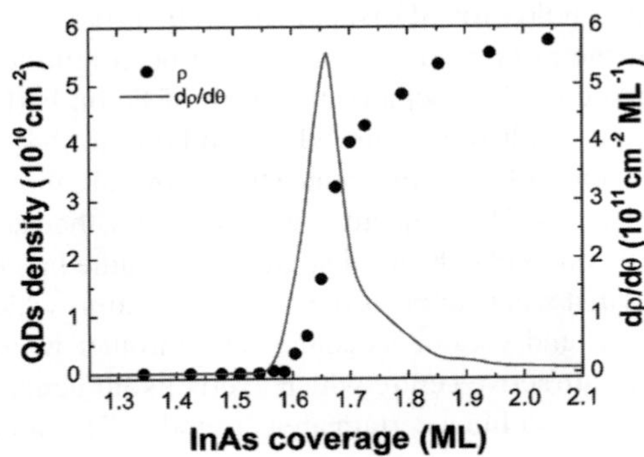

Fig. 1.16 InAs/GaAs(001) QDs number density ρ (full dots) and rate of nucleation $d\rho/d\theta$ (line) plotted as a function of coverage. Note that the nucleation process develops within 0.2 ML

is a quite fast process and, besides, the size of the QDs is rather smaller than the respective VCs (Fig. 1.17).

We have measured the size distribution for a certain values of coverage and generated the "experimental" associated VDs and the attached distributions. The superposition of the two set of experimental data is shown in Fig. 1.18. The agreement is more than satisfactory. However, this is simply the proof that the growth proceeds proportionally to the VD.

To prove the scale invariance we have performed an appropriate fit of the distributions and compared the results. The fits have been done using the gamma distribution (see Appendix)

$$f_\beta(x) = \frac{\beta^\beta}{\Gamma(\beta)} x^{\beta-1} e^{-\beta x}. \tag{1.1}$$

The received β parameters, listed in Table 1.1, demonstrate that between 1.68 ML and 1.85 ML the distribution function is, indeed, invariant. Beyond this range the islands become too large, the tessellation starts changing, moreover coalescence (impingement) begins dominating the kinetics and the scale invariance is lost.

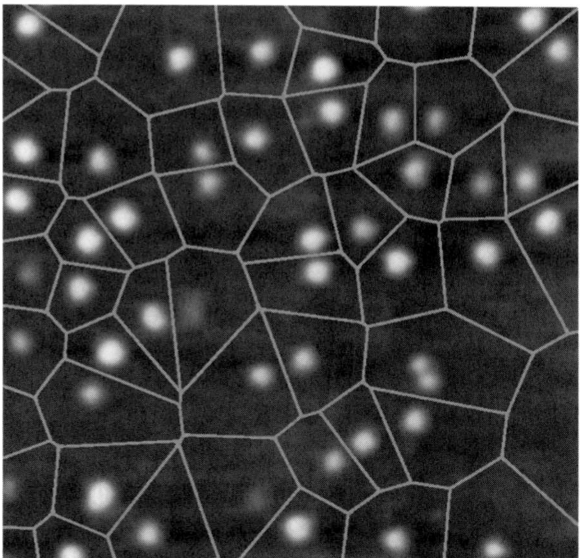

Fig. 1.17 AFM image ($350 \times 350\,nm^2$) of 1.68 ML of InAs deposited at $500\,^\circ C$ on the GaAs(001) surface.The superimposed network is the associated Voronoi tessellation calculated with respect to the QD centres

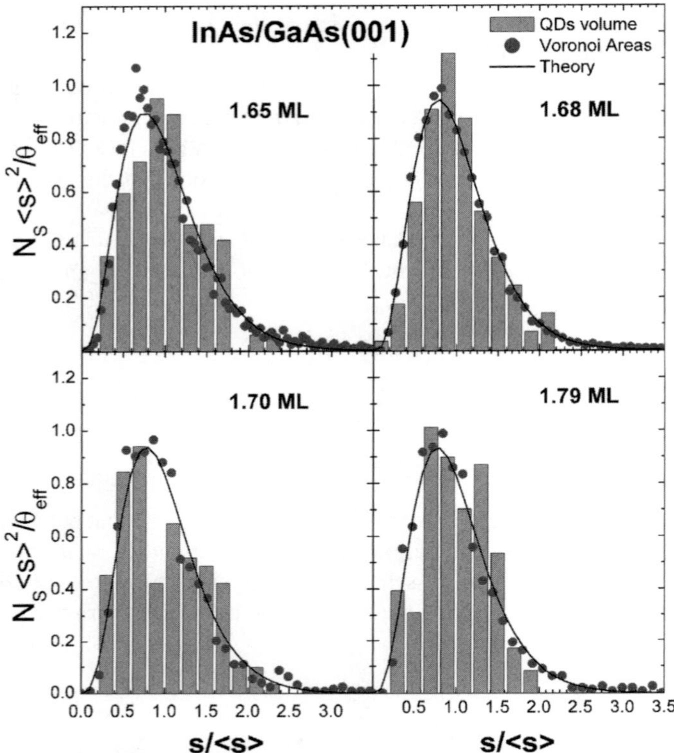

Fig. 1.18 Comparison among the experimental scaled distributions of the island volume (histograms), the Voronoi areas (full dots) and the best fit to Eq. (4) of the text (Kiang function) for coverages ranging from 1.65 to 1.79 ML

Table 1.1 Values of β as a function of coverage calculated by fitting the Voronoi areas distributions to Eq. (4)

Θ (ML)	1.65	1.68	1.70	1.73	1.79	1.85
β	4.1	4.6	4.5	4.7	4.5	4.6

1.5 Conclusions

We have reported evidence of step erosion from QDs nucleated on step edges. Such effect occurs, during the 2D–3D transition, in a narrow range of InAs coverage (1.6–1.8 ML) coincident with the sudden increase of QDs density. The total QDs volume growth rate is 4.6 times the impinging flux rate in the 1.6–1.8 ML range, and the same as the flux rate above 1.9 ML. Thus, our data show that all processes involving the WL take place in a narrow range of coverages just above the S-K transition, after which no further contributions are observed. We also found evidence that the excess QD volume is linked to the step density of the WL, suggesting that it could be related to the step erosion. The surface mass transport responsible for most of the nucleated volume of three-dimensional InAs QDs on GaAs(001) within few tenths of a ML at the critical thickness has been evidenced by observing the 2D to 3D RHEED pattern transition during growth interruption. It is speculated that the cation current is triggered by the As desorption and by the lowering of the binding energy of In atoms due to compressive strain of the substrate at the critical composition. Such condition is first met at θ_c around large dots sitting on step edges. Intermixing and segregation of In play a crucial role in the InAs/GaAs(001) heteroepitaxy, because they determine the critical surface composition that triggers the 2D–3D growth instability. Such processes could be at the basis of an energetic mechanism that triggers the formation of quantum dots in strained heteroepitaxy. Two thresholds are clearly distinguishable for the $2D$–$3D$. The first threshold occurs at 1.45 ML, where a low density of small islands nucleate at the upper step edges; the second occus at 1.59 ML, where an explosive nucleation of 3D QDs takes place, supported by the large surface mass current associated to the step erosion. Step instability and surface mass transport also influence strongly the kinetic of the 2D–3D transition [60]. Finally,we find that the QDs volume distributions scale as those of the Voronoi-cells areas. This scale invariance supports a mechanism of growth for the dots from a capture zone identified as the corresponding Voronoi cell. The key point for scale invariance is that the nucleation rate is much faster than the growth rate. Our results provide clear evidence that the InAs QDs formation on the singular (001) GaAs surface, is largely a case of low-correlated heterogeneous nucleation, which proceeds by diffusion and aggregation.

The present work has been partially supported by the FIRB project cod. num. RBNE01FSWY007 and by 2005 PRIN National Project

1.6 Appendix

The growth takes place in a 1D space where dots are distributed at random with concentration λ (see Fig. 1.19). One is interested in two stochastic variables: Y and L. The former is the distance between two successive dots, the latter is the size of the Voronoi cell; in this specific case a segment. Then $P(y < Y < y+dy) = \lambda e^{-\lambda y} dy$, while, being $2L = Y_1 + Y_2$

$$P(l < L < l + dl) = dl \int_0^{2l} \lambda e^{-\lambda y} \lambda e^{-\lambda(2l-y)} dy = 2\lambda^2 l e^{-2\lambda l} dl$$

By introducing the rescaled variable $x = \lambda l$, one ends up with

$$f_2(x) = 2xe^{-2x}, \tag{1.2}$$

which is the exact solution in 1D, as far as the Voronoi cell distribution is concerned. Kiang conjectured that it could be applied to any dimensionality, and proposed the following equation

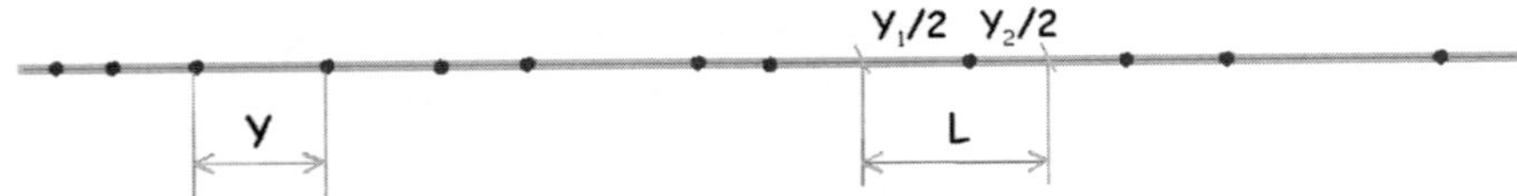

Fig. 1.19 Sketch of a Poissonian distribution of dots on a line

$$f_\beta(x) = \frac{\beta^\beta}{\Gamma(\beta)} x^{\beta-1} e^{-\beta x}.$$

which coincides with (1.2) for $\beta = 2$.

References

1. T. Walther, A.G. Cullis, D.J. Norris, and M. Hopkinson, Phys. Rev. Lett. **86**, 2381 (2001)
2. A.G. Cullis , D.J. Norris , T. Walther, M.A. Migliorato and M. Hopkinson, Phys. Rev. B **66**, 081305 (2002)
3. A.G. Cullis, D.J. Norris, M.A. Migliorato, and M. Hopkinson, Appl. Surf. Sci. **244**, 65 (2005)
4. P. Michler, A. Kiraz, C. Becher, W.V. Schoenfeld, P.M. Petroff, L. Zhang, E. Hu, and A. Imamoglu, Science **290**, 2282 (2000)
5. N.H. Bonadeo, J. Erland, D. Gammon, D. Park, D.S. Katzer, and D.G. Steel, Science **282**, 1473 (1998)
6. X.Q. Li , Y.W. Wu , D. Steel, D. Gammon, T.H. Stievater, D.S. Katzer, D. Park, C. Piermarocchi, and L.J. Sham, Science **301**, 809 (2003)
7. M. Bayer, P. Hawrylak, K. Hinzer, S. Fafard, M. Korkusinski, Z.R. Wasilewski, O. Stern, and A. Forchel, Science **291**, 451 (2001)
8. J. Stangl, V. Holy, G. Bauer, Rev. Mod. Phys. **76**, 725 (2004)
9. B.A. Joyce and D.D. Vvedensky, Mat. Sci. and Engineering R **46**, 127 (2004)
10. D. Leonard, K. Pond, and P.M. Petroff, Phys. Rev. B **50**, 11687 (1994)
11. N.P. Kobayashi, T.R. Ramachandran, P. Chen, and A. Madhukar, Appl. Phys. Lett. **68**, 3299 (1996)
12. T.R. Ramachandran, R. Heitz, P. Chen, and A. Madhukar, Appl. Phys. Lett. **70**, 640 (1996)
13. T.J. Krzyzewski, P.B. Joyce, G.R. Bell, T.S. Jones, Phys. Rev. B **66**, 121307 (2002)
14. F. Patella, A. Sgarlata, F. Arciprete, S. Nufris, P.D. Skutznik, E. Placidi, M. Fanfoni, N. Motta, and A. Balzarotti, J. Phys. Cond. Matt **16**, S1503 (2004)
15. Z. Ding, D.W. Bullock, P.M. Thibado, V.P. LaBella, K. Mullen, Phys. Rev. Lett. **90**, 216109 (2003)
16. J.M. Moison, C. Guille, F. Houzay, F. Barthe, and M. Van Rompay, Phys. Rev. B **40**, 6149 (1989)
17. F. Patella, S. Nufris, F. Arciprete, M. Fanfoni, E. Placidi, A. Sgarlata, and A. Balzarotti, Phys. Rev. B **67**, 205308 (2003)
18. J. Mirecki Millunchick, A. Riposan, B.J. Dall, B. Pearson, B.J. Orr, Appl. Phys. Lett. **83**, 1361 (2003)
19. O. Dehaese, X. Wallart, and F. Mollot, Appl. Phys. Lett. **66**, 52 (1995)
20. S. Fukatsu, K. Fujita, H. Yaguchi, Y. Shiraki, and R. Ito, Appl. Phys. Lett. **59**, 2103 (1991)
21. Y. Tu and J. Tersoff, Phys. Rev. Lett. **93**, 216101 (2005)
22. M.J. Da Silva, A.A. Quivy, P.P. Gonzlez-Borrero, Jr E. Marega, J.R. Leite, J. Crystal Growth **241**, 19 (2002)
23. G. Costantini, A. Rastelli, C. Manzano, P. Acosta-Diaz, G. Katsaros, R. Songmuang, O.G. Schmidt, O.V. Knel, and K. Kern, J. Crystal Growth **278**, 38 (2005)

24. R.L. Schwoebel and E.J. Shipsey, J. Appl. Phys. **37**, 3682 (1966)
25. D. Kandel and J.D. Weeks, Phys. Bev. B **49**, 5554 (1994)
26. R. Leon, T.J. Senden, Y. Kim, C. Jagadish, and A. Clark, Phys. Rev. Lett. **78**, 4942 (1997)
27. S. Tsukamoto, T. Honma, G.R. Bell, A. Ishii and Y. Arakawa, Small **2**, 386 (2006)
28. T. Ala-Nassila, R. Ferrando, and S.C. Ying, Adv. Phys. **51**, 949 (2002)
29. H. Bulou and J.P. Bucher, Phys. Rev. Lett. **96**, 076102 (2006)
30. F. Patella, F. Arciprete, E. Placidi, S. Nufris, M. Fanfoni, A. Sgarlata, D. Schiumarini, and A. Balzarotti, Appl. Phys. Lett. **81**, 2270 (2002)
31. O.V. Lysenko, V.S. Stepanyuk, W. Hegert, and J. Kirschner, Phys.Rev.B **68**, 033409 (2003)
32. E. Placidi, F. Arciprete, V. Sessi, M. Fanfoni, F. Patella, and A. Balzarotti, Appl. Phys. Lett. **86**, 241913 (2005)
33. P.B. Joyce, T.J Krzyzewski, G.R. Bell, B.A. Joyce, and T.S. Jones, Phys. Rev. B **58**, R15981 (1998)
34. A. Ohtake and M. Ozeki, Appl. Phys. Lett. **78**, 431 (2001)
35. E. Penev, S. Stojkovic, P. Kratzer, and M. Scheffler, Phys. Rev. B **69**, 115335 (2004)
36. W. Seifert, N. Carlsson, J. Johansson, M. Pistol, and L. Samuelson, J. Cryst. Growth **170**, 39 (1997)
37. X.Z. Liao, J. Zou, D.J.H. Cockayne, Z.M. Jiang, X. Wang, and R. Leon, Appl. Phys. Lett. **77**, 1304 (2000)
38. S.A. Chaparro, Y. Zhang and J. Drucker, Appl. Phys. Lett. **76**, 3534 (2000)
39. N. Motta, J. Phys. Cond. Matt **14**, 8353 (2002)
40. I. Kegel, T.H. Metzger, A. Lorke, J. Peisl, J. Stangl, G. Bauer, J.M. Garcia, and P.M. Petroff, Phys. Rev. Lett. **85**, 1694 (2000)
41. A.L. Barabsi, Appl. Phys. Lett. **70**, 2565 (1997)
42. F.K. Le Goues, M.C. Reuter, J. Tersoff, M. Hammar, and R.M. Tromp, Phys. Rev. Lett. **73**, 300 (1994)
43. F. Patella, M. Fanfoni, F. Arciprete, S. Nufris, E. Placidi, and A. Balzarotti, Appl. Phys. Lett. **78**, 320 (2001)
44. H. Tsuyoshi, S. Tsukamoto, and Y. Arakawa, Jap. J. Appl. Phys. **45**, L777 (2006)
45. M.C. Xu, Y. Temko, T. Suzuki, and K. Jacobi, Surf. Sci. **589**, 91 (2005)
46. E. Placidi, F. Arciprete, F. Patella, M. Fanfoni, E. Orsini, and A. Balzarotti, J. Phys. Cond. Mat. **19**, 225006 (2007)
47. Y Temko, T. Suzuki, M.C. Xu, and K. Jacobi, Surf. Sci. **591**, 117 (2005)
48. M.C. Bartelt and J.W. Evans, Phys. Rev. B **46**, 12675 (1992)
49. J.G. Amar and F. Family, Phys. Rev. Lett. **74**, 2066 (1995)
50. G.S. Bales and D.C. Chrzan, Phys. Rev. B **50**, 6057 (1994)
51. P.A. Mulheran and J.A. Blackman, Phil. Mag. Lett. 72, **55** (1995);
52. P.A. Mulheran and J.A. Blackman, Phys. Rev. B **53**, 10261 (1996)
53. J.A. Stroscio and D.T. Pierce, Phys. Rev. B **49**, R8522 (1994)
54. Y. Ebiko, S. Muto, D. Suzuki, S. Itoh, K. Shiramine, T. Haga, Y. Nakata, N. Yokoyama, Phys. Rev. Lett. **80**, 2650 (1998)
55. P.A. Venables and D.J. Bell, Proc. R. Soc. London A **332**, 331 (1971)
56. A. Okabe, B. Boots, K. Sugihara, and S.N. Chin, Spatial Tessellations. Concept and Application of Voronoi Diagrams 2nd edn., New York: Wiley, (2000)
57. M.C. Bartelt, A.K. Schmid, J.W. Evans, and R.Q. Hwang, Phys. Rev. Lett. **81**, 1901 (1998)
58. M. Fanfoni, E. Placidi, F. Arciprete, E. Orsini, F. Patella and A. Balzarotti, Phys. Rev. B **75**, 205312 (2007)
59. F. Arciprete, E. Placidi, V. Sessi, M. Fanfoni, F. Patella, and A. Balzarotti, Appl. Phys. Lett. **89**, 041904 (2006)
60. F. Patella, F. Arciprete, E. Placidi, M. Fanfoni, V. Sessi, and A. Balzarotti, Appl. Phys. Lett. **87**, 252101 (2005)

Chapter 2
Self-assembly of InAs Quantum Dot Structures on Cleaved Facets

E. Uccelli, J. Bauer, M. Bichler, D. Schuh, J. J. Finley, G. Abstreiter, and A. Fontcuberta i Morral

2.1 Introduction

Strain induced self-assembled quantum dots (QDs) have been extensively studied in the past decade. Their unique properties are mainly derived from the carrier confinement leading to atom-like energy level structure [1]. In this context, QDs are nanostructures that are indeed expected to have a great impact in optoelectronic devices. Methods to control the size and therefore the emission properties have been successfully developed, but the control of the position has shown to be a more difficult task. QDs usually nucleate randomly on a substrate by self-assembly [2], and methods to control the position are based on inducing a preferential nucleation of the self-organized dots at certain points on a surface. The different approaches range from the use of lithographically patterned substrates to vicinal surfaces [3, 4, 5, 6, 7, 8].

Alternatively, the Cleaved Edge Overgrowth technique has been used for the fabrication of atomically precise one and zero dimensional quantum structures. In this case, the quantum wires (QWRs) and dots are defined by the change in confinement potential at the intersection of one or two quantum wells (QWs), respectively [9, 10].

In this chapter, we present a different technique to deterministically control the position of self-assembled InAs QD arrays at the nanometer scale on (110) crystal surfaces. The method is a combination of self-assembly and the Cleaved Edge Overgrowth technique. In the first section, the growth technique is introduced. Then, in section 2.3 the results concerning the growth are shown. The optical properties of the self-assembled QDs are explained in Section 2.4. Finally, in Section 2.5, future prospects of the technique are presented.

2.2 Growth Technique

2.2.1 Cleaved Edge Overgrowth

The concept of Cleaved Edge Overgrowth (CEO) for the synthesis of quantum dot arrays (QDs) is illustrated in Fig. 2.1. The CEO technique was proposed by Störmer, Gossard and Wiegmann more than 20 years ago [10] and realized for the first time

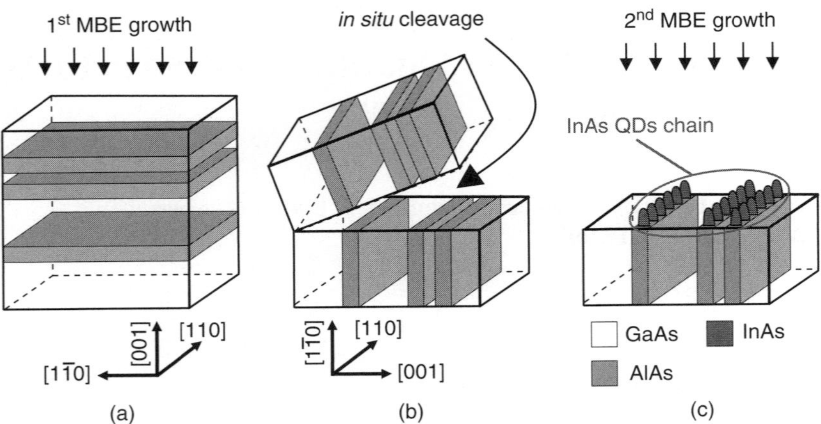

Fig. 2.1 The concept of the Cleaved Edge Overgrowth technique. (a) First MBE growth step, with the growth of AlAs/GaAs layers on a (100) wafer. (b) *in situ* cleave along the (110) plane. (c) Second MBE step with the overgrowth of InAs and formation of aligned InAs nanostructures on the cleaved facet

by *Pfeiffer et al.* at AT&T Bell Laboratories in 1990 [11]. It consists of two MBE growth steps separated by an *in situ* cleavage. The first MBE growth step is realized on a normal (001) substrate. Then, the sample is removed from the chamber and thinned to a thickness of 100–120 μm. The thin wafer is subsequently scribed and cleaved into rectangular pieces of 7 by 7 mm. A scratch along the [1–10] direction is made where the future in situ cleave is to occur. After the cleavage along the [110] direction inside the MBE chamber a new (110) surface is obtained, which is exposed to the second growth.

In our case, the first MBE step consists of the growth of an alternating series of AlAs and GaAs layers on a semi-insulating (001) GaAs oriented substrate (Fig. 2.1(a)). The substrate temperature is 650 °C, the growth rate 1 Å/s for both GaAs and AlAs and the As_4 pressure 1.25×10^{-5} mbar. In the second MBE step, the material grown is InAs from 1 to 4 nominal monolayers (MLs). As it will be shown in this chapter, InAs tends to accumulate only on the AlAs regions, resulting in ordered arrays of InAs nanostructures such as quantum dots (QDs) and nanowires (NWs) (Fig. 2.1(c)). The successful growth of InAs on the cleaved facet has been found to have a particularly small process window: surface temperatures between 440 and 470 °C, In rate lower than 0.15 Å/s and As_4 pressure between 4 and 6.5×10^{-5} mbar.

2.2.2 *Mechanisms of QD Formation on (110) Surfaces*

The substrate composition and orientation has an important effect on the morphology of epitaxially grown nanostructures. In a strained system such as InAs on GaAs, the surface energy of the underlying substrate strongly influences the way the energy is released. Typically, the growth of InAs on (001) GaAs surfaces leads to the formation of QDs by the Stranski-Krastanov growth mechanism [2]. On the contrary, the same growth conditions on the (110) GaAs surfaces do not lead to QD formation. In this case, the strain is released by the formation of misfit dislocations, leading to much bigger islands [12, 13, 14]. An Atomic Force Microscopy measurement (AFM) of the surface morphology of InAs growth on (001) and (110) GaAs surfaces is shown in Fig. 2.2. Clearly, a high density of small, relatively uniform lens-shaped InAs islands are obtained on (100) substrates, while on (110) surfaces much larger trapezoidal islands are observed.

In order to investigate how the surface energy influences the way in which the stress is released in mismatch systems, growth of InAs has been investigated on the (110) facets of different materials. This is easily implemented by the CEO technique, as the composition of the (110) surface is given simply by the design of the (001) substrate. Indeed, one starts fabricating substrates composed of several layers of different materials, such as GaAs, $In_{0.3}Ga_{0.7}As$, $Al_{0.45}Ga_{0.55}As$ and AlAs. Subsequently, in the second MBE step, InAs growth is performed on the (110) facet of these materials and the surface morphology is investigated by AFM.

The structure of the substrate is as follows: two different layer sequences (superlattices[1], in the following denoted as SL) of $In_{0.3}Ga_{0.7}As$-GaAs and AlAs-GaAs layers were grown on a semi-insulating GaAs (001) substrate. They are separated by a GaAs layer. (Sample-A). Here, SL1 consists of 5 periods of 20 nm $In_{0.3}Ga_{0.7}As$ and 40 nm GaAs, whilst SL2 is composed by 10 periods 20 nm AlAs and 20 nm GaAs. After *in situ* cleaving of the substrate in the MBE chamber, different amounts

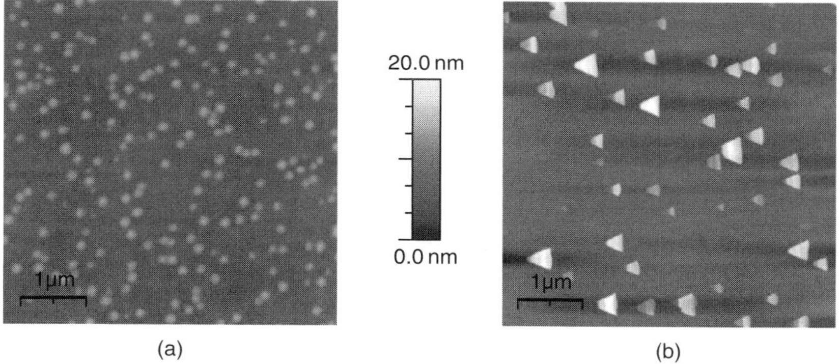

Fig. 2.2 Surface topology of InAs grown on (001) and (110) GaAs surfaces by AFM. (a) On (001) GaAs: high density formation of lens shaped QDs (b) Larger triangular and trapezoidal islands formation on (110) oriented GaAs surface

[1] For simplicity these layer sequences are called "superlattices" (SL) within this work. It has to be emphasized that this does not refer to the concept of overlapping wavefunctions and miniband formation but that this expression is only used to highlight the periodicity of the alternating layers grown.

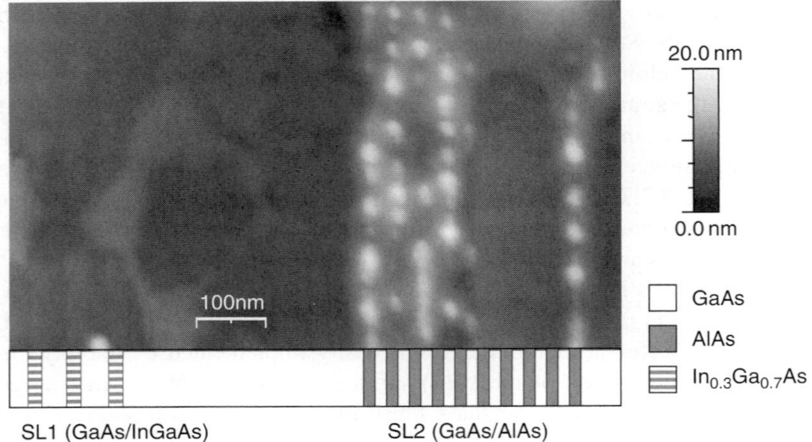

Fig. 2.3 Effect of the surface composition on the nucleation of QDs: Atomic Force Microscope measurement of the (1–10) surface of Sample A with both $In_{0.3}Ga_{0.7}As$/GaAs and AlAs/GaAs heterostructure for QDs template. Only on top of the AlAs stripes three dimensional QD like structures have been formed, after deposition of a nominal thickness of 3 ML InAs at a surface temperature of 460 °C

of InAs (namely 2–3–4 ML) were deposited directly on the exposed (1–10) surface at different growth temperatures (450–470 °C).

Figure 2.3 is a typical AFM measurement of the surface after deposition of 3 ML of InAs. As depicted, three dimensional islands or quantum dots (QDs) form on some of the stripes. QD formation is only observed on the AlAs stripes, and no QD nucleation has been observed on the $In_{0.3}Ga_{0.7}As$ stripes. The lack of QD nucleation onto $In_{0.3}Ga_{0.7}As$-GaAs SLs was a surprising observation, as it seems to contradict some published theoretical predictions and experimental works [15, 16]. However, it cannot be excluded that the MBE growth conditions were simply inappropriate ones for nucleation of QDs on (110) $In_{0.3}Ga_{0.7}As$. The same growth was tried on AlAs/$Al_{0.45}Ga_{0.55}As$ superlattices. As in the case of AlAs/GaAs, only on the AlAs stripes QDs were observed.

In the remainder of the chapter, we will focus on the growth of InAs nanostructures on the AlAs-GaAs (110) system.

2.3 Experimental Results

2.3.1 General Characteristics

To investigate the growth characteristics of InAs on AlAs stripes, a second substrate was used, composed of 5 different AlAs-GaAs SLs separated by a 1 μm GaAs layer (Sample-B). Each SL consists of 10 layers of AlAs separated by a 70 nm thick GaAs stripe. The thickness of the AlAs layers varied from 20 nm (SL1) up to 40 nm (SL5).

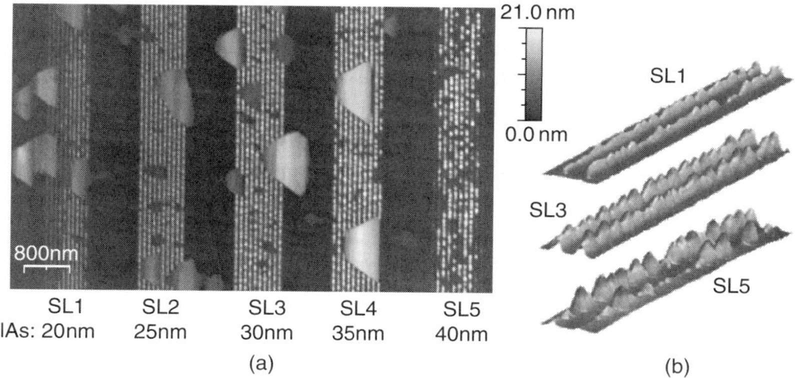

Fig. 2.4 General morphology of InAs growth on AlAs/GaAs SLs: (a) Atomic Force Microscopy of InAs QDs grown on a cleaved facet with AlAs/GaAs superlattices with 5 different AlAs stripe thicknesses (indicated on the picture). (b) 3-dimensional representation of the QDs grown on the stripes for three different stripe thicknesses

A typical Atomic Force Microscopy measurement of the cleaved-edge surface after the nominal growth of 3.2 ML of InAs is presented in Fig. 2.4(a). QDs have formed on the AlAs stripes for all SLs. For a more clear representation, a three-dimensional image of some of these arrays is presented in Fig. 2.4(b). Also visible in Fig. 2.4(a) are much larger triangular islands with height of some monolayers. These nanostructures correspond to the islands that have been already observed in MBE grown structures on (110) GaAs substrates [17].

In order to get a quantitative characterization of the morphology of the QDs, a detailed statistical analysis of the dot dimensions has been performed. As can be seen from Fig. 2.5(a), the lateral size of the QDs follows the thickness of the underlying AlAs stripe. The smallest dots are nucleated on SL1 (AlAs 20 nm) and largest dots on SL5 (AlAs 40 nm). As it will be shown in next section and according

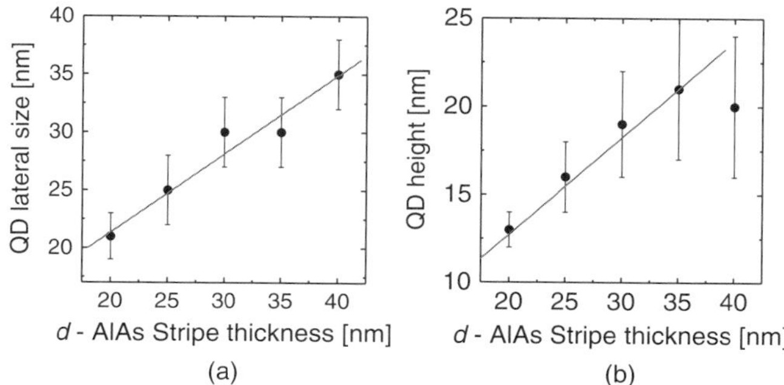

Fig. 2.5 Dimensions of the QDs as a function of the AlAs stripe thickness: (a) Lateral size (b) Height

to Fig. 2.5(b), the height of the dots increases linearly with the AlAs stripe thickness (from 13 nm on SL1 up to more than 20 nm for dots on SL4 and SL5). The average linear dot density along the AlAs stripes has also been estimated, but no specific dependency for density versus AlAs thickness has been found. For the calculations, regions with trapezoidal defects have been excluded, as these are considered to be strong perturbations to the dominant growth mode.

2.3.2 Growth Studies of the QD Formation Mechanism

In order to gain insight into the growth mechanisms of preferential migration of InAs on (110) AlAs regions and nucleation of QDs, a series of systematic growth runs were performed. The density and morphology of InAs nanostructures was studied as a function of the AlAs stripe width and nominal InAs thickness. For this particular study, the substrate (labeled as Sample-C) consisted of a series of AlAs stripes of thicknesses varying from 3 nm to 42 nm, separated by 70 nm GaAs spacers. A typical AFM measurement of a sample grown on this kind of superlattice is illustrated in Fig. 2.6. From the intensity scale, it is clear that the QD height increases with the AlAs stripe thickness. Indeed, on the thinner stripes the height does not differ much from the neighboring GaAs, which means that there seems to be almost no accumulation of InAs in these regions and no QDs are formed. In contrast, for the largest stripes the QDs reach heights up to 22 nm.

The size uniformity of QDs within a stripe was studied by measuring the height distribution from the AFM measurements. A variance of less than 17% was found in all cases. A much narrower size distribution would be highly desirable.

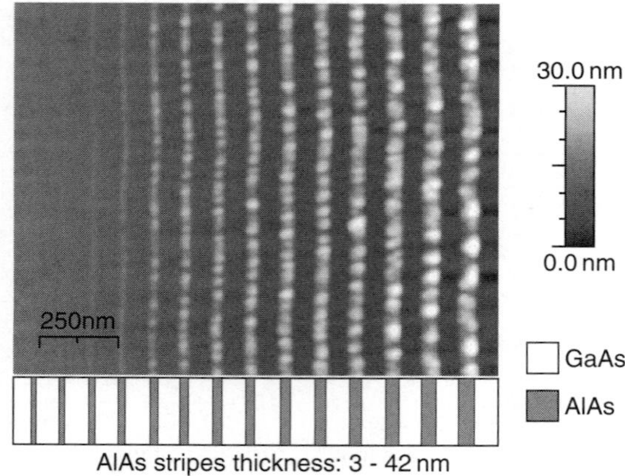

Fig. 2.6 Nucleation of InAs on AlAs stripes with increasing thickness: (a) AFM picture of Sample-C overgrown by 2.0 ML InAs. Formation of nanowires on narrow AlAs stripes as well as QDs chain on AlAs stripe equal and larger than 15 nm

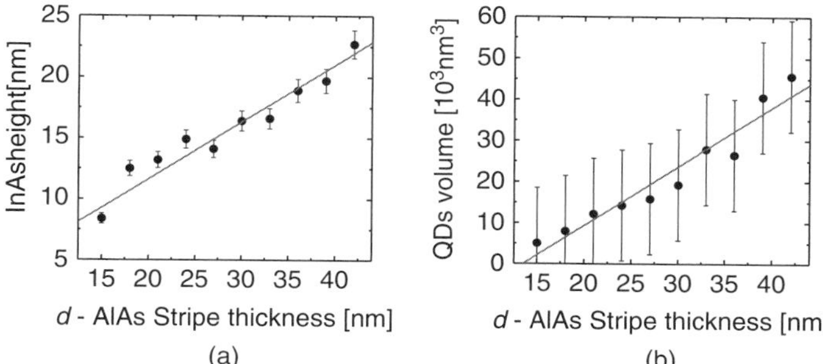

Fig. 2.7 Morphology of the InAs QDs for increasing AlAs stripe thickness relation to Fig. 2.6: (a) height and (b) volume of the QD's. The QD nucleation onset (Volume close to zero) occurs at $d = 15\,\text{nm}$

The mean height and volume of the QDs is plotted in Fig. 2.7 as a function of the stripe thickness, d. The QD volume and height increase linearly with the AlAs stripe thickness. This means that the nominal thickness is distributed unevenly through the surface. InAs tends to accumulate on AlAs stripes, increasing the total InAs effective thickness. This effect is linearly dependent on d. As it will be shown in the section 2.3.3, this is a consequence of the much lower diffusion coefficient and lifetime of the In-adatoms on (110) AlAs with respect to (110) GaAs. As the effective InAs thickness depends on d, the critical thickness needed for the nucleation of QDs, t_c, is achieved only on some of the stripes. On the thinnest stripes, the effective thickness lies below t_c and no QDs can form. This is exactly what it is observed experimentally. As shown in Figs. 2.6 and 2.7, in good agreement with this expectation, an onset of nucleation for the QDs is observed at a stripe thickness of 15 nm.

Even if the QDs lateral size and height scales with the AlAs stripe thickness, it is common sense that this can only occur up to certain values. Indeed, after a certain AlAs thickness, a transition to 2D growth should occur. In order to understand where this limit value lies, a second substrate (Sample-D) was designed with 80 and 100 nm wide AlAs stripes. An AFM measurement after growth of 3.2 ML is shown in Fig. 2.8. In this case, the single QD array is not observed anymore. However, instead of observing a transition to a random 2D arrangement, the QDs continue to grow aligned along the stripes. The difference here is that the QDs tend to nucleate next to the interface between stripes. This result may indicate that the QD formation does not follow exactly a Stranski-Krastanov growth mechanism [2] and deserves to be studied in more detail in the future.

2.3.3 Discussion of the Growth Mechanism

In this section we demonstrate, with a simple model, that the selectivity of the InAs nucleation on (1–10) surfaces can be understood as a consequence of the different

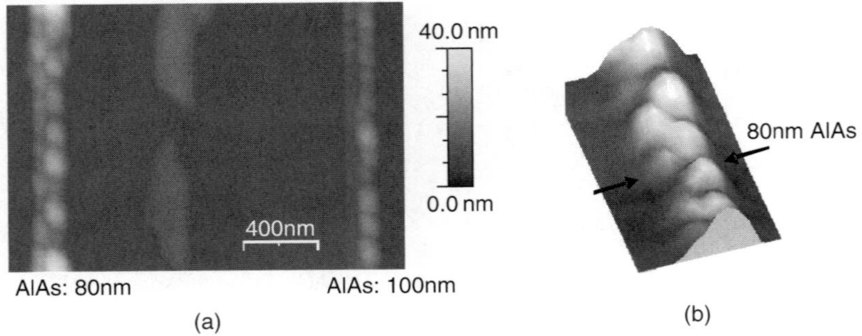

Fig. 2.8 Formation of QD pairs on thicker AlAs stripes: (a) AFM picture of Sample-D overgrown by nominally 3.2 ML of InAs. On very broad AlAs stripe (>80 nm) InAs can nucleate in bi-QD chain. (b) 3D picture of bi-QD chain

desorption rate and mobility of the impinging In-adatoms on AlAs and GaAs (110) growth surfaces [18]. As it has been previously reported, the diffusion length of the In-adatoms is more than three time larger on (001) Ga rich surfaces than on Al-rich (001) surfaces (35 nm vs. 10 nm) [19]. This ratio remains almost constant on (Ga,Al)As (110) surfaces, but the absolute diffusion length increases in each case by more than a factor of 30 [15, 20] (>1 μm vs. 350 nm). As a consequence, it is expected that In-adatoms would naturally tend to accumulate on the AlAs stripes rather than on the GaAs regions.

A very simple model can help understand how this happens. The main processes occurring at the growing surface are depicted in Fig. 2.9(a). First, Arsenic is a highly volatile atom which is continuously adsorbed and desorbed on the surface. On the contrary, after impinging on the surface, Indium diffuses along the surface before being incorporated in a film or desorbed. Depending on the temperature, surface composition and/or crystal orientation, the diffusion length of In can vary from several nm to microns. It has been shown that the In-adatoms mobility and desorption rate on Al-containing layers is lower [19, 20]. As it will be shown in the following, the difference in mobility and desorption rate leads to a higher effective mass flow towards the AlAs stripes [21]. We start by considering movement of In-adatoms in one dimension. Resulting from the combination of random walk and interaction with

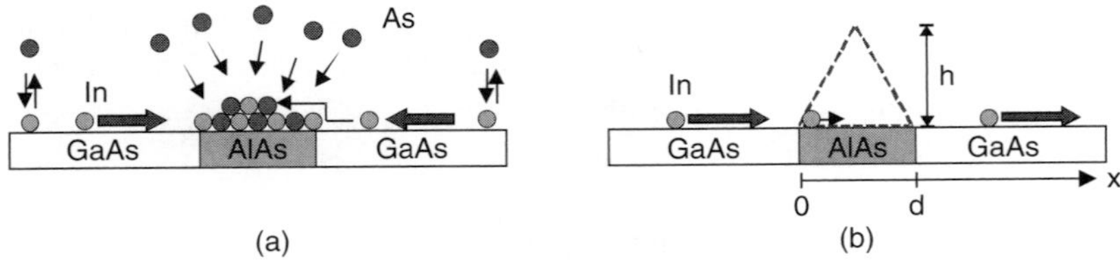

Fig. 2.9 Schematics of dynamics of In and As-adatoms on the growing surface (a) Mechanism for the preferential nucleation of InAs QDs on top of AlAs (110) stripes; (b) Schematics of the diffusion model presented

the underlying substrate, the effective movement of In-adatoms can be described by the diffusion equation with an effective diffusion coefficient, D. The diffusion coefficient depends on the type of material underneath and is significantly smaller for AlAs than for GaAs [19].

As described above, we assume that the movement of In-adatoms follows the equation

$$x(t) = x_o + \sqrt{D \cdot t} \tag{2.1}$$

where D is the diffusion coefficient that varies on top of AlAs and GaAs surface, and t is time. Per unit time, the probability of an In atom precipitating to form InAs is inversely proportional to the lifetime (τ) of In-adatom on the material. This means that for a certain time Δt, the amount of deposited In atoms will be proportional to this probability. If one now approximates the geometry of the QDs as triangular, the amount of InAs on top of the AlA region, V_{InAs}, will be:

$$V_{InAs} = \frac{\triangle t}{\tau_{AlAs}} = \frac{1}{2} h \cdot d \tag{2.2}$$

where and h and d are the height and width of the QDs.

Now, let us consider a single In ad-atom arriving at the interface between GaAs and AlAs ($x = 0$) and moving towards $x > 0$. By using the diffusion equation for the movement of In-adatoms (Eq. 2.1), one can calculate the average time $\triangle t$ that an In ad-atom would spend on the AlAs stripe. By introducing this value into Eq. 2.2 it is then possible to calculate the average height of the QDs as a function of the AlAs stripe thickness d:

$$h(d) \sim \frac{d}{D_{AlAs} \tau_{AlAs}} \tag{2.3}$$

The diffusion of In-adatoms on GaAs is not considered in this simple calculation, which is based on the assumption that the diffusion coefficient and lifetime of In-adatoms in (110) GaAs are much larger than in AlAs ($D_{GaAs} \gg D_{AlAs}$ and $\tau_{GaAs} \gg \tau_{AlAs}$). However, such simple calculations can help to understand intuitively why the effective thickness of InAs increases linearly with d and in what sense the values of D and τ play a role in the process.

2.4 Optical Properties

In this section, the optical properties of the InAs QDs on the AlAs stripes are investigated by micro-photoluminescence spectroscopy (μPL). In section 2.4.1, preliminary measurements comparing the substrate response with and without the InAs QDs will be presented. Then, in section 2.4.2 the properties of ensembles of QDs as a function of size will be presented. In section 2.4.3, the dependence of PL with the excitation will be shown for an ensemble of a few ODs. Finally, in section 2.4.4 the temperature dependence of the PL will be briefly discussed.

The samples were mounted in a vertical holder inside a liquid helium cryostat. PL was excited with a HeNe laser (632 nm), incident perpendicularly to the (1–10) surface. Due to the confocal excitation and collection geometry, the spatial resolution in both excitation and detection channels was 1.5 to 2.5 μm. The measurements were performed at temperatures from 10 to 85 K. Two different substrates (Sample-E and Sample-F) were investigated. Their structure is depicted schematically in Fig. 2.10(a–b). Sample-E consists of four spatially separated AlAs/GaAs SLs (SL1: five periods of 32 nm AlAs and 68 nm GaAs; SL2: five periods of 20 nm AlAs and 40 nm GaAs; SL3: five periods of 11 nm AlAs and 22 nm GaAs; SL4: ten periods of 20 nm AlAs and 20 nm GaAs) grown on semi-insulating (001) GaAs. Sample-F consists of a stack of SLs with different AlAs thicknesses. Each SL consists of a stack of 10 AlAs/GaAs, the thickness of AlAs varies from 20 to 40 nm, while the thickness of GaAs was kept at 70 nm.

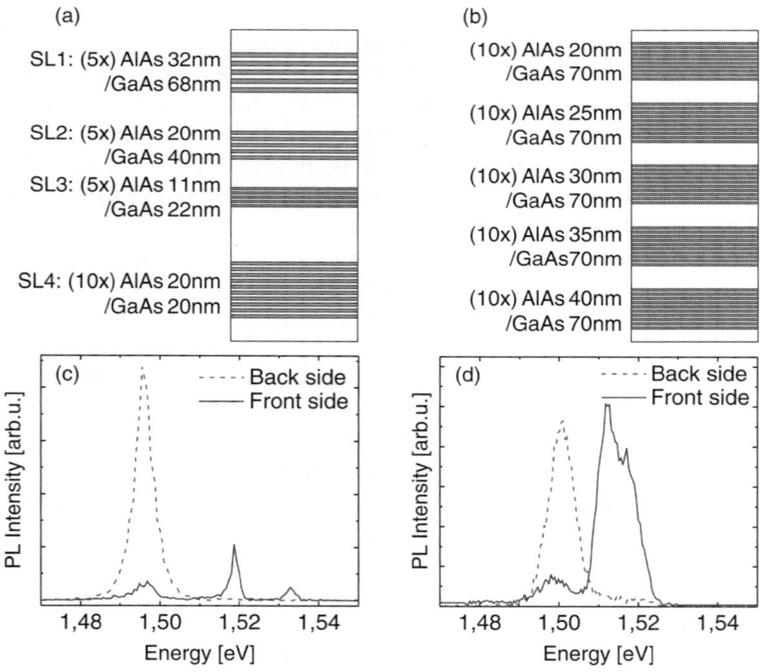

Fig. 2.10 Photoluminescence of the substrates: (a)–(b) Schematics of the substrate geometry of Sample-E and F. (c)–(d) PL-spectra from both front side (solid line) and back side (dashed line) of the substrates

2.4.1 Optical Response of the Substrates

Before studying in detail the optical properties of the InAs nanostructures, the macro-photoluminescence response of the underlying substrate was recorded, in order to have an optical reference to clearly distinguish the GaAs bulk material and the MBE grown layer sequences from the μ-PL measurements of the overgrown material.

Photoluminescence measurements of Sample-E and Sample-F are shown in Fig. 2.10(c–d). In all substrates, the epi-side was not thicker than 10 μm, much thinner than the thickness of the wafer (100–150 μm). Since the laser exciting perpendicular to the (001) surface of the substrate has a finite penetration depth [22], only material near the surface can contribute to the PL signal. Therefore, measurements on both front- and back-side of the samples can correspond respectively to the MBE grown layer sequences and the GaAs bulk material, giving a spectral fingerprint of the epitaxially grown layer.

The spectra confirm that the luminescence characteristics are strongly different. Front-side spectra show generally a stronger emission (single peak or a couple of peaks) around 1.515 eV and an emission with low intensity close to 1.50 eV. Back-side spectra show only one emission line near 1.50 eV. The peak at 1.519 eV (Fig. 2.10(c)) corresponds to GaAs energy gap [23], while we suggest that the peak at 1.533 eV (above the gallium arsenide band gap) arises from the weak quantum confinement effect in 20 nm GaAs stripe. This peak has been observed only in the Sample-E. According to [23], peaks at 1.517 eV and at 1.512 eV (Fig. 2.10(d)) arise from free and bound exciton (in GaAs). The peak around 1.499 eV (front and back side in both figures) is caused by the presence of carbon impurities in GaAs [23]. In Sample-F, the exciton peak of GaAs is significantly broadened and seems to be composed of more than one peak, shifting the whole band to lower energies. We attribute this shift to the presence of tensile stress of the GaAs sandwiched between the multiple layers of AlAs (difference in the lattice constant). To illustrate this phenomenon we have added the μPL scan of the (1–10) side of the substrate for Sample-F. It is clear from Fig. 2.11 that the shift increases in the growth direction with the number of SLs.

2.4.2 Optical Properties of Ensemble of QDs

The optical properties of QDs resulting from the growth of 3 ML of InAs capped by 50 nm of GaAs on both Sample-E and Sample-F were investigated by μ-PL spectroscopy. As has been presented in section 2.3, the size of the QDs is predetermined by the thickness of the AlAs stripe, which should be reflected directly in the optical properties. Additionally, the measurements should also give some measurement of the size homogeneity.

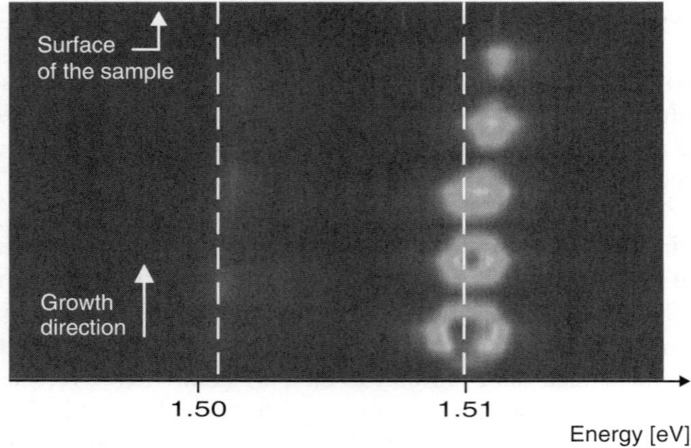

Fig. 2.11 μ-PL scan of substrate F: The excitonic peak clearly shifts to higher energies as more SLs are added

A 2.5 μm excitation spot was scanned across the (1–10) surface along the [001] direction of Sample-E in order to collect the PL response of the QDs arising from the different SLs. A strong luminescence signal coming from the excitonic recombination in InAs QDs was observed around 1.3 eV. The spectra in Fig. 12 correspond to the PL signal excited with a power of ~70m W/cm^2. In the figure, two photoluminescence spectra are presented. They were collected with the laser spot positioned on two different SLs. The peak mainly centered at 1.328 eV (labeled PL1) corresponds to the measurement realized on the QDs grown on 32 nm AlAs stripes (SL1). The second spectrum corresponds to the PL signal from the QDs grown on the 20 nm AlAs stripes (SL4) and consists of a main peak centered at 1.365 eV (labeled PL4). As expected, smaller QDs emit at slightly higher energies. In both spectra a second peak is present; whose origin is the PL emission of the QDs in the neighboring SLs. Both spectra are characterized by small line width

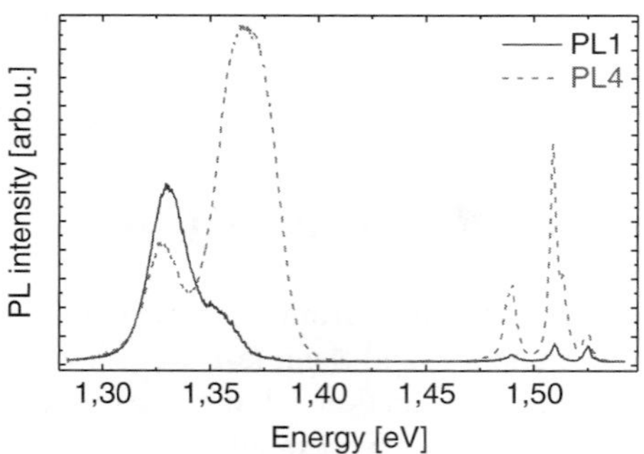

Fig. 2.12 μ-PL measurements of ensemble of QDs grown on Sample-E: spectra taken on different SLs, with AlAs stripe thickness of 32 nm (solid line) and 20 nm (dashed line). Both peaks present a clear Gaussian shape at 1.324 (1.3625) eV and show very narrow FWHM of 17.7 (25.0) meV

(∼25 meV and ∼18 meV), meaning a relatively good size homogeneity of the dots, in agreement with the AFM analysis presented above [24]. It is interesting to note that the obtained values for the line width are very close to some other very narrow measured values in InAs self-assembled QDs ensembles by other groups (∼16 meV e.g.[25,26]).

2.4.3 Power Dependence Measurements of Single QDs

In order to verify the QD nature of the InAs islands, we performed PL spectroscopy of single dots. For that, we scanned the stripes of low QD density with a 1.5 μm laser spot. In certain areas of low dot density, we found sharp line luminescence indicating PL of few individual dots. In these areas we performed power dependent μ-PL measurements. The power was varied from 1.8 to 1800 μW/cm^2. In Fig. 2.13, we show the result of such a measurement, where one quantum dot is contribuiting to the spectra. Only a single emission line at 1.308 eV is observed at low excitation power. The linear power dependence of the intensity identifies it as arising from a single exciton (X^0). In addition, this line is an emission doublet, possibly due to elongation of the quantum dots along the AlAs layer and the resulting anisotropic electron-hole exchange interaction [27]. Upon increasing the excitation power density, several sharp lines at lower (1.3044 eV) and higher (1.3135 eV, 1.3148 eV, 1.3166 eV and 1.3182 eV) energies emerge. The intensity of the emission line at 1.3044 eV increases roughly quadratically with excitation power density and dominates the spectra for the highest excitation densities investigated. This characteristic behavior identifies the peak as arising from bi-exciton (2X) recombination in the dot [28]). The other lines observed at high power arise from multi exciton complexes (mX) and possibly also charged excitons [29]. In conclusion, we have observed single dot like luminescence that confirms the QD nature of the InAs islands.

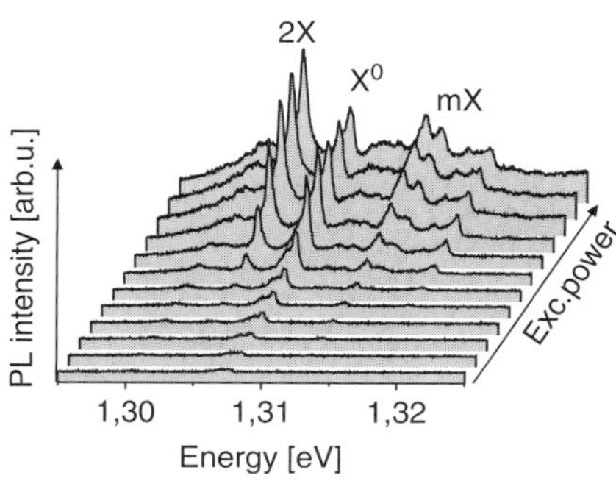

Fig. 2.13 PL spectra taken at different excitation power with spot laser localized only on few QDs

2.4.4 Temperature Dependence Measurements

A temperature series of the PL was realized on Sample-F to further elucidate the optical properties. The sample temperature was controlled by a resistive heater together with a silicon temperature diode placed near the sample in the cryostat. Care was taken to ensure that the temperature of the sample was stable. For all measurements, the excitation power ($\sim 4\,\mu\mathrm{W/cm}^2$) and the scanning position were kept constant. Figure 2.14 shows spectra recorded on the SL with AlAs stripes of 35 nm thickness of Sample-F.

The measurements clearly reveal a decreasing PL intensity as well as an increasing broadening of the PL, with rising temperature. This behavior can be explained as the enhancement of carrier escape processes from the quantum dot energy levels to the continuum states of the surrounding material at higher temperatures [1]. Due to the three-dimensional confinement, the localization energy for electron and hole is higher for QDs than for QWRs or QWs. Indeed, the integrated intensity of the PL lines I should follow the Arrhenius relation [30]:

$$I(T) = I_o \cdot \exp\left(\frac{-E_{loc}}{K_B T}\right) \tag{2.4}$$

where E_{loc} is the confinement energy, K_B the Boltzman constant and T the temperature. As a consequence, the escape rate for thermally activated carriers is smaller and hence the increase in PL line broadening is also smaller for QDs than for QWs or QWRs. This general behavior can be observed in Fig. 2.14. From the measurements, an E_{loc} of 18 meV was obtained. This seems to be a very low value (as 100 meV was expected, e.g. [31]). One possible origin could be the intermixing of the InAs dots with the surrounding materials, but more detailed investigations will be realized in the future in order to a better understanding.

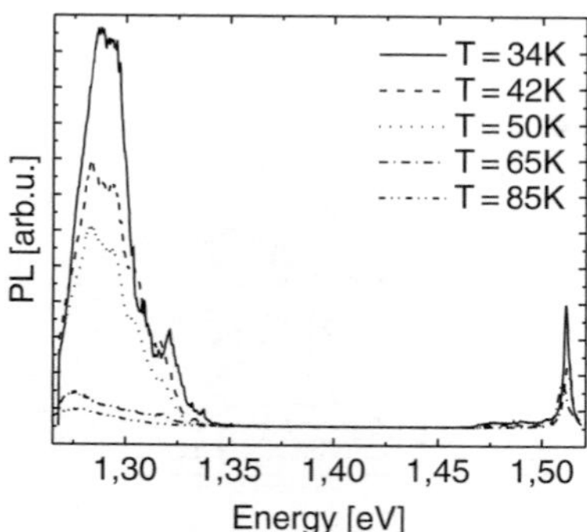

Fig. 2.14 Temperature dependent spectra taken from ensemble of QDs on Sample-F

2.5 Perspectives and Future Work

The successful control of the synthesis of QD arrays opens many perspectives in the fields of physics, electronics and information technology. It is relatively straightforward to imagine different devices that are enabled by this technique. Some of them will be briefly presented in this section.

One possibility is the use of the arrays as the channel of a transistor to investigate the transport properties of a QD chain. This can be achieved if the AlAs quantum well on which the QDs are nucleated contains a two dimensional electron gas (2DEG) and by depositing a thin gate on the (001) surface. As shown in Fig. 2.15, it is possible to separate the 2DEG into two electron reservoirs by applying a gate voltage, the distance between being defined by the gate thickness. Theoretically, the electronic transport from side to side of the 2DEG is only possible through the QD array.

Other possibilities are to use the optical functionality of the QDs for quantum information processing. Indeed, one can exploit the optical excitations in QDs to obtain coherent manipulation of two excitonic levels within one dot but also between coupled dots. The extrapolation to several dots is enabled by the work presented in this chapter and opens up a large variety of experiments where the coupling between single dots and chains can be investigated.

Finally, the structural characteristics of the QD arrays will need more careful investigation. In that respect, cross-sectional High Resolution Transmission Electron Microscopy and Raman spectroscopy measurements will be realized. These studies will provide not only information on the crystalline structure of the QDs, but also insights on the formation mechanism of the QDs. With these studies, we expect to answer the question of whether the QDs nucleation occurs via a Stranski-Krastanov or by an AlAs intermixing process.

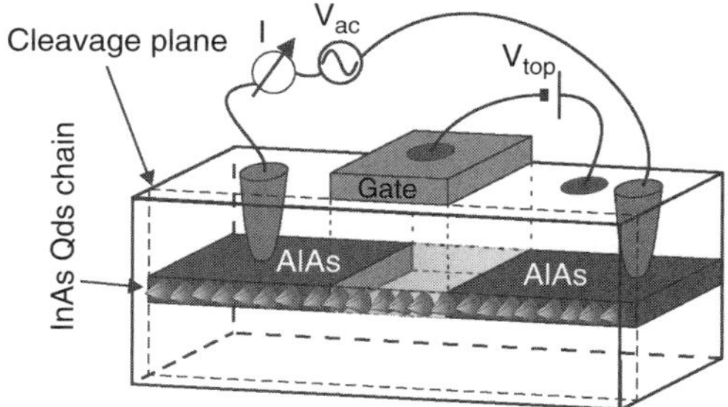

Fig. 2.15 Schematics of the configuration for a QD chain transistor device

Acknowledgments This work was financially supported by the Deutsche Forschungsgemeinschaft (DFG) in the framework of SFB 631 TP B1, by the Bundesministerium für Bildung, Wissenschaft, Forschung und Technologie (BmBF) through the Grant NanoQUIT-01BM469 and by the EC-Research Training Network COLLECT, HPRN-CT-2002-00291. The Marie Curie Excellence Grant program is also acknowledged (Project MEXT-CT-2006-042721, "SENFED").

References

1. Bimberg D, Grundmann M, Ledentsov NN (1999) Quantum Dot Heterostructures. John Wiley and Sons, Chichester
2. Stranski IN, Krastanow L (1938) Zur Theorie der orientierten Abscheidung von Ionenkristallen aufeinander. Sitzungsberichte der Akademie der Wissenschaften in Wien, Mathematisch-Naturwissenschaftliche Klasse, Abt. IIb, 146(1–10):797
3. Nötzel R, Ploog KH (2001) MBE of quantum wires and quantum dots. J. Cryst. Growth 227–228:8–12
4. Kim HJ, Park YJ, Park YM, Kim EK, Kim TW (2001) Fabrication of wirelike InAs quantum dots on 2°-off GaAs (100) substrates by changing the thickness of the InAs layer. Appl. Phys. Lett. 78:3253–3255
5. Leon Leon R, Chaparro S, Johnson SR, Navarro C, Jin X, Zhang YH, Siegert J, Marcinkevicius S, Liao XZ, Zou J (2002) Dislocation-induced spatial ordering of InAs quantum dots: Effects on optical properties. J. Appl. Phys. 91:5826–5830
6. Bhat R, Kapon E, Hwang DM, Koza MA, Yun CP (1988) Patterned quantum well heterostructures grown by OMCVD on non-planar substrates: Applications to extremely narrow SQW lasers. J. Cryst. Growth 93:850–856
7. Gerardot BD, Subramanian G, Minvielle S, Lee H, Johnson JA, Schoenfeld WV, Pine D, Speck JS, Petroff PM (2002) Self-assembling quantum dot lattices through nucleation site engineering. J. Cryst. Growth 236:647–654
8. Songmuang R, Kiravittaya.S, Schmidt OG (2003) Formation of lateral quantum dot molecules around self-assembled nanoholes. Appl. Phys. Lett. 82:2892–2894
9. Schedelbeck G, Wegscheider W, Bichler M, Abstreiter G (1997) Coupled quantum dots fabricated by cleaved edge overgrowth: From artificial atoms to molecules. Science 278: 1792–1795
10. Wegscheider W (2005) Cleaved Edge Overgrowth, T-Shaped Quantum Wires and Dots. In: Bryant GW, Solomon GS (eds) Optics of Quantum Dots and Wires. Artech House, Boston, pp 271–314
11. Pfeiffer L, West KW, Stormer HL, Eisenstein JP, Baldwin KW, Gershoni D, Spector J (1990) Formation of high quality two-dimensional electron gas on cleaved GaAs. Appl. Phys. Lett. 56:1697–1699
12. Belk JG, Sudijono JL, Zhang XM, Neave JH, Jones TS, Joyce BA (1997) Surface contrast in two dimensionally nucleated misfit dislocations in InAs/GaAs(110) heteroepitaxy. Phys. Rev. Lett. 78:475–478
13. Belk JG, Pashley DW, McConville CF, Joyce BA, Jones TS (1998) Surface morphology during strain relaxation in the growth of InAs on GaAs(110). Surf. Sc. 410:82–98
14. Joyce BA, Jones TS, Belk JG (1998) Reflection high-energy electron diffraction/scanning tunneling microscopy study of InAs growth on the three low index orientations of GaAs: Twodimensional versus three-dimensional growth and strain relaxation. J. Vac. Sci. Technol. B, 16:2373–2380
15. Wasserman D, Lyon SA (2004) Cleaved-edge overgrowth of aligned quantum dots on strained layers of InGaAs. Appl. Phys. Lett. 85:5352–5354
16. Zhao C, Chen YH, Cui CX, Xu B, Sun J, Lei W, Lu LK, Wang ZG (2005) Quantum-dot growth simulation on periodic stress of substrate. J. Chem. Phys. 123:094708–094711

17. Yoshita M, Oh JW, Akiyama H, Pfeiffer LN, West KW (2003) Control of MBE surface step-edge kinetics to make an atomically smooth quantum well. J. Cryst. Gr. 251:62–67
18. Shchukin VA, Ledentsov NN, Bimberg D (2004) Epitaxy of nanostructures, Springer-Verlag, Berlin
19. Ballet P, Smathers JB, Yang H, Workman CL, Salamo GJ (2001) Control of size and density of InAs/(Al,Ga)As self-organized islands. J. Appl. Phys. 90:481–487
20. Lobo C, Leon R (1998) InGaAs island shapes and adatom migration behavior on (100), (110), (111), and (311) GaAs surfaces. J. Appl. Phys. 83, 4168–4172
21. Bauer J, Schuh D, Uccelli E, Schulz R, Kress A, Hofbauer F, Finley JJ, Abstreiter G (2004) Long-range ordered self-assembled InAs quantum dots epitaxially grown on (110) GaAs. Appl. Phys. Lett. 85:4750–4752
22. Gustafsson A, Pistol ME, Montelius L, Samuelson L (1998) Local probe techniques for luminescence studies of low-dimensional semiconductor structures. J. Appl. Phys. 84:1715–1775
23. Pavesi L, Guzzi M (1994) Photoluminescence of $Al_xGa_{1-x}As$ alloys. J. Appl. Phys.75:4779–4842
24. Schuh D, Bauer J, Uccelli E, Schulz R, Kress A, Hofbauer F, Finley JJ, Abstreiter G (2005) Controlled positioning of self-assembled InAs quantum dots on (110) GaAs. Physica E 26:72–75
25. Costantini G, Manzano C, Songmuang R, Schmidt OG, Kern K (2003) InAs/GaAs (001) quantum dots close to thermodynamic equilibrium. Appl. Phys. Lett. 82:3194–3196
26. Yang T, Tatebayashi J, Tsukamoto S, Nishioka M, Arakawa Y (2004) Narrow photoluminescence linewidth (<17 meV) from highly uniform self-assembled InAs/GaAs quantum dots grown by low-pressure metalorganic chemical vapor deposition. Appl. Phys. Lett. 84:2817–2819
27. Bayer M, Ortner G, Stern O, Kuther A, Gorbunov AA, Forchel A, Hawrylak P, Fafard S, Hinzer K, Reinecke TL, Walck SN, Reithmaier JP, Klopf F, Schäfer F (2002) Fine structure of neutral and charged excitons in self-assembled In(Ga)As/(Al)GaAs quantum dots. Phys. Rev. B 65:195315–195337
28. Brunner K, Abstreiter G, Böhm G, Tränkle G, Weimann G (1994) Sharp-Line Photoluminescence and Two-Photon Absorption of Zero-Dimensional Biexcitons in a GaAs/AlGaAs Structure. Phys. Rev. Lett. 73:1138–1141
29. Findeis F, Zrenner A, Böhm G, Abstreiter G (2000) Optical spectroscopy on a single InGaAs/GaAs quantum dot in the few-exciton limit. Solid State Comm. 114:227–230
30. Kapteyn CMA, Lion M, Heitz R, Bimberg D, Brunkov PN, Volovik BV, Konnikov SG, Kovsh AR, Ustinov VM (2000) Room-temperature 1.3 mm emission from InAs quantum dots grown by metal organic chemical vapor deposition. Appl. Phys. Lett. 76:1573–1575
31. Chu L, Zrenner A, Bichler M, Böhm G, Abstreiter G (2001) Intersubband photocurrent spectroscopy on self-assembled In(Ga)As/GaAs quantum dots. Phys. Stat. Sol. (b) 224: 591–594

Chapter 3
InAs/GaAs Quantum Dots with Multimodal Size Distribution

Udo W. Pohl

3.1 Introduction

Semiconductor research and applications have experienced a progressive reduction in dimensionality of structures from bulk, through quantum wells and wires to quantum dots (QDs). QDs provide an ultimate limit of charge carrier confinement. Their size is in the nanometer range, i.e. of the order of the exciton Bohr radius, giving rise to fully quantized confined electron and hole states. Their electronic properties differ significantly from those of systems with higher dimensionality and offer potential applications in novel optoelectronic devices.

The most effective method for the fabrication of coherent, dislocation-free semiconductor QDs is the strained-layer epitaxy in the Stranski–Krastanow mode [1, 2]. Such self-organized transformation from two-dimensional layer-by-layer growth to a three-dimensional mode was found for many heterostructure systems. InAs quantum dots in GaAs matrix represent the best studied model system. The discrete energy states and the ease of integration into an optoelectronic device recently lead, e.g., to the presentation of electrically driven single photon sources [3]. Prerequisite for practical applications is the ability to control the electronic properties of the dot. Fabrication of quantum dots by applying Stranski-Krastanow growth is hampered by the action of entropy at growth temperature, usually leading to ensembles of dots which vary in size, shape, and composition. Consequently the energies of bound states vary within the ensemble. Since this growth mode still yields defect-free dots with superior electronic properties, the technique was recently advanced to improve the shape homogeneity within the ensemble and the integrity of the dot-matrix interfaces [4]. The development aimed to find a growth regime which simultaneously favors the formation of equilibrium-near dots with homogeneous composition and sharp interfaces. This chapter addresses the growth control of such particular kind of InAs dots in GaAs matrix and their electronic properties.

3.2 Decomposition of the Size Distribution into Subensembles

An ensemble of self-organized QDs usually contains a broad distribution of QD sizes, often accompanied by a spread of QD's compositions due to intermixture with matrix material. These fluctuations of single QD properties lead to a statistical distribution of the eigenenergies within the ensemble. Consequently, the luminescence of the QD ensemble is a convolution of sharp single QD spectra with the distribution function [5], typically yielding an inhomogeneously broadened band with a gaussian shape of up to 100 meV half width. An early study on chemical beam epitaxy of strained InAs quantum wells in InP matrix indicated the existence of a growth regime which simultaneously produces equilibrium-near non-intermixed dots and sharp interfaces [6]. The introduction of a growth interruption (GRI) after InAs deposition and prior to InP cap layer growth lead to layers which show up to eight well separated emission lines. The lines were attributed to one-monolayer (ML) thickness fluctuations of the InAs/InP well, as previously also reported for lattice-matched materials GaAs/AlGaAs and InP/GaInAs. The interpretation was improved later, stating that three-dimensional islands with heights corresponding to integral numbers of InAs MLs form under the given growth conditions [7, 8].

More recent studies on the Stranski-Krastanow growth of InAs demonstrated that such growth regime also exists for QD formation in GaAs matrix and proved the dot height quantization. Figure 3.1 shows PL spectra of two samples with thin InAs layers in GaAs, grown using metalorganic vapor phase epitaxy. In one sample the GaAs cap layer was immediately grown after the deposition of the InAs layer, in the other sample a short growth interruption was introduced prior to GaAs cap deposition, keeping other growth parameters unchanged. The first sample shows a narrow PL which corresponds to an e_0-hh_0 exciton recombination in a pseudomorhic InAs/GaAs quantum well of ~1.8 ML thickness, in good agreement with the amount of deposited material. The asymmetry at the low-energy side is well

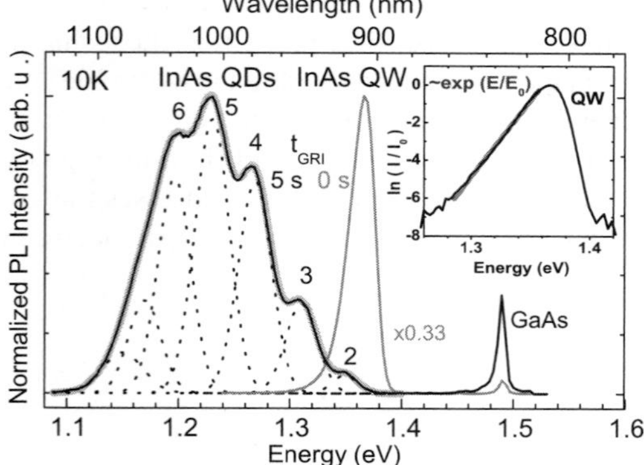

Fig. 3.1 PL spectra of InAs/GaAs samples grown without (grey curve) and with $t_{GRI} = 5$ s growth interruption. The thick grey curve is the sum of the dotted Gaussian fits, GaAs denotes matrix near band-edge emission. The inset shows the QW spectrum in a semilogarithmic scale. After [4], inset from [9]

described by an exponential function $I \propto \exp(E/E_0)$, cf. inset in Fig. 3.1. The dependence indicates quantum well (QW) states with a density-of-states tail originating from potential fluctuations due to thickness variations [9]. This proves the formation of a rough InAs/GaAs quantum well if no GRI is applied.

The sample grown with an additional growth interruption of 5s clearly shows PL of a QD ensemble peaking near 1.2 eV. The apparent modulation of the QD-ensemble emission originates from the decomposition into narrow emission lines of subensemles. The subensembles are related to InAs QDs which differ in height by one InAs ML [4, 10]. The numbers on the peaks in Fig. 3.1 mark the heights of the respective subensemble QDs in ML units. The multimodal nature of the ensemble size-distribution was concluded from a combined structural, optical, and theoretical study described in more detail in the following sections.

3.2.1 Initial Stages of Dot Formation

If InAs is deposited on GaAs instead of InP the strain is nearly doubled. The critical layer thickness for elastic strain relaxation in Stranski-Krastanow growth is hence reduced. A critical value near only 1.5 ML [2] leads to the requirement of an atomically flat GaAs surface prior to InAs deposition to obtain well-defined bottom interfaces of the dots. This is easily achieved by an appropriate process [11]. The InAs/InP islands were reported to form from a two-dimensional InAs layer during the growth interruption [6]. This means that the deposited amount of InAs was close to the critical value for the Stranski-Krastanow transition on InP. To study the respective conditions for deposition on GaAs, transients of the anisotropic reflectance were recorded in-situ during the growth of In(Ga)As layers with strain below and just above the critical value for elastic relaxation. A fixed photon enery of 2.6 eV was selected due to its high surface sensivity [12]. The Stranski-Krastanow transition was retarded by composition with Ga to reduce the strain in the layers. This allowed for optically tracing the response during deposition. Transients of reflection anisotropy-spectra (RAS) of such InGaAs/GaAs layers show oscillations, which are particularly pronounced at high Ga composition, cf. Fig. 3.2. The oscillations originate from reconstruction domains of reduced As coverage close to monolayer steps which occur in the island growth mode at the low deposition temperature used for dot growth [13]. The damping of the modulation for increasing In content is attributed to a combined effect of a more group-III-rich surface due to the reduced As-In bond strength as compared to Ga-As and that of a transition to step-flow growth due to the higher surface mobility of In species as compared to Ga.

The InGaAs layer with 65% Ga exceeds the critical strain for dot formation. The respective RAS signal shows just two monolayer oscillations. Dot formation does not lead to a clear feature in the response. The apparent drop is related to a slight depletion of As at the surface due to an interruption of As precursor supply during GRI. The RAS response during GRI basically originates from the In(Ga)As wetting

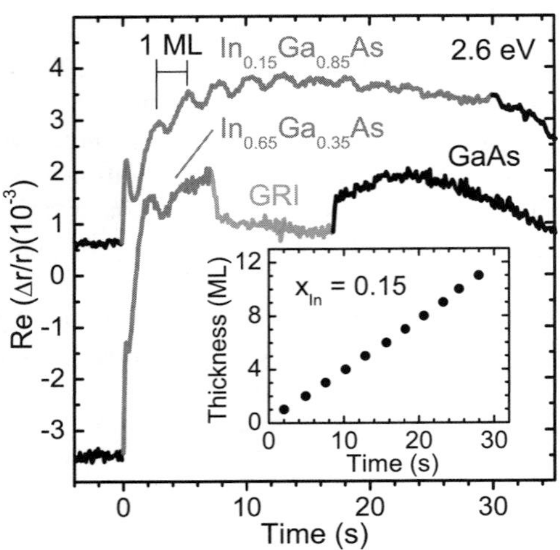

Fig. 3.2 Transients of the anisotropic reflectance recorded during metalorganic vapor phase epitaxy of InGaAs layers with different In compositions on GaAs at 485°C. GRI denotes a growth interruption without precursor supply. The inset shows the temporal thickness increase in units of monolayers. From [27]

layer, the slow changes are assigned to an optically detected coarsening due to dot ripening. The in-situ RAS study clearly demonstrates that a smooth growth front is maintained during strained layer deposition on GaAs prior to dot formation.

The in situ study indicates that comparable conditions as reported for InAs on InP are hence also given for the deposition of InAs on GaAs, encouraging growth of a multimodal QD ensemble. To induce such QD formation, ~1.9 ML thick InAs layers were deposited at typ. 490°C with a comparatively high deposition rate of 0.4 ML/s. Growth interruptions of variable duration were applied after InAs deposition and prior to GaAs cap layer deposition, showing a strong effect on the structural and optical properties of the layers [4, 9]. The PL of a sample series grown with gradually prolonged duration of short growth interruptions shows how the multimodal dot ensemble forms from the initial rough QW, cf. Fig. 3.3. Without interruption a PL maximum at 1.38 eV is found, corresponding to the exciton recombination in a QW. Comparison of the low-energy emission tail with the peak-energy separations of QD subensembles indicates significant localization in the DOS of the QW and a related strong roughness. Application of a 0.5s growth interruption leads to a small blue shift of the PL maximum and some broad emission band around 1.3 eV. These features originate from thinning of the QW and an enhanced appearance of locally thicker parts of the QW. The blue shift of the QW-related PL saturates for longer interruptions, and the emission of a multimodal QD ensemble with individual subensemble peaks evolves from the shallow localizations in the DOS tail of the rough QW. The material of the initial InAs deposition hence partially concentrates at some QD precursors of dots located on a wetting layer of contant thickness. The existence of such a constant, 1 monolayer thick wetting layer consisting of pure InAs was proved by PL excitation studies discussed below and confirmed by TEM investigations [4].

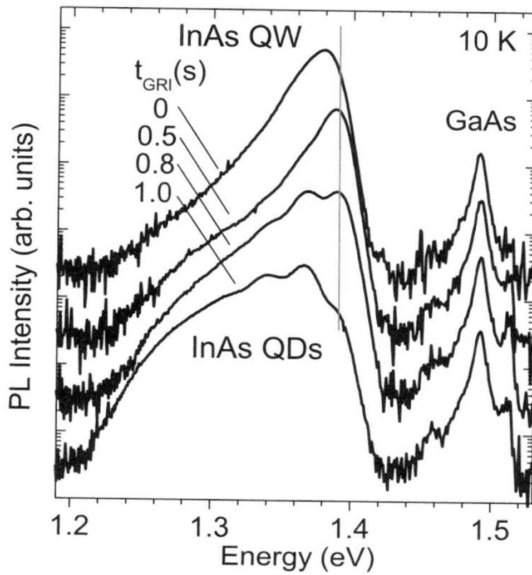

Fig. 3.3 PL spectra of InAs/GaAs layers grown with different durations t_{GRI} of short growth interruptions prior to GaAs cap layer deposition. Curves are vertically shifted by one decade for clarity, the grey vertical line is a guide to the eye. From [27]

3.2.2 Evolution of the Dot-Size Distribution

The detection of the quantization of dot heights within a multimodal ensemble provides a sensitive tool to trace the ripening of dots during the growth interruption. For longer GRI durations the PL energy of the QD ensemble shifts to lower energy due to an average dot-size increase. The emission lines of individual subensembles do *not* shift to the red. The lines rather shift to the blue as shown in Fig. 3.4 by comparing, e.g., corresponding emissions of samples grown with 2 s and 10 s GRI. Since the height of QDs in a given ensemble remains constant for varied GRIs the shift indicates a lateral dissolution of subensembles of smaller dots and a materials redistribution to form subensembles of larger dots. PL excitation spectra recorded in the respective maxima of samples grown with different GRIs show heavy hole and light hole exciton-resonances of the wetting layer QW remaining at fixed energies for all samples, cf. Fig. 3.4c. Thickness and composition of the wetting layer consequently remain unchanged during the ripening of the dots. The *hh* and *lh* exciton transition energies correspond to a QW of 1 ML thickness and pure InAs composition. This means that the evolution of the QD ensemble proceeds solely by mass transport among the dots.

The dynamics of evolution is revealed by analyzing the integral PL intensity of individual subensembles at various stages of the evolution shown in Fig. 3.4. Model calculations of excitons confined in dots of varied sizes show that the relative oscillator strength of the emission is largely insensitive to the size of the QDs [14]. The integral intensity of the subensemble peaks is hence considered a reasonable direct measure for the number of QDs with a specific corresponding height.

The temporal evolution of these intensities given in Fig. 3.5 directly shows the formation of subensembles with larger dots by dissolution of the subensembles built

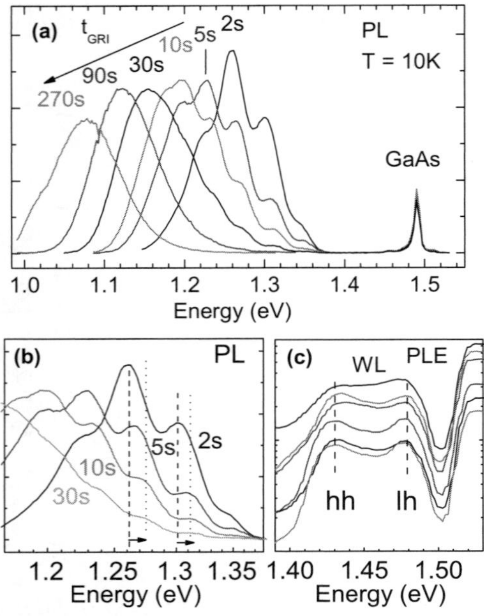

Fig. 3.4 PL (a, b) and PL excitation spectra (c) of InAs/GaAs samples with multimodal QDs grown with varied growth interruptions t_{GRI}. WL denotes the wetting layer with resonances of the heavy hole and light hole. (a) and (c) from [4]

by smaller dots. The number of, e.g., dots with 5 ML height increases during the initial 5s growth interruption on expense of those with 2 to 4 ML height. Then the number of these dots decreases – either by dissolution to provide material to increase the number of 6 ML high dots, or by growth. After 10s GRI also the 6 ML high dots start to dissolve.

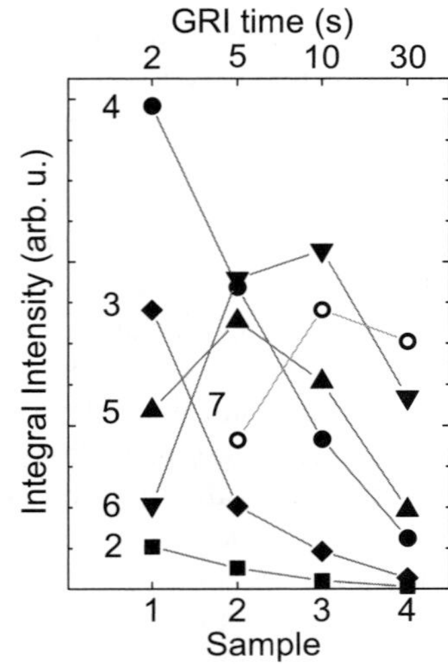

Fig. 3.5 Integral PL intensities of subensemble peaks measured from samples grown with various growth interruptions. The numbers refer to QD heights in monolayer units as indicated in Fig. 3.1. From [14]

3.2.3 Structural Properties of Multimodal InAs/GaAs Dots

The assignment of the modulation observed in the PL of the dot ensemble (Fig. 3.1) to a discretization of dot heights and a consequential decomposition of the ensemble into subensembles is substantiated by cross-sectional transmission-electron micrographs. The images shown in Figs. 3.6 and 3.7 were recorded along (100) using the strain sensitive (220) reflections, the chemically sensitive (200) reflections, and the direct beam. InAs dots then appear as bright central areas, as shown in the overview Fig. 3.6. The dark areas above the dots – for larger dots also below – correlate to strain fields in the surrounding GaAs matrix. The dots have a low aspect ratio with flat top and bottom facets. Their lateral extension is about 10 nm.

Details of the dot shape are obtained from high resolution images. The local chemical composition is obtained from Fourier-filtered images using the {200} components and the transmitted beam in the reconstruction. Figure 3.7 shows a HRTEM image of a dot and the result of such image processing. The crystal lattices of the InAs dot and the GaAs lattice are apparently different and allow for a quantitative analysis. The base and the top layer of the flat dots consist of plain, continuous InAs, forming atomically sharp interfaces to the cladding GaAs matrix. No significant InAs-GaAs intermixing is found. The dots consequently differ in height by integer numbers of InAs monolayers. The side facets are less well resolved in the images, due to high strain perturbations at the base perimeter of the QDs. They appear as steeply inclined facets, giving the QDs a shape of a truncated pyramid. Such a result was also observed in cross-section scanning tunneling images of multimodal InAs/GaAs QDs [15], see top image in Fig. 3.7. The filled-state STM image shows horizontal lines due to the III-V zigzag chains at the scanned zincblende (110) surface, one such line corresponds to two atomic monolayers. Dots appear as bright contrast with sharp interfaces to the matrix, confirming flat top and bottom facets built of pure InAs.

Theoretical studies on the shape of InAs/GaAs dots were reported particularly for uncovered or partially covered structures. Truncated pyramids with flat top and bottom facets, as found for the covered InAs/GaAs QDs studied here, agree with predictions [16, 17] and findings [18] of equilibrium shapes of uncovered dots. The observation of truncated pyramidal shape for *covered* dots is in agreement with a recent STM study on the capping of free standing InAs dots [19]. During the initial stages of cap-layer deposition the dots rapidly shrink vertically, building flat structures which are close to equilibrium shapes.

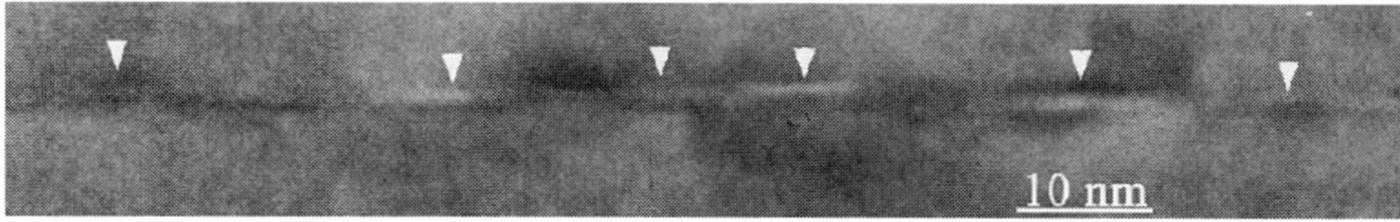

Fig. 3.6 Cross section transmission electron micrograph of InAs/GaAs dots growth with 5 s GRI, imaged along [100] using strain and chemically sensitive reflections. White arrows mark areas with contrast changes due to flat dots

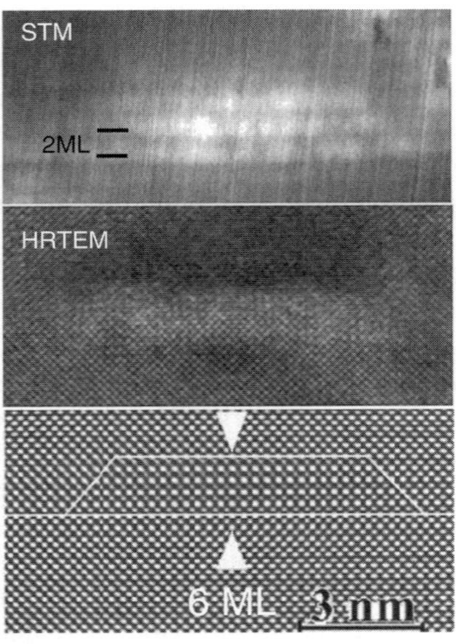

Fig. 3.7 Cross-section scanning tunneling (top) and high resolution transmission-electron micrograph (middle) of InAs/GaAs QDs with 6 ML height. Bottom: Fourier-filtered image of the HRTEM micrograph. Top image from [15], middle and bottom from [4]

The evolution of the dot ensemble by dissolution of small dots described in Section 3.2.2 is also reflected in the QD areal density. Plain view bright-field TEM images given in Fig. 3.8 show that the QD density decreases from $4.0 \times 10^{10}\,\mathrm{cm}^{-2}$ for 5s growth interruption to $1.7 \times 10^{10}\,\mathrm{cm}^{-2}$ for 90s [4]. The simultaneously increased sizes of the dots induce larger strain fields. This leads to higher contrast in the images recorded using the strain-sensitive (220) reflection and slightly off-axis two-beam conditions (images for 5s and 90s GRI). The black-white contrast in these images is oriented along the (220) diffraction vector. No dislocated islands were observed in the samples, except for the longest GRI of 270s where a high density of misfit dislocations ($1.9 \times 10^{8}\,\mathrm{cm}^{-2}$) which originate at InAs islands was found. For such samples containing dislocated islands, the investigation of strained-QD evolution is complicated due to a faster growth of relaxed islands, i.e. an increased material transfer to these regions [20]. The density of not dislocated dots has then consequently dropped, in this sample to $0.9 \times 10^{10}\,\mathrm{cm}^{-2}$. The study of coherently strained dots did hence not include such long growth interruptions.

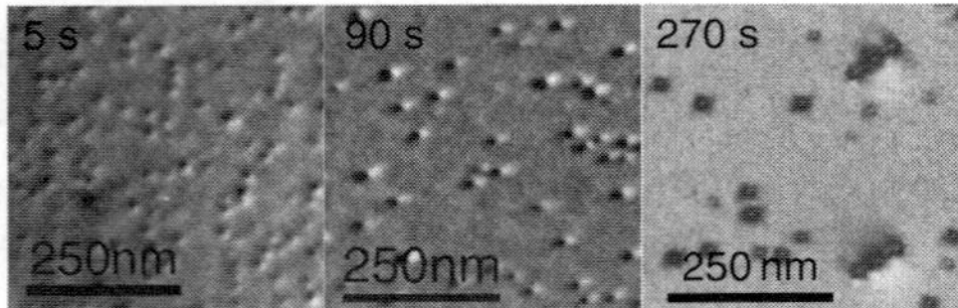

Fig. 3.8 Plan view transmission-electron micrographs of InAs/GaAs QD samples grown with different growth interruptions as indicated. From [4]

3.2.4 Kinetic Model of Multimodal Dot Ensembles

The formation of a dot ensemble with a multimodal size distribution may theoretically be described using a kinetic approach [4, 14]. The model outlined in Fig. 3.9 assumes strained dots of truncated pyramidal shape being surrounded by an adatom sea which represents the InAs wetting layer. Growth and dissolution of a dot occurs by attachment and detachment of adatoms. The kinetics is basically controlled by the stress which is concentrated at the base perimeter of a dot. There the dot's elastic energy creates a barrier for the nucleation at the side facets. The barrier increases with the height of a dot. Therefore growth by adding new layers on the top facet is favored, keeping the dot base constant. Material for growth is provided from dissolution of smaller dots.

A partially grown layer on the top facet of a dot leads to a change of the Gibbs free energy of the system, given by

$$\delta\Phi_{dot} = \delta E_{elast} + \delta E_{edge} - \mu\delta N \tag{3.1}$$

δE_{elast} is the elastic relaxation-energy change and leads to a decrease of Gibbs free energy as the number of atoms in the dot increases. δE_{edge} is the step energy of the additional partial layer on the top facet and increases with the length of its perimeter. δN is the number of atoms in the partial layer, that are transferred from the adatom sea. μ is the chemical potential of the adatom sea. $\delta\Phi_{\mathrm{dot}} > 0$ at the initial growth stage of an additional layer, but eventually becomes negative when this layer exceeds a critical size. Gibbs free energy then decreases if the dot grows. Each time a top layer of a dot is completed the Gibbs free energy has local minimum, cf. Fig. 3.10. Since the size of the top facet of the truncated pyramid decreases with height, it will at a certain height become too small to further favor formation of a subsequent layer. Growth of this dot will then stop. QDs with larger initial bases will therefore stop growing at a larger height.

For numerical modeling, the exchange of adatoms between sea and dot is described by rate equations [14]. An Arrhenius dependence in the probability of

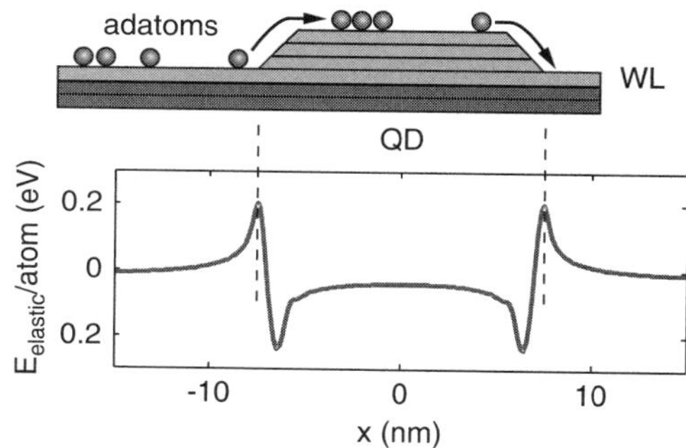

Fig. 3.9 Model of an InAs dot on a 1 ML thick wetting layer. Bottom: Elastic energy density of the dot

dot-height increase or decrease $W(b, h \rightarrow h \pm 1)$ accounts for the strain-dependent barrier,

$$W(b, h \rightarrow h \pm 1) = \omega \exp(-\delta\Phi_{barrier}/k_B T). \tag{3.2}$$

An equal prefactor ω is assumed for all dots. The temporal evolution of the dot heights h and base lenths b in the ensemble is described by the distribution function $F(b, h, t)$ which obeys a master equation

$$\frac{\partial F}{\partial t} = \sum W \times F, \tag{3.3}$$

with a sum over all height increases ($h-1 \rightarrow h$ and $h \rightarrow h+1$) and decreases ($h+1 \rightarrow h$ and $h \rightarrow h-1$). Growth or dissolution of the dots proceeds by mass exchange between the dots and the adatom sea. For the growth interruption, the mass conservation criterion reads

$$q_{\text{adatom}}(t) + \sum_b \sum_h F(b, h, t)\, V(b, h) = Q = const. \tag{3.4}$$

The number of adatoms q_{adatom} in the adatom sea decreases as a dots grows, thereby lowering the chemical potential μ due to the relation

$$q_{\text{adatom}} \propto \exp(\mu/(k_B T)), \tag{3.5}$$

and consequently raising $\delta\Phi_{\text{dot}}$. An increase of Gibbs free energy leads at an early stage of evolution (μ_1) to a larger critical size of the additional layer on top of a dot, see Fig. 3.10. At a later stage (μ_2) both growth of high dots and dissolution of small dots lead to a decrease of the free energy. The driving force for dissolution acts inverse to that of growth, i.e. predominantly at the respective topmost layer of

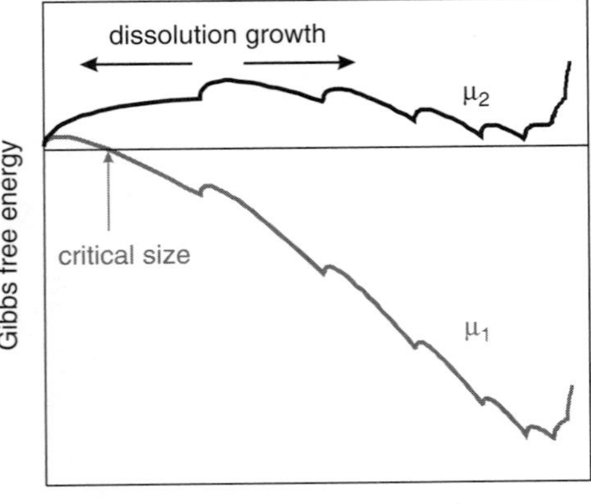

Fig. 3.10 Gibbs free energy of a dot with truncated pyramid shape due to consecutive growth of layers on the top facet. Grey and black curves refer to adatom sea chemical potentials at an early (μ_1) and a later stage ($\mu_2 < \mu_1$) of the evolution, respectively. From [4]

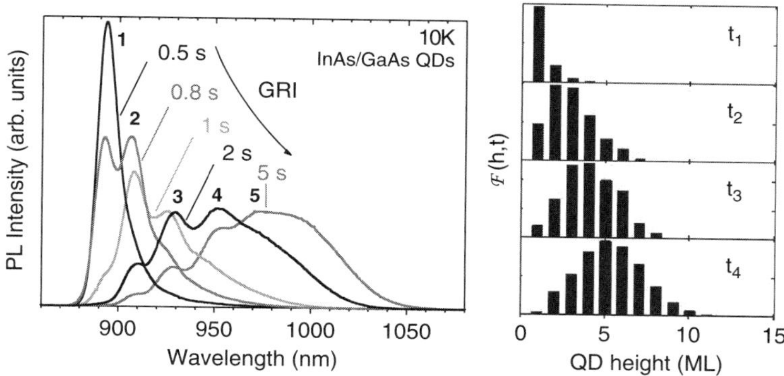

Fig. 3.11 Measured (left panel) and simulated initial stages (right panel) of the evolution of a dot ensemble with a multimodal size distribution

the top facet. Therefore small dots are expected to dissolve by a reduction of height, feeding their material to the adatom sea. Parameters to account for δE_{elast} and δE_{edge} of InAs/GaAs dots were adopted from continuum elasticity and density-functional theory, respectively.

Results of a numerical solution of the coupled set of Eqs. (3.3) to (3.5) are given in Fig. 3.11 for different stages of the evolution [14]. The simulated ensemble initially consists of 1 ML high dots and some minor contributions of higher dots. These dots have a broad distribution of base lengths b. At later stages the number of 1 ML high dots gradually decreases and the maximum of the distribution function F shifts towards higher dots by either dissolution or enlargement of more shallow dots. The model accounts for major features of the experimental data: Formation of a dot ensemble with a multimodal size distribution, an evolution by dissolution of smaller and growth of larger dots, dissolution of the dots by reduction of height, and the persistence of a flat top facet.

3.3 Electronic Properties of Multimodal InAs/GaAs Dots

The detailed knowledge about the structural properties of InAs/GaAs quantum dots in an ensemble with a multimodal size distribution provides an excellent opportunity to study the complex interplay of Coulomb interaction and external confinement for confined charge carriers. The equally shaped dots with well-defined sizes allow for a direct correlation of structural and optical properties, and an analysis of the origin of excitonic properties by realistic model calculations.

3.3.1 Size-dependent Energy Levels

Generally differences in the size of QDs lead to a spread of optical transition energies within a QD ensemble. The apparent decomposition of the ensemble PL into subensemble emissions shows that the statistical deviations around the mean values within a subensemble are smaller than the differences of the mean values of adjacent subensembles. Spectral broadening of a given subensemble due to variations of QD base length, remaining interface roughness and composition fluctuations is hence smaller than the energy shift caused by the addition of one InAs monolayer to the height. The energy shift between subsequent subensemble maxima is expected to decrease with increasing height, in qualitative agreement with the experiment. The shifts of the emissions depicted in Fig. 3.1 descease from ~40 meV for shallow dots to 30 meV with increasing height. The effect of monolayer splitting is even more pronounced if antimony is added during formation of the dots, acting as a surfactant and leading to smoother interfaces [4, 10]. In the following such a sample is used to prove that the spectral separation of subensemble emission maxima is due to a discrete variation of QD height in steps of InAs monolayers, accompanied by a simultaneous increase in base length.

The photoluminescence of an InAs/GaAs QD ensemble with a pronounced multimodal size distribution is given in Fig. 3.12 [21]. An excitation density below 4 mW/cm^{-2} ensured an average occupation below one exciton per dot, excluding the occupation of exited states. This is confirmed by PLE spectra Fig. 3.13 which do not show excitation transitions corresponding to emissions in the PL spectrum.

The PL spectrum shows excitonic ground-state emissions of eight subensembles with mutual energy spacings from 51 to 29 meV with increasing QD height. The

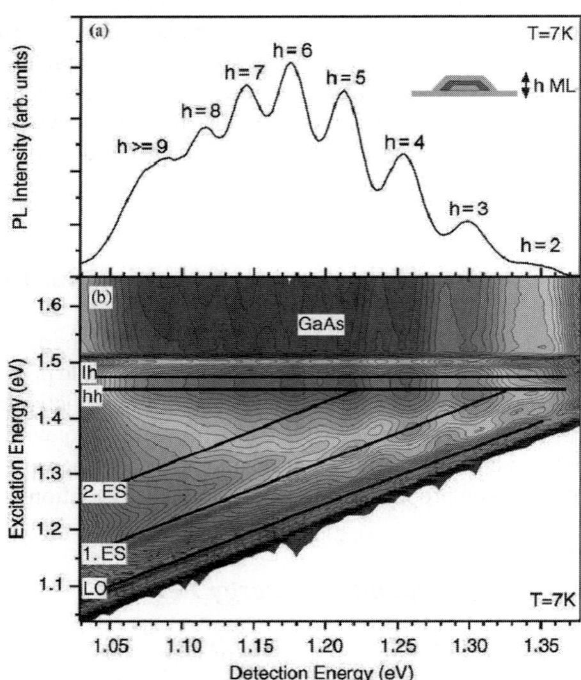

Fig. 3.12 PL spectrum (a) and PLE contour plot (b) of an InAs/GaAs QD ensemble with multimodal size distribution. Horizontal black lines indicate light-hole (*lh*) and heavy-hole (*hh*) resonances of the wetting layer, inclined lines mark the first and second excited QD level and the 1-LO phonon transition. From [21]

emission energy of a subensemble with a given height is smaller than that found in samples grown without antimony, cf. Fig. 3.1. This indicates a larger lateral extension of the QDs formed under mediation of Sb.

The excited exciton states of the QDs are assessed by PLE spectra. Figure 3.12b displays the PL intensity on a logarithmic scale as a function of both detection and excitation energies. The resonance of the heavy-hole exciton in the wetting layer at 1.45 eV marks the onset of the excitation continuum that limits the localization in the QDs. Due to this common resonance for all QDs in the ensemble exciton localization decreases from ~360 meV for high QDs to ~100 meV for shallow QDs with 2 ML height. QDs with a ground-state transition energy below 1.22 eV have at least two excited states (1.ES and 2.ES), whereas smaller QDs with transition energies between 1.22 eV and 1.32 eV have only one. Quantum dots with transition energies above 1.32 eV have no bound excited state.

Transitions into excited states are also clearly resolved in the PLE spectra given in Fig. 3.13. They reveal the first excited state about 120 meV above the ground state. Such large substate splitting supports a high InAs content in small QDs. The quantization of the very shallow QDs with 3 and 2 ML height apparently decreases. This effect is attributed to delocalization of the first excited state into the wetting layer.

Based on the structural data of the QDs presented in Section 2.3 the energy of excitons confined in such truncated pyramidal InAs dots was calculated [10, 22, 23]. The three-dimensional model used 8-band **k.p** theory and included a configuration interaction scheme. The **k.p** method is based on the envelope-function ansatz of Bloch's theorem, stating that the electron (or hole) wave function in a crystal with translational symmetry can be separated into a Bloch part and an envelope part. The Bloch part oscillates with the atomic distance as period, while the smooth envelope function only varies on mesoscopic distances. The oscillating Bloch functions can be eliminated from the electron Hamiltonian, leaving eigen-equations of the

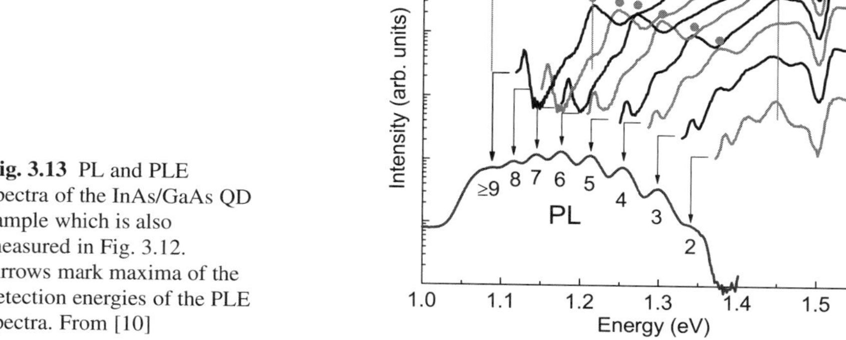

Fig. 3.13 PL and PLE spectra of the InAs/GaAs QD sample which is also measured in Fig. 3.12. Arrows mark maxima of the detection energies of the PLE spectra. From [10]

envelope functions only. In a zincblende semiconductor the single-particle wave function $\Psi(\mathbf{r})$ is given by

$$\Psi(\mathbf{r}) = \sum_{\mathrm{R}} \sum_{i=1}^{8} \psi_i(\mathbf{R}) w_i(\mathbf{r} - \mathbf{R}), \tag{3.6}$$

ψ being the 8-dimensional envelope part determining the amplitude of Ψ at the locations $\mathbf{R}$ of the atoms, and w_i is the Wannier (Bloch) part featuring the S or P character. The eight enveloppe functions of ψ belong to the electron of the lowest conduction band, the heavy hole, the light hole, and the split-off hole of the uppermost valence band for the two orientations of the spin, respectively. Two-particle exciton energies are calculated using the configuration-interaction (CI) method, including excited-state configurations in the basis set [23]. Modeling thus accounts for direct Coulomb interaction, correlation and exchange. Since both the structure and the size of the studied InAs/GaAs QDs are well known due to their multimodal nature and all prominent interactions are included, the modeling provides a realistic picture of the confined particle's properties. Modeling starts assuming shape, size and composition of a QD. Then the strain distribution and the consequential piezoelectric potential are calculated [24]. Solution of the Schrödinger equation with these potentials yields the single-particle wave-functions. Parameters for the calculation are taken from experimental values for the bulk material Γ-point band structure. It should be noted that the model does not contain adjustable parameters. Excitonic properties are finally calculated in the basis of the single-particle states and the mentioned interaction schemes.

The outlined calculation was applied to pure InAs/GaAs QDs with truncated pyramid shape, (001) bottom and top facets, and {110} side facets. In the model an additional 1 ML thick InAs wetting layer is assumed, in agreement with the heavy- and light-hole energies observed in the PLE spectra. The calculation leads to a good agreement to the experimental results if an appropriate base length is assumed. The energies of the ground and excited states in a small QD of 3 ML height are well explained, if a base length of 10.2 nm is used, see Fig. 3.14. However, keeping this base length for higher dots results in too high predicted transition energies. Similarly energies of a 9 ML high dot are well explained using 13.6 nm base length, but such base length yields too low predicted energies for smaller dots. Consequently both, height and base length actually increase for larger dots, as proved by the excellent agreement of the calculated energies connected by the black lines in Fig. 3.14. Quantum dots in the ensemble with a multimodal size distribution hence show a gradual shell-like increase in volume.

In section 3.2 the dynamics of QD evolution was assessed using the integral intensities of the subensemble emissions. This approach is justified by a calculation of the relative oscillator strength, proving that this quantity is insensitive to the size of the QDs. As shown in Fig. 3.15 the calculated ground-state wave-functions of electron and hole do not vary strongly if height and base length of a QD are changed from 3 ML / 10.2 nm to 13 ML / 15.8 nm, respectively. Consequently the overlap

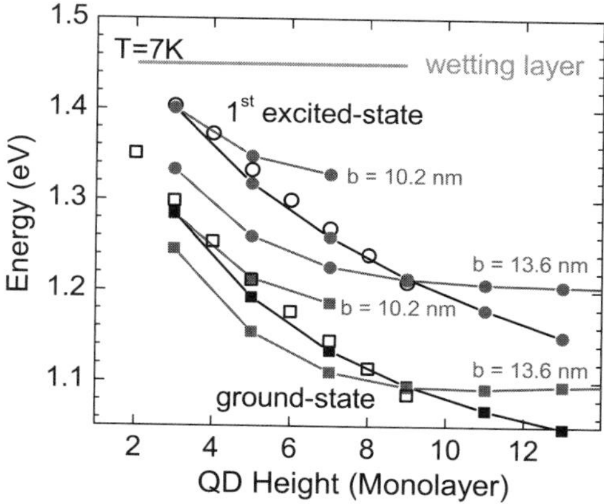

Fig. 3.14 Calculated (solid symbols) and experimental (open symbols) energies of the exciton ground state (squares) and first excited state (circles) for truncated pyramidal InAs QDs. Black lines connect calculated data for varied base length *b*. From [10]

hardly changes for QDs considered here. In the strong-confinement regime the oscillator strength of the ground-state transition is ruled solely by the ground-state electron-hole overlap [25]. Even the largest QDs still provide a strong confinement, electron and hole wave-functions are hence spatially squeezed.

3.3.2 Binding Energy of Confined Exciton Complexes

For an engineering of quantum dots as building blocks in novel applications knowledge about the electronic structure of few-particle states and the relation of struc-

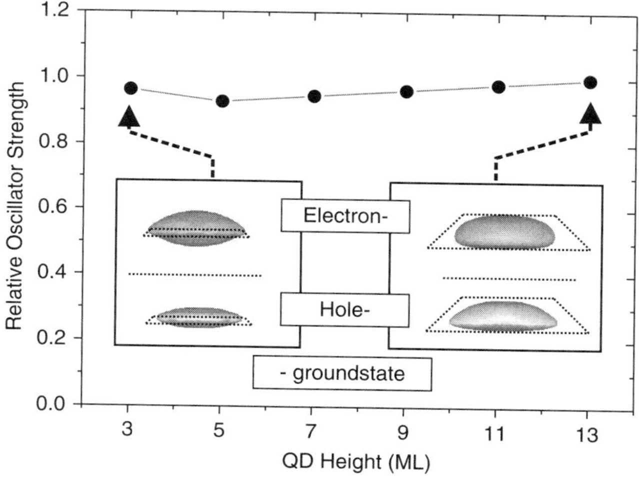

Fig. 3.15 Relative oscillator strength of the exciton emission in an InAs/GaAs QD of truncated pyramid shape. The inset shows calculated e_0 and hh_0 wave functions for 3 ML high QDs (left) and 13 ML high QDs. From [14]

tural and electronic properties is essential. Inhomogeneous broadening of the QD ensemble or subensemble is generally much larger than energy differences between states of various excitonic complexes or exchange energies. Therefore single-QD spectroscopy has been established as a powerful tool to study few-particle effects by excluding influences of inhomogeneous broadening. Results presented below were obtained using spatially resolved cathodoluminescence (CL) with an optically opaque near-field shadow-mask made of Au, which is evaporated onto the sample surface to reduce the number of simultaneously probed dots. For ~100 nm apertures and typically 4×10^{10} dots cm^{-2} about four QDs are detected. Further selection is achieved spectrally, and an unambiguous assignment of spectral lines to a single specific QD is provided by exploiting the omnipresent effect of spectral diffusion [26].

Single-dot spectra allow to gain insight into the Coulomb-affected few-particle properties. An important aspect of Coulomb interaction is the renormalization of few-particle transition energies occurring when an electron and a hole confined in a QD recombine in the presence of additional charge carriers. The biexciton binding energy is of particular interest for applications, e.g. for emitters of polarization-entangled single-photon pairs from excitonic and biexcitonic transitions. Its value is given by the energy difference between excitonic and biexcitonic recombination. Representative spectra of a large and a small InAs/GaAs QD are shown in Fig. 3.16. The different excitonic transitions within a single QD spectrum were identified via polarization and excitation dependend measurements [21]. Due to the statistical nature of carrier capture into the dots emission from neutral excitons (X), biexcitons (XX), as well as from charged excitonic complexes are observed in the studied case.

A remarkable feature of the spectra of a large and a small dot in Fig. 3.16 is the reversed order of exciton and biexciton emission. Another apparent feature discussed in Section 3.3 is the large fine-structure splitting of the large dot's exciton emission in the upper spectrum compared to the small splitting of the small dot.

For a systematic investigation of the relation between excitonic binding energies and structural properties of the QDs many single-dot spectra were recorded all over

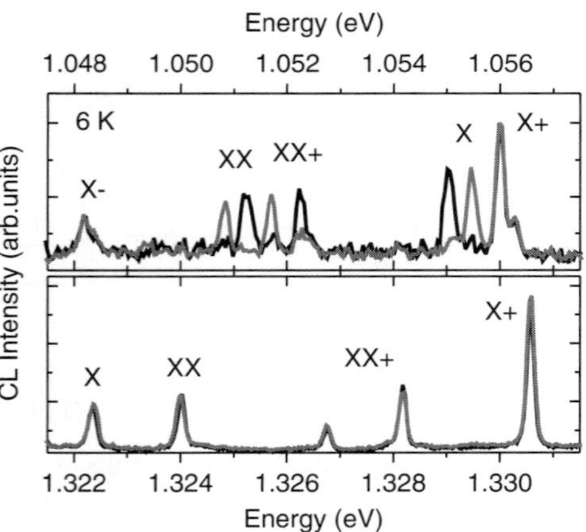

Fig. 3.16 Catholdoluminescence spectra of a single large (top) and a small dot (bottom), linearly polarized along $[\bar{1}10]$ (π^+, black spectra) and [110] (π^-, grey spectra). From [27]

the inhomogenously broadened ensemble peak. The resulting binding energies for the biexciton and the two trions are plotted in Fig. 3.17 as a function of the neutral exciton recombination energy. There exist apparent characteristic trends and energy regimes for the three excitonic complexes. The negatively charged exciton (also termed negative trion) has always a positive binding energy, i.e. the emission is shifted to the red, $E(X) - E(X^-) > 0$. In contrast to this behavior, the binding energies of the biexciton and the positive trion decrease as the exciton energy increases, i.e. the dot gets smaller. The energy of the biexciton even changes from binding to antibinding.

The binding energies originate from the Coulomb interaction between the confined charge carriers. This interaction is discussed in terms of direct Coulomb interaction and correlation to understand the observed trends. The energy regimes of the two trions can be explained by the direct Coulomb interaction alone [28]. The negative trion X^- consists of two electrons and one hole confined in the QD. Its binding energy directly depends on the difference between the two direct Coulomb terms $C(e,h)$ and $C(e,e)$. The binding energy of the positive trion X^+ correspondingly depends on $C(e,h)$ and $C(h,h)$. Due to the larger effective mass of holes and the small size of the QDs the wave function of the hole is stronger localized than that of the electron, cf. wave functions in Fig. 3.15. Consequently $|C(e,e)| < |C(e,h)| < |C(h,h)|$, and the negative trion has a positive binding energy, while the positive trion has a negative binding energy.

As to the trends of the binding energies the effect of correlation is to be considered. To model the impact of correlation the number of confined QD states for electrons and holes included in the calculation was varied [28]. The procedure is motivated by the observation in Fig. 3.12 that the number of confined states decreases for increasing ground-state recombination energy, i.e. for smaller dots. In the numerical model the number of bound states is expressed by the size of the function basis, i.e. the number of electron and hole levels used to build the exciton basis-states for

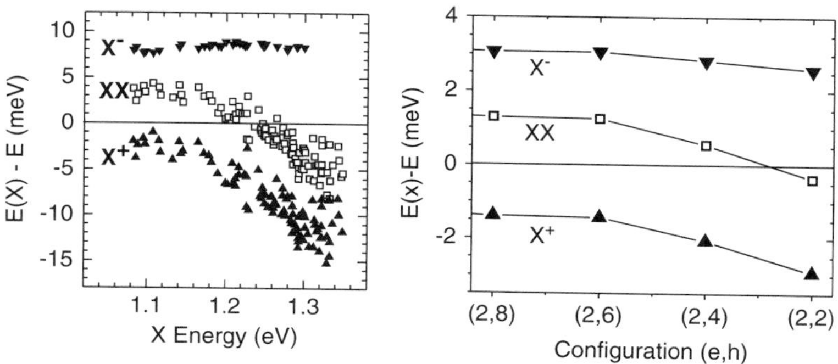

Fig. 3.17 Left: Energy distances between the recombination energy of confined exciton complexes and that of the neutral single exciton as a function of the single exciton emission energy. Right: Calculated energy distances of the recombination energy of confined exciton complexes with respect to that of the neutral single exciton as a function of the number of states included in the configuration interaction. After [28]

the CI calculation. By comparing calculations for different configurations it became obvious that the number of bound *hole* levels is the main factor governing the trend of the binding energies. The right panel of Fig. 3.17 shows the result of calculations where the number of electron states is kept constant at two and the number of hole levels is varied from eight for large QDs to two for small QDs. Obviously the trends and energy regimes observed in the experiment are well reproduced.

The study on single quantum dots demonstrates that the order of binding energies of confined few-particle complexes is controlled by the mutual direct Coulomb interaction. The magnitude of the binding energies depends on the size of the quantum dots and is governed by the number of bound states.

3.3.3 Anisotropic Exchange Interaction

The polarized spectra of a large and a small QD in Fig. 3.16 showed a splitting of the neutral exciton and biexciton transitions. Both emissions are split into two lines of equal intensity, and in both cases the order of their polarization with respect to the transition energy is reversed. The splitting of these emissions is identical. This more clearly recognizable in Fig. 3.18 which shows the respective emissions of two other single QDs. The splitting is defined to be positive if the (110) polarization appears at lower energy in the exciton doublet, i.e. the X line at lower energy is π^+ polarized. In Fig. 3.18 positive and negative splittings can be observed. The other polarization-dependend lines appearing in Fig. 3.16 which originate from the recombination of the positively charged biexciton and the positive trion change intensity as a function of polarization direction. All these features originate from the anisotropic part of the exchange interaction among the confined electrons and holes and are discussed in the following.

The identical, so-called fine-structure splitting (FSS) of the exciton and biexciton emissions results from a lifting of the degeneracy of the bright exciton state, which

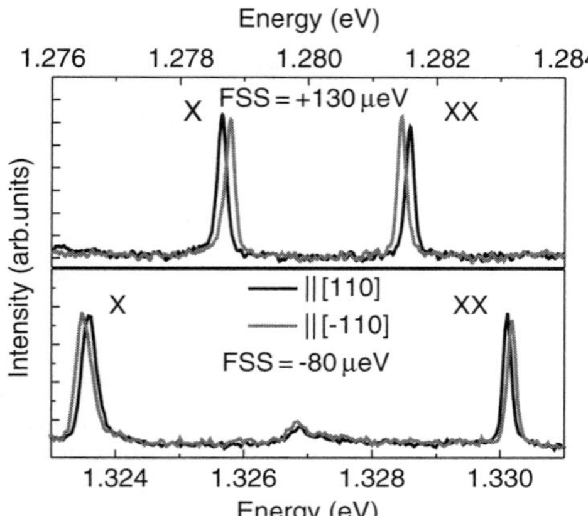

Fig. 3.18 Polarized spectra of two quantum dots with reversed order of polarization in the exciton emission

is the final state of the biexciton decay and the initial state of the exciton decay. Its origin is the exchange interaction, which couples the spins of electrons and holes, and their Zeeman interaction with internal and external magnetic fields [29]. Since the Zeeman contribution is generally negligible in absence of an external magnetic field it will not be considered here. We focus on the exciton ground state in crystals with zincblende structure like InAs and GaAs which have T_d point group symmetry.

The exciton consits of an electron and a hole. Only heavy-hole states are considered for simplification, though the calculations presented below include light-hole and split-off hole contributions. The ground state is then composed of angular-momentum projections $s_z = \pm 1/2$ and $j_z = \pm 3/2$, yielding four exciton states with momentum projections M_z equal to ± 1 and ± 2. In absence of Coulomb interaction and band- and spin-coupling effects, the exciton ground state is hence fourfold degenerate due to four combinations of products $e_i h_j$ with $i, j \in \{1, 2\}$. The two-particle Hamiltonian can be expanded into a basis of antisymmetrized product wave functions

$$e_i(\mathbf{r}_1)h_j(\mathbf{r}_2) \to \frac{1}{\sqrt{2}}\left(e_i(\mathbf{r}_1)h_j(\mathbf{r}_2) - h_j(\mathbf{r}_1)e_i(\mathbf{r}_2)\right) \quad i, j \in \{1, 2\}. \tag{3.7}$$

The degeneracy of the four exciton states with M_z equal to $(|1\rangle, |-1\rangle, |2\rangle, |-2\rangle)$ is lifted by the Coulomb interaction as depicted in Fig. 3.19. The Coulomb interaction consists of direct and exchange terms and reads in the basis of Eq. (3.7)

$$(C \cdot \mathbf{1} + \begin{pmatrix} 0 & \Delta_B & 0 & 0 \\ \Delta_B & 0 & 0 & 0 \\ 0 & 0 & -K & \Delta_D \\ 0 & 0 & \Delta_D & -K \end{pmatrix})u = E_i u. \tag{3.8}$$

C is the direct Coulomb term $\langle e_i h_j | H_{Coul} | e_i h_j \rangle$ and K is the diagonal exchange term $\langle e_i h_j | H_{Coul} | h_j e_i \rangle$. Solving this equation results in the eigenvectors and eigenstates

$$(E_1, E_2, E_3, E_4) = (\underbrace{C - K - \Delta_D, C - K + \Delta_D}_{D}, \underbrace{C - \Delta_B, C + \Delta_B}_{B}). \tag{3.9}$$

The index D marks states with $|M_z| = 2$. These states do not couple to photon fields and are usually not intermixed with the $|M_z| = 1$ states in absence of magnetic fields. Therefore they are generally not observed in luminescence experiments and denoted *dark* states. The two dark states are split by $E_2 - E_1 = 2\Delta_D$. The splitting is accompanied by the formation of symmetric and antisymmetric linear combinations $|2\rangle + |-2\rangle$ and $|2\rangle - |-2\rangle$ of the pure spin states 2. The states with indices B are correspondingly formed by linear combinations $|1\rangle + |-1\rangle$ and $|1\rangle - |-1\rangle$ and are referred to as *bright* states. The difference $E_4 - E_3 = 2\Delta_B$ is called bright splitting or fine-structure splitting. If the confining potential has a high symmetry (at least C_{4v}) then Δ_B is equal 0 and the wave functions are pure spin states 1. This is not the case for a confinement in a quantum dot, even if the shape of the dot is that of a square-based pyramid like those considered in this chapter. Piezoelectric effects lead to a reduction of the symmetry to C_{2v} [22, 24]. The experimentally observed splitting shown in Fig. 3.16 with linear polarizations oriented along ⟨110⟩ demonstrates that the actual dot symmetry is C_{2v}. The symmetry lowering also causes the pure spin states in C_{4v} symmetry to mix, and the corresponding circularly polarized transitions become linearly polarized in C_{2v} as shown in Fig. 3.19. The two directions of polarization are perpendicular to each other and reflect the symmetry characteristics of the confining potential. The biexciton state corresponds to $M_z = 0$ and is therefore not split by exchange interaction. Since the single exciton is the final state in the XX → X transition, this emission also reflects the fine-structure splitting and mixing of the bright-exciton spin states.

The single dot spectra given in Figs. 3.16 and 3.18 suggest that the magnitude of exchange interaction is related to the size of QDs. For a systematic analysis spectra of a large number of QDs with different sized were studied. Figure 3.20 compiles fine-structure splittings as a function of the single-exciton recombination energy which is directly related to the size of the dots. The exciton fine-structure splitting increases from small negative values for small QDs to positive values as large as 520 μeV for large QDs within the multimodal dot ensemble, as indicated by bars on top. Such large splittings reported in Ref. [30] were usually not observed for InAs/GaAs QDs by other groups [e.g.[29, 31, 32, 33]]. The dependence shown in Fig. 3.20 clearly confirms that the magnitude of the exciton fine-structure splitting scales with QD size.

The action of the anisotropic exchange interaction is also reflected in the recombination of the positively charged biexciton and the positive trion. Their emissions were shown in Fig. 3.16 to change intensity and the direction of polarization for different single QDs. The systematic effect of size on the action of the anisotropic exchange in the exciton fine-structure splitting raises the question if such dependence also exists for the charged biexciton and trion. A scheme of the charged-biexciton decay is shown in Fig. 3.21.

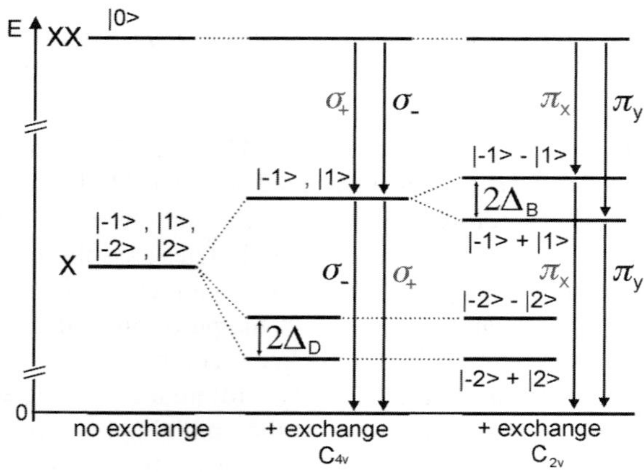

Fig. 3.19 Energy scheme of exciton and biexciton states, illustrating the effect of exchange interaction and symmetry of the confining potential. $2\Delta_D$ and $2\Delta_B$ mark the splittings of dark ($|\pm 2\rangle$) and bright ($|\pm 1\rangle$) states, $\sigma_{+/-}$ and $\pi_{x/y}$ denote circular and linear polarization, respectively

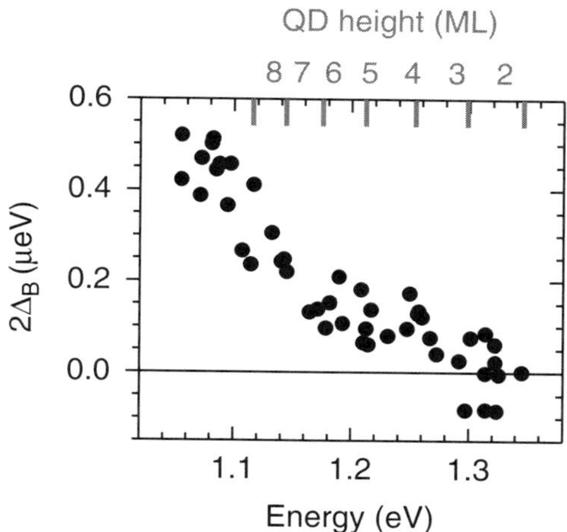

Fig. 3.20 Magnitude of the exciton fine-structure splitting $2\Delta_B$ for a number of InAs/GaAs quantum dots with different sizes. PL maxima of subensembles are marked by grey bars on top

The initial XX^+ state of the cascade is degenerate, analogous to the case of the neutral biexciton. The final state of the positive trion (X+*) comprises one singlet and three triplet states which are all twofold degenerate [34]. Contrary to the exciton case, this degeneracy is not lifted under exchange interaction. The exchange interaction leads to a mixing of the triplet states and alters the initially pure circular polarization to elliptic polarization. The degree of polarization of these lines hence reflects the degree of intermixing of the different charged trion states and therefore the magnitude of the anisotropic exchange interaction. The change of the polarization of the two XX^+ recombination lines from circular for pure states to elliptic, i.e. the degree of mixing of the trion states, is defined by the polarization degree

$$p = \frac{I^{\pi+} - I^{\pi-}}{I^{\pi+} + I^{\pi-}}. \tag{3.10}$$

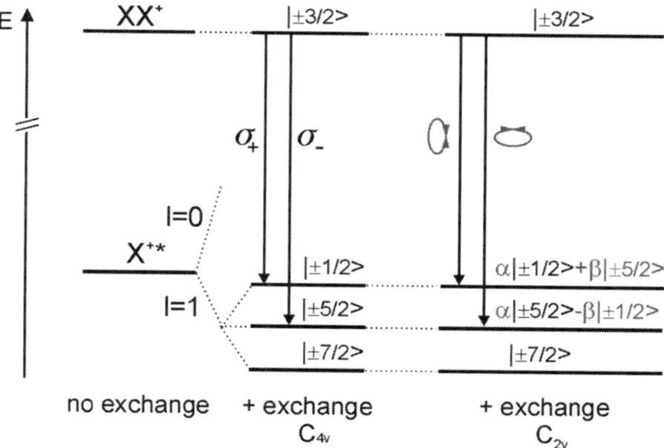

Fig. 3.21 Energy scheme of the charged-bieciton to charged-exciton decay. Grey features indicate changes by the anisotropic exchange interaction, ellipses indicate elliptic polarization. X^{+*} singlet states are omitted for clarity

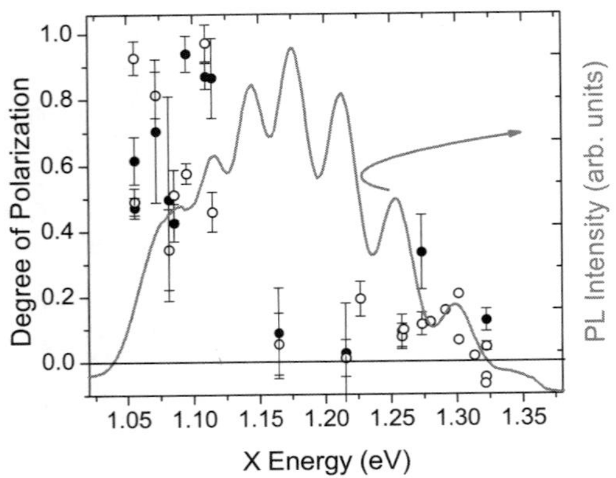

Fig. 3.22 Degree of polarization of the two emission lines of the positive-biexciton decay. The grey spectrum represents the photoluminescence of the QD ensemble. Circles and dots refer to the two different polarizations of the split emission lines. From [37]

Data of the polarization degree as measured all over the inhomogeneously broadened QD ensemble emission are given in Fig. 3.22 [37]. They show that also the polarization of the positive-biexciton to trion decay follows a systematic trend with the size of QDs.

The measurements clearly demonstrate the impact of the anisotropic part in the exchange interaction on electrons and holes confined in a QD. Such anisotropy arises in the dots, when the symmetry of the confining potential is lower than C_{4v}. Sources for such symmetry lowering are structural anisotropy of the dots, piezoelectricity induced by strain [22, 24], and atomistic symmetry anisotropy [35]. Structural elongation of the studied quantum dots can be ruled out as a main source for the observed exchange effects. TEM measurements outlined in Section 2.3 do not show a significant anisotropy, and numerical modeling fails to reproduce experimental results [30]. Moreover, the data demonstrate that the effect of exchange scales with the size of the QDs. Piezoelectricity provides a possible explanation for the observed trend [30]. Its magnitude is proportional to the occurring shear strain in the QDs. Due to the lattice mismatch between GaAs and InAs, the shear strain is larger for larger QDs [24]. Larger QDs with stronger shear strain components hence lead to stronger piezoelectric fields. Consequently such QDs have larger values of the fine-structure splitting, and the XX^+ lines show a larger degree of polarization. Though the experimentally observed scaling trend of the fine-structure splitting was well reproduced by first-order calculations of piezoelectricity [30], the magnitude was not. Furthermore, the importance of second order terms in the piezoelectric effect was recently pointed out [36]. Linear and quadratic terms were shown to have an opposite effect, and the quadratic term may dominate for large strains. The role of atomistic symmetry anisotropy is not yet studied in detail. A major contribution cannot be excluded in a complete treatment of exchange interaction in quantum dots.

3.4 Conclusion

Control of strained-layer deposition in the Stranski-Krastanow mode can be employed to fabricate equally-shaped InAs quantum dots in GaAs matrix with a well-defined multimodal distribution of sizes. Growth studies demonstrate that such dots form from a rough quantum well and ripen by a material exchange between dots of different subensembles. The evolution is theoretically described by an interplay of the strain energy in the dots and the chemical potential of the adatom sea.

The well-defined structure of the InAs/GaAs quantum dots has enabled a systematic study of the complex interplay between Coulomb interaction and confining potential for confined charge carriers. An apparent dependence of excitonic properties on the size of the quantum dots is found. The bright-exciton fine-structure splitting changes monotonously from negative values to more than 0.5 meV and the biexciton binding energy varies from -6 meV to 4 meV, i.e. from antibinding to binding, as the height of the truncated pyramidal dots increases from 2 to above 9 InAs monolayers. The binding energies of charged and neutral excitons increase due to correlation by the gradually increasing number of bound states for increasing dot size. The increasing magnitude of the fine-structure splitting with dot size is caused by the anisotropic part of the exchange interaction, with major contributions from the strain-induced piezoelectricity. The identification of key parameters allows to tailor exciton properties, providing a basis for applications of quantum dots in novel devices.

References

1. M. Grundmann (Ed.), *Nano-Optoelectronics*, Springer, Berlin (2002).
2. D. Leonard, K. Pond, P.M. Petroff, Phys. Rev. B **50**, 11687 (1994).
3. A. J. Bennett, D. C. Unitt, A. J. Shields, P. Atkinson, K. Cooper, D. A. Ritchie, Appl. Phys. Lett. **86**, 181102 (2005).
4. U.W. Pohl, K. Pötschke, A. Schliwa, F. Guffarth, D. Bimberg, N.D. Zakharov, P. Werner, M.B. Lifshits, V.A. Shchukin, D.E. Jesson, Phys. Rev. B **72,** 245332 (2005).
5. M. Grundmann, J. Christen, N.N. Ledentsov, J. Böhrer, D. Bimberg, S.S. Ruvimov, P. Werner, U. Richter, U. Gösele, J. Heydenreich, V.M. Ustinov, A.Yu. Egorov, A.E. Zhukov, P.S. Kop'ev, Zh.I. Alferov, Phys. Rev. Lett. **74,** 4043 (1995).
6. J.F. Carlin, R. Houdré, A. Rudra, M. Ilegems, Appl. Phys. Lett. **59**, 3018 (1991).
7. A. Gustafsson, D. Hessmann, L. Samuelson, J.F. Carlin, R. Houdré, A. Rudra, J. Crystal Growth **147,** 27 (1995).
8. S. Raymond, S. Studenikin, S.-J. Cheng, M. Pioro-Ladrière, M. Ciorga, P.J. Poole, M.D. Robertson, Semicond. Sci. Technol. **18,** 385 (2003).
9. U.W. Pohl, K. Pötschke, M.B. Lifshits, V.A. Shchukin, D.E. Jesson, D. Bimberg, Appl. Surf. Sci. **252**, 5555 (2006).
10. R. Heitz, F. Guffarth, K. Pötschke, A. Schliwa, D. Bimberg, N.D. Zakharov, P. Werner, Phys. Rev. B **71**, 045325 (2005).
11. R. Sellin, F. Heinrichsdorff. Ch. Ribbat, M. Grundmann, U.W. Pohl, D. Bimberg, J. Crystal Growth **221,** 581 (2000).

12. U. W. Pohl, K. Pötschke, I. Kaiander, J.-T. Zettler, D. Bimberg, J. Crystal Growth **272,** 143 (2004).
13. J.-T. Zettler, J. Rumberg, K. Ploska, K. Stahrenberg, M. Pristovsek, W. Richter, M. Wassermeier, P. Schützendübe, J. Behrend, L. Däweritz, Phys. Stat. Sol (a) **152,** 35 (1995).
14. U.W. Pohl, K. Pötschke, A. Schliwa, M.B. Lifshits, V.A. Shchukin, D.E. Jesson, D. Bimberg, Physica E **32**, 9 (2006).
15. R. Timm, H. Eisele, A. Lenz, T.-Y. Kim, F. Streicher, K. Pötschke, U.W. Pohl, D. Bimberg, M. Dähne , Physica E **32**, 25 (2006).
16. N. Moll, M. Scheffler, E. Pehlke, Phys. Rev. B **58**, 4566 (1998).
17. Q. K. K. Liu, N. Moll, M. Scheffler, E. Pehlke, Phys. Rev. B **60**, 17008 (1999).
18. G. Costantini, C. Manzano, R. Songmuang, O.G. Schmidt, K. Kern, Appl. Phys. Lett. **82**, 3194 (2003).
19. G. Costantini, A. Rastelli, C. Manzano, P. Acosta-Diaz, R. Songmuang, G. Katsaros, O.G. Schmidt, K. Kern, Phys. Rev. Lett. **96**, 226106 (2006).
20. J. Drucker, Phys. Rev. B **48**, 18203 (1993).
21. S. Rodt, R. Seguin, A. Schliwa, F. Guffarth, K. Pötschke, U.W. Pohl, D. Bimberg, J. Lumin. **122-123**, 735 (2007).
22. O. Stier, M. Grundmann, D. Bimberg, Phys. Rev. B **59,** 5688 (1999).
23. O. Stier, R. Heitz, A. Schliwa, D. Bimberg, Phys. Stat. Sol. (a) **190**, 477 (2002).
24. M. Grundmann, O. Stier, D. Bimberg, Phys. Rev. B **52,** 11969 (1995).
25. M. Sugawara, Phys. Rev. B **51**, 10743 (1995).
26. V. Türck, S. Rodt, O. Stier, R. Heitz, R. Engelhardt, U.W. Pohl, D. Bimberg, R. Steingrüber, Phys. Rev. B **61** (2000) 9944.
27. U.W. Pohl, R. Seguin, S. Rodt, A. Schliwa, K. Pötschke, D. Bimberg, Physica E **35**, 285 (2006).
28. S. Rodt, A. Schliwa, K. Pötschke, F. Guffarth, D. Bimberg, Phys. Rev. B **71,** 155325 (2005).
29. M. Bayer, G. Ortner, O. Stern, A. Kuther, A.A. Gorbunov, A. Forchel, P. Hawrylak, S. Fafard, K. Hinzer, T.L. Reinecke, S.N. Walck, J.P. Reithmaier, F. Klopf, F. Schäfer, Phys. Rev. B **65**, 195315 (2002).
30. R. Seguin, A. Schliwa, S. Rodt, K. Pötschke, U.W. Pohl, D. Bimberg, Phys. Rev. Lett. **95**, 257402 (2005).
31. K. Kowalik, O. Krebs, A. Lemaître, S. Laurent, P. Senellart, P. Voisin, J.A. Gaj, Appl. Phys. Lett. **86,** 041907 (2005).
32. A.S. Lenihan, M.V. Gurudev Dutt, D.G. Steel, S. Gosh, P.K. Bhattacharya, Phys. Rev. Lett. **88,** 223601 (2002).
33. R.J. Young, R.M. Stevenson, A.J. Shields, P. Atkinson, K. Cooper, D.A. Ritchie, K.M. Groom, A.I. Tartakovskii, M.S. Skolnick, Phys. Rev. B **72**, 113305 (2005).
34. K.V. Kavokin, Phys. Stat. Sol. (a) **195**, 592 (2003).
35. G. Bester, S. Nair, A. Zunger, Phys. Rev. B **67**, R161306 (2003).
36. G. Bester, X. Wu, D. Vanderbilt, A. Zunger, Phys. Rev. Lett. **96**, 187602 (2006).
37. R. Seguin, S. Rodt, A. Schliwa, K. Pötschke, U.W. Pohl, D. Bimberg, Phys. Stat. Sol. (b) **243**, 3937 (2006).

Chapter 4
Carrier Transfer in the Arrays of Coupled Quantum Dots

Yuriy I. Mazur, Georgiy G. Tarasov, and Gregory J. Salamo

4.1 Introduction

A quantum dot (QD) can be imagined as a giant artificial atom while consisting of thousands of real atoms. It is a structure that spatially confines electrons to a region on the order of the De Broglie electron wavelength dramatically affecting the quantum states of a system. This enables major advances in the study of the fundamental properties of zero-dimensionality systems and creates the opportunity for a variety of novel device applications. These applications range from novel lasers to quantum computing to new medical devices. QD fabrication techniques include molecular beam epitaxy (MBE), electron beam lithography, and colloidal synthesis. In this review we focus on semiconductor QDs that were produced by MBE via the Stranski-Krastanov growth technique. In addition, we focus on developing an understanding of the underlying physics that gives rise to the QD electronic structure and dynamical carrier processes both of which are important for device applications [1]. That is the electronic structure and dynamical processes depend significantly on QD size and shape, on the QDs imbedded surrounding, and on whether the QDs are randomly spatially distributed or spatially organized in arrays.

One exciting application of QDs depends on coherent manipulation of charge and spin for solid-state based implementations of quantum logic. In this regard, excitonic excitations of individual QDs have been used to demonstrate Rabi oscillations [2, 3], conditional two-qubit gate based on exciton-biexciton transitions [4], strong coupling to photons [5–7] and dipolar couplings in a pair of QDs [8]. Moreover, optically induced spin manipulation in quantum dots is currently pursued in various research laboratories [9, 10] and most recently single spin Rabi oscillations have been demonstrated with radio-frequency excitation [11]. These first steps towards quantum information processing are significant since QD systems offer–at least conceptually–the potential of implementing scalable arrays of quantum bits.

Meanwhile, ensembles of semiconductor QDs allowed fabrication of high-performance lasers [12, 13], infrared optical detectors [14–17], and white light sources [18, 19]. Despite early predictions, realization of QD lasing occurred to be a much more challenging problem then corresponding quantum well devices.

Z. M. Wang, *Self-Assembled Quantum Dots.*

This difficultly was overcome recently [20] due to the use of self-assembled growth techniques that made possible dense arrays of coherent islands, reasonably uniform in shape and size, and relatively defect-free. For example, 1.3 μm lasers with low threshold and high power have been successfully demonstrated on a GaAs substrate with InGaAs multi-stacked QDs. Currently, state-of-the-art for 1.3 μm QD lasers have been used to deliver more than 10 GHz of 3 dB-modulation bandwidth [21].

In addition to lasers, InAs/GaAs QD systems have the potential to be the leading technology for infrared detectors that could be used for many military and domestic applications [14]. QD infrared photodetectors yield high operating temperatures, low dark currents, and good wavelength tunability. In contrast to quantum wells (QWs), they can also be used for the detection of normal-incidence light. Moreover, InAs QDs embedded in InP matrix revealed room-temperature electroluminescence in a very wide wavelength range from 950 to 2200 nm [18]. The wide range of emission indicates that the QDs have the potential for broadband optical gain in the infrared region at room temperature which could be used as efficient optical amplifiers for 1.0–1.6 μm fiber communication wavelength region.

While these results are encouraging, the fabrication of QD arrays suitable for application, is challenging, due to size and composition fluctuations in QD ensembles as well as the difficulty of achieving controlled spatial ordering of QDs. In this regard different strategies have been proposed for generating vertical or lateral ordering of QDs, vertical strain correlation [7, 22–24], templating [25] or lithography [26]. Most recently, droplet epitaxy self-assembly using lattice-matched semiconductors has emerged as a novel, promising technique for fabricating geometrically controlled pairs, rings and even more complex QD arrays of high optical quality [27–30].

Usage of semiconductor QDs in high-performance lasers [12, 13], quantum computations [31] or single electron transistors [32] normally implies carrier transfer from a continuum of states into the discrete atomic-like states of QD. Efficiency of such transfer is determined to a great extent by the strength of coupling between the QD array and a carrier reservoir. For example a quantum well (QW) [33] or QW wetting layer [34] that collects carriers can act as a carrier reservoir from which lateral diffusion leads to capture by QDs. In fact, QW nanostructures with a QD array grown on top (QDs:QW), is considered as a good candidate for such a reservoir for QDs [35–37]. Due to a controllable QW thickness the size distribution of the strain-induced QDs in such systems becomes significantly narrower and while the QDs:QW are formed inside the QW layer, the QD capture rate for photogenerated carriers is expected to be improved. In view of the significantly complicated energy structure of the QDs:QW system, a clear picture of carrier relaxation has to be carefully constructed.

Adding to this complication, QD arrays used in application are normally dense and generally not separated spatially by long distances or infinite potential barriers. As a result, the electronic wavefunctions of adjacent QDs can overlap, thus allowing carriers to travel from one QD to another due to interdot coupling. This carrier transfer, even if slow, can affect carrier dynamics after optical excitation.

In what follows we examine the effect of coupling, both laterally and vertically, in In(Ga)As/GaAs bilayer structures. Different mechanisms of carrier relaxation in coupled QD systems are discussed and demonstrated experimentally.

4.2 Engineering of Quantum Dots Sizes, Density and Coupling in In(Ga)As/GaAs Quantum Dots Arrays

4.2.1 In(Ga)As/GaAs Arrays with Lateral Coupling

It is well known that a morphological transition from two dimensional (2D) to three-dimensional (3D) growth of highly strained InAs layers on GaAs substrates by MBE technique can result in the appearance of three-dimensional islands, or QDs when the InAs layer thickness exceeds a critical value. This process, which on the (001)-oriented InAs/GaAs system proceeds via the Stranski-Krastanov growth mode, results in a thin InAs "wetting layer" covered with InAs (or, more generally InGaAs) QDs, whose size and density is dependent on various aspects of the growth conditions, including substrate temperature, arsenic over-pressure, InAs deposition rate, and growth interruptions.

After numerous studies, a picture has emerged on how the transition from 2D to 3D growth takes place. 3D growth resulting in QDs occurs over a relatively large temperature range between 400 and 520 °C. Results based on photoluminescence (PL), photoluminescence excitation spectroscopy (PLE), and transmission electron microscopy (TEM) of structures prepared with the relatively low growth temperature T_G of 420 °C indicate that dot formation begins with an InAs coverage of less than 1 monolayer (ML) [38, 39] and the dot size grows continually with coverage, reaching 8 nm with 2 ML of coverage and up to 37 nm for 6 ML coverage [40]. Results from in situ atomic force microscopy (AFM) indicate that, at least at the relatively high growth temperature T_G of 500 °C the growth is 2D and the InAs builds a pseudomorphically strained film up to about 1.57 ML InAs coverage [41, 42]. Beyond 1.57 ML, dots appear whose size is relatively independent of further coverage up to 2.2 ML when the large density results in coalescence. Furthermore, in addition to the QDs, larger patches of one extra ML high appear on the WL, as well as quasi-3D patches 2–4 ML high, for intermediate InAs coverage. These quasi-3D patches appear to nucleate the quantum dots. A later in situ scanning tunneling microscopy (STM) study [43] reveals that, at least at higher growth temperatures, the deposited InAs grows mostly two dimensionally up to 1.57 ML, then the 2D layer decreases in average thickness with further coverage as it is consumed to form first the quasi-3D patches [42] and then QDs with relatively constant size of 16 nm. The size distribution, for an InAs coverage greater than 1.65 ML, is about 4 nm. Furthermore, the development of the quasi-3D patches occurs twice [44]: first as small patches which disappear and then as larger patches which may evolve into the QDs. PL from both the small and large quasi-3D patches was also measured. A bimodal size distribution for low InAs coverage was also reported for the relatively low growth temperature of 450 °C [45]. In that study, the WL luminescence was still dominant and two distinct size clusters were attributed to QDs and to patches of 2-ML-high regions of the WL, which are thought to serve as dot precursors. Other work shows that at both medium and high growth temperatures, different size QDs or quasi-3D patches coexist near the critical thickness for dot formation, which was determined to be in the 1.4–1.7 ML range, depending on growth temperature [46]. Low-temperature growth with thicker InAs coverage has also resulted in a bimodal size distribution on the GaAs(211)B substrate surface [47]. The authors have also observed a bimodal size distribution using AFM in InAs quantum dots grown at 420 °C [48]. The evolution of size distributions with emphasis on achieving narrow distributions is also described [49, 50].

The QD size distribution with an intermediate InAs coverage has been studied [51]. It is shown that, although growth at relatively high substrate temperatures results in large dots (base size approximately 14 nm) for higher InAs coverage, the first dots which appear are much smaller. These smaller dots co-exist with the appearance of the larger ones for intermediate InAs coverage, and then disappear for larger InAs coverage. Thus the size distribution at intermediate InAs coverage is not described by a single maximum, but is bimodal. At lower substrate temperature, bimodal or multimodal distributions appear for intermediate InAs coverage. This behavior is explained in terms of at least two distinct thresholds for dot formation, one being the minimum InAs coverage which can lead to islanding and the other being that which supports an optimal dot size. The temperature dependence of the PL further demonstrates presence of strong coupling of dots from different size classes and thermally activated electron transfer from dots in the smaller size class to those in the larger one.

The samples investigated in the study [51] were GaAs superlattice structures containing thin layers of InAs grown on semi-insulating (001)-oriented GaAs substrates in a Riber 32-P gas-source molecular-beam epitaxy (GSMBE) system with growth rates of 2.5, 3.125, and 0.2 Å /s for GaAs, AlGaAs, and InAs, respectively. All samples consisted of 10 periods of 60 Å $Al_{0.2}Ga_{0.8}As$, 30 Å GaAs, 60 Å GaAs:Si ($N_D = 2 \times 10^{18}\,cm^{-3}$), 60 Å GaAs, 4 to 8 Å InAs, 60 Å GaAs, 60 Å GaAs:Si, 30 Å GaAs, 60 Å $Al_{0.2}Ga_{0.8}As$, subsequently capped with a 50 Å GaAs:Si layer. The dot containing layers were sufficiently separated such that no significant vertical correlation is expected. Similar reference structures containing a single InAs layer were also investigated and were found not to be significantly different from the superlattice samples described here. There were two sets of samples: those with a relatively high substrate temperature T_G during the InAs growth (T_G = 505 °C) and those with a relatively low T_G (T_G = 420 °C) [51]. The growth temperature of all structures was 530 °C, except of the InAs and the adjacent GaAs layers which were grown at T_G after a short growth interruption allowing the substrate to reach this lower temperature. The samples grown at T_G = 420 °C included growth interruptions; specifically, the 40 s of InAs deposition was divided into five successive depositions, separated by 5 s interruptions. Double crystal x-ray diffraction together with a simulation of the double x-ray rocking curves was used to verify the completed structures. Even in the case of the InAs existing in the form of islands, the experimental x-ray diffraction results are characteristic only of the average coverage as long as the InAs being measured is not too close to the surface [52]. We found a linear and 1:1 dependence of the nominal InAs coverage d_{InAs} as determined by x-ray analysis on the InAs deposition time [52]. (This time excludes growth interruption during the dot formation). The samples investigated in the study [51] were grown near both the high- and low-temperature limits for the self-formation of InAs QDs in GaAs [41] resulting in dots of different lateral sizes and areal density (D_a) dependent on InAs coverage (d_{InAs}) and growth temperature. The QDs grown at the higher T_G have a more homogeneous size distribution, whereas the QDs grown at the low-temperature limit demonstrate a more distinct redistribution among different size families [48]. Reference samples with a single InAs layer were studied in a plan-view TEM for the characterization of shape, size, and density of the InAs QDs and the majority-sized QDs for the given growth conditions described.

More precise information about the size distribution, including that of different size families, was derived from the AFM measurements. Figure 4.1a shows an AFM image of a reference sample with $d_{InAs} = 1.79$ ML, but only one InAs layer at the surface.

The AFM image demonstrates the presence of two different families of QDs: one with base length of $b = 8$ nm and $D_a = 1.7 \times 10^{11}$ cm^{-2} and the other family with $b = 14.5$ nm and $D_a = 3 \times 10^{10}$ cm^{-2}. The histogram of the QD base length (Fig. 4.2a) clearly shows this bimodal size distribution. Figure 4.1b shows the AFM image for another reference sample with $d_{InAs} = 2.46$ ML, but, again, with only one InAs layer. It was grown at $T_G = 420\,^\circ$C and with growth interruption. In this image, three families of sizes of dots are visible: the larger sized QDs with base lengths of both 8 nm and 14 nm coexist with a much larger density of smaller sized (b = 6 nm) QDs. Figure 4.2b shows a QD density ratio of approximately 12:3:1 for the 6 nm:8 nm:12 nm dots. QDs on the surface of uncapped samples are not identical to those embedded in the matrix material, meaning that the size distributions shown in Fig. 4.2 are not necessarily identical to those in superlattice samples. In particular,

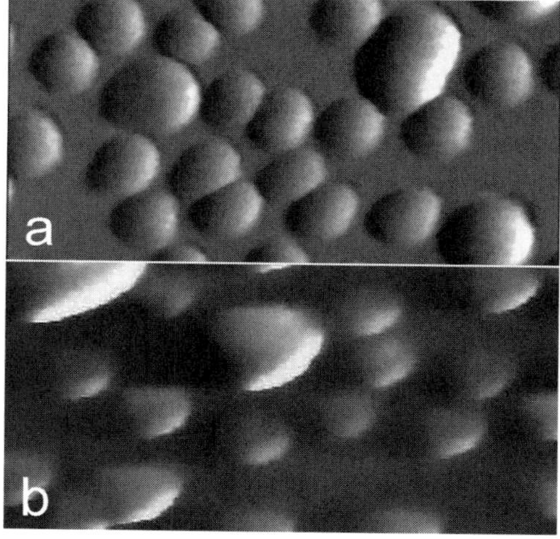

Fig. 4.1 Embossed and shaded atomic force micrographs of QDs grown at: a) $T_G = 505\,^\circ$C and (b) $T_G = 420\,^\circ$C with an InAs coverage of 1.79 ML and 2.46 ML, respectively. The AFM pictures showing the surface morphology are of a field of a) 70×35 nm^2 and b) 50×25 nm^2

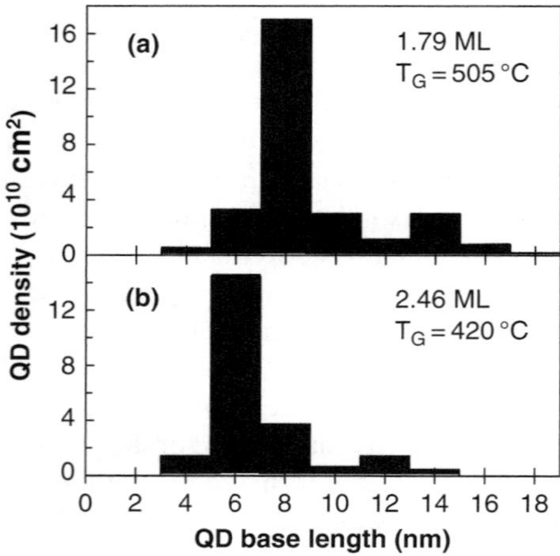

Fig. 4.2 Size distribution of InAs QD samples grown at (a) $T_G = 505\,^\circ C$ and (b) $T_G = 420\,^\circ C$ with an InAs coverage of 1.79 and 2.46 ML, respectively. The AFM pictures of these two samples are shown in Fig. 4.1

the growth kinetics during the overgrowth of the QDs by the capping material is complex and could result in material redistribution between the WL and QDs [53]. The growth interruption and low T_G can also alter the surface morphology. Despite these caveats, the AFM images will provide some insight into the size distributions of the samples with buried dots.

The post growth treatment of the QD samples, as well as, QD capping significantly affect the sample morphology. The effects of annealing and quenching on InAs/GaAs(001) QDs grown at a low growth rate by MBE have been studied by means of STM [54]. Significant changes in QD density, average volume (of more than one order of magnitude), WL morphology, and QD volume density have been observed after in situ annealing for up to 90s at a growth temperature of 485 °C providing direct evidence for a QD ripening process resembling that observed in other heteroepitaxial growth systems. Inefficient sample removal from the growth chamber and cooling (quenching) can lead to unintentional annealing and ripening of the QDs and/or the appearance of clusters on the sample surface. The appearance of these clusters can have a significant impact on any statistical analysis of QD samples. The results show [54] that MBE-grown InAs/GaAs(001) QDs are kinetically controlled structures even at low growth rates, and underline the importance of the quench method in minimizing postgrowth changes in the characteristics of the uncapped QD array and preserving the in situ surface morphology for ex situ surface studies.

A microscopic picture for the GaAs overgrowth of self-organized InAs/GaAs(001) quantum dots is developed in Ref. [55]. STM measurements reveal two capping regimes: the first being characterized by a dot shrinking and a backward pyramid-to-dome shape transition. This regime is governed by fast dynamics resulting in island morphologies close to thermodynamic equilibrium. The second regime is marked by a true overgrowth and is controlled by kinetically limited surface diffusion processes. Observed structural changes were addressed the energetic minimization driven by lattice mismatch and alloying. The atomic scale morphology of InAs/GaAs QDs capped with $In_{0.1}Ga_{0.9}As$ layers of different thickness has been studied using STM and compared to the effects of capping with a pure GaAs layer [56]. It is found that the QDs capped with a 5 nm $In_{0.1}Ga_{0.9}As$ layer exhibit a longer PL emission wavelength than those capped with GaAs. STM studies show that the QDs capped with $In_{0.1}Ga_{0.9}As$ retain their height during the initial stages of capping (up to ~2 nm), whereas the GaAs-capped QDs collapse as material migrates from their tops onto the cap surface. After deposition of a 25 nm GaAs cap the surface is still far from flat, whereas the $In_{0.1}Ga_{0.9}As$ capping layer is planar after just 5 nm deposition. High-resolution STM images, supported by reflection high-energy electron diffraction measurements, reveal a $(4 \times 3)/c(4 \times 6)$ reconstruction for the $In_{0.1}Ga_{0.9}As$ cap, whereas the GaAs cap layer rapidly exhibits a $c(4 \times 4)$ reconstruction after the first few monolayers of deposition. The planar morphology is controlled by the enhanced In adatom diffusion on the InGaAs alloy surface.

The substrate orientation-dependence of InAs/GaAs QD growth has been studied also [57]. It is shown that abrupt change from 1D- to 3D- charge confinement takes place in InAs deposited on (100) GaAs substrate. On the tilted (311)B substrates, the QD morphology is different, resulting in a weaker charge confinement that gradually increases with the amount of deposited InAs. At 1.9 ML, the QD confinement on this substrate orientation is as effective as for the (100) oriented substrates. By studying the confinement of the charges in samples with QDs at different stages of development, it is possible to get insight into the QD formation process.

Thus we can see that by adjusting the QD growth conditions as well as the post growth treatment of the samples, it is possible to engineer the QD sizes, shapes and interdot distances. All together this determines the lateral QD coupling due to overlap of the electronic wavefunctions of adjacent QDs in the In(Ga)As/GaAs QD arrays. The interdot coupling significantly affects the interdot carrier transfer rate and allows to create peculiar QD complexes for various applications.

4.2.2 In(Ga)As/GaAs Arrays with Vertical Coupling

Under most growth conditions the formation of QDs is controlled by In surface migration. In case of a single QD layer this leads to interdependent QD size and density, and hence to a substantial inhomogeneity of QD size distribution. In multilayered samples, the surface strain introduced in the capping layer by an initial set of buried QDs leads to strain-driven adatom migration and hence to a vertical self-organization of QD stacks [22, 58]. A low energy shift and a decreasing full width at half maximum (FWHM) of the ground-state PL in multiply stacked samples evidence the formation of coherent superlattice-like states [58] and the altered strain distribution in QD stacks [59]. This strain-driven vertical alignment of QDs in multilayered samples has opened new possibilities for application of coupled QD structures. For example, the ability to trap, localize, store, and transfer carriers within QD stacks creates the potential for memory applications [60–62]. In fact, in this case the inherent disadvantage of a broad QD size distribution can even be considered as an advantage offering the potential for multiple storage bits [63].

Driven by both potential application and the interesting interplay between QD layers, there have been several investigations on vertically aligned InGaAs/GaAs bi-layers separated by a GaAs spacer of variable thickness [59, 64, 65]. Control of the deposition in each of the two layers has enabled engineering of the relative size and density of the QDs in each layer. For example, in bi-layer samples, the first (seed) layer with deposition Θ_1 is used to control the island density in the subsequent layer via stress-induced, vertical self-organization [22], and the InAs deposition Θ_2 in the second layer then determines the average island size in the second layer. In this way, the correlation between the relative QD size and density in the two layers can be systematically explored.

Another interesting reason for the study of bi-layers is that QD stacks can also be utilized to produce a more homogeneous QD size distribution. Cross sectional TEM images clearly reveal a growing uniformity from layer-to-layer in multilayered QD stacks. This uniformity results in much more narrower PL band in comparison with the monolayer QD array of appropriate quality [59].

Typically InAs/GaAs QD bi-layer samples are grown in a solid-source MBE chamber coupled to an ultrahigh vacuum scanning tunneling microscope (STM) following the procedure: first a GaAs (001) substrate is grown, followed by a 0.5 μm GaAs buffer layer, a 28 nm $Al_{0.3}Ga_{0.7}As$ layer and 10 min annealing at 580 °C that allows to get a nearly defect-free atomically flat surface [66]. Then a seed QD layer is grown by depositing InAs layer of needed thickness Θ_1 with a growth rate of 0.1 ML/s, As_4 partial pressure of 8×10^{-6} Torr, and substrate temperature of 500 °C. This is followed by the GaAs spacer layer deposited on top of the seed QD layer. The second QD layer is then added. The InAs deposition coverage Θ_2 in the second layer is varied (from 1.8 to 2.7 ML for different samples used in Ref. [66]). Each sample is finally capped with 20 nm GaAs and 28 nm $Al_{0.3}Ga_{0.7}As$ layers.

The samples are structurally characterized by plan-view STM and cross-sectional transmission electron microscopy (XTEM). Figure 4.3 shows the XTEM image of the bi-layer with 1.8 ML/2.4 ML InAs deposition on the seed and second layers and with a GaAs spacer layer ($d_{sp} = 30$ ML) between them [66]. This sample demonstrates almost fully vertical correlation between the arrays of QDs. For other samples with a thicker spacer layer, it is observed a partly correlation. For example, the statistical analysis of the 1.8/2.7 ML sample proves that the InAs QDs in the resulting sample are weakly vertically correlated ($\sim$50%) for the 50 ML GaAs spacer thickness and the QDs in the second layer are nearly twice the volume of those in the seed layer due to larger amount of the InAs deposited in the second layer [66]. The samples with partly vertical correlation between QD arrays are of prime interest for the study of the interlayer carrier transfer.

Recently the research efforts are aimed to improve the surface morphology of stacked InAs/GaAs QD system. A simple, but effective way, to achieve improvement for stacked 1.3 μm InAs/GaAs QDs, grown by metal-organic chemical vapor deposition, is to have the GaAs middle spacer and top confining heterostructure layers deposited at a low temperature of 560 °C in order to suppress a post growth

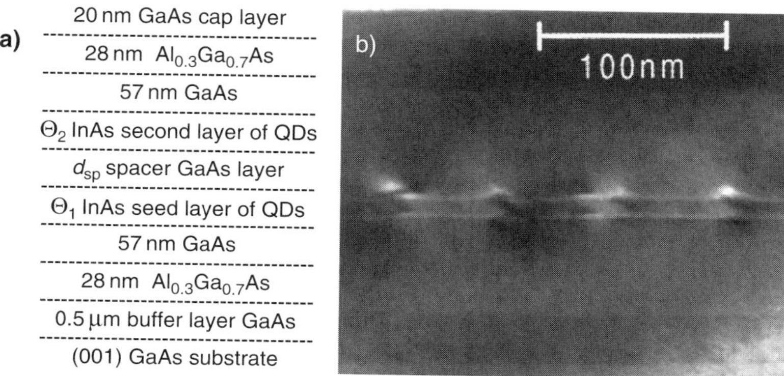

Fig. 4.3 XTEM image of the bi-layer with $\Theta_1 = 1.8$ ML, $\Theta_2 = 2.4$ ML and $d_{sp} = 30$ ML

annealing effect [67]. In this case, the annealing processes, just after depositing the GaAs spacer layers, leads to a significantly improved surface morphology of the top GaAs layer. For a structure of five-layers of QDs, the surface roughness, with the introduced annealing processes, is reduced to about 1.3 nm ($5 \times 5\,\mu m^2$ area), much less than the roughness of 4.2 nm without annealing. Furthermore, PL measurements do not reveal any changes in emission wavelength after inserting the annealing steps. This dramatic improvement in surface morphology results from the improved GaAs spacer surfaces due to the introduced annealing processes.

Another highly effective reported growth technique has made it possible to both dissolve large islands and to prevent further defect propagation in closely spaced (15 nm) stacked InAs/GaAs QDs, while maintaining an emission wavelength $>1.3\,\mu m$ [68]. Island dissolution is accomplished via an In flush using an AsH_3 pause inserted into the growth sequence just after each QD layer is capped. The low V/III ratio enables the flushing of surface In atoms from defect sites while the fully capped QDs remain intact. This technique eliminates the need for in situ annealing that activates the In flush in other growth scenarios and results in a large emission blueshift. Strain propagation within the closely spaced QD stacks is reduced by GaP strain-compensation layers. AFM and TEM confirm an improved surface morphology and crystalline quality of stacked QDs [68].

The effect of the growth temperature of the spacer layer on multilayer InAs/GaAs QD structures has also been investigated [69]. Change of the growth temperature of the spacer layers is used to try to obtain identical layers for multilayer QD structures. A 5-layer 1.3 μm InAs/GaAs QD structure with 50 nm GaAs spacer layers served as a model system. It is found that the growth temperature of the GaAs spacer layer has pronounced effects on both the structural and optical properties of the InAs QDs. For GaAs spacer layers grown at a low temperature of 510 °C, dislocations are observed in the second and subsequent layers, a result of significant surface roughness in the underlying spacer layer. However by increasing the growth temperature to 580 °C for the final 35 nm of the 50 nm GaAs spacer, a much smoother surface is achieved, allowing the fabrication of essentially identical, defect free QD layers. The suppression of defect formation enhances both the room-temperature PL efficiency and the performance of 1.3 μm multilayer InAs/GaAs QD lasers. An extremely low continuous-wave room-temperature laser threshold current density of 39 A/cm^2 is achieved for an as-cleaved 5-layer device with emission at 1.306 μm and ground state operation up to 100 °C [69].

For InAs/GaAs QD bilayers with 11.2 nm GaAs spacer layers it has been found that the emission of the second QD layer is blueshifted with respect to the first one [70]. The two PL peaks can be made coincident by increasing the amount of InAs deposited in the second layer and also by annealing the spacer layer. STM measurements show that QDs in both layers have the same composition (pure InAs) before capping. However, QDs with similar shapes and volumes exhibit a blueshift in PL when grown in the second layer compared to a single layer. This is attributed [70] to enhanced intermixing during the capping stage of the second-layer QDs, and is a consequence of these dots being more strain-relaxed due to the strain fields associated with the first QD layer. The blueshift is smaller for annealed spacer layers due to a lesser degree of strain relaxation as a result of the change of surface morphology induced by the annealing process.

The above overview of the techniques used recently for the growth of layers of vertically correlated arrays of QDs gives evidence that we can produce high quality QD arrays with controllable parameters and interlayer coupling. Our focus now moves to investigating and influencing both the intralayer and interlayer carrier transfer in QD arrays.

4.3 Mechanisms of Interdot Carrier Transfer

Usually optically excited (hot) carriers in a In(Ga)As/GaAs QD system arising after light is absorbed in the GaAs barrier, relax to the energy states of the InAs WL and finally become trapped in the discrete atomic-like states of QDs. It is an effective way to populate the QD states. Characteristic times of this carrier transfer are rapid in comparison with the carrier radiative lifetimes (~1 ns) both in the excited and in the ground QD states and significantly vary with the QD density D_a. For example, it has been found that the time for the carrier to reach a QD state increases from 2 to 20 ps for the self-assembled InGaAs/GaAs QD structures with D_a decreasing from $10^{10}\,cm^{-2}$ to $10^{8}\,cm^{-2}$ [71]. The temperature and photoexcited carrier density dependencies of the carrier transfer times suggest that potential barriers at WL and QD interfaces hinder carrier capture in low-density QD structures. Further studies [72] have demonstrated that the carrier transfer time can be different for modulation-doped self-assembled InAs/GaAs QDs changing from 5 to 6 and 12 ps for the doped and undoped samples, respectively. The experiments [73] suggest that in all samples the carrier capture into the highest QD levels is accompanied by LO phonon scattering. The significant difference in the transfer times is attributed to different relaxation mechanisms for the subsequent processes of intradot carrier relaxation. In doped samples, the available built-in carriers in the QDs lead to efficient electron-hole scattering, while in the undoped sample relaxation progresses in a cascaded manner by the emission of LO phonons. Additionally, experimental results show decreased carrier lifetimes in the doped structures, which is attributed to nonradiative recombination at doping-induced recombination centers in the vicinity of the QD layers. The described processes are related to the carrier transfer from the barrier or WL states into the states of QDs. However, being trapped in the QD states carriers can further recombine radiatively or escape from the QD state before recombination reducing the QD radiative yield. After escape the carriers can be trapped by available defects in the QD structure and decay nonradiatively or re-trapped by other QDs resulting finally in the interdot carrier transfer and emission in the another spectral region. Let us consider a few of mechanisms of interdot transfer.

4.3.1 Temperature Induced Transfer

The idea of temperature induced carrier transfer in self-assembled InAs/GaAs QDs samples is based on an analysis of broad and often non-Gaussian PL bands observed in dense QD ensembles. Unusual temperature dependence of both the peak energy position and the FWHM of PL bands were observed earlier for InAs/GaAs QDs [73] have been consistently explained in terms of thermally activated carrier transfer among QDs belonging to different size families [74]. These dot families were identified through decomposition of the complicated PL band shape into a set of Gaussians. Analysis of the temperature dependence of the integrated PL intensity measured in the temperature range 10–180 K for the self-assembled InAs/GaAs QDs with InAs coverage ranging from 1.6 to 3.2 ML has revealed two thermally activated processes responsible for this dependence: i) quenching of PL emission due to the escape of carriers from the QDs to non-radiative recombination centers, through the GaAs barriers; and ii) electron transfer between neighboring QDs with different sizes, through intermediate states of the WL. The energy difference between the WL and the QD electron states is derived to range from a few meV to 30–40 meV for QDs emitting in the 1.35–1.23 eV spectral range [74]. Therefore at finite temperatures electrons can be thermally excited into propagating states of the WL, travel in real space and be trapped by other remote QDs, thus jumping between dots of different size before radiative recombination takes place. This mechanism favours carrier transfer from small to large dots, which have a deeper confining potential, and, consequently, lower energy levels. A similar idea was used also for explanation of temperature-dependent PL in a bimodal InAs/GaAs self-assembled QD systems [75]. The changes in the PL spectra were interpreted in terms of carrier thermal redistribution between two families or "modes" of QDs. The ratio of PL intensities ascribed to these two QD modes is analyzed based on a model in which the WL acts as the carrier transfer channel between dots.

Unusual temperature behavior of the PL for asymmetrical QD pairs in a vertically aligned double-layer InAs/GaAs QD structure also provides clear evidence for carrier transfer from smaller to larger QDs by means of a nonresonant multiphonon-assisted tunneling process in the case of interlayer transfer and through carrier thermal emission and recapture within one layer [76, 77].

Extensive studies [78–80] of the temperature influence on the PL properties of the QD systems have led to the following understanding: i) The PL intensity quenching with increasing temperature is commonly attributed to the thermal escape of carriers from QDs to a wetting layer and/or to a barrier, and/or to the nonradiative recombination centers; ii) The tunneling of carriers between dots via the wetting layer is responsible for the unusual reduction of linewidth with increasing temperature and for the faster redshift of emission energy than expected for the band gap of the dot material. In what follows we show that immediate tunneling between QDs is also important for a more complete understanding.

4.3.2 Coulomb Scattering

Photoexcited electron-hole pairs in the barrier rapidly relax to the WL states due to the carrier thermalization within the GaAs barrier during $\sim$1 ps and subsequent capture by the WL. Final energy relaxation to the QD states of about 40–60 meV is very fast (within 2–60 ps) and is happened due to three alternative mechanisms: multiphonon-emission, Auger carrier-carrier scattering, and polaron decay.

Considering Auger carrier-carrier scattering in InAs/GaAs self-assembled QD structures it is generally assumed that carriers are effectively injected directly into the WL, and then they are captured by the QDs or recombine in the WL [81]. Two processes are taken usually into account: 1) a 2D electron or 2D hole in the WL collides with a 2D electron and is captured by the QDs, whereas the other 2D electron scatters into a 2D state of higher energy in the WL. Similar processes of this type can involve 2D holes; 2) a 2D hole is captured by the QD due to Coulomb scattering with the electron previously captured by this QD. As a result this electron is excited into a 2D state in the WL. Analysis shows that Auger capture due to scattering by electrons is more effective than that involving hole scattering [34, 81]. The capture rates are larger for smaller dot sizes and these rates decrease very quickly with the dot size. It is shown also that these rates depend very strongly on ΔE, the energy level separation between the WL band edge and the QD level into which capture occurs. When the single-phonon processes are energetically allowed the single-phonon capture rates can dominate the Auger capture rates. However, while the ΔE value significantly varies due to the inhomogeneous broadening in self-assembled dots the multiphonon- and Coulomb-mediated processes could both contribute to carrier capture in QDs.

Various theoretical approaches have been developed for the analysis of the Coulomb scattering in different QD systems [33, 81–84]. In case of the QDs coupled to QW through barrier layer, the system very prospective for the laser application [33], Auger-assisted carrier transfer from QW to QDs competes with phonon-assisted tunneling. It is shown that an average carrier transfer time is defined by the carrier concentration in the QW. Depending on the barrier width between the

QDs and QW, both mechanisms can result in a carrier transfer time of a few to several hundred picoseconds. A consistent picture of electron relaxation within QDs in this structure appears to demand two relaxation mechanisms: electron-hole Auger scattering and polaron (phonon-assisted) decay.

The Coulomb scattering influences mainly the carrier exchange between WL and QDs system rather than the immediate interdot carrier transfer. It could be very effective at high excitation density creating high carrier density in the WL, however in case of moderate and low electron densities in the WL that is in focus of our study other mechanisms are responsible for the interdot carrier dynamics.

4.3.3 Phonon-assisted Tunneling

Carriers trapped by the QD can escape due to tunneling in the WL or to other QDs or defects, thus contributing to the conductivity or carrier transfer. The effectiveness of interdot tunneling is defined by the overlapping of electronic wavefunctions of coupled QDs. Therefore double QDs arising due to vertical correlation in bilayer QDs structures are considered as very suitable for the demonstration of the quantum mechanical tunneling process. There exist extended theoretical and experimental studies developing various aspects of electronic coupling and carrier tunneling in different QD systems [85–96]. Intuitively it is clear that in case of vertically and laterally coupled InAs/GaAs self-assembled QDs the electron states will follow the artificial molecule analogy, whereas the coupling of hole states is much more complicated. Usually self-assembled QD molecules are intrinsically asymmetric with inequivalent dots resulting from imperfect control of crystal growth. Vertically aligned pairs of InAs/GaAs quantum dots have been grown by MBE, introducing intentional asymmetry that limits the influence of intrinsic growth fluctuations and allows selective tunneling of electrons or holes [85]. Interdot barrier thickness can be easily changed thus the tunneling energies vary over a wide range permitting engineering the electron and hole tunneling with asymmetric InAs/GaAs QD molecules, important for the systematic design of complicated QD nanostructures. In such system different defects can arise during MBE growth affecting the carrier dynamics. Indeed deep level transient spectroscopy (DLTS) and capacitance-voltage (C-V) studies of InAs/GaAs vertically coupled QDs inserted in an active region of laser diode reveal point defects located in regions close to the QDs [86]. The point defects create deep levels and carriers can be trapped in the localized states of these levels. The Coulomb interaction between carriers captured in the QDs and on the point defects form the dipoles. It has been demonstrated that in the dipoles the carriers from deep levels can tunnel into the states of QDs from which they were previously thermally evaporated [86]. Similar process we observed in bilayer InAs/GaAs QD structures under temperature lowering [66].

Different experimental techniques are used to reveal the contribution of the tunneling in physical phenomena related to QDs excitation. High resolution (Laplace) transient spectroscopy of electronic states of ensembles of self-assembled

InAs/GaAs QDs allows monitoring the *s* and *p* state occupations as a function of time starting from the moment when all electrons are captured by the QDs with different Fermi level positions under various thermal excitation [87]. The increase in resolution of Laplace transient spectroscopy over conventional experiments reveals quite specific rates of carrier loss, which are attributed to tunneling at low temperatures and a combination of thermal emission and tunneling as the temperature is increased.

Combined PLE and time-resolved PL and cathodoluminescence (CL) studies of the carrier dynamics in a high-density ensemble of $Cd_xZn_{1-x}Se/ZnS_ySe_{1-y}$ QDs reveal lateral carrier transfer among single QDs [88]. Under nonresonant excitation of the QD system a significant probability of independent capture of electrons and holes in separate QDs is observed. The subsequent lateral migration of carriers between adjacent QDs forms a slow decay component of the exciton ground-state PL. At low temperatures the lateral carrier transfer is reduced to the phonon-assisted inter-QD tunneling, resulting in migration times of the order of several nanoseconds. The role of independent carrier capture is suppressed at high excitation densities or increased temperatures, enabling thermally activated migration [88]. Completely opposite picture of the carrier capture and tunneling we observed recently in intentionally designed strained InAs: $In_{0.3}Ga_{0.7}As/GaAs$ QDs:QW structure [89]. The carrier transfer occurs through the excitonic trapping from the QW states to the QD states. This transfer is very efficient at low excitation densities and low temperatures and explains the excitation density and nonmonotonic temperature dependences of the QW photoluminescence.

Interdot coupling can be effectively tuned using hydrostatic pressure [97], applied voltage [98] or magnetic field [99] thus providing the way for engineering of electron and hole tunneling in QD nanostructures. The strength of coupling can be increased if the distance between the dots becomes small enough. It is shown recently by means of the PL spectroscopy under high hydrostatic pressure that electronic coupling due to quantum-mechanical tunneling in stacks of InGaAs/GaAs QDs can be effectively increased [97]. Varying the hydrostatic pressure one influences the interdot-barrier height for electronic coupling between self-assembled QDs in vertical stacks. The magnitude of the coupling can be tuned in a controllable way by applying high hydrostatic pressure. A sufficiently high pressure can also quench the coupling.

The application of an external bias along growth direction in vertically stacked InAs/GaAs QD arrays substantially impacts their PL properties [98]. The time resolved PL data are consistently interpreted in terms of the bias-induced tuning the energetic states of the QDs with respect to each other. Tuning through resonances between the ground states of QDs significantly increases the probability of carrier tunneling along the stacked QDs and substantially modifies PL intensities as well as rise and decay times in time-resolved PL. This bias sensitivity paves the way for new photonic applications of such structures.

In general, the magnetic field drives the system of vertically coupled QDs into a strongly correlated regime by modulating the single-particle gaps [99]. In coupled QDs different components of the magnetic field, either parallel or perpendicular to the tunneling direction, affect single-dot orbitals and tunneling energy, respectively. This allows us to tune the QDs system and strongly influence the rate of carrier tunneling.

Summarizing this Section we have shown that in reality various mechanisms of carrier transfer have an impact on the optical and transport properties of the QD system. For distinguishing and investigating the transfer mechanism we should provide special experimental conditions, grow suitable QD structures and apply a corresponding external field. In what follows we describe efforts to explore this behavior of carrier dynamics in dense QD arrays.

4.4 Temperature Effects in Photoluminescence of In(Ga)As Quantum Dots Arrays

4.4.1 Line-shape Analysis of Temperature Dependent Photoluminescence Spectra of Dense In(Ga)As/GaAs Arrays with Lateral Coupling

Let us consider the temperature behavior of the PL spectra measured for the dense InAs/GaAs QDs arrays. Figure 4.4 shows these spectra for the sample with an InAs coverage Θ of 2.46 ML, the AFM picture of which is plotted in Fig4.1 (b) [100]. PL spectra excited by the 514.5 nm line of a cw Ar^+ laser have been studied under various conditions. The temperature was changing from $T = 6$ K up to room temperature, the excitation intensity was increasing from 0.01 to 20 W/cm^2, and the

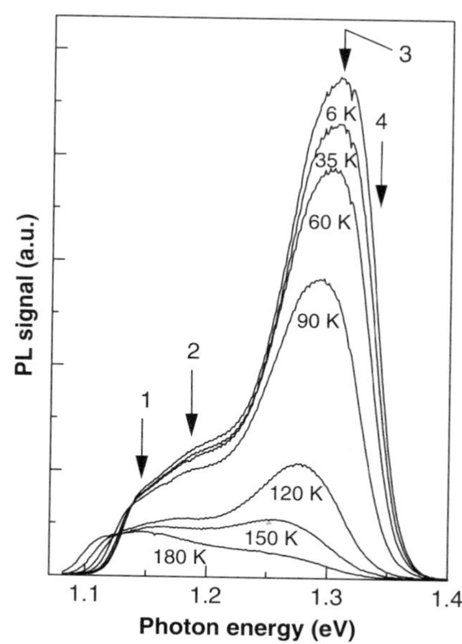

Fig. 4.4 Temperature evolution of the PL spectrum from InAs/GaAs QD sample ($\Theta = 2.46$ ML, $T_G = 420\,^\circ$C). Positions of Gaussian peak energies for the low temperature spectrum ($T = 6$ K) are shown by arrows

magnetic field B was varying from $B = 0\,\mathrm{T}$ to $7\,\mathrm{T}$. The study in magnetic field shows no effects on both the PL peak line shape and the PL peak position, as it was expected for the small-sized QDs ($\sim$10 nm) and such a broad PL feature [101]. Variation of the excitation intensity within the chosen interval shows a remarkable reproducibility of the PL line shape, excluding the contribution of the spectral components, which stem normally from the excited states of the QDs at least at the highest excitation densities used for the cw PL study. The PL maximum depends linearly on the excitation density over 3 orders of magnitude, when the excitation intensity decreases from the moderate level of 5 W/cm^2 to the low level of 0.01 W/cm^2.

Dramatic changes modify the PL line shape under the temperature variation. Figure 4.4 demonstrates these changes. It is clearly seen that the high-energy side of the PL band decays rapidly with temperature increase, whereas the low-energy side preserves its shape up to temperatures of about 200 K. In order to make more definite assignment the PL spectrum was deconvoluted into set of four Gaussians (cf. arrows and notations 1–4). Figure 4.5 depicts the temperature dependence of all characteristic features of these Gaussians: photon energies [Fig. 4.5 (a)], FWHM [Fig. 4.5 (b)], and peak magnitudes [Fig. 4.5 (c)]. The observed behavior gives strong indication of a multimodal distribution in QD sizes. The modes assembled from the smaller sized QDs are responsible for the high energy PL peaks and the

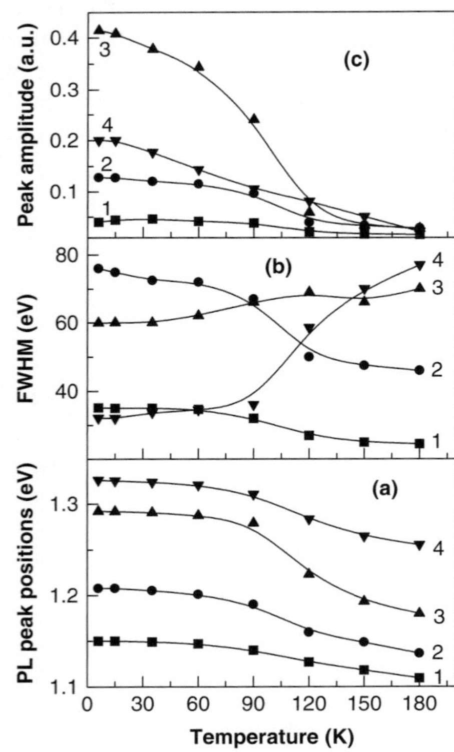

Fig. 4.5 Temperature dependence of PL parameters in InAs/GaAs QD sample for Gaussian components derived from the PL spectra deconvolution: (a) energy peak positions; (b) FWHM; (c) peak amplitudes

modes aggregated from larger sized QDs give rise to the low energy PL peaks. The temperature dependent spectral position of each PL component follows the shift of the InAs band gap [74, 102]. We can clearly see that amplitudes of low-energy peaks (1–2) are practically invariant up to 90 K, whereas those of high energy peaks decrease substantially with the temperature. The FWHM of all components varies slightly within the temperature interval of 6–90 K, but then decreases for low-energy components as well as grows for line 4 at the highest photon energy. Similar behavior is also observed for integrated intensities of all four contributions. Since the excitonic mechanism of QD recombination, the observation of higher luminescence efficiency from QDs of smaller sizes indicates an additional confinement of InAs excitons in the lateral direction. The high PL efficiency also implies that the phonon bottleneck effect is not efficient for the present case. Up to a temperature of ~60 K the spectral positions are almost invariable. One of the reasons is that the InAs band gap does not change (less than 10 meV) strongly within the temperature range $0\,K < T < 60\,K$. Because the PL linewidth is more than 75 meV we cannot detect the expected redshift.

Another reason is the exciton localization in a definite potential minimum. When the temperature exceeds 60 K the peak energy position starts to move toward lower energies. We assume that this can be connected with carrier escape from QDs of the smaller size family via tunneling to adjacent QDs. Actually, the unipolar and excitonic thermal escape is still a question in dispute even in quantum wells. When the k_BT becomes comparable to the exciton binding energy, free carriers arise because after exciton dissociation they can escape from the QDs at least in part due to tunneling between the nearby QDs. The thermalized carriers can relax over a long distance and find a lower local-energy minimum, resulting in shrinkage of the FWHM. The tunneling will be more likely for the QDs with sizes in the center of the corresponding Gaussian distribution. This also can contribute to the reduction of FWHM observed for peaks 1 and 2 [Fig. 4.5 (b)]. With further temperature increasing (above 150 K) the electron–phonon scattering and thermal re-distribution of carriers are of importance, giving rise to an increasing FWHM. Figure 4.5 (b) shows that the electron–phonon scattering contributes strongly to the high-energy component 4, the FWHM value of which increases by more than factor 3 within the temperature range 90–180 K. The FWHM behavior for peaks 1–3 rather reflects the coexistence of excitonic recombination with the carrier transfer between the QDs with different energy levels and QDs of different modes. Indeed the electron-hole pairs created due to light absorption in the 2D InAs regions and in the GaAs barriers are effectively captured into the QDs, where they afterwards recombine. Both the carrier tunneling between InAs QDs and carrier percolation through the 2D InAs layer may contribute to the carrier transfer. The efficiency of these processes is expected to be high in the assemblies with small average distance between the QDs. Moreover, the 2D InAs layer connects QDs of different sizes and consequently of different emission energies. Taking into account the excitonic nature of the PL, the excitonic thermal activation, carrier transfer, including the tunnel transfer among the individual QWs, we are able to explain the temperature behavior of peaks 1 and 2. Indeed, the relaxation of carriers and their radiative recombination occurs for all separate size modes. However, the nonradiative decay of the PL for peaks 3 and 4 serves as a complementary channel of feeding for the QDs assembled in larger size modes. Being of a lower density these latter QDs become saturated due to the carrier flux from the decaying radiative states belonging to smaller size modes. This saturation is clearly manifested by the temperature behavior of peaks 1 and 2. Both their magnitude and FWHM change very slightly for temperature variation in contrast to strong quenching of the PL signal from the QDs belonging to the small size mode. The multimodal behavior of the PL described above represents a clear demonstration of thermally induced carrier transfer among various size families in a multimodal QD size distribution, supporting and extending the picture outlined by Brusaferri *et al* [74] for a very similar system of self-organized InAs QDs. We give evidence of the stabilization of PL characteristics, caused by a strong coupling between modes, due to feeding of the radiative transitions through the nonradiative decay and carrier transfer from decaying excitonic states of the small sized QD modes.

Summarizing we have demonstrated the stabilization caused by strong lateral coupling between the QD modes of the PL magnitude and FWHM of large QD modes within a wide temperature interval 50–150 K due to nonradiative decay and lateral carrier transfer arising from decaying excitonic states of small size QD modes and tunneling between the different QD modes.

4.4.2 Temperature Induced Carrier Transfer in Vertically Coupled In(Ga)As/GaAs Quantum Dot Arrays

In this Section we present a detailed study of the carrier transfer between two InAs/GaAs QD families with different size distribution but separated from each other by a thin GaAs layer with the thickness d_{GaAs} of 16 nm [76]. The particular QD system under investigation is a vertically aligned double-layer InAs/GaAs QD structure with different sized QDs in the first layer compared to the second layer. The PL spectrum from such QDs system (Figure 4.6) reveals a pronounced double-peak structure. It can be attributed to the total contribution in PL signal from QD ensembles of both layers. Indeed, a line shape analysis shows that the total PL signal is well reproduced by a convolution of two Gaussian-shaped bands and a third contribution arising from the WL (see bottom curve in Fig. 4.6). The high energy QD peak 2 is ascribed to the dots in the seed layer ($\Theta_1 = 1.8$ ML) while the low-energy peak 1 originates from dots in the second layer ($\Theta_2 = 2.4$ ML). Figure 4.6 plots also the temperature dependence of the PL emission from this sample measured at excitation intensity of 1 W/cm^2. There exist two different temperature regions: i) low temperature region (10–80 K), where PL spectra are quite stable and exhibit no significant changes. Such behavior of PL spectra is characteristic for excitonic recombination; ii) high temperature region (100–200 K) where dramatic changes in the PL spectra are observed. Beginning at around 110 K peak 2 (small dots in seed layer) experiences a strong redshift of about 2 meV/K (see inset of Fig. 4.6).

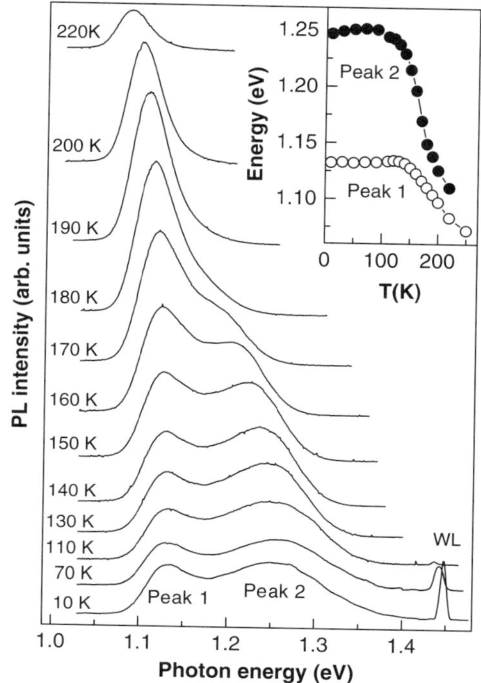

Fig. 4.6 PL spectra of stacked bilayer InAs/GaAs QD structure with different sized QDs at different temperatures ($\Theta_1 = 1.8$ ML, $\Theta_2 = 1.8$ ML and $d_{GaAs} = 16$ nm). The inset shows the temperature dependence of the two PL peaks energies

Simultaneously with the shift, a slight increase of the integrated PL intensity in the temperature range of 110–160 K is observed with a subsequent decrease as the temperature rising to 250 K. Such a strong redshift of the PL signal in an InAs/GaAs QD system has not been observed before. The slight increase of peak 2 in the range of 100–150 K is a real effect. It is due to a transfer of carriers with a short lifetime in small dots of the seed layer to the states with a long lifetime in large dots of the seed layer, resulting in a higher equilibrium density of excited QDs [103]. Over the same temperature range the behavior of peak 1 (large dots) is basically different. The spectral redshift starting at $T \sim 150$ K is much smaller than for peak 2 while the integrated PL intensity begins to grow dramatically with increasing temperature above 120 K. As shown in Fig. 4.7 (a) peak 1 reaches a maximum at 190–200 K and then drops rapidly between 200 and 250 K.

We propose an explanation of the observed temperature behavior based on the interlayer carrier transfer between dots due multiphonon-assisted tunneling. This tunneling is ineffective at low temperature but become efficient at temperatures higher than 100 K. At the same time basing on previous investigations of stacked QDs [104] we believe that the quality and thickness of the GaAs barrier layers limits the role of thermionic emission in explaining our observations of the bilayer QD structure in the 100–200 K range. According to previous investigations [100, 103], one can suggest a time hierarchy of carrier relaxation processes. The fastest times are for intralayer and intradot carrier relaxation τ_I, followed by interlayer carrier

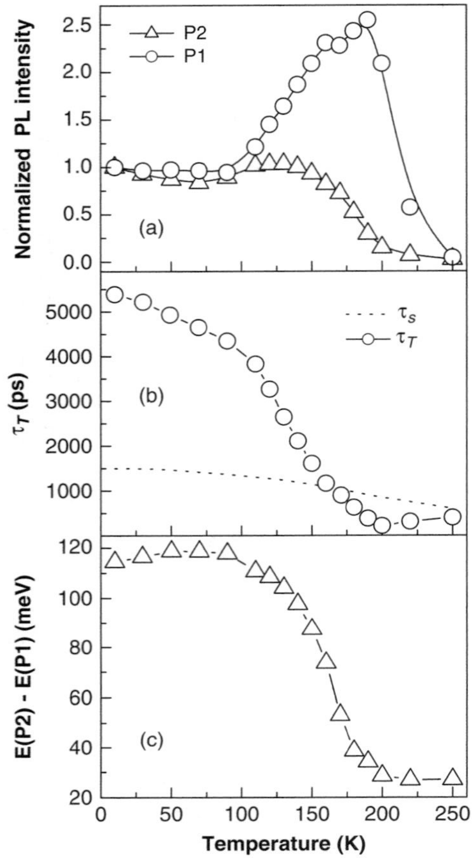

Fig. 4.7 Temperature dependence (a) of the normalized integrated PL intensity for peaks 1 and 2 shown in Fig. 4.6; (b) estimated from Ref. [103] radiative ground state lifetime τ_S of seed layer and interlayer carrier transfer time τ_T ; (c) energy difference between PL peak positions from QDs of the seed and second layers

transfer time τ_T, and finally the radiative ground state lifetime of seed layer dots τ_S and second layer dots τ_L. For low temperatures a ground state lifetime (τ_S) of ~1500 ps has been determined for the 1.8 ML single layer [103]. Following Ref. [94], the yield (η_T) of excitation transfer from small dots of the seed layer to large dots of the second layer may be determined by $\eta_T = \tau_S/(\tau_S + \tau_T)$ and estimated from the ratio of the integrated PL intensity of peaks 1 and 2 (Fig. 4.6). For $T = 10$ K the estimated τ_T is ~6000 ps. This large value of τ_T in comparison with ground-state lifetime τ_S is in a good agreement with results in Ref. [94]. Figure 4.7 (b) depicts τ_T derived using η_T and utilizing the τ_S (T) estimated from Ref. [103]. One can see that τ_T sharply decreases starting at 100 K and become smaller than the ground-state lifetime τ_S at about 150 K indicating that the process of interlayer carrier transfer begins to play a more essential role and becomes dominant at $T > 160$ K. It should be noticed, that the estimated temperature dependence of τ_T gives only an upper limit, since we did not take into account nonradiative recombination channels, which can be activated at high temperatures. The physical reason for a greater interlayer carrier transfer at temperatures higher than 120 K is that nonresonant tunneling is becoming increasingly more resonant. Indeed, at low temperature [Fig. 4.7 (c)] the average energy difference ΔE between ground states of dots from the seed and second layers is about 120 meV. Without going into the nature of actual transfer mechanism, one can emphasize that at least four LO phonons ($E_{\mathrm{LO,InAs}} \sim 29$ meV) are required. With increasing temperature, however, the carrier thermal energy in the ground states of the small QDs becomes comparable to the exciton binding energy, so that the exciton dissociates into free carriers into the WL and GaAs barrier and then relaxes into a different dot. Carrier hopping between dots favors a transfer of carriers to the dots having a higher binding energy, and hence, a lower emission energy, resulting in a narrowing and spectral redshift of the PL spectrum. Thus, with elevating temperature carrier redistribution takes place and for high temperatures most carriers are in the ground states of the largest dots. For example, in the temperature range of 180–220 K [Fig. 4.7 (c)], ΔE is only $\sim$30 meV, a value which is close to the LO phonon energy. In this case, the yield of nonresonant tunneling must be significantly higher.

Summarizing we have uncovered a more complicated picture of energy and carrier transfer in multimodal and high-density QD systems that will provide for greater control of the PL behavior.

4.5 Photoluminescence of In(Ga)As/GaAs Quantum Dot Systems Under High-Density Optical Excitation

High-density optical excitation of the In(Ga)As/GaAs QD system reveals new features in the PL response related to the peculiarities of hot carrier dynamics. Among them there are the state filling and PL of excited states in self-assembled QDs [105–107]. Due to discrete density of states for the QDs the peculiarities of intersublevel relaxation lead to the observation of PL from excited-state transitions at comparatively low excitation intensities. Intersublevel relaxation occurs with participation of few longitudinal acoustic or optic phonons (so-called phonon bottleneck effect), due to multiphonon processes, or Coulomb interactions and the intersublevel relaxation rates are much faster than the $\sim$1 ns interband dynamics of the QD ground states. Evidence of this fact follows from the observation of the inhomogeneously broadened Gaussian PL line shape ascribed to the statistically distributed ground states of self-assembled QDs at low excitation densities. Discrete energy spectrum of each QD state results in a state filling effect due to exclusion principles when only a few carriers can populate the lower states. This will also lead to hindered intersublevel dynamics, and, as a result, to the observation of excited-state interband transitions as the excitation intensity is increased. The emission peaks related to excited-state transitions are observed at higher energy than the ground-state QD emission, but should not be confused with the higher-energy multipeak emission assigned to the segregated inhomogeneous broadening caused by ML fluctuation in the QW [108]. The above three effects (phonon bottleneck, state filling, and segregated inhomogeneous broadening) giving rise to higher-energy emission peaks can be distinguished due to different physics nature. In case of the state filling effect under low intensity excitation of the QD system only the ground-state emission is observed because of the fast intersublevel relaxation. As the intensity increases the intersublevel carrier relaxation toward the lower levels is slowed due to the reduced number of available final states in the QD and their progressive filling resulting in growth of population of excited states. This latter process accompanies with the appearance of PL peaks originating from the excited-state interband transitions. What is important, the relative magnitudes of the excited state PL peaks grow with the excitation density increase. In contrast to state filling effect, the phonon bottleneck effect will permit excited-state interband transitions even under low excitation conditions because the intersublevel and interband relaxation dynamics are comparable [109]. In the case of segregated inhomogeneous broadening the observed multiple peaks have the same relative amplitude over several orders of magnitude of excitation intensity [108]. Their energy positions follow the energy predicted for a QW with a fluctuation of a few ML from its average thickness. Excitons localized in a thin QW with ML fluctuations are confined by a much shallower potential than in the case of self-assembled QDs and therefore can be easily distinguished under temperature increase. The onset of thermionic emission and the thermal PL quenching will be observed at much lower temperatures in case of segregated inhomogeneous broadening than in the self-assembled QDs while QDs possess much deeper confining potentials.

Figure 4.8 shows the PL spectra of self-assembled InAs/GaAs QDs grown by MBE using the Stranski-Krastanow growth mode [110]. The sample was grown on a GaAs (001) substrate, with a 0.5 mm GaAs buffer layer and 10 min annealing at 580 °C to provide nearly defect-free atomically flat surface. QDs were then grown by depositing 2.1 ML of InAs at a substrate temperature of 500 °C and finally capped with a 150 ML GaAs layer. The cw PL is measured using the 514.5 nm line of an Ar^+ laser for GaAs excitation, spanning excitation densities from 0.01 to 2000 W/cm^2. At low excitation density (~5 W/cm^2) only one emission peak is observed at 1.13 eV. As the power is increased, four peaks subsequently develop (Fig. 4.8), including the ground state emission. These peaks are separated by ~40 meV and their FWHM ranges from 30 to 40 meV, as deduced from a multiple Gaussian fit. The temperature behavior of PL band evidences for excitons localized in deep confining potentials.

Let us discuss this PL experiment. We consider that carriers after excitation into GaAs barrier rapidly relax to the 2D continuum states of the WL and then are trapped by a QD. As a result the carriers initially populate any of the available discrete bound QD states, with no need to occupy all the excited states sequentially. However, while the number of available higher-energy levels is large the majority of carriers will enter the dot via an excited state. This imposes that intersublevel relaxation times are much shorter than the excited-state radiative lifetimes, otherwise excited-state luminescence would be observed at low excitation intensity (phonon bottleneck effect), which is not seen in Fig. 4.8. The carrier population and therefore the PL intensity of a given excited state will be proportional to $\tau_{int}/(\tau_{int}+\tau_R)$, where

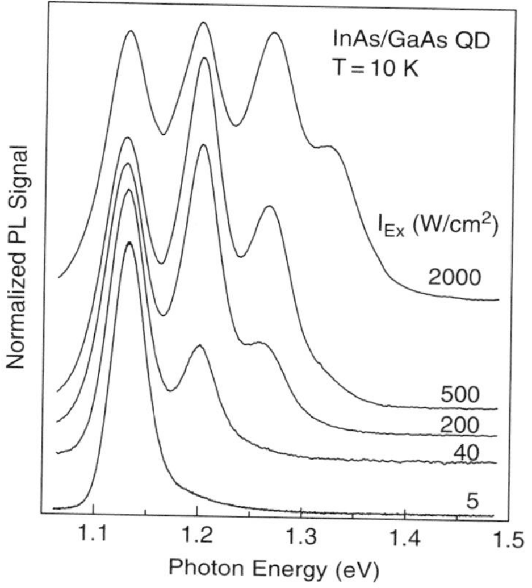

Fig. 4.8 Low-temperature (T = 10 K) PL spectra of InAs/GaAs self-assembled QDs, displaying excited-state radiative recombination under elevating the excitation density [110]

τ_{int} is the intersublevel relaxation time and τ_{R} is the radiative lifetime [105]. If $\tau_{\mathrm{int}} << \tau_{\mathrm{R}}$ this factor is negligible and excited-state PL does not develop. However, if τ_{int} is comparable to τ_{R}, the excited-state emission should be detectable. While no excited-state emission is observed at low excitation intensities, the condition $\tau_{\mathrm{int}} << \tau_{\mathrm{R}}$ is held for empty QDs, and we therefore interpret the longer decay time as being the radiative lifetime (~1 ns). However, since excited-state PL is observed at high excitation intensities, it is clear that intersublevel relaxation is slowed due to the reduced number of available final states caused by the state-filling effect. As the intersublevel relaxation rate decreases and approaches the interband radiative recombination rate for a given excited-state transition, the spectra display a progressive saturation of the lower-energy transitions (seen in Fig. 4.8) and appearance of additional emission peaks related to the upper excited states.

The filling and saturation of the QD states become even more pronounced in the system of vertically coupled QDs if the excitation density increases. We can observe these effects for the ground state QD emission in the excitation intensity range far beyond that needed to develop the excited-state transitions. It is related again with the restricted number of the available QD ground states and with the existence of additional channel for the carrier relaxation in the QD ground state, the interdot (interlayer) tunneling. In order to demonstrate these effects the samples with vertically coupled QDs in two InAs layers separated by the GaAs spacer layer of variable thickness have been grown [111]. In this case the seed QD layer was grown by depositing 1.8 ML of InAs on the GaAs substrate, then 30, 40, 50 or 60 ML of GaAs spacer (d_{sp}) was deposited on top of the seed QD layer, and, finally, the second QD layer was added through deposition of 2.4 ML InAs. Each sample for optical studies was capped with a 150 ML GaAs layer. The variation of the spacer thickness permits to control of the interlayer coupling resulting in the pair dot binding. The STM statistical analysis indicates a size distribution for the QDs with (4 ± 1.5) nm for the height, (20 ± 3) nm for the width, and a dot density of about 4.5×10^{10} cm^{-2} in the seed layer. The dot density in the second layer is variable over the range from 2.5×10^{10} cm^{-2} to 4×10^{10} cm^{-2} depending on the value of d_{sp}. Meanwhile, the second layer 3D islands are nearly twice in volume of the seed islands (for $d_{sp} = 30$ ML) due to additional deposition, as well as the influence of the strain field from the seed layer. The resulting bi-layer QD structure contains different sized dots in each layer. However, the interlayer dot correlation is not complete, since a part (α) of the QDs of the seed layer, as well as, part of the QDs of the second layer, are still uncoupled. The XTEM statistical analysis proves the appearance of asymmetric QD pairs (AQDP) [112]. The AQDP fraction (the correlation degree ($1-\alpha$)) in total QDs density of the seed layer depends on the d_{sp} value: $(1-\alpha) = 0.95, 0.70, 0.50$ and 0.10 for the $d_{sp} = 30$, 40, 50, and 60 ML, respectively.

PL spectra of grown bilayer samples are shown in Fig. 4.9 as a function of d_{sp} at a low excitation density (I_{ex}) of about 0.1 mW/cm^2. As expected [64] the PL spectrum for $d_{sp} = 30$ ML shows a single, slightly asymmetric Gaussian shaped peak (P_{sec}) at 1.09 eV, which is the emission energy expected from the large QDs (LQDs) in the second layer. In this case the tunneling time (τ_T) from the seed layer dots to the second layer is apparently so fast compared to the radiative lifetime of the small QDs (SQDs) (τ_{R1}) (τ_T is less than τ_{R1}) that very little PL from the seed layer is observed. With increasing d_{sp}, however, an additional PL peak becomes evident on the high-energy side at 1.27 eV, resulting in a clear double peak structure for the 50 and 60 ML spacer samples. This high-energy peak (P_{seed}) coincides with the

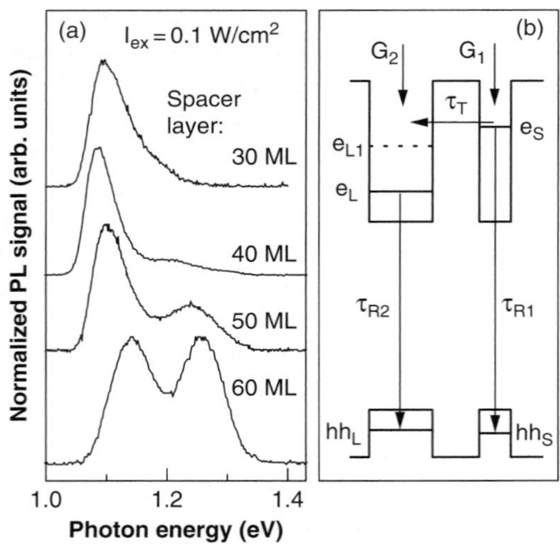

Fig. 4.9 (a) Low excitation density cw PL spectra at $T = 10$ K for 1.8/ 2.4 ML AQDP samples with various spacer thickness. (b) Schematic representation of processes contributing to carrier populations in SQDs and LQDs ensembles. Photoexcitation of carriers is represented by G_1 and G_2, tunneling time for electrons is τ_T, and recombination lifetimes are τ_{R1} and τ_{R2} for SQDs and LQDs, respectively [111]

QD PL peak observed for single-layer samples with a 1.8 ML InAs deposition [76] and is therefore attributed to the emission from SQDs. The apparent increase in the PL yield of the SQDs with increasing spacer thickness is also expected due to the decreasing electronic coupling and, consequently, decreasing carrier transfer probability between AQDPs.

Perhaps less expected is the fact that the cw PL spectra as a function of d_{sp} reveal a significantly different PL behavior between the SQD and LQD peaks as a function of the excitation density I_{ex}. Investigation of PL spectra in our samples over wide range of excitation densities covering four orders of magnitude as well as thorough line-shape analysis convince us that we are still below the conditions of higher levels excited state emission becomes visible in the PL spectra thus a contribution of excited states of LQDs can be ignored in our cw PL measurements. Figure 4.10 (a) plots normalized PL spectra, as a function of I_{ex}, for the sample with $d_{sp} = 40$ ML [110]. The peak intensity ratio I_{LQD}/I_{SQD}, as a function of I_{ex} [Fig. 4.10(b)], shows a weak dependence on I_{ex} for $d_{sp} = 60$ ML when the seed and second QD layers are nearly decoupled, but a significant influence for a d_{sp} of 40 ML, when the coupling is strong. For the strong coupling case, at high I_{ex}, the filling of the energy levels in the LQDs understandably reduces the carrier transfer from the ground state of the SQDs thus increasing the population of this state and decreasing the ratio I_{LQD}/I_{SQD} as it is seen in Fig. 4.10(b). Thus, for AQDPs we attribute the decrease in carrier transfer probability with elevating I_{ex} to a decrease of available free ground states in the larger QDs in the second layer.

In order to interpret our data we assume a schematic band energy diagram for the AQDP structure presented in Fig. 4.9(b) and apply a simple three-level model for this structure in order to uncover the behavior of the tunneling time τ_T and radiative lifetimes of the SQDs and LQDs, τ_{R1}andτ_{R2}, respectively. Within this model the observed decrease in carrier transfer probability can be understood as state filling due to increasing I_{ex}. In addition, for the samples with $d_{sp} \leq 50$ ML, when τ_T becomes smaller than τ_{R1} and $\tau_{R1} \leq \tau_{R2}$, carrier tunneling from a SQD to the LQD is temporarily blocked for a time of order τ_{R2}. This scenario can be treated within a three-level model [see Fig. 4.9(b)] approximation, using the following rate equations [110]:

$$\begin{aligned}\frac{dn_1(t)}{dt} &= -\frac{n_1(t)}{\tau_{R1}} - \frac{n_1(t)\left(N_2 - n_2\left(t\right)\right)}{N_2\tau_T} + G_1\left(N_1 - n_1\left(t\right)\right)/N_1,\\ \frac{dn_2(t)}{dt} &= -\frac{n_2(t)}{\tau_{R2}} + \frac{n_1(t)\left(N_2 - n_2\left(t\right)\right)}{N_2\tau_T} + G_2\left(N_2 - n_2\left(t\right)\right)/N_2.\end{aligned} \tag{4.1}$$

Here $n_1(t)$ and $n_2(t)$ represent the carrier densities at time t in the lowest energy levels of the SQDs and the LQDs, respectively. N_2 is the total number density of LQDs that are coupled to an equal number of SQDs N_1 in the AQDPs. The factor $n_1(t)(N_2 - n_2(t))/N_2$ defines the fraction of SQDs for which tunneling within AQDP takes place since the corresponding LQDs are empty. G_1 and G_2 are the carrier generation rates for the SQDs and the LQDs, respectively. The set of equations

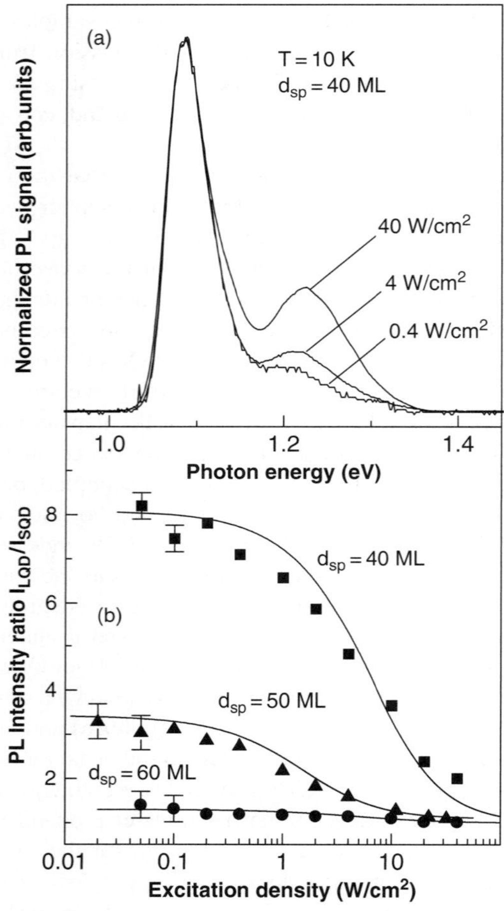

Fig. 4.10 (a) The normalized PL spectra of bilayer sample with $d_{sp} = 40$ ML for different I_{ex}. (b) PL peak intensity ratio I_{LQD}/ I_{SQD} of the LQD and SQD PL spectra vs excitation level for various barrier thicknesses. The fit with Eq. (4.2) is shown by solid lines [110]

(4.1) takes into account the saturation of the carrier trap into the SQDs and LQDs states, respectively.

Applying Eq. (4.1) for the steady state case, $dn_1/dt = 0, dn_2/dt = 0$, we get the solutions

$$\begin{cases} n_1^S = N_1 - \tau_{R1}\left(-B - \sqrt{B^2 - 4AC}\right)/2A \\ n_2^S = N_2 - \tau_{R2}\left(\gamma - \delta\left(N_1 - n_1^S\right)/\tau_{R1}\right), \end{cases} \tag{4.2}$$

with $A = \dfrac{\tau_{R1}\tau_{R2}}{N_2\tau_T}\delta$, $B = -1 - G_1\tau_{R1}/N_1 - (\gamma + \delta N_1/\tau_{R1})\,A/\delta$ and $C = \gamma\,\dfrac{N_1\tau_{R2}}{N_2\tau_T} + \dfrac{N_1}{\tau_{R1}}$, where $\gamma = \left(\dfrac{N_1}{\tau_{R1}} + \dfrac{N_2}{\tau_{R2}}\right)/\,(1 + G_2\tau_{R2}/N_2)$ and $\delta = (1 + G_1\tau_{R1}/N_1)\,/\,(1 + G_2\tau_{R2}/N_2)$.

Equations (4.2) describe saturation in the AQDP system resulting from the blockage of the carrier transfer out of the SQDs into the LQDs as the population of the LQDs ground state becomes large. This effect becomes pronounced under a substantial elevation of the pumping level.

For our excitation wavelength, we assume equal excitation rates for both the SQDs and LQDs [94]. In the range of low excitation densities the states of the LQDs populate substantially faster than states of the SQDs in the AQDPs due to the additional flow of carriers from the SQDs. When tunneling takes place the ratio n_2^S/n_1^S tends to the limiting value of $\tau_{R2}\left(\tau_T + 2\tau_{R1}\right)/\left(\tau_{R1}\tau_T\right)$ as $G_1, G_2 \to 0$ (the limit of extremely low excitation densities). If the tunneling carrier transfer is negligible ($\tau_T \to \infty$) this ratio tends to the value τ_{R2}/τ_{R1}, while in the case of rapid carrier escape from the SQDs states ($\tau_T \to 0$) the ratio n_2^S/n_1^S tends to the limit $2\tau_{R2}/\tau_T$, which can be very large. On the other hand, if the excitation densities are high enough ($G_1, G_2 \to \infty$), the populations n_1^S and n_2^S are saturated for both the SQDs and the LQDs. Equation (4.2) allow quantitatively describe the experimental data [Fig. 4.10 (b)], thus verifying the microscopic picture of carrier relaxation in the AQDP system described in this Section.

Summarizing we have demonstrated the elaborated picture of carrier relaxation in various QDs under high excitation level. This relaxation involve both carrier thermalization and non-thermal carrier transfer through the interdot tunneling. Peculiarities of the density-of-states for the QD system allow observation of states filling and saturation of the radiative transitions in the QD PL spectra under comparatively low excitation densities.

4.6 Time-Resolved Photoluminescence Spectroscopy of In(Ga)As/GaAs Quantum Dot Systems

4.6.1 Hierarchy of Relaxation Times in Laterally Coupled In(Ga)As/GaAs Quantum Dot Systems

Though there are numerous studies of the optical properties of QD arrays in the literature, carrier transfer processes in dense QD arrays are not understood in detail. The PL properties of dense arrays differ substantially from those of low-density ensembles partly due to the distinctly different-bimodal or even multimodal QD size distribution [see Fig. 4.1, 4.2]. Time-resolved PL studies of dense QD arrays revealed mostly a monoexponential or biexponential decay of PL intensity on a time scale between several hundreds of picoseconds and several nanoseconds [59, 105, 113]. Such exponential behavior has been assigned to PL contributions from ground states of different QD sizes and/or to ground- and excited-state contributions. PL decay times τ_d derived from the observed exponential transients have been used to describe the dynamic behavior of the population $n_i(t)$ of the i-th QD. This analysis, however, does not account for carrier transfer and coupling between different QDs which should result in finite rise-times of PL emission and/or a substantially modified decay behavior. We demonstrate the occurrence of such transfer processes, strongly affecting the overall picosecond PL kinetics, in particular at very low excitation densities [114]. A range of carrier densities in which lateral interdot carrier transfer saturates will be identified. This transfer involves transitions from higher-lying ground states of SQDs into lower-lying states of LQDs. For this demonstration the InAs/GaAs QD sample ($\Theta = 1.79$ ML, $T_G = 420\,^{\circ}$C) shown in Fig. 4.1, 4.2 with bimodal size distribution is chosen. Low temperature PL spectrum of this sample is shown in Fig. 4.11. Broad PL band (the FWHM is of ~120 meV) preserves its spectral shape if the excitation density is raised at least from 1 to 20 W/cm^2. A line-shape analysis of the spectrum shows that the PL signal is well reproduced by a convolution of two Gaussian-shaped peaks and a third contribution arising from the WL (dotted lines in Fig. 4.11). Taking into account the size distribution and areal densities of both QD families [Fig.4.1, 4.2], the stronger peak at 1.28 eV is attributed to the larger number of smaller QDs, having an average base length of 8 nm, while the lower-energy and weaker peak at 1.22 eV refers to larger dots with $b = 14$ nm.

Transient PL measurements were performed with sub-100-fs pulses from a Ti:sapphire laser ($\lambda = 732$ nm, $\tau \approx 80$ fs, f = 82 MHz) allowing population of the InAs QDs by exciting the GaAs barrier. Different detection energies within the broad PL spectrum (Fig.4.11) and in a wide range of excitation intensities were used. Up to excitation densities of 10^{11} photons/(pulse/cm^2), the PL emission at the different detection energies rises within the time resolution of our experiment of 10 ps and displays a monoexponential decay characterized by a decay time τ_d. In Fig. 4.12 τ_d, determined from a least-square fit of the transients, are summarized for different excitation densities [114, 115]. The values are almost constant below 10^{11} photons/(pulse/cm^2). For higher excitation densities [$>10^{11}$ photons/(pulse/cm^2)] τ_d decreases, nonexponential transients appear, and finally a well-pronounced biexponential decay is observed at least in the spectral region of the QD emission band. In Fig.4.12, the largest time constant measured, i.e. the one that is most likely to be

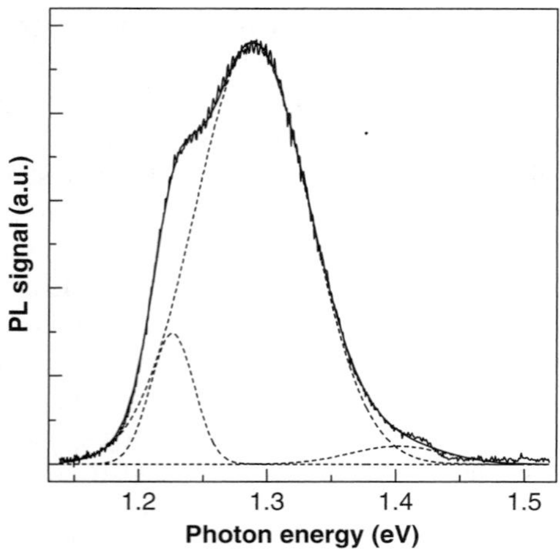

Fig. 4.11 Steady-state PL spectrum for the InAs/GaAs QD sample with $d_{\mathrm{InAs}} = 1.79$ ML at $T = 10$ K. A line-shape analysis of the low temperature spectrum proves that the PL signal is a convolution of two Gaussian-shaped peaks and a third smaller contribution arising from the WL [114]

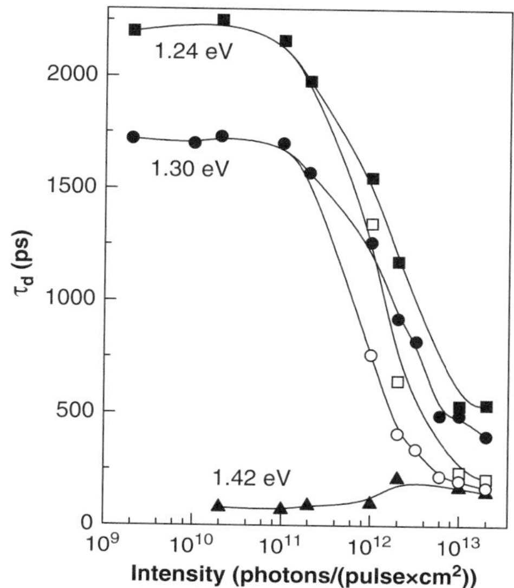

Fig. 4.12 PL decay time (τ_d) data for InAs/GaAs QD sample shown in Fig. 4.11 determined from a least-square fit of transients for various excitation densities. Full symbols mark the long transients, whereas open symbols denote the faster transients that exclusively appear for high excitation densities [114, 115]

governed by ground-state recombination, is marked by full symbols, whereas open symbols of the same type denote the additional faster components measured at the same photon energy for high excitation densities. The threshold for the appearance of biexponential decay is the same for both spectral positions corresponding to the QD PL maximums (see Fig. 4.12). This latter fact can be interpreted in terms of the intradot relaxation with the participation of excited QD states. However, the presence of faster components in the PL decay related to the excited states does not give evidence of the phonon bottleneck existence in our dense QD system, while excited states do not develop at the elevation of cw exciation. Besides switching from mono-exponential to biexponential decay at higher excitation densities, a strikingly strong decrease of the PL decay time for the QD states is observed at higher excitation densities. Such a behavior is quite unusual in self-assembled QD systems. Indeed, the saturation effects expected under an elevation of the excitation density tend to produce longer carrier decay times in view of the exclusion principle. However, it can be wrong in a system of strongly coupled QDs with a large range of base lengths b. In this latter case, the excited states of large QDs can be energetically close to the ground states of the small QDs. While the wave functions of the excited states are weakly localized in comparison with the ground state wave functions, the overlap of excited- and ground-state wave functions for neighboring QDs is stronger than the ground state wave-function overlap. In this case, the saturated states of large QDs can relax through the ground states of smaller QDs. The carriers will escape from large QDs into states with smaller relaxation time Σ (smaller QDs) Π thus providing an effective reduction of the decay time through a complementary channel. This channel of energy relaxation can be of importance just for high-density QD systems.

We now present the interesting results measured at very low excitation densities which give evidence of a finite rise time of the PL at low detection energies, i.e., for the LQDs [114]. In Fig. 4.13 (a), we plot the PL time evolution at a detection energy of 1.20 eV for excitation fluxes between 10^9 and 2×10^{11} photons/(pulse/cm^2). For the lowest excitation densities, the PL signal was integrated over about 10^{11} subsequent excitation pulses. The data at low excitation clearly display a maximum shifted to positive times, a behavior which is indicative of a delayed rise to PL. With increasing excitation density, i.e., saturation of the QD ensemble, this maximum shifts to shorter times. For the highest density of 2×10^{11} photons/(pulse/cm^2), the signal rises within the time resolution of 10 ps. As a measure for the delayed rise, we analyzed the temporal position of the maximum PL in each transient by fitting the whole curve. Figure 4.13 (b) shows this temporal shift of the PL maximum versus excitation density. The position of the PL transient peak for 2×10^{11} photons/(pulse/cm^2) was set as $t = 0$. The line in Fig. 4.13 (b) corresponds to a saturation curve with $t = t_0/(1 + I/I_0)$, with $t_0 = 300$ ps and $I_0 = 1 \times 10^{11}$ photons/(pulse/cm^2).

The delayed rise of PL for weak excitation is consistent with the model of interdot carrier transfer that considers the ground-state relaxation in a system of coupled QDs. In our model, carriers in the ground state of a QD can relax by radiative recombination giving rise to PL and, in addition, carriers populating the ground states of smaller QDs can be transferred into levels of larger QDs being even lower in energy. This represents an extra depopulation mechanism for the smaller sized dots and therefore the overall PL decay time τ_d becomes faster for smaller QDs, i.e., for increasing photon energy in the overall PL spectrum. Such a behavior was found in the experiments described above. Correspondingly, PL from large QDs exhibits different kinetics: First, there is the population increase caused by accumulation of carriers that relax from smaller QDs, giving rise to a delayed rise of PL. Second, there is a decrease of the number of energy levels below the considered one into which the carriers might relax, resulting in longer PL decay times.

Theoretically these experimental data can be treated in the terms of rate equations. For weak excitation density of the QD ensemble and under the assumption that the carrier relaxation rate from an energy level in a QD is proportional to the number of vacant lower-energy levels in adjacent QDs, the rate equation for a particular quantum level E_i can be written as [114]

$$\frac{dn_i}{dt} = -\frac{n_i}{\tau_0^i} - \sum_{i>j} \frac{n_i D_j}{\tau_t^{ij}} + \sum_{i<j} \frac{n_j D_i}{\tau_t^{ji}}, \tag{4.3}$$

where n_i and τ_0^i are the population and the total ground-state recombination lifetime in the ith ground state, respectively; τ_t^{ij} is the interdot carrier transfer time between E_i and E_j states, and D_j is the density of the (final) Ej states. Equations (4.3) have an analytical solution allowing a straightforward analysis of $n_i(t)$. In order to find this solution let us rewrite Eq. (4.3) in more convenient form

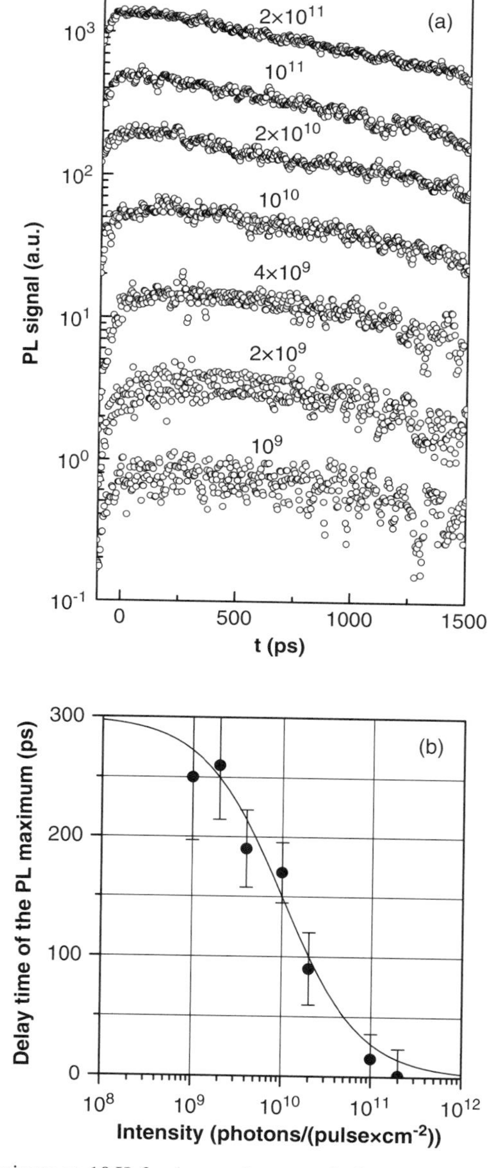

Fig. 4.13 (a) PL transients at 10 K for low and extremely low excitation densities. The detection energy is 1.2 eV. The intensity as parameter is given in photons/(pulse/cm^2). (b) Shift of the PL maximum vs excitation density. The position of the PL transient peak for 2×10^{11} photons/(pulse/cm^2) was set to zero [114]

$$\frac{dn_i}{dt} = -\lambda_i n_i + \sum_{j=1}^{i-1} a_{ij} n_j. \tag{4.4}$$

with

$$\lambda_i = \left(\frac{1}{\tau_0^i} + \sum_{j=i+1}^{N} \frac{D_j}{\tau_t^{ij}} \right) \text{ and } a_{ij} = \frac{D_i}{\tau_t^{ji}}.$$

The index i labelling the levels runs from 1 for the one with the highest ground state energy, over all N states considered. The parameter λ_i determines the excitation decay into the state i and the corresponding a_{ij} value quantifies the energy transfer from higher energy states.

The solution of equation set (4.4), satisfying the initial conditions $n_i = n_i^0$, can be written as follows

$$n_i = \left(n_i^0 + \sum_{j=1}^{i-1} \varphi_{ij} \right) \exp\{-\lambda_i t\} - \sum_{j=1}^{i-1} \varphi_{ij} \exp(-\lambda_j t). \tag{4.5}$$

The parameters of equation (4.5) are calculated from the recurrent procedure,

$$\begin{aligned}
&\varphi_{11} = n_1^0, \\
&\varphi_{ij} = \frac{a_{ji}}{(\lambda_j - \lambda_i)} \Theta_{ij} \varphi_{jj}, \; i > j, \quad a_1 = 0, \\
&\Theta_{ij} = \begin{cases} 1 + \sum\limits_{m=1}^{n=i-j-1} \left[(-1)^m \sum\limits_{k=1}^{P_{mn}} \left\{ \mathrm{T}(k,m) \prod\limits_{l=1}^{m} \frac{a_{j\rho(k,l)}}{\left(\lambda_j - \lambda_{\rho(k,l)}\right)} \frac{a_{\rho(k,l)r}}{a_{jr}} \right\} \right], & if \;\; i \geq j+2 \\ 1, \; if \;\; i < j+2 & \end{cases} \\
&\varphi_{ii} = n_i^0 + \sum_{k=1}^{i-1} \varphi_{ik}.
\end{aligned} \tag{4.6}$$

Here $P_{nm} = \frac{n!}{m!(n-m)!}$ is the number of combinations from n elements by m elements, $\mathrm{T}(k,m)$ is the ordering operator that arranges all integer numbers $\rho(k,l))$ within the interval $j+1 \leq \rho(k,l)) \leq i-1$in descending order for every combination determined by pair (k,m) in equation (4.6), the r index is equal to i for $\max\{\rho(k,l)\}$ in any (k,m) combination and to the preceding value of $\rho(k,l))$ for every subsequent term in the ordered product (k,m). Equation (4.5) is the analytical solution for the set of equations (4.4) allowing a straightforward analysis of n_i(t). Nevertheless it occurs not very informative in view of huge number of input parameters τ_0^i and τ_t^{ij} typical for ground states of QDs system of high density. Therefore in order to illustrate the characteristic features of QDs kinetics it worthy to simplify system assuming a single time τ_0 for recombination lifetime in the i-th ground state and a single time τ_t for the inter-dot carrier transfer. In what follows we point out the physical reason for such simplification even in the real case of high density QDs arrays. Under this assumption equation (4.6) includes the only modification

$$\Theta_{ij} = \begin{cases} \prod_{l=j+1}^{i-1} \left(1 - \frac{a_i}{\lambda_j - \lambda_l}\right) & if \quad i \geq j+2 \\ 1, \quad if \quad i < j+2, \end{cases} \tag{4.6a}$$

and it can be easily applied to the transient PL analysis. It should be noticed here that the only assumption in Eq. (4.3) is the existence of lower (upper) lying states with respect to a certain state i in (from) which a carrier can appear (escape). Therefore, the model can be applied also to the temporal evolution of the PL in a system without interdot coupling since the physical meaning of τ_t could be taken as an interdot transfer time as well as a nonradiative scattering time.

Possessing n_i(t) the PL decay time τ_d is now calculated according to

$$\tau_d = -n_i(t) \left[\frac{dn_i(t)}{dt}\right]^{-1}. \tag{4.7}$$

In fact, the proposed rate equations model of equation (4.4) holds for a dense QD system independent of the number of modes of its size distribution. The presence of more than one distinct dot size distribution within a large QD ensemble can be taken into account simply by introducing additional Gaussians into Eq. (4.5), resulting in several steplike variations of τ_d with emission energy. As a result, we find a staircase like spectral dependence of the ground-state PL time constants τ_d [103]. This dependence is shown in Fig. 4.14 [103, 115].

Let us discuss the spectral dependence of τ_d, which exhibits the stair-case like shape mentioned above. The shortest τ_d values characterize the relaxation within the wetting layer ($\tau_d < 100$ ps) and the GaAs matrix ($\tau_d < 200$ ps). This is caused by rapid trapping of carriers by the QDs and/or nonradiative decay caused by the interfaces. Furthermore, there are two steps in the spectral region of QD emission, namely, at 1.22 and 1.28 eV. These are exactly the spectral positions of the cw PL peaks for both QD families seen in upper part of Fig. 4.14. The correlation of the two features in both spectra survives up to room temperature. For InAs/GaAs QDs appearing in a monomodal QD size distribution, the spectral dependence of τ_d has been studied by Tackeuchi *et al* [90]. The correlation between the cw PL peak and the observed single step we interpret by taking into account carrier transfer within one QD family. If the ground state of a smaller dot relaxes, the energy transfers to the larger-sized QDs (within the family). This process has a cascade-like nature described by Eq. (4.4). The probability to escape from the ground state of small QDs is determined by the number of available unoccupied states with lower energies supplied by the larger-sized QDs. Thus, this probability will be determined by the integral over the Gaussian distribution up to the photon energy of the PL emission. The integrated Gaussian is, of course, the error function erf(E). Since τ_d is inversely

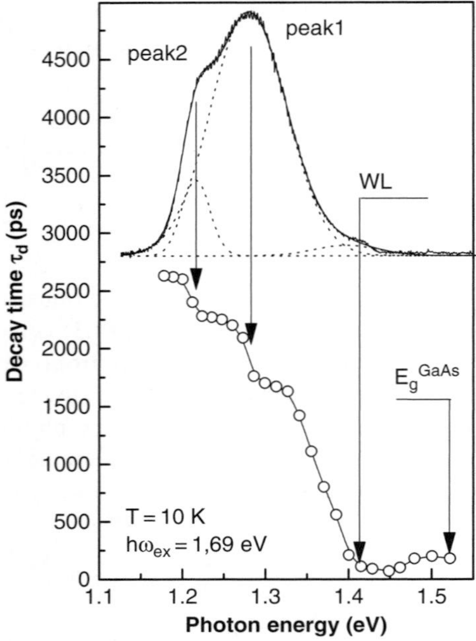

Fig. 4.14 Steady-state PL spectrum (upper part) and PL decay time τ_d (bottom part) for high-density InAs/GaAs QDs sample with $d_{\mathrm{InAs}} = 1.79$ ML measured at $T = 10$ K. Spectral positions of the QD PL bands, wetting layer PL, and GaAs PL are marked by arrows. Excitation density is 2×10^{10} photons/(pulse/cm^2)

proportional to the probability of escape, $\tau_d(E)$ is expected to be step-like shaped with the strongest change in the center of the Gaussian. This is exactly what we see twice, both at 1.22 and 1.28 eV, in our InAs/GaAs QDs appearing in a bimodal QD size distribution, as shown in Fig. 4.14. Consequently, we demonstrate an additional criterion on how the modal structure of QD distribution can be revealed using the time-resolved PL measurements. The experimental $\tau_d(E)$ dependence exactly follows that one calculated from Eqs. (4.5) and (4.7) thus completely justifying the applicability of the rate equations for the description of the carrier dynamics in dense multimodal QD arrays. Even more pronounced $\tau_d(E)$ dependence is observed in the InAs/GaAs QDs sample with $d_{\mathrm{InAs}} = 2.46$ ML shown in Fig. 4.1 (b) and Fig. 4.4. This QD sample demonstrates multi-modal size distribution what is immediately reflected by its $\tau_d(E)$ dependence shown in Fig. 4.15 [116].

A realistic description of carrier dynamics in dense QD arrays has to include the case of saturation, i.e., situations in which the number of nonequilibrium carriers n_i reaches the number of available QD ground states N_j in the jth dot distribution. In order to include saturation effects, Eq. (4.3) have to be modified to

$$\frac{dn_i}{dt} = -\frac{n_i}{\tau_0^i} - \sum_{i>j} \frac{n_i(N_j - n_j)D_j}{\tau_t^{ij}} + \sum_{i<j} \frac{n_j(N_i - n_i)D_i}{\tau_t^{ji}}. \tag{4.8}$$

The equations of set (4.8) are nonlinear differential equations, and do not have a simple solution. Therefore they were solved numerically and the results of such solution predicted a well pronounced peak that can develop itself in the PL transient in the

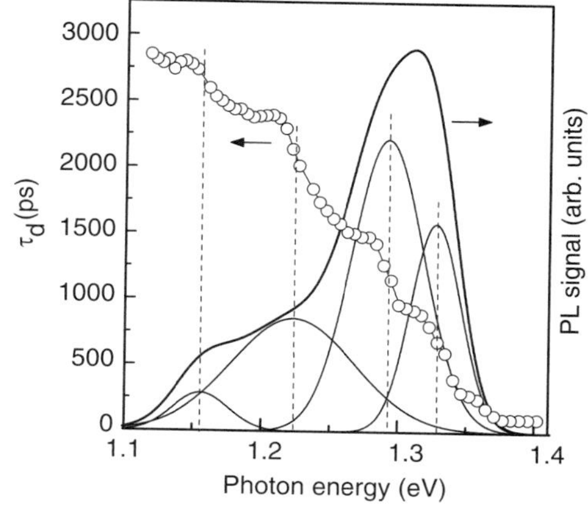

Fig. 4.15 PL spectrum (line) and decay time τ_d (circles) for the InAs/GaAs QDs sample with $d_{\mathrm{InAs}} = 2.46\,\mathrm{ML}$ shown in Fig.4.4 at $T = 10\,\mathrm{K}$. τ_d dependence fully reproduces the multi-modal QD distribution

case of weakly filled QDs ground states when the radiative recombination time (τ_0) and that of the carrier transfer (τ_t) caused by the interdot coupling are of a comparable value. In the saturation case the PL decay is again close to a single exponential. The exponentially decaying parts of the transients, e.g., the 1500–2000 ps range, show a roughly constant slope resulting in about the same τ_d value independent of the initial ground-state filling. It is evident that saturation rather leads to exponential behavior, whereas the PL kinetics in the absence of saturation may also be described by the simpler model described by Eq. (4.6).

Applying developed above rate-equation model to the actual experimental situation we can explain the details of QD kinetics. Indeed, carrier transfer from smaller to larger QDs results in a delayed buildup of population in the latter, a behavior reflected in the PL kinetics [114]. With an increasing excitation flux, saturation of the popuation in such low-energy states occurs, leading to a faster rise of PL. This experimental result is in full agreement with the predictions of the rate-equation models. For low excitation, we reach a regime where a simple model without saturation (Eq. (4.3)) applies. For strong excitation, saturation leads to a much faster population increase in the emitting QDs and a subsequent monoexponential decay of the PL intensity. Applying the simple model for the low-density case and using $t_0 = 300\,\mathrm{ps}$ we derive a value of $\gamma = \tau_t/\tau_0 = 1.3$. It is clear that the simple model seems not to be completely adequate to the real QD system by reason of the strong scattering of τ_0^i and τ_t^{ij} values that can be expected for the ground states of small-sized QDs. However, in high-density QD systems with a strong overlap of the wave functions and a great importance of QD size fluctuations and scattering of the interdot distances, the individual characteristics of a single QD becomes substantially smeared, and an averaging over the relaxation times for both radiative recombination and carrier transfer in the QD ensemble becomes even of more physical meaning than those introduced for every isolated (or free-standing) QD. Taking a value $\tau_0 \sim 2.6\,\mathrm{ns}$ corresponding to the PL decay time of the largest QDs where carrier transfer is almost absent, one finds an "averaged carrier transfer" time $\tau_t > 3.4\,\mathrm{ns}$ for an individual transfer channel. Note that the much faster decay of

PL from small QDs represents the sum over all possible transfer channels from such QDs into states at lower energies resulting in a short overall decay time.

Finally, we conclude that a saturation of QD ground states takes place. In this case, however, one could expect a lineshape variation of the low-power cw PL on the basis of the saturation of the interdot carrier transfer mechanism. A thorough PL study at extremely low excitation densities (less than 1 W/cm^2) reveals such a modification [114]. It was clearly demonstrated that increased excitation power leads to an enhanced low energy part of PL spectrum, and a decay of the high-energy part. We found that the low-energy part reaches the saturation at 1 W/cm^2. A further power increase does not change the line shape in this spectral region. The decay of the PL signal at the high-energy side of the PL band at very moderate excitation densities leads to a shrinking of the total PL spectrum. This effect can be assigned, e.g., to a carrier release from shallow traps within the WL with subsequent transfer of them to the QDs when the pumping level increases. When the excitation density reaches 1 W/cm^2 no further change of the PL spectrum is observed.

In summary, we have learned that transient PL data for dense InAs/GaAs QD arrays are consistent with the model of carrier transfer from small to large QDs within the ensemble directly influencing the PL kinetics. For low excitation densities, the low-energy states of large LQDs display a delayed rise of PL due to the delayed accumulation of carriers transferred from smaller QDs. With an increasing excitation flux, the carrier population saturates, resulting in a fast rise of PL and a mono-exponential decay. From a rate equation analysis of the time-resolved data, we derive a time constant of carrier transfer of the order of 3.5 ns. Interdot tunneling of carriers is considered as the main transfer mechanism at low temperatures. The behavior found here is highly relevant for the physics of dense arrays of coupled QDs.

4.6.2 Detection of Interlayer Coupling in Time-Resolved Photoluminescence of In(Ga)As/GaAs Quantum Dot Systems

In Section 4.5 we have considered the InAs/GaAs QDs in bilayer structures with interlayer coupling varying through the change of the GaAs barrier thickness. The cw PL spectra allowed to get the set of parameters for these structures however this set is incomplete. For example, although the cw PL spectra as a function of d_{sp} provides significant insight into tunneling from the SQD layer to the LQD layer, they do not give any information about of the τ_{R1} and τ_{R2} times thus preventing immediate usage of Eq. (4.2) for the comparison with cw PL results. Moreover cw PL measurements do not allow a direct determination of the tunneling time, which may explain the large spread in the values among different reports. For the determination of these important physical parameters, transient PL measurements are of crucial importance.

Figure 4.16(a) gives the PL transient curves for SQDs and LQDs in AQDP sample with $d_{sp} = 30\,\mathrm{ML}$ shown in Fig. 4.9 while Fig. 4.16(b) gives transient PL

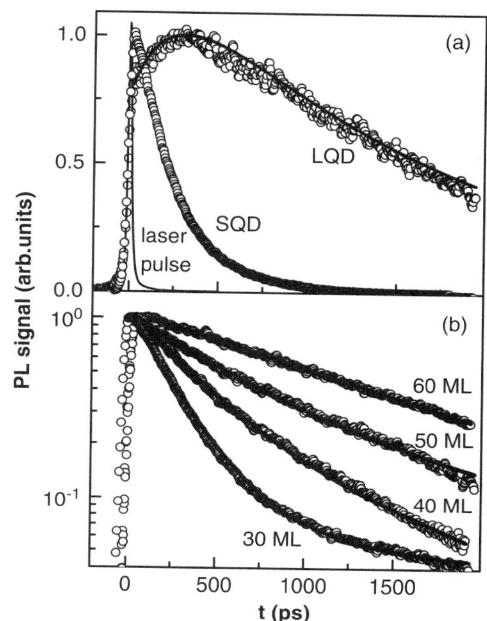

Fig. 4.16 (a) Normalized PL transients for SQDs and LQDs in an AQDP sample (see Fig. 4.9) with a spacer thickness of 30 ML. The detection energies are 1.25 and 1.094 eV for SQDs and LQDs, respectively. (b) Normalized PL transients for SQD in AQDP samples with various spacer thicknesses. The detection energy is at 1.25 eV. The calculated dependences shown with solid lines where found from the least-squares fit by Eqns. (4.9) and (4.10) with the correlation parameters $(1-\alpha)$ taken to be 0.95, 0.75, 0.55, and 0.10 for the $d_{sp} = 30$, 40, 50, and 60 ML, respectively [110, 111]

signals of the fundamental exciton transition in the SQDs for different d_{sp}. It is of interest that at the excitation levels required for the transient spectra we could not distinguish the excited state emission from the LQDs for the 30 ML spacer sample at 1.25 eV. Optical excitation for the transient PL measurements was provided by 80-fs pulses at $\lambda = 732$ nm from a mode-locked Ti:sapphire laser producing an optical pulse train at 82 MHz and an excitation density that was varied between 10^9 and 2×10^{13} photons/(pulse$\times$cm^2). For all samples investigated here, the PL transients were measured at the position of the PL maximum for both the SQDs and the LQDs. In addition, we chose a low I_{ex} at 5×10^9 photons/(pulse$\times$cm^2) as a compromise between desired the low-excitation levels needed to avoid excited-state PL contributions and provide sufficient high signal-to-noise ratios. As seen in Fig. 4.16 (b), the transients for SQDs of different samples cannot uniformly be fitted by a single decay time that decreases with increasing d_{sp}. In addition, the data for LQDs show a very different behavior in comparison with the SQD PL transient. The delayed rises in PL transients of QDs can be observed from single dot layers also, as it was discussed in Section 4.5. These delays are a result of relaxation from the excited states to the ground state combined with Pauli blocking or intralayer carrier transfer between the ground states of coupled QDs in dense QD array. In our case of low excitation densities and low areal QD density taking into account that the LQD PL increases with approximately the same time constant as the decrease of the SQD PL we relate the delayed rise of LQD PL to inter-dot tunneling transferring carriers from seed and second layers. With increasing d_{sp}, and correspondingly increasing τ_T, this delayed maximum becomes much less pronounced and practically vanishes for $d_{sp} = 60$ ML. These experimental observations are in full agreement with the predictions of Eq. (4.1), which can be modified for the case of a δ-like excitation and extremely low population of the QD ground states ($n_1 << N_1, n_2 << N_2$). In order to apply Eq. (4.1) we assume that intra-layer carrier capture into localized QD states is significantly faster (~5 ps) than the excitation rate. The intra-layer and intra-dot carrier relaxation (transferring the captured carriers in the ground QD states) is also assumed faster than the excitation transfer processes and radiative recombination [94]. In this case the radiative recombination and the excitation transfer between localized QD states determine the observed PL behavior. Tunneling originates in the ground state of SQDs with the tunneling time τ_T. For a very thin spacer layer (<30 ML) the estimated inter-dot transfer time is comparable with the time of intra-dot relaxation. In this case, the calculated τ_T value gives only an upper limit. However, under these approximations the rate equations (4.1) transform into:

$$\begin{aligned} \frac{dn_1(t)}{dt} &= -\frac{n_1(t)}{\tau_{R1}} - \frac{n_1(t)}{\tau_T}, \\ \frac{dn_2(t)}{dt} &= -\frac{n_2(t)}{\tau_{R2}} + \frac{n_1(t)}{\tau_T} \end{aligned} \tag{4.9}$$

These equations can be immediately integrated under the initial conditions of n_1 $(t = 0) = n_1(0)$ and $n_2(t = 0) = n_2(0)$, as provided by a δ- like excitation at $t = 0$, to yield the expressions of interest:

$$\begin{aligned} n_1(t) = n_1(0)e^{-t(1/\tau_{R1}+1/\tau_T)} \text{ and } n_2(t) &= [n_2(0) + Dn_1(0)]e^{-t/\tau_{R2}} \\ -Dn_1(0)e^{-t(1/\tau_{R1}+1/\tau_T)} \text{ where } D &= \frac{\tau_{R1}\tau_{R2}}{\tau_{R1}\tau_{R2} + \tau_T(\tau_{R2} - \tau_{R1})}. \end{aligned} \tag{4.10}$$

The PL decay time for the SQDs τ_{PL} is therefore given by $\tau_{PL} = \tau_{R1}\tau_T/(\tau_{R1} + \tau_T)$. Meanwhile, the recombination lifetime τ_{R1} of the SQDs can be found experimentally since it can be taken equal to that measured τ_{R1}^{ref} from the reference single QD layer sample of appropriate QD density.

Up to now describing the AQDP kinetics and cw experiments (Section 4.5) we considered a fully correlated system where all of the SQDs of seed layer form the AQDPs with the LQDs of second layer. In practice there exist a certain fraction,α, of the SQDs that are not coupled to any LQD in the second layer. Likewise, there will be a certain fraction β of LQDs that are not coupled to any SQD in the seed layer. Intuitively it is clear that $\alpha \geq \beta$ in bi-layer structures and the magnitude of α and β must depend on the d_{sp} thickness. Indeed, XTEM statistical analysis gives the correlation factor (1-α) of ~0.95, ~0.70, ~0.50, and ~0.10 for $d_{sp} = 30$ ML, 40 ML, 50 ML, and 60 ML, respectively. The carriers in non-correlated QDs in each layer are assumed to relax independently of the AQDPs and each other (no lateral coupling) due to the low QD densities in the samples under investigation. Therefore in what follows we add the PL contribution of the non-correlated QDs additively to the PL yield from AQDPs. Then one finds for the SQDs:

$$n_1^{total}(t) = n_1^{total}(0)\left\{\alpha e^{-t/\tau_{R1}} + (1-\alpha)\, e^{-t(1/\tau_{R1}+1/\tau_T)}\right\}, \qquad (4.11)$$

where n_1^{total} and n_2^{total} are the densities of optically excited QDs from the total dot densities N_1^{total} and N_2^{total} measured by STM in the seed and second layers, respectively.

Measuring the transient PL for the seed layer at the maximum of the emission band of SQDs and taking into account the independently determined $\tau_{R1} = \tau_{R1}^{ref}$ and (1-α) values we can calculate the τ_T value for all of the samples under investigation and uncover the relation $\tau_T = f(d_{sp})$. It is worth noting that the procedure developed here of τ_T extraction from the transient PL data is correct in essence and gives τ_T values that are substantially different from those cited in analogous bi-layer experiments without accounting for the contribution from non-correlated QDs.

Using the determined values of τ_T, τ_{R1} and taking τ_{R2} as a fitting parameter for the experimentally measured transients for LQDs [see Fig. 4.16(a)], the relaxation time τ_{R2} is determined. The result is most accurately done for the $d_{\text{sp}} = 30\,\text{ML}$ sample where the contribution of the non-correlated LQDs is negligibly small.

Possessing τ_T, τ_{R1} and τ_{R2} for all samples we can calculate the dependence of $I_{\text{LQD}}/I_{\text{SQD}}$ on the excitation density and compare with the result from the cw measurements. Equation (4.2) shows good agreement with the experimental data [Fig.4.10(2b)], especially if one considers that the experimental accuracy in determining the SQD PL is limited at low excitation densities by poor PL signal-to-noise ratio and by the asymmetric high energy tail of the LQD emission.

We have also analyzed τ_T as a function of d_{sp} from the transient PL spectroscopy data. This dependence, shown in Fig. 4.17 [110, 111] on a semi logarithmic scale, can be fitted using a straight line and follows a simple expression for the tunneling time for barrier penetration assuming a square barrier (semi classic Wentzel-Kramers-Brillouin (WKB) approximation developed for tunneling processes in coupled QWs):

$$\tau_T \propto \exp[2d_{sp}\sqrt{(2m^*/\hbar^2)(V - E_{sn})}]. \qquad (4.12)$$

Here m^* is the effective mass in the spacer layer, V is the band discontinuity of the conduction band, and E_{sn} is the lowest confinement energy level in SQD energy. The assumption of a square barrier, especially for small d_{sp}, can be incorrect since indium segregation during the capping phase will produce a spacer with an InGaAs alloy composition rather than GaAs. This would decrease the height of the barrier and the effective mass, both of which will decrease τ_T.

Although segregation is expected during GaAs capping, its amplitude will not be significant for InAs deposition at 500 °C and a GaAs capping rate of 1.0 ML/ s. In fact, a sharp interface between InAs QDs and GaAs barriers under these growth conditions is often assumed in order to achieve reasonable agreement between calculations and optical characterizations.

The dependence shown in Fig. 4.17 supports a model for the observed dependence of τ_T on d_{sp} that is based on carrier tunneling. Naturally one would consider

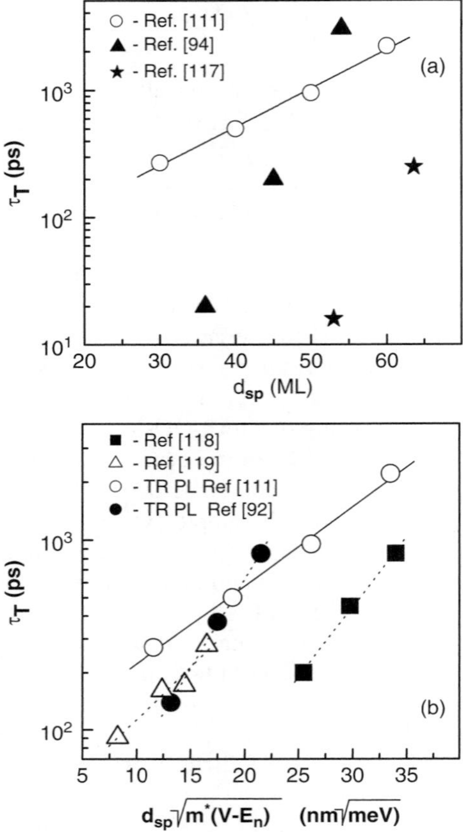

Fig. 4.17 (a) Tunneling time (open circles) deduced from the SQD PL transient as a function of barrier width. The solid line is fitted to tunneling times by the least-square method. The data of tunneling times for the similar AQDP systems (closed triangles: Ref. [94]; stars: Ref. [117]) estimated from cw PL measurements are shown. (b) Comparison of nonresonant electron tunneling times as a function of the effective barrier width observed in $Al_xGa_{1-x}As/GaAs$ ADQWs (closed squares: Ref. [118]; open triangles: Ref. [119]), the InAlAs/GaAs/InAs AQDP closed circles: Ref. [95] together with data from Ref. [111] (open circles). Dotted lines are guides for the eye

this as non-resonant tunneling since neither the electron nor the hole energy levels in the SQDs and the LQDs are aligned. According to the WKB approximation one can refer to the greater propensity for electrons to tunnel compared with holes due to the large difference in their effective masses. Equation (4.12) predicts that for a d_{sp} of about 8 nm the tunneling time for the holes is more than three orders of magnitude larger than that of the electrons. In the PL experiments, since carriers with a shorter lifetime govern decay time, the experimentally observed τ_T should be attributed to electron tunneling. In the same Fig. 4.17(a), the values of τ_T versus d_{sp} are shown for similar InAs/GaAs AQDPs estimated from the ratio of the integrated cw PL intensities of SQDs and LQDs, assuming equal ratio of the excitation rates for both QD layers, as reported in Refs. [94, 117]. The difference in the obtained values of τ_T between our data and the data in Ref. [94] might be caused by several reasons: First, the difference in radiative lifetime τ_{R1}^{ref} for the single layer samples, which were used as τ_{R1} of the seed layer (1400 ps in our sample in comparison with 600 ps in Ref. [94]). Second, the reduced accuracy of exact determination of integrated PL intensity from SQDs for small values of d_{sp}. For the cw PL investigation of AQDP grown at the specific temperature conditions in Ref. [117], the estimated τ_T according to Ref. [94] led to an extremely fast value of $\tau_T \sim 16$ ps even for a considerably thick $d_{sp} \sim 15$ nm. This led the authors of Ref. [117]–unlikely for the size distribution of QDs in both layers–to suggest a model of carrier resonant tunneling from the ground state of the SQD to the second excited state of the LQD, followed by with a following rapid relaxation to lower states. Third, the contribution of the non-correlated QDs had not been taken into account in Ref. [92, 94, 117], which would lead to an overestimation of the true tunneling time τ_T. Fourth, comparing our results with Ref. [94] and [117] we should take into account the difference in growth conditions. While the growth conditions for every group are always different, it is worth to compare them for the possibility to reveal some common features. While the size of QDs in Ref. [94] may be similar to what we have, the QDs grown at lower growth rates in Ref. [117] are significantly larger and one might expect a rather different variation with spacer layer thickness.

A comparison with InAlAs/GaAs/InAs AQDPs [92] as well as $Al_xGa_{1-x}As/GaAs$ ADQWs [118, 119], in which electron tunneling was observed, is displayed in Fig. 4.17(b). Measured values of τ_T are plotted as a function of quantity $D = d_{sp}\sqrt{(2m^*/\hbar^2)(V - E_{sn})}$ in order to account for the different structure parameters in InAs, GaAs and AlInAs. For our structure we used the value of the energy difference $V - E_{sn}$ between the electron ground state in SQD and the GaAs conduction-band edge to be about 110 meV following the data in Ref. [120]. An important comment should be made on the mechanisms underlying inter-dot carrier transfer. In contrast to the ADQW structures, for the vertically aligned AQDPs three-dimensional shape of the InAs dots makes the definition of d_{sp} somewhat difficult. The uniformity of the dots leads to statistical variations (especially in the case of weakly correlated QD pairs) which requires us to introduce an effective tunnel barrier thickness d_{sp}^{eff} defined as the average distance between the top of QDs in the seed layer and the bottom of the second layer. So in presenting our data in Fig. 4.17(b), we used d_{sp}^{eff} by taking into account the average dot height of ~4 nm determined from STM images of uncapped single layer 1.8 ML samples [65].

As can be seen in Fig. 4.17(b), the transient data for the InAlAs/GaAs/InAs AQDP [92] (the only electron tunneling time τ_T directly measured in AQDP to our knowledge) and the ADQW [118, 119] (one of the several transient measurements with similar results) indicate a difference in tunneling time by one-order-of-magnitude. (Our data revise the results of Ref. [92] over an even broader range of d_{sp}^{eff}.) The origin for this difference is not well understood. One explanation [92] is that a longitudinal optical phonon assists the electron tunneling and, therefore, results in a phonon bottleneck for the AQDP. Here we suggest an alternative explanation. For ADQWs the tunneling rate must be calculated over a significant number of final degenerated and nearly degenerated states while for AQDPs the final state is well defined. This alone could explain the longer tunneling time for the AQDP and is consistent with the fact that in our experiments the tunneling time was insensitive to the energy mismatch between the SQDs and the LQDs.

Let us turn to a discussion of the underlying tunnel mechanisms. The tunneling is logically considered non-resonant tunneling, since electrons (holes) are clearly localized in both the SQDs and LQDs and their corresponding energy levels are not aligned significantly. In principle it could be that both the separate transfer of electrons (holes) and the actual transfer of excitons play a role. Since the energy separations between excited states and ground states of QDs are large ($\sim$70 meV), non-resonant tunneling by emission of optical phonons rather than acoustical phonons takes place. Our data gives evidence of electron transfer, while the tunneling time of heavy holes is estimated to be at least one order of magnitude larger than that of electrons due to the larger effective mass. This electron transfer can be interpreted in terms of a transfer from a direct $X_{e_1hh_1}$ into an indirect "cross" exciton state $X_{e_2hh_1}$ (in real space), where an electron is located in the e_2 state of LQDs. Meanwhile, the heavy hole occupies the hh_1 state of SQDs. However, the low recombination effectiveness, as well as the substantially longer transfer time for the hole, efficiently suppresses PL from the indirect exciton state. One would expect rapid relaxation of the $X_{e_1hh_1}$ exciton into the direct $X_{e_2hh_2}$ exciton due to the Coulomb potential that is attracting the hole into LQD state. This is the case for a comparatively thin barrier. In our case the charged carriers attract carriers of the opposite sign more efficiently in their corresponding layers, thus creating a new pair of direct excitons placed in adjacent layers. These excitons are taken into account by the third term in Eq. (4.1). Appearance of an exciton in the LQD of the second layer makes the probability of tunneling of an additional electron from SQD significantly smaller, thus for a moment decoupling the AQDP. This process is taken into account by the second term in Eq. (4.1). Together, this simple microscopic model is in a good agreement with the experimental data both for the cw PL data and for the time-resolved PL and does a good job describing the real physical picture in the bi-layer system of coupled QDs.

4.6.3 Contribution of Excited States to the Carrier Transfer in Vertically Stacked In(Ga)As/GaAs Quantum Dot Systems

Let us consider the contribution of excited states to the carrier transfer in vertically stacked In(Ga)As/GaAs quantum dot systems. In what LQD state does a carrier occur after tunneling through barrier between seed layer and second layer? Is it excited or ground QD state? What is a contribution of QD excited states in carrier kinetics? In order to clarify these issues we modified our model including the possibility to tunnel into excited state of LQDs. Thus the rate equations take the form

$$
\begin{aligned}
\frac{dn_1(t)}{dt} &= -\frac{n_1(t)}{\tau_{R1}} - \frac{n_1(t)\,(N_2 - n_2(t))}{N_2 \tau_T} - \frac{n_1(t)\,(N_2^{exc} - n_2^{exc}(t))}{N_2^{exc} \tau_T^{exc}} + \\
&\quad G_1\,(N_1 - n_1(t))\,/N_1, \\
\frac{dn_2^{exc}(t)}{dt} &= -\frac{n_2^{exc}(t)}{\tau_{R2}^{exc}} - \frac{n_2^{exc}(t)\,(N_2 - n_2(t))}{N_2 \tau_2^{int}} + \frac{n_1(t)\,(N_2^{exc} - n_2^{exc}(t))}{N_2^{exc} \tau_T^{exc}} + \\
&\quad G_2^{exc}\,(N_2^{exc} - n_2^{exc}(t))\,/N_2^{exc}, \\
\frac{dn_2(t)}{dt} &= -\frac{n_2(t)}{\tau_{R2}} + \frac{n_1(t)\,(N_2 - n_2(t))}{N_2 \tau_T} + \frac{n_2^{exc}(t)\,(N_2 - n_2(t))}{N_2 \tau_2^{int}} + \\
&\quad G_2\,(N_2 - n_2(t))\,/N_2.
\end{aligned} \tag{4.13}
$$

Additionally to Eq. (4.1), $n_2^{exc}(t)$ represents here the carrier density at time t in the excited energy levels of the LQDs and N_2^{exc} is the total density of first excited states (generally degenerate) in LQDs that are coupled to SQDs in the AQDPs. Third term of first equation in the set (4.13) describes the carrier transfer from the ground states of SQDs to the excited states of LQDs due to tunneling τ_T^{exc}. The factor $(N_2^{exc} - n_2^{exc}(t))/N_2^{exc}$ defines the fraction of unoccupied excited states of LQDs onto which tunneling from ground states of SQDs within AQDP is actual. Middle equation of the set (4.13) describes the population change in the excited states of the LQDs. Here radiative lifetime of excited state is defined as τ_{R2}^{exc} and τ_2^{int} is the time of inter-level relaxation in the LQDs due to which the carriers from excited states transfer to the ground states of the LQDs. G_2^{exc} is the carrier generation rate in the excited states of the LQDs. The set of equations (4.13) takes into account the state filling in the excited and ground states of the LQDs as well as the ground state of the SQDs. Let us consider the case of low excitation densities, $n_2 << N_2$, $n_2^{exc} << N_2^{exc}$, and $n_1 << N_1$, and rapid inter-level relaxation in LQDs, $\tau_2^{\text{int}} << \tau_{R2}^{exc}$ [121]. Then set of equations (4.13) can be reduced to the set of equations (4.1) and the relaxation of the SQDs and the LQDs in ground states is described by three-level model (4.1) with parameters $\tau_T^{ef} = \frac{\tau_T \tau_T^{exc}}{\tau_T + \tau_T^{exc}}$ and $G_2^{eff} = G_2 + G_2^{exc}$. These effective parameters have been defined in our experiments. Under restrictions implied by given inequalities three-level model (4.1) adequately describes AQDPs independently on excited states and state degeneracy in QD. Thus our analysis correctly reproduces the interlayer carrier transfer in bi-layer asymmetric InAs/GaAs quantum dots.

In summary, the dependence of the tunneling time on the thickness of the separation layer is determined. Controlling the spacer thickness within the limits imposed for noticeable vertical correlation it has been possible to tune the carrier dynamics in the SQDs within wide limits from ~2500 ps for 60 ML down to ~250 ps for the 30 ML GaAs spacer sample. Analysis of the experimental data is performed in terms of rate equations taking into account the peculiarities of the tunneling transfer in the system of AQDPs. It has been shown that it is important to account for the contribution of non-correlated QDs both in the second layer and in the seed layer of bi-layer InAs/GaAs QD structure that significantly corrects the estimated τ_T value. The dependence of the tunneling times on the barrier thicknesses is in good agreement with the behavior of the tunneling time for barrier penetration calculated in the WKB approximation developed for tunneling processes in coupled QWs. The data support a proposed microscopic model of tunneling including the non-resonant electron transfer from a direct into a cross exciton state, with subsequent generation of two direct excitons in adjacent QD layers. This process temporarily blocks the tunneling in the AQDPs leading to a saturation of the tunneling channel and enhancement of the SQD PL signal.

4.7 Role of Continuous States in the Interdot Carrier Transfer

4.7.1 Origins of Continuous States in Quantum Dots Systems

Recent studies on the optical properties of single In(Ga)As QDs coupled to their WL have revealed continuum states spread in energy below the lowest quantum state of the quantum dot barrier [122–125]. As a result, coupling between these continuum states and the discrete QD states can strongly affect intra-dot relaxation and provides de-coherence channels that can be harmful to the application of QDs to quantum information processing. One might expect, therefore, that there exist a basic understanding of the nature and role of the continuum states to further development of QD technology. However, while the origin of the continuum states as well as their effect on QD optical properties have been partly explored in various systems [122–128], many important aspects are remain unresolved.

Studies of the continuum states in QD systems have been carried out using several different techniques, i.e., near-field PLE [122], polarized PL [126], and infrared absorption measurements [127]. These investigations have established the important point that the continuum states couple the barrier states to the discrete QDs states providing efficient carrier relaxation from the lowest barrier state into the QD states. As a result, carriers excited above the WL absorption edge can very quickly relax through the continuum states losing their energy by producing longitudinal acoustic phonons. In this way the carriers easily trickle down rapidly reaching the energy region of a QD excited state before emitting an integral number of longitudinal optical (LO) or localized phonons and transitioning to the QD ground state. While the self-assembled QDs are coupled to the WL the density of continuum states is suggested to be a very gradual crossover from QW to QD [122] and it is an intrinsic feature of the single QD and cannot be ascribed to the QD ensemble.

Further micro-PL measurements under cw excitation have also demonstrated a strong PL up-conversion from single InAs/GaAs self-assembled QDs and also from the InAs WL [123]. In fact, identical features from the WL and single QDs have been observed using up-converted PL signals, evidencing that the WL continuum states could originate from the deep states localized at the rough interfaces of the

WL quantum well. In this scenario, depending on the roughness of the WL the band tail of localized states can penetrate sufficiently deeply in the region of discrete lines of the QDs spectra.

In addition to this possibility, several alternative explanations for the origin of the continuum states have been proposed. For example, low-energy phonon satellites of the WL states have been proposed to result in a continuous background intensity which linearly increases with the WL population [129]. Another explanation depends on crossed electron-hole states [123] that generate a continuous absorption background from single QDs without the need for participation of sample imperfections. This theoretical model [123] allows for the unambiguous identification of the origin of the continuum transitions in the absorption spectra of single self-assembled $In_{0.5}Ga_{0.5}As$ QDs [125]. For example, it has been demonstrated that the QD spectra of this system contains a region of low-energy transitions up to 50 meV above the exciton ground state with sharp atom-like spectra followed by the onset of the continuum. The PL features of the continuum region have significantly increased FWHM values, strong temperature dependence, and demonstrate field dependent broadening under the applied electric field. This behavior has given direct evidence for coupling of the confined QDs states and the WL states.

The results discussed above apply mainly to single QDs and little consideration has been paid to the ensembles of QDs, in which inhomogeneity of size obscures the contribution of the continuum background to the behavior of single QDs. Nevertheless, recently it has become important to account for the continuum states in more complicated quantum systems. For example, for single InGaAs self-assembled quantum rings embedded in a field effect structure it has been found that optical transitions among electronic shells of unpaired symmetry are favored by the presence of a nearby Fermi sea. In this case, crossed transitions between localized and delocalized electronic states have to be considered in order to explain the observed PL and PLE data [130]. As another example, for InAs/InP QD superlattices emitting at 1.55 μm, it has been shown [131] that the increase of QD density on the WL leads to enhanced splitting and mini-band effects for the QD states and also induces fragmentation of the WL density of states. Finally, for an ensemble of self-assembled (In,Ga)As QDs on GaAs (100), it has been shown that in addition to the well-known temperature behavior of the PL peak position and width around 90 K caused by carrier redistribution among the QDs through the WL and phonon assisted tunneling [100], a similar redistribution takes place at much lower temperatures around 30 K through low-energy continuous states between the WL and QDs [133].

While all of the examples above point to a significant role for the continuum states an unambiguous explanation for their existence is still wanting. In what follows we identify unambiguously the crossed transitions from the WL valence band to the electron ground states of self-assembled $In_{0.4}Ga_{0.6}As$ QDs in the cw and in time-resolved PL of QD ensemble. These crossed transitions develop themselves as an additional PL band in the spectral region ~50 meV above the PL band of the exciton ground states. The peculiar temperature dependence of the QD PL band measured under different excitation intensities is found to be consistent with a model of continuum of states with a band tail [123, 133].

4.7.2 Manifestation of Continuous States in Photoluminescence and Time-Resolved Spectroscopy of In(Ga)As/GaAs Quantum Dot Arrays

In order to reveal continuous states a set of samples with $In_xGa_{1-x}As$ ($0.3 \leq x \leq 0.4$) QDs has been grown under appropriate growth conditions [134]. For instance, the $In_{0.4}Ga_{0.6}As$ QDs were grown on a GaAs (100) substrate in a Riber 32 MBE system by deposition of 10 ML $In_{0.4}Ga_{0.6}As$ at the substrate temperature of 540 °C and a growth rate of 0.38 ML/s for InGaAs. After a 10 s growth interruption, a 20 nm GaAs capping layer was grown on the (In,Ga)As layer before the sample was heated up to 580 °C for growth of an 80 nm GaAs layer. Finally, the (In,Ga)As QDs layer was again deposited at 540 °C on top of the GaAs surface for morphology characterization by AFM. Statistically the QD ensemble is characterized with an average QD diameter, height, and density to be of ~40 nm, ~8 nm, and $\sim 4 \times 10^{10}$ cm^{-2}, respectively. In the case of low-density excitation one can expect a single PL band broadened due to the size distribution in the QD ensemble.

Figure 4.18 shows three typical PL spectra measured for randomly buried $In_{0.4}Ga_{0.6}As$ QDs excited with different wavelengths at low-temperature (T = 10 K). The bottom PL spectrum excited with the 532 nm line ($E_{exc} = 2.33$ eV) from a doubled Nd:YAG laser under low excitation density of ~250 mW/cm^2 indeed follows a Gaussian distribution centered at $E_{max} = 1284$ meV with the FWHM value $\Gamma = 38$ meV. The line-shape of two other spectra excited at $E_{ex} = 1.65$ eV and $E_{ex} = 1.53$ eV cannot be fitted by a Gaussian due to reasons discussed below.

The PLE spectra, also shown in Fig. 4.18, are measured at two different energy positions: $E_{det} = 1284$ meV corresponding to the PL maximum in bottom spectrum; and $E_{det} = 1340$ meV corresponding to the energy position of low-energy PLE feature seen in the PLE spectrum measured at the $E_{det} = 1284$ meV. Both PLE spectra distinctly show the heavy and light hole absorptions of the WL giving the excitation resonances at 1424 and 1486 meV, respectively. These transition energies are typical for our growth conditions and are similar to those found for the WL in Ref. [113]. Both PLE spectra are characterized by intense background absorption detected below the WL absorption band edge but subsiding within the energy band gap. Let us discuss the PLE feature ($E_{det} = 1284$ meV) distinctly seen at an energy ~50 meV above the exciton ground state that marks the onset of the observed continuum states. In earlier work this energy region was explained as crossed transitions that involve one bound electronic state of the QDs and one delocalized 2D state of the WL valence band [123, 124]. These crossed transitions should follow a Gaussian distribution similar to the energy distribution of the electronic ground states of the QD ensemble. In fact, the PLE distribution in the crossed transition spectral region in our experiment is nicely described by a Gaussian function.

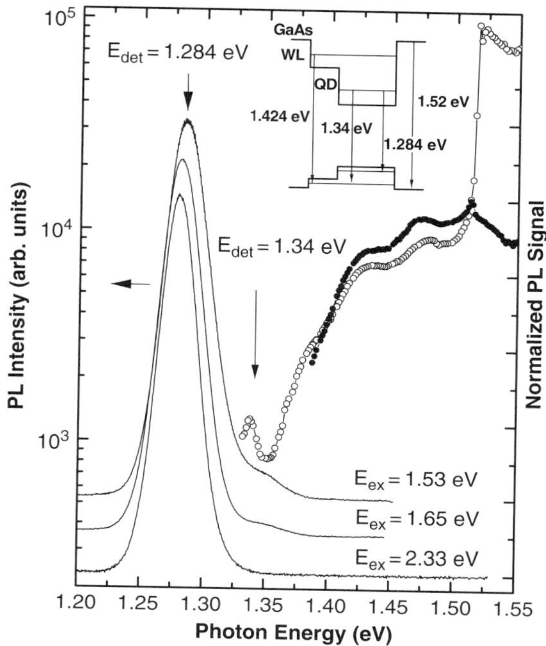

Fig. 4.18 Low temperature (T = 10 K) PL and PLE spectra of buried $In_{0.4}Ga_{0.6}As$/GaAs QDs. PL spectra are measured for the different excitation energies. Excitation density $I_{exc} = 250\,mW/cm^2$. Arrows down show the detection energies for PLE spectra. Empty circles stand for $E_{det} = 1.284\,eV$, filled circles–for $E_{det} = 1.34\,eV$. Inset: Scheme of optical transitions derived from PLE and PL spectra [134]

In cw PL the crossed transitions are expected to be of much lower intensity in comparison with the intensity of the excitonic ground state transitions and consequently to our knowledge have never been directly observed. In order to change their relative contributions and clearly distinguish the crossed transitions contribution in the PL spectrum, the PL excitation energy should be decreased to below the energy of the WL transition. Figures 4.18 and 4.19 demonstrate the effect of the excitation energy reduction on the PL spectrum. As the energy of the exciting quanta approaches the energy of the GaAs band-gap an additional feature begins to develop on the high-energy side of the PL band. Appearance of this band in the PL spectra under very low excitation density excludes its assignment to the excited QD states. Indeed, the PL signal from the excited QDs states is strongly dependent on the excitation intensity. In case of Fig. 4.18 the excitation density is evaluated to be of $\sim 250\,mW/cm^2$. The additional feature, distinctly seen at this excitation intensity, persists in the PL spectrum under intensity lowering by two orders of magnitude. Such a behavior is atypical of the PL from excited QD states while in this low-power excitation condition the state filling effect in QDs is negligible.

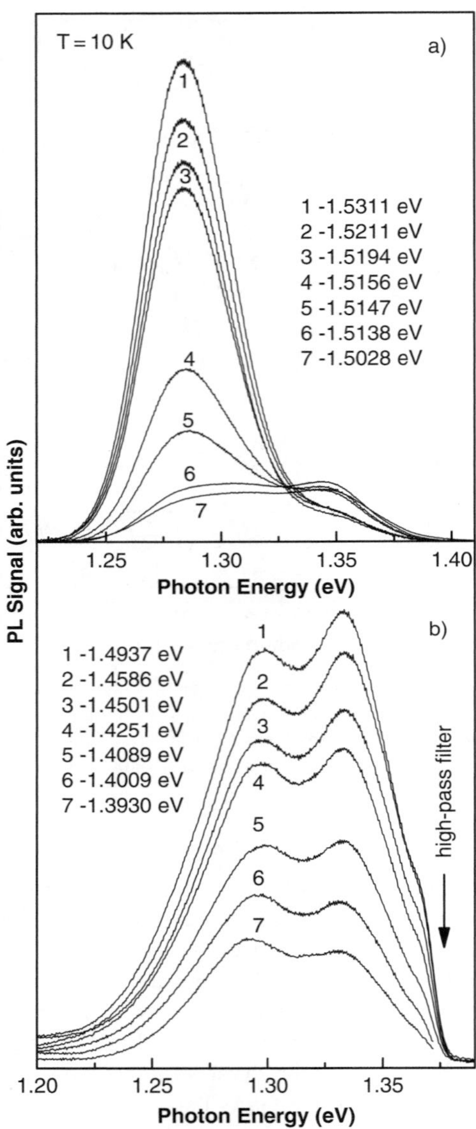

Fig. 4.19 Low temperature PL spectra of buried $In_{0.4}Ga_{0.6}As$/GaAs QDs for excitation energies crossing the energy of GaAs excitonic transitions a); for the below GaAs barrier excitation b). Distinct two-peak structure develops. $I_{exc} = 250\,mW/cm^2$. Low energy PL band corresponds to the QD excitonic transitions, high energy PL band is contribution of crossed transitions [134]

One can assume that all the phenomenology presented here might stem from extrinsic (defect dependent) effects. Indeed, as far as the band around 1.34 eV is concerned, a defect band has been revealed in the same energy range, whose intensity was strongly enhanced by sub-gap excitation in In(Ga)As/GaAs QD samples [135]. Such defect can effectively compete in carrier capture with the QDs, when the excitation is performed in the WL, thus giving rise to fluctuations in the intensity ratio between the QD and the defect emissions. Resulting from a detailed optical characterization [135] the 1.356 eV PL band was ascribed to the radiative transition between the conduction band and the doubly ionized Cu_{Ga} acceptor in GaAs. A peculiar PL behavior was observed [135] in all investigated QDs samples caused by the acceptor presence: i) a resonant quenching of the QDs PL when exciting on the excited level of this deep defect; ii) the PLE spectrum of the 1.356 eV emission turned out to be almost specular to the QD PLE. This correlation between the PL efficiency of the QDs and the Cu centers evidences a competition in the carrier capture arising from a resonant coupling between the excited level of the defect and the electronic states of the WL on which the QDs nucleate. At least three important features differ our observation of the additional PL band under sub-band excitation from that found in the QDs systems in Ref. [135]. First, the energy position of the 1.34 eV PL band varies with the indium content in our samples whereas the position of the L1 band is fixed in all QDs samples from Ref. [135]. Second, the PLE spectrum detected at the position of the QDs PL maximum shows pronounced peak at the excitation energy of 1.34 eV (see Fig. 4.18) in contrast to absence of such peak in the QD PLE spectra in Ref. [135]. This evidences about the energy transfer from the states suited in the spectral range of the 1.34 eV to the ground states of QDs in our samples. Third, in our samples we do not see a specular behavior of the PLE spectrum of the 1.34 eV emission in respect with the QD PLE observed in Ref. [135] (see Fig. 4.18). Thus, assigning a defect or interface nature to the additional PL band seen in our experiment has to be excluded. As a result, we assign this new band to the crossed transitions earlier us detected in the PLE spectrum. Based on these arguments the optical transitions in the cw PL spectra are depicted in the inset of Fig. 4.18.

In addition to the PLE and PL spectra, further experiments have reinforced this assignment [134]. Figure 4.19(a) shows the PL spectral modifications observed as the excitation energy passes the energy of the GaAs excitonic transitions. The intensity of the QD emission drops by an order of magnitude, whereas the strength of the crossed transitions does not change appreciable and even grows. Under excitation above and within the WL states (Fig. 4.19b), the QD PL and the PL caused by the transitions between localized QD electronic ground states and the 2D states of the WL are already comparable by intensity. If the excitation energy decreases from $E_{exc} = 1.5311$ eV down to $E_{exc} = 1.3930$ eV (see Fig. 4.19a and Fig. 4.19b) the QD PL maximum shifts non-monotonically. Initially it moves towards higher energies. Then (see Fig.4.19b) it shifts slightly (~5 meV) towards lower energies. Such behavior can be explained by peculiarities of energy relaxation for hot carriers under above-GaAs barrier excitation. Carriers possessing excess kinetic energy are preferably trapped by larger QDs realizing deeper potential wells. If the excitation energy approaches the GaAs band-edge the carriers possessing smaller kinetic energy can be trapped by smaller QDs realizing shallower potential wells. Thus the excitation energy decrease favors the more effective population of smaller QDs, emission of which forms the high-energy side of the QD PL band. Effectively, such redistribution of the QD excitation effectiveness will leads to the total shift of the QD PL band to the blue side with the excitation energy decrease seen in Fig.4.19a. Further decrease of the excitation energy below the GaAs and wetting layer barrier energies (Fig. 4.19b) makes actual the QD excitation through the process $E_{QD} = E_{exc} - n\hbar\omega_{LO}$. Due to this process the larger QDs are excited more effectively if the E_{exc} value decreases resulting in total red shift of the QD PL band seen in Fig. 4.19b. The energy position of additional PL feature does not reveal a significant variation under the E_{exc} decrease what is completely consistent with the behavior expected from a model of crossed transitions. These transitions, which give rise to the continuum of states in the absorption spectrum of single QDs, transform into a noticeable peak in the PLE spectrum of the QD ensemble and become directly visible in the PL spectrum with below GaAs barrier excitation. The characteristic energy interval of ~50 meV above the energy of the excitonic transitions found in both the PLE and PL spectra of our self-assembled $In_{0.4}Ga_{0.6}As/GaAs$ QD system and in the absorption spectra of single self-assembled $In_{0.5}Ga_{0.5}As$ QDs (Ref. [125]) is defined mainly by the structure of the WL valence band and QD hole states.

Entirely different continuum of states related to structural effects such as overlapping strain fields can be actual in our high-density QD ensemble as well. Indeed, the QD PL band (see. Fig. 4.18) excited with the 532 nm line from a doubled Nd:YAG laser reveals a peculiar behavior under temperature variation. This behavior can be partly interpreted in terms of continuum states. Figure 4.20 presents the results of measurements of the PL peak position (Fig. 4.20a), integrated intensity (Fig. 4.20b), and FWHM (Fig. 4.20c) at various temperatures ranging from 10 K to 170 K and three different excitation densities [134]. These results are consistent with the results of Ref. [133] for the self-assembled $In_{0.36}Ga_{0.64}As/GaAs$ QDs. It is seen that the PL maximum significantly (~50 meV) shifts towards lower energies, the integrated PL intensity non-monotonically decreases and the FWHM value changes with two pronounced minimums with increasing temperature. The shift of the PL peak energy (Fig. 4.20a) does not follow the well-known Varshni's law for the energy gap temperature dependent change in bulk semiconductors [102] and partly can be explained in terms of thermally induced carrier transfer from smaller QDs with higher optical transitions energies into larger QDs possessing lower energies of inter-band transitions [100]. Thus resultant shift of the PL maximum is contributed by semiconductor band gap shrinking and by the predominant excitation of larger QDs in the QD size distribution. When temperature increases up to 90 K. The carrier transfer is responsible also for the substantial reduction of the PL FWHM (~7 meV) observed as the temperature is increased to T ~90 K (Fig. 4.20c). At low temperature the PL lineshape is determined by the inhomogeneous QD distribution, i.e. the PL FWHM arises from the sum of emission of QDs of many different sizes from Gaussian distribution. When temperature increases the thermalized carriers predominantly repopulate the QDs of larger sizes representing deeper potential wells. Thus the

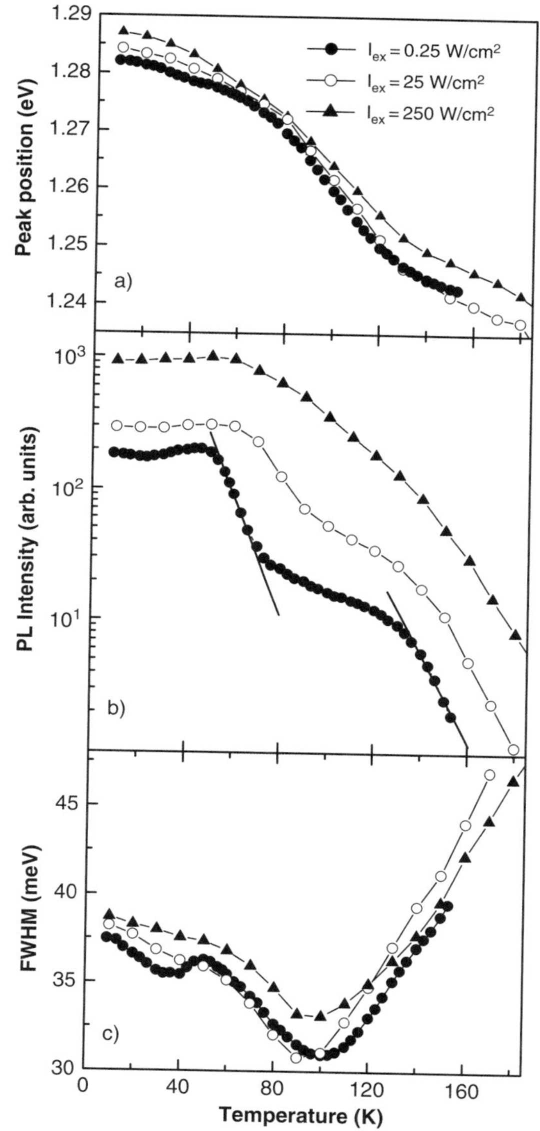

Fig. 4.20 Temperature dependences of the QD PL band for buried $In_{0.4}Ga_{0.6}As/GaAs$ QDs: a) PL maximum energy; b) Integrated PL intensity; c) Full width at half-maximum, FWHM, measured under three different excitation densities. Two–minimum FWHM (T) dependences are observed. Solid lines in b) show the extrapolated parts used for the determination of the activation energies [134]

PL linewidth is expected to be reduced. When the temperature is high enough, the effect of electron-phonon scattering becomes a dominant contribution and the PL line-width starts to increase with increasing temperature resulting in the FWHM minimum seen at 90 K in Fig. 4.20c.

It is significant to note that the temperature dependence of the FWHM (Fig. 4.20c) demonstrates an additional minimum at T~30 K for the lowest excitation density, which washes out at high excitation density. The origin of this minimum in the FWHM temperature dependence was attributed in Ref. [133] to inter-dot carrier transfer through the continuous states located above the QD ground-state energy. It is not clear whether or not the density of these continuous states approaches zero above the QD ground-state energy. However the temperature behavior of the FWHM gives evidence of a thermal-like activation over a small energy gap. In this case the carriers could efficiently transfer at a low temperature corresponding to a gap. In Ref. [133], a rather small FWHM reduction of only 1 meV observed in the temperature region of ~30 K, gives evidence that the carriers excited to the continuum are only locally redistributed among the QDs. Thus these continuum states span over only a limited area around the QDs. It has also been assumed that tunneling [73, 76] cannot be considered as a mechanism of inter-dot carrier transfer and yet give the PL spectra temperature behavior observed in Fig. 4.20c. As an argument it was offered [133] that in case of tunneling all carriers would transfer to the largest QDs at low temperatures. Thus an increase of FWHM is expected rather than a reduction of FWHM when temperature grows from 10 K to 30 K. Nevertheless we argue for the tunneling as a mechanism of interdot carrier transfer basing on speculation as follow. In fact the times of interdot tunneling measured by means of time-resolved PL in our systems occur to be the order of ~1 ns [100]. They are comparable with the radiative times ~1 ns in QDs [105] and thus the carrier distribution even at the lowest temperatures is determined by the excitation density as well. It means that even at the lowest temperatures carriers are distributed over whole ensemble of QDs due to balance between tunnel carrier transfer and carrier excitation. The interdot tunneling is non-resonant and increasing temperature favors increasing carrier transfer due to both tuning of the electronic levels of adjacent QDs and the thermal excitation of QD electrons into the local continuum states. In this last case the QD electronic wave function has a greater space spread and its overlap with the electron wave function of adjacent QDs becomes stronger increasing the probability of interdot tunneling. Thus the presence of local continuum states nearby QD facilitates the interdot carrier transfer through tunneling.

The activation energies derived from the temperature dependence of the integrated PL (Fig. 4.20b) measured at the lowest excitation density and around the temperatures ~60 K and ~140 K are found to be $E_{WL}^{hh} = 40$ meV and $E_{GaAs}^{el} = 160$ meV, respectively. These energies roughly correspond to the hole excitation from the QDs into the heavy-hole-valence band of the WL and to the electron excitation from the QD into the conduction band of the GaAs barrier, as determined from the PLE and PL measurements. Thus analysis of the temperature dependence of the QD PL provides independent evidence of coupling between the QD states and the WL states further supporting the claim of the observation of crossed transitions.

The behavior of the FWHM temperature dependence under the elevation of excitation density, seen in Fig 4.20c, is also consistent with the model of thermally induced carrier transfer from smaller QDs into larger QDs. Indeed, the increase of PL excitation density leads to increased population of larger QDs, and the rate of carrier transfer from smaller QDs reduces due to Pauli blocking. This results in change of the temperature dependence of FWHM for higher excitation density and shallow minimum at 30 K related to carrier transfer disappears.

Finally we observed the crossed transitions inherent to the joint nature of the WL valence-to-QD electronic density of states in PL of self-assembled $In_{0.4}Ga_{0.6}As$/GaAs QDs. The strength of these transitions becomes comparable with the excitonic ones with below GaAs-barrier excitation and decays significantly with below WL excitation. An additional minimum in the temperature dependence of PL FWHM and its behavior with the excitation density increase is explained in terms of inter-dot carrier transfer through the continuum states related to the WL morphology and phonon-assisted processes. Therefore we distinguish two different contributions to the continuum states caused by QD–WL coupling: crossed transitions between 0D electron states in QDs and the 2D states of the WL and the continuum of electronic states due to influence of QD environment. These latter states can easy originate from inhomogeneous strain in QD, variation of composition and geometry of QD, overlap of strain fields, etc.

Continuum states discussed above can serve as the intermediated states for the carrier relaxation in the QD system. We present results of time-resolved PL measurements for the spectral range of the continuum states in InGaAs/GaAs QD structures uncovering the transfer of carriers through these states. For these measurements the $In_{0.4}Ga_{0.6}As$/GaAs QDs are grown on (100) and (311)B GaAs under similar conditions depositing of 10 ML $In_{0.4}Ga_{0.6}As$ at the substrate temperature of 540 °C and growth rate of 0.38 ML/s for InGaAs.The AFM analysis indicates a comparatively narrow size distribution in both $In_{0.4}Ga_{0.6}As$/GaAs QD ensembles grown on (100) and (311)B GaAs substrates. The average QD diameter, height, and density are found to be of ~40 nm, ~8 nm, and $\sim 4 \times 10^{10}$ cm^{-2}, respectively. These narrow distributions are reflected in the low temperature (10 K) PL spectra shown in Fig. 4.21 for (311)B (Fig. 4.21a) and (100) (Fig. 4.21b)) GaAs samples [136]. A single-peaked PL spectrum with a maximum at $E_{\max} = 1.284$ eV and full width at half maximum (FWHM) of $\Gamma = 38$ meV is detected in the (100) sample and a $E_{\max} = 1.334$ eV and $\Gamma = 25$ meV in the (311)B sample. A blue shift of ~50 meV and a FWHM reduction of ~10 meV of the QD PL band in the (311)Bsample relative to the (100) sample is attributed to a differing amount of strain in the structure and/or stronger confinement in the growth direction. PLE spectra, shown in Fig. 4.21(a, b) distinctly reveal the heavy-hole and light-hole absorptions of the WL giving the excitation resonances at ~1.42 eV and ~1.48 eV, respectively. For both samples the PLE spectra develop a significantly strong background due to the rough interfaces of the WL and the interaction of the QD states and transitions with their surrounding environment. In order to explore the kinetic properties of these

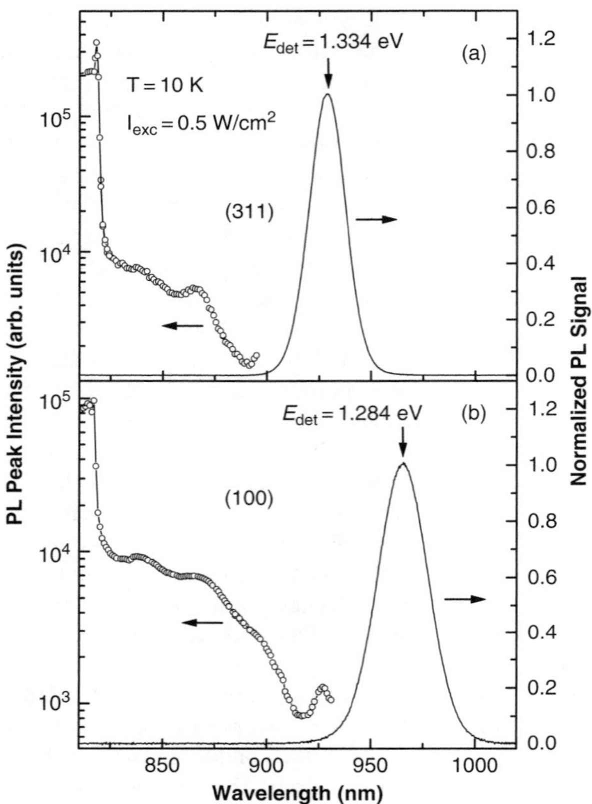

Fig. 4.21 a) Low temperature PL and PLE spectra: a) for the InGaAs/GaAs(311)*B* QD sample; b) for the InGaAs/GaAs(100) QD sample [136]

below WL states, transient PL measurements were carried out within the energy region ranging from the energy of the WL absorption band-edge to the energy of the ground states excitonic transition in the QDs.

Figure 4.22 presents the PL transients measured for the InGaAs/GaAs (100) QD sample at T = 10 K using pulsed laser excitation above the GaAs bandgap ($\lambda_{\mathrm{exc}} = 752\,\mathrm{nm}$) [136]. In particular, the detection energy (E_{det}) scans at 4 nm steps are given for the vicinity of the WL states (Fig. 4.22a), just below WL transitions (Fig. 4.22b), and with 8 nm steps in the vicinity of the inherent QD transitions. The data show that the shape of PL transients significantly changes as the E_{det} deepens into the WL energy gap. The characteristic rise time for PL transients I_{PL} (t) does not exceed ∼20 ps. Nevertheless all I_{PL} (t) dependences measured below the WL absorption edge (Fig. 4.22a) demonstrate a delayed rise and convex shape of the PL decay curve that is obviously non-exponential. A part of the I_{PL} (t) dependence can be seen to follow a mono-exponential decay within the wavelength interval from ∼848 nm to ∼926 nm as shown in Fig. 4.22b. The time interval of mono-

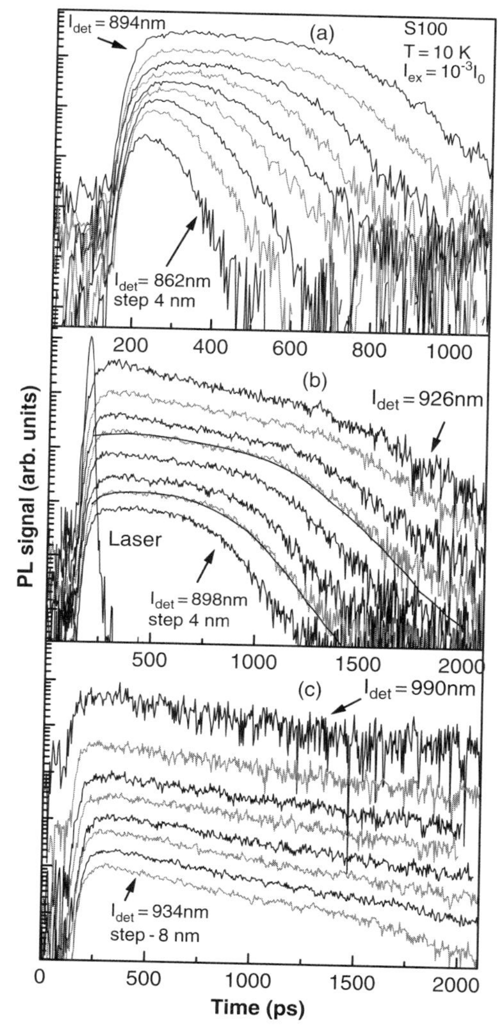

Fig. 4.22 PL transients measured within different wave length intervals for PL detection in the InGaAs/GaAs(100) QD sample : a) from 862 nm to 894 nm; b) from 898 nm to 926 nm; c) from 934 nm to 990 nm. Distinct convex shapes of transients are seen within the region of the sub-WL transitions. PL transients calculated using the set of Eqns. (4.14–4.16) are shown by solid lines in Fig. 4.22b [136]

exponential decay becomes longer reaching ~1500 ps as the E_{det} value approaches the spectral region of the QD ground state transitions. Scanning the QD spectral region, as shown in Fig. 4.22(c), we see that we can describe the PL decay as a mono-exponential process at least within the time interval as large as 2000 ps. The decay time τ_d for this mono-exponential relaxation is found to be dependent on the

PL excitation density varying from $\tau_d = 2.9\,\text{ns}$ to $\tau_d = 1.5\,\text{ns}$ as the $I_{\text{PL}}(0)$ is varied over three orders of magnitudes. The similar results have been obtained for the (311)B GaAs sample also.

The observed behavior of PL decay for the below WL states and QD states is typical for a system with cascade carrier relaxation through the set of saturable states. Indeed, optically injected carriers in QD systems or devices initially populate the continuum states of the WL and/or the surrounding barriers. After an energy dissipation processes, these carriers diffuse further into the below WL continuum and/or become captured by the QDs and relax down to the ground state before radiative recombination occurs. The efficiency of this relaxation cascade directly affects the efficiency of the carrier capture by the QD states as well as device performance. Our transient PL measurements prove that besides capture into the QDs, the carriers live in the below WL states for a time comparable with the time of radiative recombination from the QD states. While the density of the below WL states significantly reduce into the WL energy gap the rate of energy relaxation can depend on the continuum population. In this case the carrier dynamics in the region of below WL states could be governed by rate equations like

$$\frac{dn(\varepsilon,t)}{dt} = -\frac{n(\varepsilon,t)}{\tau_r(\varepsilon)} - \int_{\varepsilon_{QDs}}^{\varepsilon} n(\varepsilon,t)\frac{(n_0(\varepsilon') - n(\varepsilon',t))}{\tau_{tr}(\varepsilon-\varepsilon')\,n_0(\varepsilon')}d\varepsilon' + \int_{\varepsilon}^{\varepsilon_{WL}} n(\varepsilon'',t)\frac{(n_0(\varepsilon) - n(\varepsilon,t))}{\tau_{tr}(\varepsilon''-\varepsilon)\,n_0(\varepsilon)}d\varepsilon'', \tag{4.14}$$

$$n(\varepsilon,t) \le n_0(\varepsilon), \tag{4.15}$$

$$n(\varepsilon,0) = f(\varepsilon, I_0). \tag{4.16}$$

Here $n(\varepsilon,t)$ defines the electron density in the state with energy ε at time t; $\tau_r(\varepsilon)$ is the time of radiative recombination. The second term in the right part of Eq. (4.14) describes the carrier transfer from the state ε down the all unoccupied low-lying energy states of continuum; $n_0(\varepsilon')$ defines the density of these states at the energy ε' and $\tau_{tr}^{-1}(\varepsilon-\varepsilon')$ determines the rate of carrier transfer from the state ε to the state ε'. The third term in the right part of Eq. (4.14) describes the carrier augmentation in the state ε due to their transfer from the upper lying energy states ε'' to unoccupied states at the energy ε. Inequality (4.15) defines the saturation limit due to state filling. $n(\varepsilon,0)$ determines the initial conditions and defines the state ε population at the moment $t = 0$. The $n(\varepsilon,0)$ value depends on the excitation intensity I_0. In order to apply Eqns. (4.14–4.16) to the QD system we assume that after the laser pulse, the carriers excited into GaAs barrier rapidly relax into QDs and below WL states creating an initial population of these states during a short time of $\sim$20 ps as indicated by transient PL measurements. At this time scale the high density hot carriers, through a number of scattering processes, including electron–electron scattering, electron-hole scattering, multi-phonon emission, populate initially empty states in the below WL energy region. After this time, taken as a time origin, much slower processes of carrier redistribution and their diffusion into lower-lying energy states take place. As a result, this cascade feeds these states elongating the PL decay. Approximating the density of continuum states $n_0(\varepsilon')$ by an exponential tail, and taking the $\tau_r(\varepsilon) = \tau$, $\tau_{tr}(\varepsilon - \varepsilon') = \tau_1$, and $\tau_{tr}(\varepsilon'' - \varepsilon) = \tau_2$ as fitting constants independent of energy it is possible to solve Eqns. (4.14–4.16) and the results of the fit are shown with lines in Fig. 4.22b. Such good fit with a simplified choice of fitting parameters supports the model presented by Eqns. (4.14–4.16). The set of parameters $\tau = 1.4$ ns, $\tau_1 = 1.2$ ns, and $\tau_2 = 0.8$ ns proves the existence of long-lived states in the below WL energy region. These states could be related to the WL morphology and/or the interaction of the buried QDs with the environment [72, 125, 137].

Thus time-resolved PL measurements carried out for the InGaAs QDs quantum dots grown on (001) and (311) oriented GaAs substrates allow us to explore the continuum states below the WL. For this purpose detection energy scans of the spectral region from the energy of the QD excitonic transition up to the WL absorption edge. The convex-shaped decay of the PL signal in this region gives evidence of carrier relaxation through the continuum states below the WL similar to carrier diffusion. Due to the exponential reduction of the density of states into the WL energy gap, the low-lying states can be blocked temporarily for further carrier transfer due to saturation. In this case the convex shape of the PL decay develops. Strong dependence of the decay time on the excitation density observed for the QD ground-state PL is explained in terms of a suggested model of carrier transfer from the continuum states below the WL to the QD states. In this case, the dynamics of carrier capture into the QDs can be better explained by a high density of states below the WL and carrier relaxation through the continuum states than by more conventional phonon mediated processes.

4.8 Summary

In conclusion, we have shown that dense QD arrays typically used for the optoelectronic applications demonstrate complex carrier dynamics changeable in different excitation conditions, depending on growth conditions and applied external fields. Numerous relaxation channels enrich the possibility of carrier transfer between adjacent dots or 2D electronic systems and interplay within wide temperature and excitation density intervals. Knowledge of these channels is of great importance for producing highly efficient QD devices with the lowest threshold characteristics and exploration of further application. We have demonstrated the ways of engineering the QD sizes, density, lateral and vertical coupling in the In(Ga)As/GaAs QD arrays affecting the carrier transfer rate. The coupling in fact creates net QD system where QDs form complex aggregates and modify surrounding in a way allowing very efficient carrier transfer for a long distance. Further advances are recently promoted due to fabrication of the nanostructures by so called "droplet epitaxy" allowing creation the quantum dots molecules, quantum dot wires, rings, etc [138]. Investigation of the carrier transport in such systems is under way.

Acknowledgments We are appreciated many people contributed this work, Zh. M. Wang, H. Kissel, Vas. P. Kunets, B. L. Liang, D. Guzun, W. T. Masselink, J. W. Tomm, V. G. Talalaev, T. Elsaesser, C. Walther, Z.Ya. Zhuchenko, Min Xiao for their skill and participation.

References

1. M. S. Skolnick and D. J. Mowbray, Annual Review of Materials Research **34**, 181 (2004).
2. T. H. Stievater, X. Q. Li, D. Gammon, D. S. Katzer, D. Park, C. Piermarocchi, and L. J. Sham, Phys. Rev. Lett. **81**, 133603 (2001).
3. A. Zrenner, E. Beham, S. Stufler, F. Findeis, M. Bichler, and G. Abstreiter, Nature **418**, 612 (2002).
4. X. Q. Li, Y. W. Wu, D. Steel, D. Gammon, T. H. Stievater, D. S. Katzer, D. Park, C. Piermarocchi, and L. J. Sham, Science **301**, 809 (2003).
5. J. P. Reithmaier, G. Sek, A. Loffler, C. Hofmann, S. Kuhn, S. Reitzenstein, L. V. Keldysh, V. D. Kulakovskii, T. L. Reinecke, and A. Forchel, Nature **432**, 197 (2004).
6. T. Yoshie, A. Scherer, J. Hendrickson, G. Khitrova, H. M. Gibbs, G. Rupper, C. Ell, O. B. Shchekin, and D. G. Deppe, Nature **432**, 200 (2004).
7. A. Badolato, K. Hennessy, M. Atature, J. Dreiser, E. Hu, P. M. Petroff, and A. Imamoglu, Science **308**, 1158 (2005).
8. T. Unold, K.Mueller, C. Lienau, T. Elsaesser, and A. D. Wieck, Phys. Rev. Lett. **94**, 137404 (2005).
9. M. Kroutvar, Y. Ducommun, D. Heiss, M. Bichler, D. Schuh, G. Abstreiter, and J. J. Finley, Nature **432**, 81 (2005).
10. A. Greilich, D. R. Yakovlev, A. Shabaev, A. L. Efros, I. A. Yugova, R. Oulton, V. Stavarache, D. Reuter, A. Wieck, and M. Bayer, Science **313**, 341 (2006).
11. F. H. L. Koppens, C. Buizert, K. J. Tielrooij, I. T. Vink, K. C. Nowack, T. Meunier, L. P. Kouwenhoven, and L. M. K. Vandersypen, Nature **442**, 766 (2006).
12. H. Shoji, Y. Nakata, K. Mukai, Y. Sugiyama, M. Sugawara, N. Yokoyama, and H. Ishikawa, Appl. Phys. Lett. **71**, 193 (1997).
13. D. Bimberg, M. Grundmann, and N. N. Ledentsov, Quantum Dots Heterostructures, 1st ed. (Wiley, New York, 1999).
14. P. Bhattacharya, A. D. Stiff-Roberts, S. Krishna, and S. Kennerly, International Journal of High Speed Electronics and System **12**, 969 (2002).
15. S. Krishna, Journal of Physics D **38**, 2142 (2005).
16. E. Towe and D. Pan, IEEE Journal of Selected Topics in Quantum Electronics **6**, 408 (2000).
17. C. Zhonghui, K. Eui-Tae, Y. Zhengmao, J. C. Campbell, A. Madhukar, Conference Digest of the 2004 Joint 29th International Conference on Infrared and Millimeter Waves and 12th International Conference on Terahertz Electronics (IEEE Cat. No.04EX857), p. 237 (2004).
18. R. Oga, W. S. Lee, Y. Fujiwara, and Y. Takeda, Appl. Phys. Lett. **82**, 4546 (2003).
19. H. A. Jones-Bey, Laser Focus World **42**, 10 (2006).
20. N. N. Ledentsov, M. Grundmann, F. Heinrichsdorff, D. Bimberg, V. M. Ustinov, A. E. Zhukov, M. V. Maximov, Zh. I. Alferov, and J. A. Lott, IEEE Journal of Selected Topics in Quantum Electronics **6**, 439 (2000).
21. S. M. Kim, Proceedings of the SPIE–The International Society for Optical Engineering **4999**, 423 (2003).
22. Q. H. Xie, A. Madhukar, P. Chen, N. P. Kobayashi, Phys. Rev. Lett. **75**, 2542 (1995).

23. M. Schmidbauer, S. Seydmohamadi, D. Grigoriev, Z. M. Wang, Y. I. Mazur, P. Schafer, M. Hanke, R. Kohler, and G. J. Salamo, Phys. Rev. Lett. **96**, 066108 (2006).
24. T. Mano, R. Notzel, G. J. Hamhuis, T. J. Eijkemans, and J. H. Wolter, Appl. Phys. Lett. **81**, 1705 (2002).
25. B. Krause, T. H. Metzger, A. Rastelli, R. Songmuang, S. Kiravittaya, and O. G.Schmidt, Phys. Rev. B **72**, 085339 (2005).
26. S. C. Lee, L. R. Dawson, K. J. Malloy, and S. R. Brueck, Appl. Phys. Lett. **79**, 2630 (2001).
27. T. Mano, T. Kuroda, S. Sanguinetti, T. Ochiai, T. Tateno, J. Kim, T. Noda, M. Kawabe, K. Sakoda, G. Kido, and N. Koguchi, Nano Lett. **5**, 425 (2005).
28. Zh. M. Wang, K. Holmes, J. L. Shultz, and G. J. Salamo, Phys. Status Solidi A **202**, R85 (2005).
29. B. L. Liang, Zh. M. Wang, J. H. Lee, K. Sablon, Yu. I. Mazur, and G. J. Salamo, Appl. Phys. Lett. **89**, 043113 (2006).
30. J. H. Lee, Zh. M. Wang, N. W. Strom, Yu. I. Mazur, and G. J. Salamo, Appl. Phys. Lett. **89**, 202101 (2006).
31. M. Friesen, P. Rugheimer, D. E. Savage, M. G. Lagally, D. W. van der Weide, R. Joynt, and M. A. Eriksson, Phys.Rev. B **67**, 121301 (2003).
32. K. K. Likharev, Single Electron Devices and their applications, Proc. IEEE **47**, 606 (1999).
33. S. Chang, S. Chuang, and N. Holonyak, Phys. Rev. B **70**, 125312 (2004).
34. I. Magnusdottir, S. Bischoff, A. V. Uskov, and J. Mørk, Phys. Rev. B **67**, 205326 (2003).
35. M. Braskén, M. Lindberg, M. Sopanen, H. Lipsanen, and J. Tulkki, Phys. Rev. B **58**, 15993 (1998).
36. G. Walter, N. Holonyak, Jr., J. H. Ryou, and R. D. Dupuis, Appl. Phys. Lett. **79**, 1956 (2001).
37. S. L. Chuang, P. Littlewood, G. Walter, and N. Holonyak, in Conference on Lasers and Electro-Optics/Quantum Electronics and Laser Science Conference, Baltimore Convention Center, Baltimore, 2003.
38. A. Polimeni, A. Patane, M. Capizzi, F. Martelli, L. Nasi, and G. Salviati, Phys. Rev. B **53**, 4213 (1996).
39. A. Patane, M. G. Alessi, F. Intonti, A. Polimeni, M. Capizzi, F. Martelli, L. Nasi, L. Lazzarini, G. Salviati, A. Bosacchi, and S.Franchi, J. Appl. Phys. **83**, 5529 (1998).
40. G. S. Solomon, J. A. Trezza, and J. S. Harris, Appl. Phys. Lett. **66**, 991 (1995).
41. Q. Xie, N. P. Kobayashi, T. R. Ramachandran, A. Kalburge, P. Chen, and A. Madhukar, J. Vac. Sci. Technol. B **14**, 2203 (1996).
42. N. P. Kobayashi, T. R. Ramachandran, P. Chen, and A. Madhukar, Appl. Phys. Lett. **68**, 3299 (1996).
43. T. R. Ramachandran, R. Heitz, P. Chen, and A. Madhukar, Appl.Phys. Lett. **70**, 640 (1997).
44. R. Heitz, T. R. Ramachandran, A. Kalburge, Q. Xie, I. Mukhametzhanov, P. Chen, and A. Madhukar, Phys. Rev. Lett. **78**, 4071 (1997).
45. I. Kamiya, I. Tanaka, and H. Sakaki, Physica E 2, 637 (1998).
46. A. S. Bhatti, M. Grassi Alessi, M. Capizzi, P. Frigeri, and S.Franchi, Phys. Rev. B **60**, 2592 (1999).
47. S. P. Guo, A. Shen, Y. Ohno, and H. Ohno, Physica E **2**, 672 (1998).
48. C. Walther, R. P. Blum, N. Niehus, W. T. Masselink, and A. Thamm, Phys. Rev. B **60**, R13962 (1999).
49. L. Chu, M. Arzberger, G. Böhm, and G. Abstreiter, J. Appl. Phys. **85**, 2355 (1999).
50. M. Colocci, F. Bogani, L. Carraresi, R. Mattolini, A. Bossacchi, S. Franchi, P. Frigeri, S. Taddei, and M. Rosa-Clot, Superlattices Microstruct. **22**, 81 (1997).
51. H. Kissel,U. Müller, C. Walther, W. T. Masselink, Yu. I. Mazur, G. G. Tarasov, and M. P. Lisitsa, Phys. Rev. B **62**, 7213 (2000).
52. A. A. Darhuber, P. Schittenhelm, V. Holy, J. Stangl, G. Bauer, and G. Abstreiter, Phys. Rev. B **55**, 15 652 (1997).
53. J. M. Garcia, G. Medeiros-Ribeiro, K. Schmidt, T. Ngo, J. L. Feng, A. Lorke, J. Kotthaus, and P. M. Petroff, Appl. Phys. Lett. **71**, 2014 (1997).

54. T. J. Krzyzewski and T. S. Jones, J. Appl. Phys. **96**, 668 (2004).
55. G. Costantini, A. Rastelli, C. Manzano, P. Acosta-Diaz, R. Songmuang, G. Katsaros, O. G. Schmidt, and K. Kern, Phys. Rev. Lett. **96**, 226106 (2006).
56. W. M. McGee, T. J. Krzyzewski, and T. S. Jones, J. Appl. Phys. **99**, 043505 (2006).
57. J. Maes, M. Hayne, V. V. Moschalkov, A. Patane, M. Henini, L. Eaves, and P. C. Main, Appl. Phys. Lett. **81**, 1480 (2002).
58. G. S. Solomon, J. A. Trezza, A. F. Marshall, and J. S. Harris, Phys. Rev. Lett. **76**, 952 (1996).
59. R. Heitz, A. Kalburge, Q. Xie, M. Grundmann, P. Chen, A. Hoffmann, A. Madhukar, and D. Bimberg, Phys. Rev. B **57**, 9050 (1998).
60. G. Yusa and H. Sakaki, Appl. Phys. Lett. **70**, 345 (1997).
61. J. J. Finley, M. Skalitz, M. Arzberger, A. Zrenner, G. Böhm, and G. Abstreiter, Appl. Phys. Lett. **73**, 2618 (1998).
62. A. Nakajima, T. Futatsugi, K. Kosemura, T. Fukano, and N. Yokoyama, Appl. Phys. Lett. **70**, 1742 (1997).
63. W. V. Schoenfeld, T. Lundstrom, P. M. Petroff, and D. Gershoni, Appl. Phys. Lett. **74**, 2194 (1999).
64. R. Heitz, I. Mukhametzhanov, J. Zeng, P. Chen, A. Madhukar and D. Bimberg, Superlattices and Microstructures **25**, 97 (1999).
65. Yu. I. Mazur, Z. M. Wang, G. J. Salamo, Min Xiao, G. G. Tarasov, Z. Ya. Zhuchenko, W. T. Masselink, and H. Kissel, Appl. Phys. Lett. **83**, 1866 (2003).
66. Yu. I. Mazur, Zh. M. Wang, H. Kissel, Z. Ya. Zhuchenko, M. P. Lisitsa, G. G. Tarasov, and G. J. Salamo, Semicond. Sci. Technol. **22**, 1 (2007).
67. T. Yang, J. Tatebayashi, M. Nishioka, and Y. Arakawa, Appl. Phys. Lett. **89**, 081902 (2006).
68. N. Nuntawong, S. Huang, Y. B. Jiang, C. P. Hains, and D. L. Huffaker, Appl. Phys. Lett. **87**, 113105 (2005).
69. H. Y. Liu, I. R. Sellers, M. Gutierrez, K. M. Groom, W. M. Soong, M. Hopkinson, J. P. R. David, R. Beanland, T. J. Badcock, D. J. Mowbray, and M. S. Skolnick, J. Appl. Phys. **96**, 1988 (2004).
70. P. B. Joyce, E. C. Le Ru, T. J. Krzyzewski, G. R. Bell, R. Murray, and T. S. Jones, Phys. Rev. B **66**, 075316 (2002).
71. S. Marcinkevièius and R. Leon, Appl. Phys. Lett. **76**, 2406 (2000).
72. J. Siegert, S. Marcinkevièius, and Q. X. Zhao Phys. Rev. B **72**, 085316 (2005).
73. D. I. Lubyshev, P. P. Gonzalez-Borrero, E. Marega, E. Petitprez, N. LaScala, and P. Basmaji, Appl. Phys. Lett. **68**, 205 (1996).
74. L. Brusaferri, S. Sanguinetti, E. Grilli, M. Guzzi, A. Bignazzi, F. Bogani, L. Carraresi, M. Colocci, A. Bosacchi, P. Frigeri, and S. Franchi, Appl. Phys. Lett. **69**, 3354 (1996).
75. Y. C. Zhang, C. J. Huang, F. Q. Liu, B. Xu, J. Wu, Y. H. Chen, D. Ding, W. H. Jiang, X. L. Ye, and Z. G. Wang, J. Appl. Phys. **90**, 1973 (2001).
76. Yu. I. Mazur, X. Wang, Z. M. Wang, G. J. Salamo, and M. Xiao, H. Kissel, Appl. Phys. Lett. **81**, 2469 (2002).
77. Yu. I. Mazur, Zh. M. Wang, G. G. Tarasov, Vas. P. Kunets, G. J. Salamo, Z. Ya. Zhuchenko, and H. Kissel, J. Appl. Phys. **98**, 053515 (2005).
78. Z. Y. Xu, Z. D. Lu, X. P. Yang, Z. L. Yuan, B. Z. Zheng, and J. Z. Xu, Phys. Rev. B **54**, 11528 (1996).
79. A. Polimeni, A. Patane, M. Henini, L. Eaves, and P. C. Main, Phys. Rev. B **59**, 5064 (1998).
80. Y. T. Dai, J. C. Fan, Y. F. Chen, R. M. Lin, S. C. Lee, and H. H. Lin, J. Appl. Phys. **82**, 4489 (1997).
81. A. V. Uskov, J. McInerney, F. Adler, H. Schweizer, and M. H. Pilkuhn. Appl. Phys. Lett. **72**, 58 (1998).
82. U. Bockelmann and T. Egeler, Phys.Rev. B **46**, 15574 (1992).
83. S. Raymond, K. Hinzer, S. Fafard, and J. L. Merz, Phys.Rev. B **61**, R16331 (2000).
84. G. A. Narvaez, G. Bester, and A. Zunger, Phys. Rev. B **74**, 075403 (2006).

85. A. S. Bracker, M. Scheibner, M. F. Doty, E. A. Stinaff, I. V. Ponomarev, J. C. Kim, L. J. Whitman, T. L. Reinecke, and D. Gammon, Appl. Phys. Lett. **89**, 233110 (2006).
86. M. M. Sobolev, A. R. Kovsh, V. M. Ustinov, A. Yu Egorov, A. E. Zhukov, M. V. Maximov, and N. N. Ledentsov, Materials Science Forum **258–263**, 1619 (1997).
87. A. R. Peaker, S. W. Lin, A. M. Song, M. Missous, I. D. Hawkins, B. Hamilton, and O. Engstrom, Materials Science & Engineering C, Biomimetic and Supramolecular Systems **26**, 760 (2006).
88. S. Rodt, V. Türck, R. Heitz, F. Guffarth, R. Engelhardt, U. W. Pohl, M. Straÿburg, M. Dworzak, A. Hoffmann, and D. Bimberg, Phys. Rev. B **67**, 235327 (2003).
89. Yu. I. Mazur, B. L. Liang, Zh. M. Wang, D. Guzun, G. J. Salamo, Z. Ya. Zhuchenko, and G. G. Tarasov, Appl. Phys. Lett. **89**, 151914 (2006).
90. A. Tackeuchi, Y. Nakata, S. Muto, Y. Sugiyama, T. Usuki, Y. Nishikawa, N. Yokoyama, and O. Wada, Jpn. J. Appl. Phys. **34**, L1439 (1995).
91. F. Adler, M. Geiger, A. Bauknecht, D. Haase, P. Ernst, A. Dörnen, F. Scholz, and H. Schweizer, J. Appl. Phys. **83**, 1631(1998).
92. A. Tackeuchi, T. Kuroda, K. Mase, Y. Nakata, and N. Yokoyama, Phys. Rev. B **62**, 1568 (2000).
93. R. Heitz, I. Mukhametzanov, H. Born, M. Grundmann, A. Hoffmann, A. Madhukar, and D. Bimberg, Physica B **272**, 8 (1999).
94. R. Heitz, I. Mukhametzhanov, P. Chen, and A. Madhukar, Phys. Rev. B **58**, R10 151(1998).
95. V. López-Richard, S. S. Oliveira, and G.-Q. Hai, Phys. Rev. B **71**, 075329 (2005).
96. J. Lehmann and D. Loss, Phys. Rev. B **73**, 045328 (2006).
97. S. I. Rybchenko, I. E. Itskevich, M. S. Skolnick, J. Cahill, A. I. Tartakovskii, G. Hill, and M. Hopkinson, Appl. Phys. Lett. **87**, 033104 (2005)
98. V. G. Talalaev, J. W. Tomm, A. S. Sokolov, I. V. Shtrom, B. V. Novikov, A. T. Winzer, R. Goldhahn, G. Gobsch, N. D. Zakharov, P. Werner, U. Gösele, G. E. Cirlin, A. A. Tonkikh, V. M. Ustinov, and G. G. Tarasov, J. Appl. Phys. **100**, 083704 (2006)
99. D. Bellucci, M. Rontani, G Goldoni, and E. Molinari, Phys. Rev. B **74**, 035331 (2006).
100. G. G. Tarasov, Yu. I. Mazur, Z. Ya. Zhuchenko, A. Maaydorf, D. Nickel, J. W. Tomm, H. Kissel, C. Walther, and W. T. Masselink, J. Appl. Phys. **88**, 7162 (2000).
101. A. Wojs and P. Hawrylak, Phys. Rev. B **55**, 13 066 (1997).
102. Y. P. Varshni, Physica (Utrecht) **34**, 149 (1967).
103. Yu. I. Mazur, J. W. Tomm, V. Petrov, G. G. Tarasov, H. Kissel, C. Walther, Z. Ya. Zhuchenko, and W. T. Masselink, Appl. Phys. Lett. **78**, 3214 (2001).
104. S. Sanguinetti, M. Padovani, M. Gurioli, E. Grilli, M. Guzzi, A. Vinattieri, M. Colocci, P. Frigeri, and S. Franchi, Appl. Phys. Lett. **77**, 1307 (2000).
105. S. Raymond, S. Fafard, P. J. Poole, A. Wojs, P. Hawrylak, S. Charbonneau, D. Leonard, R. Leon, P. M. Petroff, and J. L. Merz, Phys. Rev. B **54**, 11548 (1996).
106. I. E. Itskevich, M. S. Skolnick, D. J. Mowbray, I. A. Trojan, S. G. Lyapin, L. R. Wilson, M. J. Steer, M. Hopkinson, L. Eaves, and P. C. Main, Phys. Rev. B **60**, R2185 (1999).
107. M. Grundmann, N. N. Ledentsov, O. Stier, D. Bimberg, V. M. Ustinov, P. S. Kop'ev, and Zh. I. Alferov, Appl. Phys. Lett. **68**, 979 (1996).
108. M. V. Marquezini, M. J. S. P. Brasil, J. A. Brum, P. Poole, S. Charbonneau, and M. C. Tamargo, Surface Science, **361–362**, 810 (1996).
109. K. Brunner, U. Bockelmann, G. Abstreiter, M. Walther, G. Böhm, G. Tränkle, and G. Weimann, Phys. Rev. Lett. **69**, 3216 (1992).
110. Yu. I. Mazur, G. G. Tarasov, and G. J. Salamo (unpublished)
111. Yu. I. Mazur, Zh. M. Wang, G. G. Tarasov, Min Xiao, G. J. Salamo, J. W. Tomm, V. Talalaev, and H. Kissel, Appl. Phys. Lett. **86**, 063102 (2005).
112. G. G. Tarasov, Z. Ya. Zhuchenko, M. P. Lisitsa, Yu. I. Mazur, Zh. M. Wang, G. J. Salamo, T. Warming, D. Bimberg, and H. Kissel, Semiconductors **40**, 79 (2006).
113. R. Heitz, M. Veit, N. N. Ledentsov, A. Hoffmann, D. Bimberg, V. M. Ustinov, P. S. Kop'ev, and Zh. I. Alferov, Phys. Rev. B **56**, 10 435 (1997).

114. J. W. Tomm, T. Elsaesser, Yu. I. Mazur, H. Kissel, G. G. Tarasov, Z. Ya. Zhuchenko, and W. T. Masselink, Phys. Rev. B **67**, 045326 (2003).
115. Yu. I. Mazur, J. W. Tomm, G. G. Tarasov, H. Kissel, C. Walther, Z. Ya. Zhuchenko, and W. T. Masselink, Physica E **13**, 255 (2002).
116. Z. Ya. Zhuchenko, J. W. Tomm, H. Kissel, Yu. I. Mazur, G. G. Tarasov, and W. T. Masselink, Inst. Phys. Conf. Ser. N_o. **174**: Section 3, 165 (2003).
117. E. C. Le Ru, P. Howe, T. S. Jones, and R. Murray, Phys. Rev. B **67**, 165303 (2003).
118. T. H. Wang, X. B. Mei, C. Jiang, Y. Huang, J. M. Zhou, X. G. Huang, C. G. Cai, Z. X. Yu, C. P. Luo, J. Y. Xu, and Z. Y. Xu, Phys. Rev. B **46**, 16 160 (1992).
119. D. H. Levi, D. R. Wake, M. V. Klein, S. Kumar and H. Morkoç, Phys. Rev. B **45**, 4274 (1991).
120. N. Horiguchi, T. Futatsugi, Y. Nakata, N. Yokoyama, T. Mankad, and P. M. Petroff, Jpn. J. Appl. Phys., Part 1 **38**, 2559 (1999).
121. S. Malik, E. C. Le Ru, D. Childs, and R. Murray, Phys. Rev. B **63**, 155313 (2001).
122. Y. Toda, O. Moriwaki, M. Nishioka, and Y. Arakawa, Phys. Rev. Lett. **82**, 4114 (1999).
123. C. Kammerer, G. Cassabois, C. Voisin, C. Delalande, Ph. Roussignol, and J. M. Gérard, Phys. Rev. Lett. **87**, 207401 (2001).
124. A. Vasanelli, R. Ferreira, and G. Bastard, Phys. Rev. Lett. **89**, 216804 (2002).
125. R. Oulton, J. J. Finley, A. I. Tartakovskii, D. J. Mowbray, M. S. Skolnick, M. Hopkinson, A. Vasanelli, R. Ferreira, and G. Bastard, Phys. Rev. B **68**, 235301 (2003).
126. Y. Toda, S. Shinomori, K. Suzuki, and Y. Arakawa,Phys. Rev. B **58**, R10 147 (1998).
127. S. Sauvage, P. Boucaud, J.-M.Gerard, and V. Thierry-Mieg, J. Appl. Phys. **84**, 4356 (1998).
128. R. Ferreira and G. Bastard, Nanoscale Res. Lett. **1**, 120 (2006).
129. K. Král and P. Zdenìk, Physica E **17**, 89 (2003).
130. B. Alén, J. Martínez-Pastor, D. Granados, and J. M. García, Phys. Rev. B **72**, 155331 (2005).
131. C. Cornet, C. Platz, P. Caroff, J. Even, C. Labbé, H. Folliot, A. Le Corre, and S. Loualiche, Phys. Rev. B **72**, 035342 (2005).
132. E. W. Bogaart, J. E. M. Haverkort, T. Mano, T. van Lippen, R. Nötzel, and J. H. Wolter, Phys. Rev. B **72**, 195301 (2005).
133. T. Mano, R. Nötzel, Q. Gong, T. V. Lippen, G. J. Hamhuis, T. J. Eijkemans, and J. H. Wolter, Jpn. J. Appl. Phys. **44**, 6829 (2005).
134. Yu. I. Mazur, B. L. Liang, Zh. M. Wang, G. G. Tarasov, D. Guzun, and G. J. Salamo, J. Appl. Phys. **101**, 014301 (2007).
135. P. Altieri, M. Gurioli, S. Sanguinetti, E. Grilli, M. Guzzi, P. Frigeri, and S. Franchi, Eur. Phys. J. B **28**, 157 (2002)
136. Yu. I. Mazur, B. L. Liang, Zh. M. Wang, D. Guzun, G. J. Salamo, G. G. Tarasov, and Z. Ya. Zhuchenko, J. Appl. Phys. **100**, 54316 (2006).
137. E. G. Lee, M. D. Kim, and D. Lee, J. Appl. Phys. **98**, 073709 (2005).
138. Zh. M. Wang, K. Holmes, Yu. I. Mazur, K. A. Ramsey, and G. J. Salamo, Nanoscale Res. Lett.**1**, 57 (2006).

Chapter 5
Dynamics of Carrier Transfer into In(Ga)As Self-assembled Quantum Dots

Saulius Marcinkevičius

Abstract The chapter reviews ultrafast dynamics of carrier transfer into InAs/GaAs and InGaAs/GaAs quantum dots of different doping, densities and interlevel energies. Results from theoretical modeling and time-resolved optical studies are discussed. Considered effects include carrier transport in the barriers, capture into the dots and relaxation in the dots. Several carrier capture and relaxation mechanisms, such as longitudinal optical (LO) phonon emission, carrier-carrier scattering, capture through defect states and relaxation through barrier/wetting layer continuum states are analyzed. LO phonon-assisted capture appears to be a rather universal capture mechanism. For the carrier relaxation in doped quantum dots and/or at high carrier densities, the carrier-carrier scattering-assisted relaxation dominates. In undoped dots and low excitation levels, relaxation through LO phonon emission is found to be the relevant process. Whatever the mechanism, the carrier transfer in QD structures is, as a rule, fast, occurring in 1 to 20 ps time range.

5.1 Introduction

Since the development of femtosecond lasers, and, in particular, self mode-locking Ti:sapphire lasers, ultrafast carrier dynamics have been a very intense and fast-evolving research area of semiconductor physics [1, 2]. A broad variety of effects, such as carrier dephasing, thermalization, tunneling, cooling, trapping, thermal emission and transport could be experimentally observed is a rather direct way. The sheer excitement of observation of effects occurring on a picosecond and even on a femtosecond time scale was supported by importance of the carrier dynamics for electronic, optical and optoelectronic device performance. To a large extent, these devices include semiconductor quantum structures of reduced dimensionalities, such as quantum wells, wires and dots. For instance, understanding of ultrafast carrier dynamics in quantum wells has lead to optimization of a number of devices, such as modulation-doped field effect transistors, quantum well lasers and modulators, and ultrafast saturable absorbers. Naturally, since realization of self-assembled semiconductor quantum dots (QDs), ultrafast carrier dynamics have acquired an significant place in research of these zero-dimensional structures. Development of relatively mature devices, such as QD lasers, and future

Z. M. Wang, *Self-Assembled Quantum Dots.*

applications, including single electron transistors, single photon sources as well as nanophotonic devices, strongly depends on understanding of ultrafast carrier dynamics in the QDs.

The field of ultrafast carrier dynamics in quantum dot structures is very broad. It includes carrier and spin dephasing and relaxation, carrier capture into and emission out of the dots, vertical and lateral transport, tunneling, trapping, radiative, nonradiative and stimulated recombination. Research is being performed on single quantum dots and QD molecules, coupled dots and randomly oriented QD ensembles, with each system showing its own peculiarities. Furthermore, manifestation of ultrafast effects experiences differences for free carriers, neutral and charged excitons, and biexcitons. In this chapter, only a fraction of this multitude of phenomena is reviewed. The considered processes include those that are responsible for the carrier transfer from the barriers into the QD ground state in a quantum dot structure. The effects include vertical carrier transport in the barriers, capture into the QDs and relaxation there; all of them of utmost importance for operation of QD lasers and optical amplifiers. The carrier transfer is studied in the system of In(Ga)As/GaAs self-assembled quantum dots, which is probably the most technologically mature and most practically applied QD platform. Besides, we concentrate on free carrier effects and large quantum dot ensembles, which are most relevant for QD laser applications. Certainly, investigations on single QDs would present additional information on the carrier dynamics since optical response would not be smeared by the inhomogeneous broadening. So far, however, time resolved measurements on single quantum dots, because of experimental difficulties of measuring weak optical signals with a high temporal resolution, were focused on slower effects, such as carrier recombination [3–5] or lateral transport [6]. Ultrafast measurements on single QDs, albeit not self-assembled but formed by interfacial fluctuations, considered coherent effects [7, 8] with quantum computing applications in mind.

One should note that research in the area of QD carrier dynamics has been very active during the last decade, which has resulted in a large number of scientific publications. A number of previous reviews [9–15] as well as several chapters of this book deal with different aspects of carrier dynamics in the QDs.

5.2 Theoretical Studies of Carrier Transfer in Quantum Dot Structures

Carrier transfer from the barriers into the ground state of the QDs can be divided into transport in the barriers, capture into the dots and relaxation in the dots. (Fig. 5.1) The last two processes often proceed simultaneously, for instance, if the carriers are captured directly into the QD ground state. In theoretical studies, however, these effects are often treated separately.

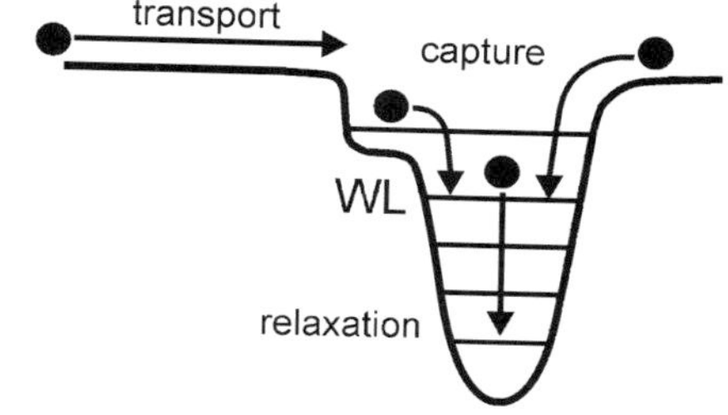

Fig. 5.1 Schematics of processes constituting electron transfer into QDs

5.2.1 Carrier Capture

Carrier capture accounts for the process then the carriers are transferred from the three-dimensional barrier or the two-dimensional wetting layer (WL) states to the discrete QD states. The capture process occurs via scattering, either with lattice vibrations or other charge carriers, and is accompanied by reduction of the potential energy of the captured carrier. In theoretical calculations, two mechanisms of carrier capture are being considered. These are the carrier capture assisted by carrier-carrier (Auger) scattering, and the capture with emission of longitudinal optical (LO) phonons. In nanostructures of higher dimensionality with a continuous spectrum of states, such as quantum wells, the LO phonon-assisted capture with its efficient energy dissipation is the main carrier capture mechanism. In QDs, however, complications related to the discrete QD energy level spectrum and the small LO phonon dispersion arise. On the other hand, carrier capture by carrier-carrier scattering, with a broad spectrum of final states, is not restricted by the limitations of energy conservation.

Modeling of carrier dynamics in a QD of a realistic shape is very complicated, besides, the shape, strain and composition distribution are not always known. Therefore, modeling is usually performed on QDs of an idealized shape, e.g. a sphere [16], a cone and a truncated cone [17, 18], a cylinder [19], or a truncated pyramid [20].

Calculations of the Auger capture in cylindrically-shaped QDs have been performed by Uskov *et al.* using the rate equation model [19]. Two processes have been considered: capture of an electron (hole) from the two-dimensional WL after interaction with a hole (electron) residing in the WL, and capture of a hole after scattering with an electron that is in the QD. The Auger capture coefficients were found to be strongly oscillating functions of the dot diameter. For the WL carrier densities of the order 10^{11}–$10^{12}\,\mathrm{cm}^{-2}$, the capture times were found to be in the range from 1 to 100 ps.

Phonon assisted capture was also found to produce strong resonances depending on the energy separation between the WL and the QD states and, consequently, the dot size [17]. It was noted, however, that hot carrier distribution, which is likely to occur in optical experiments after excitation higher-up in the barriers, eases resonance requirements. Zhang and Galbright [20] elaborated on this subject and found that, indeed, for a hot carrier temperature of 400–500 K, the resonances in the

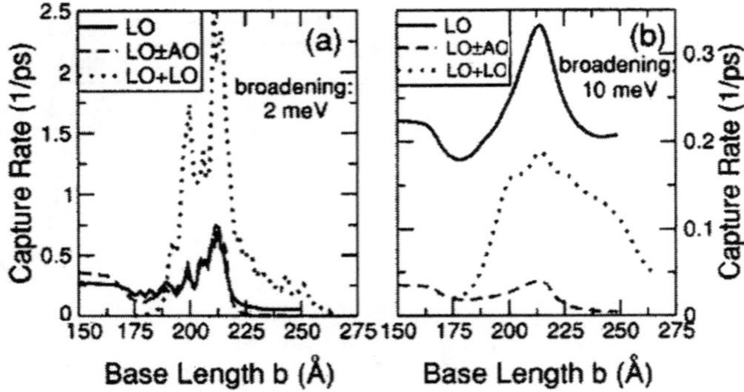

Fig. 5.2 Electron capture rates as a function of base length in truncated pyramid QDs due to various single- and two-phonon capture processes. The level broadening is (a) 2 meV and (b) 10 meV. The carrier and lattice temperatures are 450 and 300 K, respectively; the carrier density is 10^{12} cm^{-2} (Reused with permission from J.-Z. Zhang and I. Galbraith, Applied Physics Letters, 89, 153119 (2006). Copyright 2006, American Institute of Physics)

phonon-assisted capture rate dependence on the dot dimensions are considerably widened, and the phonon-assisted capture is very efficient (Fig. 5.2). In their work, the capture with one and two LO as well as LO and acoustic phonon participation has been considered, and the shortest capture times of subpicosecond duration were obtained.

Magnusdottir with coworkers studied both carrier capture mechanisms, phonon- [21] and Auger-assisted [18]. The capture rates from the WL into the QDs were calculated following the Fermi's golden rule. Similarly as in other works, strong resonances in the dependence of the phonon-assisted capture times on the QD dimensions were found, both for single and two LO phonon processes. The calculated capture times as a function of the WL carrier density are shown in Fig. 5.3(a). For a given carrier density, the single phonon process is about an order of magnitude faster, however, at moderate to high carrier densities both types of capture are very short. The capture time dependence on the carrier density has its origin in the Fermi filling of the energy levels of interacting carriers. In the experiments, however, the capture time, especially at low carrier densities, was found to be density independent; in fact, this independence was assumed as a criterion distinguishing the phonon- and Auger-assisted carrier capture [22]. The temperature dependence of the capture rate (Fig. 5.3(b)) occurs via phonon and carrier (Fermi) distributions. At low temperature, both quantities are close to zero; the subsequent temperature dependence is mainly determined by changes in the carrier distribution. At temperatures above ~50 K, the capture rate is nearly constant. For capture via carrier-carrier scattering with both carriers initially in the WL, the calculations have revealed several tendencies. Firstly, scattering by electrons has been found to be more efficient than scattering by holes because Coulomb interaction matrix elements favor small changes in the momentum transfer of the scattered carriers. Electrons in the WL have stronger dispersion than holes, consequently, for the same

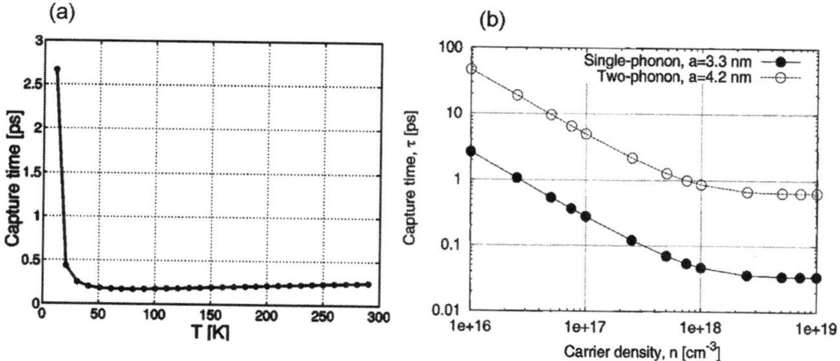

Fig. 5.3 (a) Calculated temperature dependence of the single-phonon capture time into the excited QD state for a carrier density of 5×10^{17} cm^{-3} and (b) density dependence of the single- and two-phonon capture times for two different dot sizes (Reused with permission from I. Magnusdottir, A. V. Uskov, S. Bischoff, B. Tromborg, and J. Mørk, Journal of Applied Physics, 92, 5982 (2002). Copyright 2002, American Institute of Physics)

energy exchange (corresponding to the energy between the WL and QD states), scattering by electrons involves a smaller change in momentum. In addition, the capture rate is strongly dependent on the energy separation between the WL and the QD states. With increased separation, the Auger coefficients for all types of carrier-carrier scattering (*e-e*, *e-h*, *h-h*, *h-e*) decrease. This indicates than in QDs with multiple levels, carrier capture into the highest excited states is the most efficient. The capture rate due to the carrier-carrier scattering was also compared to the LO phonon-assisted capture (Fig. 5.4). Only at the highest carrier concentrations these rates become similar; at lower carrier densities the LO phonon-assisted capture is more efficient. These different capture mechanisms experience different dependence on the sheet carrier density. At low and moderate carrier densities, the phonon assisted capture rate is proportional to the sheet carrier density n because only one carrier participates in the capture process. The carrier-carrier scattering is proportional to n^2 since two WL carriers are involved.

5.2.2 Carrier Relaxation

From all the mechanisms of carrier-phonon scattering in polar semiconductors, charge carriers strongly interact only with LO phonons because they produce an electric field [23]. However, LO phonons have small dispersion, and relaxation via emission of LO phonons has limitations stemming from the energy conservation requirement. Carrier relaxation with acoustic phonon emission is considered to be slow because of the week deformation potential interaction. The calculated slow-down of the carrier relaxation in QDs is commonly referred to as the phonon bottleneck effect [24, 25]. However, discrepancy between the phonon bottleneck effect and most of the experimental results showing fast relaxation suggest that

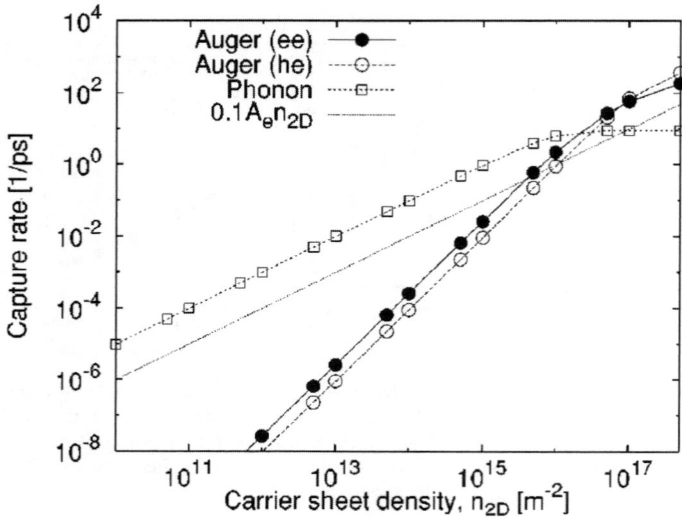

Fig. 5.4 The Auger carrier capture rates for electrons and holes for scattering with electrons compared to the electron capture with emission of one LO phonon. The dotted line indicates the approximate level for the two-phonon capture (Reprinted figure with permission from [18]. Copyright (2002) by the American Physical Society)

another relaxation mechanism, namely, carrier-carrier scattering, should play an important role.

For the carrier-carrier scattering that involves two-dimensional carriers in the WL and carriers in the QDs, carrier relaxation times were calculated by Uskov *et al.* using dipole approximation and taking into account screening by carriers located in the barriers [26]. Relaxation times of the order of 1 to 10 ps for carrier densities of 10^{11}–10^{12} cm^{-2} were obtained. Dependence of the Auger coefficient on temperature was found to be weak. Similar carrier relaxation times due to carrier-carrier scattering were also calculated in Ref. [17].

A comprehensive analysis of the role that the carrier-carrier and carrier-LO phonon scattering play in the capture and relaxation has been presented by Nielsen and coauthors [27]. Their calculations were not restricted by the Fermi's golden rule but also took into account population effects. In- and out-scattering were considered as well as screening in the WL/QD system. The carrier-carrier scattering-assisted relaxation time was of the order of 1 ps; for resonant conditions, the LO phonon-assisted scattering was of comparable duration. The carrier relaxation was found to be faster than the capture, and processes involving holes were faster than the corresponding effects for electrons.

Recently, electron relaxation rates in InGaAs/GaAs QDs have been analyzed by Narvaez *et al.* [28]. In this work, realistic QD energy levels and wave functions were computed using atomistic pseudopotential-based approach. This allowed determining Coulomb scattering matrix elements, which, using the Fermi's golden rule, led to interlevel scattering rates. Time-dependent occupation numbers were found using the rate equations. The calculated interlevel scattering rates vary between 1 and 7 ps depending on the QD band gap. Decrease of the relaxation time with the decreased band gap was attributed to increased density of the QD hole states in dots with a deeper potential.

Li *et al.*, in an attempt to reexamine the phonon bottleneck effect, explored carrier relaxation via LO phonon scattering [29]. In their work, the authors took into account the anharmonic decay of LO phonons into bulk acoustic phonons. It appears that the LO phonon lifetime of 2.5–7 ps has a remarkable effect on the relaxation process: a window of efficient carrier relaxation by the LO phonon emission is extended to tens of meV around the LO phonon energy discarding the phonon bottleneck prediction of the slowed-down relaxation and stressing the importance of the phonon scattering process for the carrier relaxation. Importance of the anharmonic LO phonon decay on the carrier relaxation was also pointed out in Ref. [30].

In an attempt to resolve the discrepancy between the experimental data and the phonon bottleneck effect, Sercel and coworkers introduced yet another relaxation mechanism, namely, defect-assisted carrier relaxation [31, 32]. According to the model, the relaxation occurs in the following sequence: first, an electron makes a transition from a higher QD state to a defect level, the defect relaxes by multiphonon emission, and the electron makes a transition back into the QD, now to a lower level. Since the process involves tunneling between the QD and defect states with the transition matrix element decreasing exponentially with the QD–defect separation, the defect-assisted relaxation is efficient only for defects that are located in the close vicinity of the dots. This, however, is not unusual for the QDs since interface defects are often present even in high quality QD structures [23, 33]. The calculations of the electron relaxation rates were carried out for spherical InGaAs/GaAs dots and typical GaAs defect states. Relaxation times on a picosecond time scale were obtained suggesting that the defect-assisted relaxation can be an efficient process. It was found that the mechanism does not require a perfect alignment between defect and QD level energies: windows of tens of meV within which the electron transfer between the dot and the defect is fast have been found.

5.3 Experimental Details

5.3.1 Quantum Dot Structures

Experiments reviewed in this chapter have been performed on InAs and InGaAs QDs grown on GaAs substrates by Stransky-Krastanov self-assembly method. Carrier dynamics in charged QDs [34] were studied on modulation-doped QD structures prepared using molecular beam epitaxy. The structures, grown on semi-insulating GaAs substrates, consist of a 500 nm thick nominally undoped GaAs buffer layer, a QD layer and a 200 nm cap layer. QDs were formed by depositing 3.5 monolayers of InAs at 520 °C. The QDs were lens-shaped with typical sizes of 45 nm diameter and 8 nm height, and densities of $3 \times 10^{10}\,cm^{-2}$ (Fig. 5.5). At a distance of 10 nm

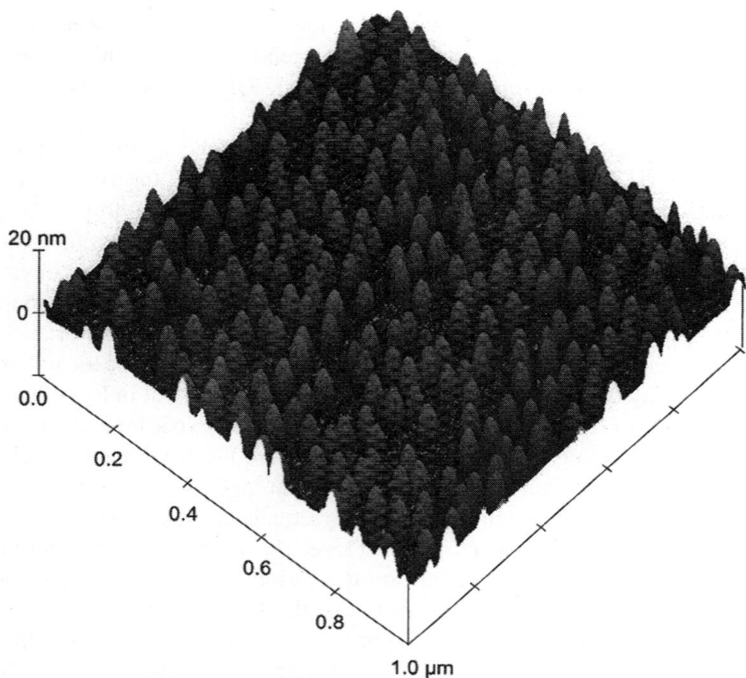

Fig. 5.5 $1 \times 1\,\mu m^2$ atomic force microscopy topographic image of the undoped InAs/GaAs QD sample prior to capping (From Siegert et al. [34])

below the dot layer, 10 nm thick Si (*n* type) or Be (*p* type) modulation-doped layers were introduced during the growth. Samples with low ($5 \times 10^{16}\,cm^{-3}$), moderate ($5 \times 10^{17}\,cm^{-3}$ for *n* and $1 \times 10^{18}\,cm^{-3}$ for *p*), and high ($5 \times 10^{18}\,cm^{-3}$ for *n* and $1 \times 10^{19}\,cm^{-3}$ for *p*) doping concentrations were prepared. For comparison, identical undoped QD sample was grown.

InGaAs/GaAs QD structures, used for investigations of carrier dynamics in dots of different densities and confinement potentials, and proton-implanted dots, were grown by metalorganic chemical vapor deposition (MOCVD). Details of the growth procedure can be found in [35–37]. The nominal QD ternary composition was $In_{0.6}Ga_{0.4}As$. The quantum dots were capped with 100 nm of GaAs. The MOCVD-grown QDs were also lens-shaped; different sets of samples had dot diameters from 25 to 46 nm. Average height/diameter ratio was typically 1:6. Some of the samples were treated by rapid thermal annealing, which was performed in argon atmosphere at temperatures from 700 to 950 °C for 30 s.

Time-integrated photoluminescence (PL) spectra of the undoped InAs/GaAs QD sample, measured at 80 K for excitation power densities ranging from 0.23 to $2300\,Wcm^{-2}$, which corresponds to 8×10^9 to 8×10^{13} photoexcited electron-hole pairs per pulse per cm^2, are shown in Fig. 5.6(a). The high excitation spectra for the *p*-doped sample is also included. Below the WL transition at 1.38 eV, four well-separated peaks, related to transitions within the quantum dots, develop with

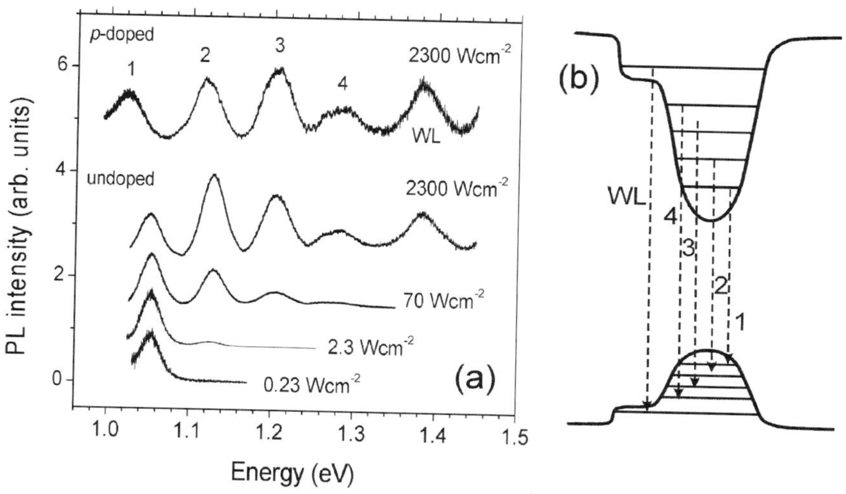

Fig. 5.6 (a) 80 K time-integrated PL spectra of the p-doped and the undoped samples obtained at excitation conditions indicated in the graph. The spectra are normalized to the ground state peak emission. (b) Schematics of the corresponding optical transitions (From Siegert et al. [34])

increased excitation power. The average energy separation between the peaks for the undoped and the n-doped structures is 77 meV, for the p-doped sample it is 85 meV.

We assign the QD PL peaks to transitions between electron and hole single particle levels with the same quantum number [38] named $e1$-$e4$ and $h1$-$h4$ (Fig. 5.6(b)). In QD level calculations, these levels are often referred as shells $(s, p, d \ldots)$ with the higher shells containing several energetically close (within a few meV) levels [39]. Such fine level splitting can not be distinguished in QD ensemble PL spectra; in the shell notation, our PL peaks-voltage would correspond to transitions between carriers located in shells with the same quantum number.

For interpretation of carrier relaxation experiments, the interlevel energy differences in the conduction and valence bands ΔE_c and ΔE_v, respectively, are very important. A PL spectrum provides a sum of these interlevel energies; to determine them individually, other types of experiments are employed. Schmidt *et al.* [40] measured the interlevel splittings in lens-shaped InAs dots using capacitance (C-V) and photoluminescence spectroscopy and found the electron level spacing $\Delta E_c = 50$ meV and the ratio between electron and hole interlevel energies $\Delta E_c/\Delta E_v = 2$. Similar results, $\Delta E_c = 49$ meV and $\Delta E_v = 25$ meV, were obtained for InAs QDs from transmission measurements in the far and near infrared spectral regions [41]. Interlevel energies in lens-shaped $In_{0.5}Ga_{0.5}As$ [42] and InAs [43] QDs have been studied by Chang *et al.* with C-V and PL spectroscopy. Again, the ratio $\Delta E_c/\Delta E_v \approx 2$ was found. For energetic intervals between the WL and the top QD levels, a close value of 1.6 has been reported [44]. These coherent results encourage using the ratio $\Delta E_c/\Delta E_v = 2$ for our dots as well. In that case, the average interlevel spacings for the InAs/GaAs QDs are about 54 and 27 meV for the conduction and valence bands, respectively. This is close to two and one InAs LO phonon energies (30 meV) with a broadening of ± 3 meV induced by strain, alloy composition variations and interfaces [45, 46]. The sum of the energy difference between the WL states and the highest electron and hole levels $e4$, $h4$ in the QDs, estimated from the PL spectra, is 109 meV. With the 2:1 ratio for the conduction and valence bands, 73 and 36 meV energy separations are obtained. These numbers match the energies of two and one GaAs LO phonons (36 ± 3 meV).

5.3.2 Experiment

Carrier dynamics were measured by time-resolved PL. A tunable mode-locked Ti:sapphire laser (pulse length 130 fs, repetition rate 76 MHz, wavelength tuning range 700–1040 nm) was used for excitation. The measurements were performed using a streak camera and an upconversion technique. The synchroscan streak camera combined with a 0.25 m spectrometer provided a temporal resolution of 3 ps. In the upconversion setup (temporal resolution 150 fs), the sum frequency signal, which results from mixing of the laser pulse and the PL in a $LiIO_3$ nonlinear crystal, was detected either by a photomultiplier and a single photon counter, or with a nitrogen-cooled low-noise CCD mounted behind a 0.5 m spectrometer [47]. Our experiments were performed in the temperature range 80 K–300 K.

Typical PL transients measured for the modulation doped and undoped QD structures are shown in Fig. 5.7. Rise of the PL signal reflects carrier arrival at a particular level, PL decay is determined by carriers leaving the level, either by relaxation, thermal escape or recombination. To find the PL rise and decay times, the PL transients are approximated using a commonly used equation

$$I(t) \propto [\exp(-t/\tau_r) - \exp(-t/\tau_d)]/(\tau_r - \tau_d),$$

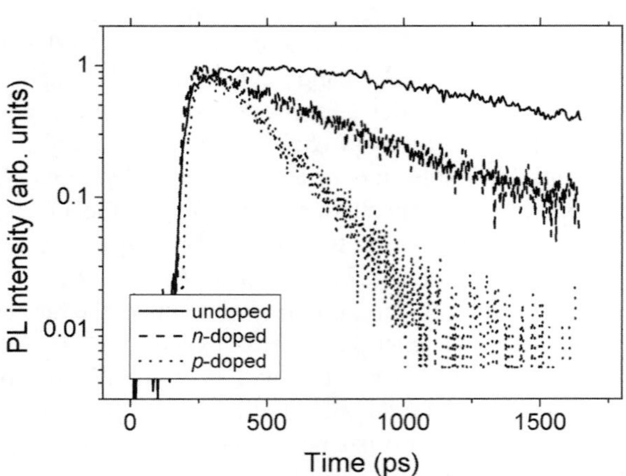

Fig. 5.7 Normalized photoluminescence transients corresponding to the ground state transition in doped and undoped QD samples (From Siegert et al. [34])

where τ_r and τ_d are the PL rise and decay times, respectively.

Some of the reviewed results were obtained using degenerate or dual-color pump-probe experiments. In such measurements, a strong pump pulse excites nonequilibrium carrier population the evolution of which is monitored by measuring transmission or reflectivity of a weaker probe pulse arriving at the sample after a controllable delay. One should note that the time-resolved PL and the pump-probe are complementing experimental techniques for studies of carrier dynamics: the luminescence signal is proportional to the product of the electron and hole distribution functions, $I_{PL} \propto f_n f_h$ while the differential transmission is proportional to their sum, $\Delta T/T \propto f_e + f_h$ [2].

5.4 Experimental Studies of Carrier Transfer into Quantum Dots

5.4.1 Free Carrier Versus Excitonic Carrier Transfer

Photoexcited carrier capture and relaxation is typically treated in a single particle or a correlated electron and hole pair (exciton) picture. In most cases, the carrier capture and relaxation are regarded to proceed as free carrier processes, with individual electron and hole capture and relaxation times. In some works, however, exciton rather than free carrier relaxation is considered [13]. Strong electron-hole pair confinement to a small volume of a QD and the corresponding strong Coulomb interaction are considered as the grounds for the excitonic approach. Exciton energy levels are represented by a combination of electron and hole levels [13]. This has serious implications on the allowed mechanisms of carrier relaxation: the much denser exciton energy spectrum would allow more alternative paths for the phonon emission, which, in effect, would influence the relaxation rate.

It would be fair to assume that application of either the free carrier or the exciton model depends on the experimental conditions. For the carrier excitation in the barriers, which is the most popular way of photoexcited carrier generation in In(Ga)As/GaAs QD structures, carriers may be captured into the QDs faster than excitons in the barrier are formed. The exciton formation time is of the order of 10 ps [48], which is similar or longer than the capture and relaxation times. On the other hand, the excitonic relaxation model might be more appropriate for the direct carrier excitation in undoped QDs when the electron–hole pairs are correlated from the excitation moment [13]. Yet, there is no unambiguous evidence indicating advantage of the excitonic relaxation model. Studies devoted to the exciton vs. free carrier problem have mostly considered thermal carrier emission rather than capture and relaxation.

For instance, the question whether the thermal emission occurs via uncorrelated carriers or excitons has been addressed in Ref. [49]. Analysis based on the rate equations has revealed that these models give very different results for the thermionic emission. An exciton has larger confinement energy than a free carrier making the threshold for exciton thermal emission higher and efficiency lower. For InAs QDs with the band gap at 1.15–1.25 eV (10 K), the temperatures at which significant differences between both models occur are $\geq$100 K. Comparison between the experiment and theory has clearly favored the free carrier model, toning down the relevance of the exciton dynamics at these elevated temperatures. Also for the intradot carrier excitation into the ground state, separate electron and hole thermal activation processes have been identified [50]. In some works, analysis of correlated vs. uncorrelated carrier activation was based on comparison between QD band offsets and the thermal activation energy, which is different for single carrier and exciton emission. In a detailed investigation of intermixed InAs/GaAs QDs, Le Ru *et al.* extracted activation energies corresponding to the uncorrelated electron and hole escape [51]. Besides, a superlinear PL intensity dependence on the excitation power, reported in the same work, supports the uncorrelated rather than the excitonic capture into the QDs. Independent electron and hole capture and relaxation have also been identified from the pump-probe data [52].

Carrier transfer effects considered in this chapter are often measured after excitation in the barriers and at elevated temperatures; thus, the free carrier picture will be used.

5.4.2 Carrier Transport

After optical excitation in the barriers at a wavelength slightly over the GaAs band gap, the majority of carriers are excited within one micrometer from the surface. Considering that the QDs are usually situated close to the sample surface, typically at a distance of ~100 nm, the absorption length, $L \cong 0.8\,\mu\text{m}$, is the distance that the photoexcited carriers have to cover before being captured into the QDs. The travel time can be measured by the time-resolved PL by selecting the detection wavelength. Figure 5.8 shows the rising parts of the PL transients for the modulation-doped and undoped InAs QDs measured at the ground QD transition after excitation in the barriers. Figure 5.9(a) collects the PL rise times measured at different transition energies in modulation doped and undoped QD structures. The PL rise time measured at the GaAs band gap energy reflects hot carrier thermalization and relaxation, and is below a picosecond. When monitored at the WL band gap, the PL rise time describes transport in the barriers and capture into the WL. This time is equal to ~2 ps and does not depend on doping. This time constant is much too short to be explained by ambipolar diffusion in the barriers, not to mention an additional picosecond or so for capture into the WL. Ambipolar diffusion, using typical GaAs diffusion coefficient at 80 K $D_{amb} = 55\,\text{cm}^2/\text{s}$, would provide diffusion time $\tau = L^2/D_{amb} \cong 130$ ps. Modulation doping, due to a partial transfer of majority carriers into the QDs, introduces potential variations and built-in fields in the vicinity of the QDs [53], which might speed-up the transport of minority carriers. However, the same short transport time is observed for all, doped and undoped samples, indicating that modulation doping does not modify carrier transport in the barriers. Heitz *et al.* have attributed such an ultrafast transport to a long-range attractive potential caused by the strain field surrounding the QDs,

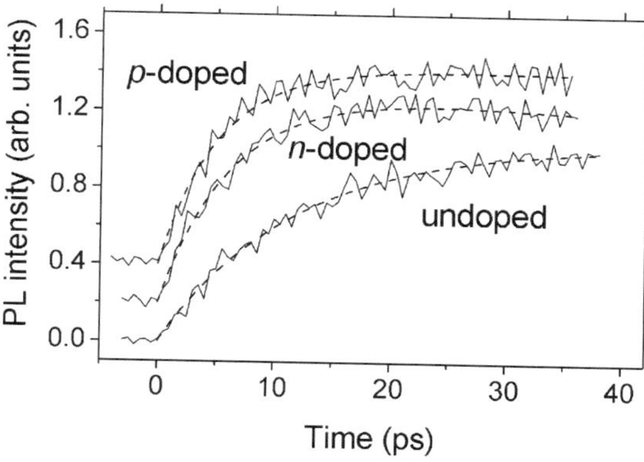

Fig. 5.8 Rising PL transients for the QD ground state transition ($e1$-$h1$) for different samples at high excitation power. Dashed lines indicate curve fits from which the PL rise times are extracted (From Siegert et al. [34])

thus, the transport is considered to proceed by drift rather than by diffusion [54]. An alternative mechanism, though not as fast as the drift, is the step-drive diffusion, observed in a GaN/AlGaN superlattice [55] and graded potential quantum well laser structures [56]. However, whichever the mechanism, the photoexcited carriers reach the QD region very rapidly, and transport in the barriers does not inhibit the overall carrier transfer.

5.4.3 Carrier Capture

Carrier transfer from states located in the vicinity of the dots to the ground QD state proceeds via carrier capture from the continuum states to the bound QD states and relaxation across the bound states. In theoretical calculations, these processes are often explored separately (Section 5.2). Experimentally, however, the capture and relaxation are usually measured together photoexciting carriers in the barriers or the WL and monitoring their arrival at the QD ground state (see Fig. 5.8). In specially designed experiments, though, the pure capture can be assessed. For instance, this can be done by monitoring carrier arrival at the highest QD states after excitation in the WL, or measuring the rate at which carriers leave the WL [57]. Sometimes, as will be discussed below, the experiments show very similar carrier arrival times for different QD states indicating that the relaxation is much shorter than the capture, which allows attributing the measured transfer times to the capture process.

In Fig. 5.9(a), the WL rise times account for the transport and capture into the WL and are 1.7, 1.8 and 2.2 ps for the p–, n-doped and undoped samples, respectively. The PL rise times for the QD transitions $e4$-$h4$ are 4.9, 5.4 and 6.1 ps. The difference between the WL and the $e4$-$h4$ transition rise times places the capture times between 3.1 and 3.9 ps. To measure the capture time unaffected by transport (but then including relaxation, since no signal for the highest QD transitions could

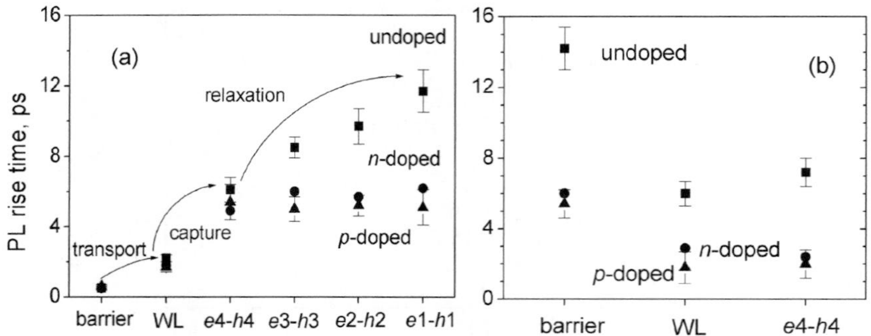

Fig. 5.9 Photoluminescence rise times of undoped (squares), n-doped (circles) and p-doped (triangles) samples measured at (a) different transition energies after excitation in the barriers at high excitation power; (b) ground state PL rise times for different excitation energies measured in the low excitation regime (From Siegert et al. [34])

be observed at low photoexcited carrier densities), carrier excitation directly into the WL was performed. In this case, the PL rise times for the ground QD transition are 2.4, 2.9 and 6.0 ps for the $p-$, n-doped and undoped samples, respectively (Fig. 5.8b). These times can be compared to a capture time <2ps measured for the transition between the WL and the QDs [58]. Carrier capture time of even shorter, subpicosecond duration has been measured in time-resolved PL experiments at low temperature in dense $In_{0.5}Ga_{0.5}As/GaAs$ QDs [59]. In this work excitation in the WL and probe in the first excited and ground levels provided the PL rise times of ~1 ps. Suggested cascade relaxation with the interlevel scattering time of 250 fs pushes the capture time to the 100 fs time scale.

The carrier capture was studied not only by the time-resolved PL, but also by the differential transmission. A capture time of 8 ps was measured by Sosnowski *et al.* at 10 K for the transition between the barrier and high-energy QD states in an undoped InAs QD structure [52]. Transport and capture times have been measured in $In_{0.4}Ga_{0.6}As/GaAs$ QDs with two confined electron levels by Urayama and coworkers [60]. Carrier excitation in the barriers at a low carrier density and probing at the excited level provided characteristic times varying from 2 to 2.5 ps in the temperature range 40–290 K. Differential two-color transmission measurements with the probe pulse tuned to the WL band gap revealed a capture time of 3 ps, independently of carrier density [61]. A capture time of the order of a picosecond has been confirmed by another pump-probe experiment in which the pump pulse created carriers in the barriers, and the IR probe pulse, tuned to the QD to WL transition, probed the occupation of the QD levels [46]. In InAs/InP QDs, a capture time of about 2 ps have been measured independently of the QD transition energy [62].

Carrier capture into QDs may proceed with the help of several mechanisms. As has been mentioned in Section 5.2.1, the capture time via emission of a resonant LO phonon at moderate carrier densities is below a picosecond; a two-phonon emission process is somewhat longer, in the range of 1 to 10 ps [21]. Experimentally, even carrier capture and relaxation with multiple phonon (4 LO + 1 LA) emission was found to be a fast process occurring in $\sim$10 ps [63]. Advantage of the phonon-assisted carrier capture, as compared to the relaxation, is that the initial state is in the continuum of either two dimensional states in the WL or three dimensional states in the barriers. Besides, the photoexcited carriers prior to their capture are typically hot, which relaxes the requirement for the initial and final state energy difference being in resonance with the energy of a single or multiple LO phonons. Capture via carrier-carrier scattering, on the other hand, at low carrier densities is slow. It strongly depends on the carrier density and reaches the rate of the capture via single LO phonon emission at a density of $\sim 10^{12}\,\mathrm{cm}^{-2}$ (see Fig. 5.4). At carrier densities of the order of 10^8 to $10^{10}\,\mathrm{cm}^{-2}$, this capture rate, compared to the phonon-assisted capture, is orders of magnitude lower. In the experiments described above, the PL rise times at the QD ground state energy do not experience any major dependence on the excitation power increasing from 11.7 to 14.2 ps for the undoped sample and staying at around 5 (6) ps for the $p-$ ($n-$) doped samples when the average excitation power density is decreased from 230 to 1 Wcm^{-2} ($8 \times 10^{12} - 4 \times 10^{10}$ carriers/cm^2). Such a behavior is characteristic for the phonon-assisted capture [22]. Considering peculiarities of the WL and QD level structure discussed in Section 5.3.1, the capture takes place with emission of one (two) GaAs-like LO phonons for the capture of a hole (electron) from the WL.

A feature that is often omitted in studies of carrier capture is the capture rate dependence on the dot density. Theoretical models usually consider capture into a single dot, thus, a possible dependence on the density is not taken into account. However, experiments show that the capture rate is different in the high and low QD density structures. To investigate this effect in detail, InGaAs QD structures in which in QD density changes by nearly two orders of magnitude while the dot size and the InGaAs alloy composition are similar have been used [64, 65, 66]. The QD structures were grown by using slight variations in GaAs (001) substrate miscut angle, from 0.0 to 2.0° towards (110). Miscut provides different concentrations of monolayer steps, which act as energetically favorable sites for island nucleation on the growth surface (Fig. 5.10). The dot densities for these QD structures range from 7×10^8 to $7.3 \times 10^9\,\mathrm{cm}^{-2}$. For comparison, high surface density ($2.4 \times 10^{10}\,\mathrm{cm}^{-2}$) randomly distributed QD sample was used. In all the samples, the dot diameters were 25 ± 5 nm. For the low density samples, the spectra show four well-defined QD-related peaks corresponding to transitions between different levels; with increased dot density, the peaks shift to the blue and energetic distances between them decrease until everything merges into one peak (Fig. 5.11). Such spectral changes have been attributed to modifications of the QD potentials by strain fields [65].

Carrier capture and relaxation in these QDs were evaluated through the PL rise times after excitation in the barriers. The PL rise times for the ground level transitions are shown in Fig. 5.12. For excited QD levels, the PL rise times agree with that of the ground state within 0.5–1.5 ps, thus, the PL rise times, at least at low excitation conditions where they reach 16 ps, can be interpreted as the capture times. With increasing excitation power and temperature, the rise times decrease down to an asymptotic value of 2–3 ps. The variation of the PL rise time is much

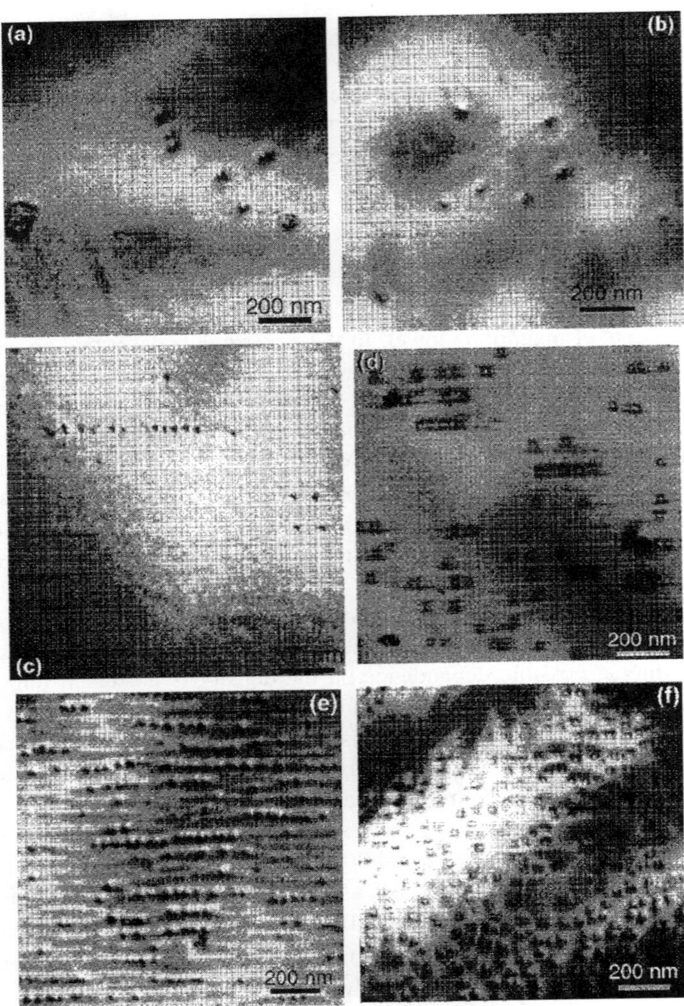

Fig. 5.10 Representative images of varying QD concentrations and spatial arrangements in InGaAs/GaAs QD structures obtained in simultaneous growth on different substrates. Substrate miscut angles from the (001) plane, degrees: (a) 0.25 ± 0.25, (b) 0.00 ± 0.25, (c) 0.75 ± 0.25, (d) 1.25 ± 0.25, (e) 2.00 ± 0.25, and (f) 0.00 ± 0.25 under growth conditions that give maximum island coverage (From Leon et al. [65])

more pronounced for the low QD density samples. The strong dependence of the PL rise times on the dot density is attributed to the carrier capture, since the vertical carrier transport and relaxation in otherwise identical QD structures should not be too different. A likely candidate hindering the capture in the low density dots is potential barriers at the QD interfaces occurring due to tensile strain induced by the lattice mismatch between GaAs and InGaAs. Such barriers have been suggested

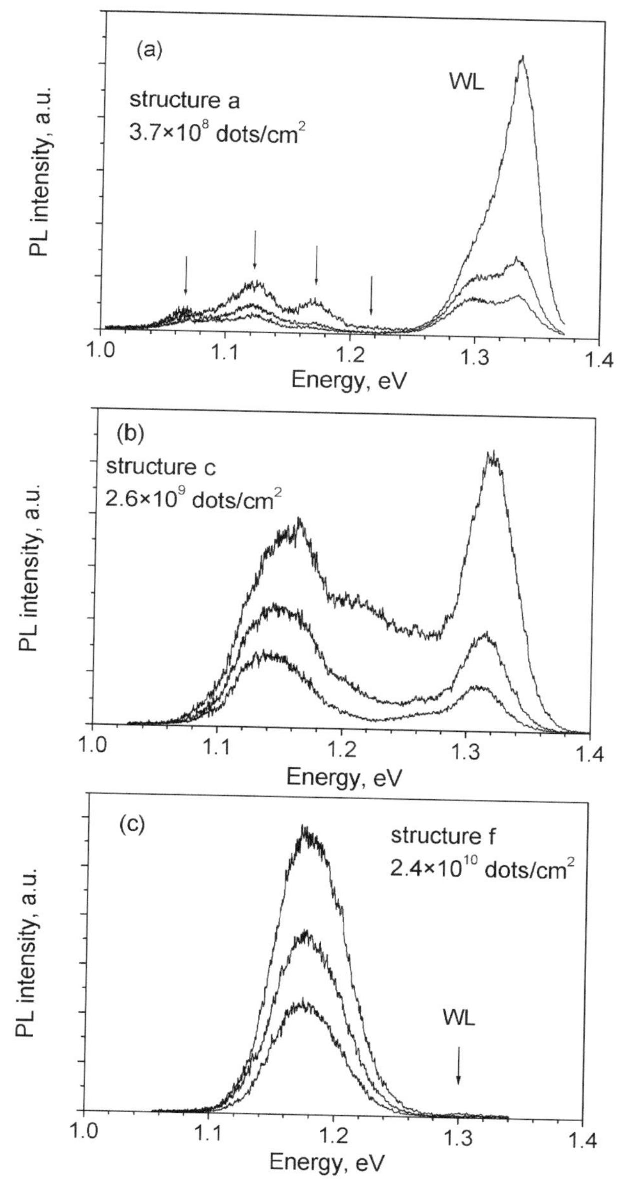

Fig. 5.11 PL spectra of InGaAs QD structures with different QD densities measured at 78 K and at 100, 840 and 1670 ps after the excitation (From MarcinkeviCius and Leon [66])

theoretically [67, 68] and observed by the deep level transient spectroscopy [69]. Recently, potential barriers at the QD interfaces have also been employed for explanation of PL data for InGaAs QDs in a quantum well [70]. The dot density, temperature and excitation power dependencies of the PL rise times are in accordance with

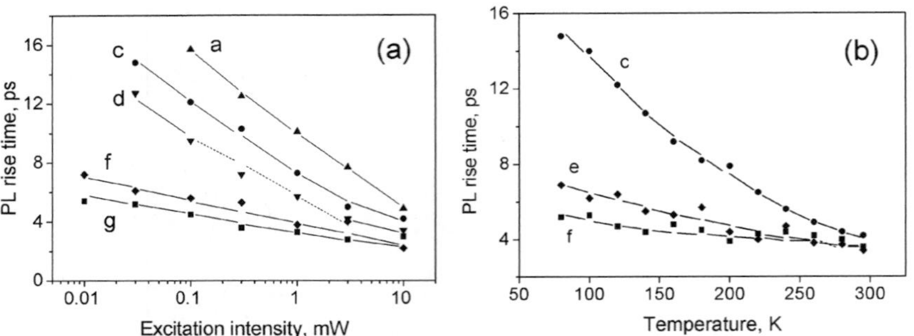

Fig. 5.12 Excitation intensity (a) and temperature (b) dependence of the 78 K PL rise times for the structures a, c-f (Following MarcinkeviCius and Leon [66])

the assumption that the capture is affected by the interfacial potential barriers. For the low density dots, the barriers should be well-defined. With increasing quantum dot density, the average inter-dot distance decreases, down to 10 nm for the highest density samples, making the potential barriers overlap, effectively diminishing their height for the individual QDs (Fig. 5.13). With increased temperature and excitation power, the capture times decrease (Fig. 5.12). The temperature dependence of the capture time allows estimating the height of these potential barriers since the rate of the carrier transfer over the barriers depends exponentially on the carrier temperature. From Fig. 5.12(b), the barrier height at the barrier/QD interface for the low QD density sample can be evaluated as 13 meV. This, however, should be considered as the lower limit of the barrier height since, before the capture; the carrier temperature should still be higher than that of the lattice (the electron-hole pairs are excited with an excess energy of 80 meV, and their temperature after 10 ps should still be higher than the lattice temperature by ~100 K [2]). It should be noted, though, that these results could also be explained by potential fluctuations in the WL, which would trap carriers and prevent them from being captured into the QDs.

Indirectly, the influence of potential barriers on the carrier capture has been confirmed by investigations of proton-irradiated QDs. The same low density InGaAs/GaAs QDs that have shown the prolonged capture times along with a few other structures were measured by continuous wave (cw) and time-resolved PL

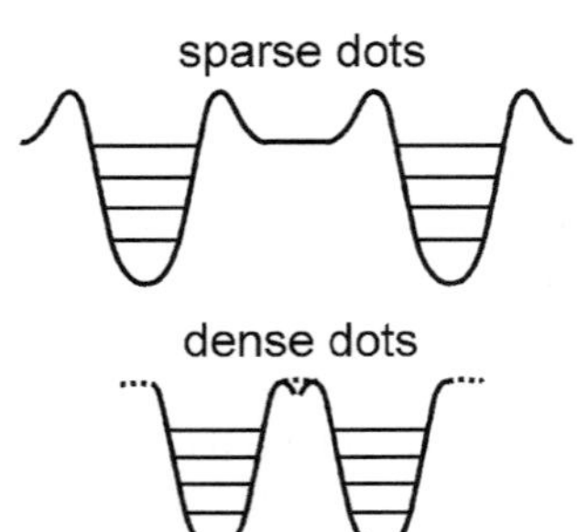

Fig. 5.13 Schematics of potential barriers in low and high density QDs

after irradiation [71, 72]. The samples were irradiated at room temperature using 1.5 MeV protons at five different doses ranging from 1.3×10^{11} to 3.5×10^{13} cm^{-2} with a dose rate of 6×10^{12} protons/sec. Figure 5.14(a) shows cw PL spectra for the as-grown QDs and dots irradiated at a low proton fluence. Figure 5.14(b) depicts dependence of the peak PL intensity measured by the time-resolved PL on the proton fluence. One can notice an increase in the QD PL intensity for the (100) InGaAs/GaAs QDs (most prominent, up to 50%, in the low-density dots) for small irradiation doses as compared to the nonirradiated QDs. In addition to PL and time-resolved PL experiments performed with excitation into the barriers, cw PL experiments were performed by varying the energy of the optical excitation. With 930 nm wavelength, carriers were excited in the WL and the QDs but not in the barriers. Direct carrier generation in the dots was achieved using 970 nm radiation. This experiment provides an answer whether the enhanced PL intensity in the irradiated samples is due to a more efficient carrier capture or relaxation. The measurements show that for 920 nm excitation a similar PL intensity increase as in the case of above-band gap excitation occurs. For the intradot excitation, the PL intensity for low to moderate irradiation fluences is irradiation-independent.

To understand the surprising enhancement of the PL intensity from the QD ground state after moderate irradiation, (which has subsequently been reported in Ref. [73]), one should consider several possible effects. The first possibility is an increase of volume from which the carriers are collected into the QDs. Such mechanism has been assumed in [73] on the grounds that proton irradiation may passivate native defects thus reducing nonradiative recombination in the barriers. However, such a mechanism would not explain the enhanced PL efficiency for excitation in the WL. Another explanation for the increased PL intensity could be a faster carrier relaxation through the defect states and an eventual reduction of the phonon bottleneck effect [31, 32] enhancing carrier concentration at the QD ground state. This

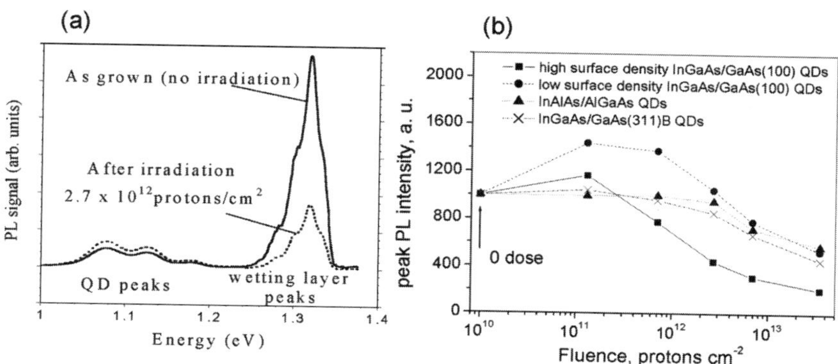

Fig. 5.14 (a) Comparison of initial (solid line) and post irradiation (dotted line) cw PL spectra at a proton dose $2.7 \times 10^{12}/\text{cm}^2$ of low density InGaAs/GaAs QDs (3.5×10^8 dots per cm^2). The spectra were obtained at constant excitation and show simultaneous emission form QD and wetting layer states. (b) Changes in PL peak intensities for different QD structures as a function of proton fluence (From Leon et al. [71])

mechanism, however, would show similar PL intensity enhancement for excitation in the barriers, the WL and the highest QD states, which contradicts the experimental data. This leaves a more efficient capture as the effect explaining the experimental observations. The more efficient carrier capture would occur if the potential barriers at the QD interfaces could be bypassed. This could happen if the carriers were trapped from the WL into irradiation-induced defect centers and then tunnel from these centers into the dots. For such a process, the defects should be located in the barriers/WL in the vicinity of the QDs or at the interfaces. Besides, the defects should have energy levels between the WL and the QD states. Studies of irradiation-induced defects in GaAs suggest that a possible candidate could be level E2, formed by an As vacancy and an interstitial [74]. The fact that we observe enhancement of the PL intensity in some, but not all types of QDs, like InGaAs/GaAs QDs grown on (311)B GaAs, suggests a resonant character of the effect. It is noteworthy that the experiments performed on the irradiated dots allow resolving the question raised above, whether it is the potential barriers or WL fluctuations that inhibit the carrier transfer in the low density dots. It is quite clear that for the enhanced carrier transfer from the WL to the QDs through the defect states, the hinder should be at the QD interfaces, which contradicts the WL potential fluctuation model and favors the interfacial barriers.

The described results fit into a rather coherent picture for the ultrafast carrier capture in InAs/GaAs QDs. In the high density QDs, the capture occurs in the time range from a subpicosecond to a few picoseconds. In low density dots, the capture is slower, which is attributed to the presence of potential barriers at the QD and WL interfaces. Relatively weak dependence of the capture time on temperature and carrier density induced either by doping or photoexcitation in the high QD density structures suggests LO phonon assisted capture as a rather universal capture mechanism. Hot electron distribution in the barriers or the WL relaxes the energy conservation requirements for such a phonon-assisted process.

5.4.4 Carrier Relaxation

In the early experimental research on carrier relaxation in QDs long relaxation times have been attributed to the phonon bottleneck effect, which implies that rapid relaxation of a hot carrier in a QD can occur only when the interlevel energy spacings are strictly in resonance with one or several LO phonons [75]. In case of carrier interaction with acoustic phonons, the relaxation is slow, typically of the order of a nanosecond [24]. However, most of the carrier relaxation experiments have failed to prove this prolonged relaxation. In fact, several experiments have shown carrier transfer times below a picosecond [76, 77]. This controversy has partially been resolved by realizing that the requirement for the strict match between the interlevel spacings and the phonon energies is, in fact, relaxed because of participation of energetically different barrier- QD- and interface-related phonons [45] as well as phonon broadening [29], which occurs due to strain inhomogeneities and anharmonic coupling between LO and acoustic phonons. In addition to the enhanced efficiency of the phonon-related relaxation, carrier-carrier scattering was suggested to play the crucial role in speeding-up the relaxation process [78]. Thus, the slow carrier relaxation in QDs because of the phonon bottleneck effect is rather an exception than a rule. It has been shown that it can be unambiguously observed only under special experimental conditions. Such an experiment has been designed by Uruyama and coworkers [79], who performed time-resolved differential transmission experiments at very low excitation powers, well below one electron-hole pair per dot. In such a situation, some of the dots capture both an electron and a hole and relaxation, occurring via carrier-carrier scattering, is fast, 7 ps. However, some dots capture only an electron or a hole and no electron-hole scattering can occur. In this case, the transmission change at the excited QD state transition energy has a long 750 ps time constant, which is attributed to the relaxation with acoustic phonon scattering and which is the experimental proof for the phonon bottleneck effect.

However, when there is more than a single carrier in a QD, the carrier relaxation typically proceeds on a time scale below 10 ps. The relaxation mechanism, however, is very much dependent on the properties of the QD and the experimental conditions. A number of mechanisms for the carrier relaxation have been theoretically proposed and experimentally observed. These include relaxation via carrier-carrier scattering, optical and acoustic phonon emission, relaxation through continuum states and relaxation through defect states.

We have studied carrier relaxation by carrier-carrier and carrier-LO phonon scattering by the time-resolved PL in modulation-doped and undoped InAs/GaAs QD structures described in Section 5.3.1. It is clear that for the doped dots the scattering would be influenced by the doping-induced carriers; for the undoped structures and low photoexcited carrier densities, the phonon-assisted relaxation may be important. The PL was measured at the energies of different QD transitions after excitation in the barriers. Besides, ground QD state PL was recorded after excitation into the WL and the excited QD states. The PL rise times are gathered in Figs. 5.9(a) and 5.9(b). For the doped structures, the PL rise times reflect minority carrier arrival to a particular level; for the undoped QDs, the PL signal is determined by photoexcited electron-hole pairs. Thus, for the n- and p-doped structures, hole and electron relaxation, respectively, is measured.

For the PL rise times measured for the different transitions, the major difference between the different samples occurs at the QD transition energies. For the undoped sample, the rise time increases with decreasing transition number (Fig. 5.9(a)). In the doped structures, the rise times are about equal for all the QD transitions and, furthermore, about twice as short as for the ground state of the undoped structure. Besides, the fast relaxation time for the doped structures is independent on the doping level. The rise time difference evidences different relaxation mechanisms in the doped and the undoped QDs. We discuss these differences considering carrier-carrier and carrier-LO phonon scattering. The faster relaxation in the doped samples indicates that carriers built-in by doping play an important role.

The carrier-carrier scattering may proceed along different lines. We can distinguish between the various relaxation paths on the grounds of the results obtained in excitation and detection wavelength-selective experiments. Before the scattering event, the carrier to be up-scattered may reside in the dot or in the WL (processes 1, 2 or 3, respectively, Fig. 5.15). The WL situation has been considered theoretically, and scattering times between 1 and 10 ps for moderate carrier densities were obtained [26]. Scattering by holes was found to be less efficient than scattering by electrons because of the larger change of the hole wave vector required for energy conservation. On the other hand, holes have much more final states to scatter to because of the higher level density determined by a larger effective mass [28].

In another case of the carrier-carrier scattering, both carriers prior to the scattering event are in the QD. Here, again, we have two different options with the final state of the up-scattered carrier being in the dot or in the barrier/WL (processes 1 or 2, Fig. 5.15). In the first case, the energy conservation requires matching between the interlevel energies in the conduction and valence bands, which, in a general case, is hardly probable. However, in the dots with $\Delta E_c / \Delta E_v = 2$ (Section 5.3.1), the energy conservation conditions can be met, especially taking into account energy level and phonon broadening.

Excitation into the WL and directly into the QDs allows distinguishing whether it is the carriers in the WL or in the QDs that play the main role in the carrier-carrier scattering. In the case of excitation into the WL, the carriers are present in this layer; for excitation at longer wavelengths, they are only in the QDs. Similar PL rise times for both excitation energies (Fig. 5.9(b)) point to the carriers built-in by doping as those responsible for the fast photoexcited carrier relaxation. The small relaxation time difference for low and high excitation powers when exciting into the barriers supports this interpretation.

The similar carrier scattering rates into all QD levels in the doped samples indicate that the carriers are simultaneously scattered into all QD levels and/or relaxation through the levels is extremely fast. Theoretical analysis of Auger capture has shown decreasing scattering rates with increasing exchange energy [18]. Calculations of Auger relaxation have not given such a clear dependence; the relaxation rate was rather found to decrease with increased excess energy of the up-scattered carrier in

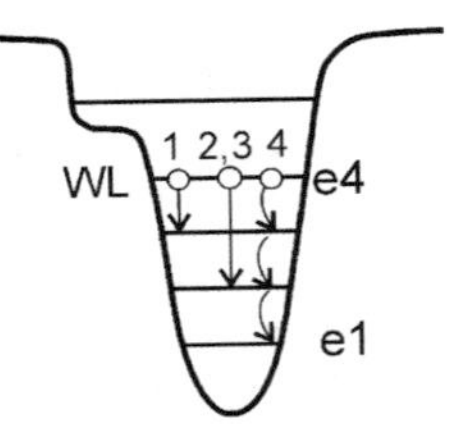

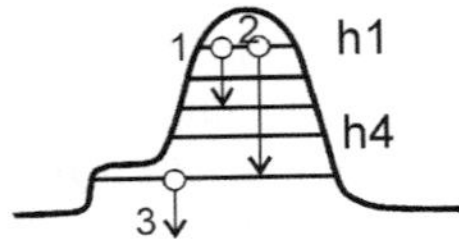

Fig. 5.15 Schematic energy level diagram of a quantum dot structure, illustrating different carrier relaxation processes

the WL [17]. Similar (and very short) scattering times for relaxation into different levels observed in our experiments require transitions in which the up-scattered carriers end up with small excess energies. Taking into account peculiarities of the QD structures, this requirement can by fulfilled in a number of different ways [34].

Ultrafast carrier capture and relaxation monitored by the time resolved PL at the QD ground state after excitation in the barriers was also reported by Gündoğdu and coauthors [77]. The characteristic carrier transfer times were 0.45 ps for the p-doped and 1.4 ps for the n-doped QD structures. These times were much shorter than the PL rise time for the undoped dots, 4.8 ps. The relaxation times in the doped structures have shown practically no temperature or photoexcited carrier density dependence, proving that carrier relaxation proceeds via scattering with the doping-induced carriers, and that the scattering rate is extremely high.

Carrier relaxation by phonon emission is also a viable relaxation mechanism in the doped structures, especially for holes for which, considering the level scheme, a single phonon emission is possible. Ultrafast hole relaxation in undoped samples has been observed in several works and was attributed to the close separation of the hole levels as well as availability of various energy-broadened phonons [46, 52]. Measurements performed on the doped and undoped samples allow distinguishing the most efficient relaxation mechanism for the holes in the n-doped QD structures. In case of a more efficient hole-LO phonon scattering, the undoped QDs would be filled with holes very soon after the capture, and the sample would resemble the p-doped structure. This would make the subsequent electron relaxation (and increase of the PL signal) equally fast in the p-doped and undoped samples. Our experiments show, however, a considerable difference in the PL rise times allowing to rule out subpicosecond hole relaxation in the undoped structure. Thus, in the doped structures, the carrier-carrier scattering-assisted relaxation is faster than the LO phonon-assisted, even for the interlevel energies nearly in resonance with the LO phonon energies. We have observed that the rate of this process basically does not depend on the number of carriers present in the dot. Experiments performed on samples with high and moderate doping provide very close relaxation times, indicating that, as soon as there are carriers confined in the QDs prior to scattering, the carrier-carrier scattering-assisted relaxation is the most efficient process.

Different PL rise times for different QD levels in the undoped sample indicate that carrier relaxation proceeds via cascade process (process 4, Fig. 5.15). At low photoexcited carrier densities, the relaxation time, i.e. the PL rise time measured after direct excitation into the QDs, is equal to 7 ps (Fig. 5.9(b)). A similar value, estimated as a difference between the highest-energy and the ground level PL rise times, is also obtained for the barrier excitation at high densities (Fig. 5.9(a)). The weak relaxation time dependence on the photoexcited carrier density suggests that for the undoped structure the prevailing carrier relaxation mechanism is emission of LO phonons. The model of relaxation by LO phonon emission is supported by the temperature dependence of the relaxation time, which at 180 K is 5.5 ps and at 80 K is 7 ps. This is consistent with the expected increase of the relaxation rate with increased temperature [63].

In general, the weak dependence of the relaxation time on the photoexcited carrier density is considered as an indication of phonon-assisted carrier relaxation. In the literature, relaxation involving various types of phonons have been discussed, including acoustic [23], optical, and a combination of those [63]. Interband pump and intraband probe experiments evidenced cascade nature of the relaxation with the interlevel relaxation time of 2.5 ps [46]. A study based on low temperature time-resolved PL experiments reported an electron relaxation time between the two adjacent levels equal to 1.5 ps [23]. In the case of electron interlevel energy of 90 meV (three InAs LO phonons), infrared absorption studies showed a somewhat longer, but still short, interlevel electron relaxation time of 3 ps [80]. Efficiency of LO phonon-assisted relaxation has also been confirmed by phonon resonances observed in PL excitation spectra of QDs [23, 54, 81].

The relevance of phonon-assisted relaxation has also been tested by time-resolved PL in undoped InGaAs/GaAs low density QDs with different interlevel energetic spacings [82]. In a number of QD structures subjected to rapid thermal annealing for 30 s between 700 and 900 °C, the QD energy levels were tuned by thermal compositional disordering of the interface [37]. This allowed gradually changing interband transition and intraband level energies. For samples annealed at different temperatures, the difference between QD PL peaks was between 26 and 51 meV (Fig 5.16), which, accepting the 2:1 interlevel energy distribution between the conduction and valence bands, provides electron interlevel energies varying between 17 and 34 meV, i.e. below and around InAs and GaAs LO phonon energies.

Figure 5.17(a) shows temporal transients for the QD ground state emission for several samples. These were measured at low excitation intensity before carrier-carrier relaxation is set into play. The carrier transfer in the 700 °C sample is faster,

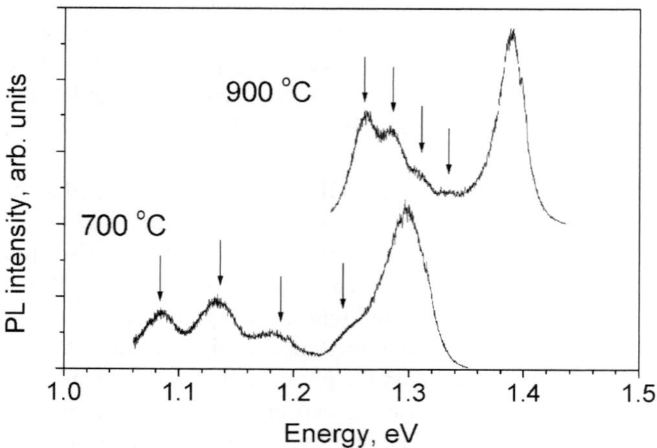

Fig. 5.16 Photoluminescence spectra measured 40 ps after the excitation for the QD samples annealed at 700 and 900 °C. The arrows mark peaks related to transitions in the QDs (From MarcinkeviCius et al. [82])

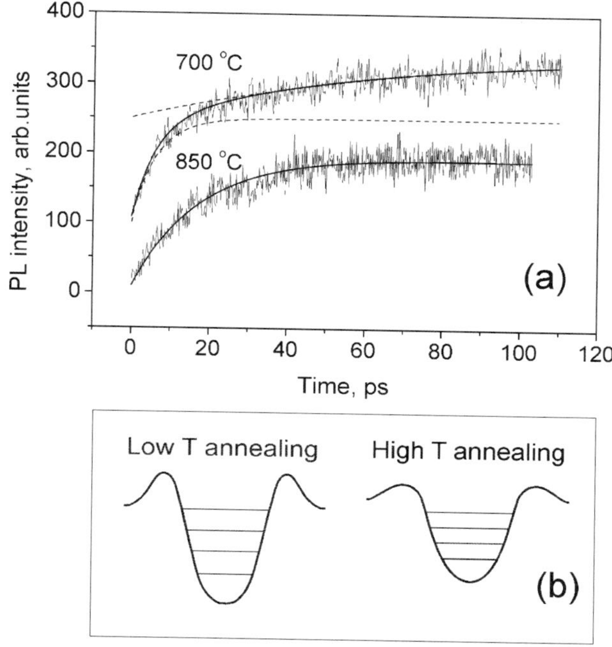

Fig. 5.17 (a) Photoluminescence transients and fits for the ground state transitions for the QD samples annealed at 700 and 850 °C. Two components of the fit for the 700 °C sample are shown as broken curves; (b) a drawing of the conduction band potential profiles for QDs intermixed at low and high temperatures (From MarcinkeviCius et al. [82])

besides, the PL transient for this sample exhibits a double-exponential rise in contrast to the samples annealed at higher temperatures. The PL transient fits provide a shorter rise time component (10–20 ps) and a longer time constant (50–100 ps). The weight of the longer component continuously decreases with increase of the annealing temperature. Transients measured for the samples annealed at 850 °C and higher show only a single-exponential rise. The vanishing long rise time component correlates with decreasing decay time for the WL PL: from 110 ps down to 24 ps for the samples annealed at 700 °C and 900 °C, respectively.

The shorter rise time component reflects a rapid transfer into the dots for the majority of carriers, the longer transient points to a sample-dependent hindrance, which slows down transfer into the dots for some of the excited carriers. Since the longer components are too long to have their origin in the carrier transport, one should consider the interfacial barriers, discussed above. These potential barriers are more efficient in blocking carrier transfer at low carrier densities because with increased photoexcitation intensity the mean carrier energy increases making the barriers easier to overcome. The potential barriers decrease with increased intermixing because the interfaces become more diffuse, see Fig. 5.17(b).This explains the diminishing weight of the long rise time component with increased annealing temperature. The majority of carriers, however, is captured into the dots while still hot and is, therefore, little affected by the barriers, which corresponds to the short component of the QD PL rise time.

The short components of the ground state PL rise times for all the samples measured at low excitation power densities are summarized in Fig. 5.18. For the lowest intensities, the rise times are intensity independent confirming that the carrier-carrier scattering regime is not reached. As can be seen from the figure, the low intensity PL rise times for the samples, annealed at 700 and 800 °C, are about half the values of that for the other samples. With fast carrier capture, which should not be different for the discussed structures, the difference in the PL rise times should be ascribed to the difference of carrier relaxation via phonon emission.

According to calculations of electron relaxation by LO phonon emission [29], the carrier relaxation time from the first excited state to the ground state increases with detuning from the LO phonon energy (solid line in Fig. 5.18). Direct comparisons of the PL rise times with the calculated values are complicated by inhomogeneous broadening of the QD structures and by the fact that the PL experiments with above barrier excitation do not allow distinguishing carrier capture from relaxation. Nevertheless, assuming that the carrier transport and capture account for the major part of the shortest QD PL rise time observed for the 700 and 800 °C samples (the WL PL rise times are from 7 to 8 ps), the measured data is in qualitative agreement with the calculations. For large detuning energies, reduced PL rise times with respect to the calculation results may be partially attributed to inhomogeneous broadening. Some of the dots would have energy spacings resonant with the phonon energies even for large average detuning, which would increase the response rate from the whole ensemble. Carrier relaxation time dependence in intermixed InAs/GaAs QDs has also been observed in Ref. [82]. In these

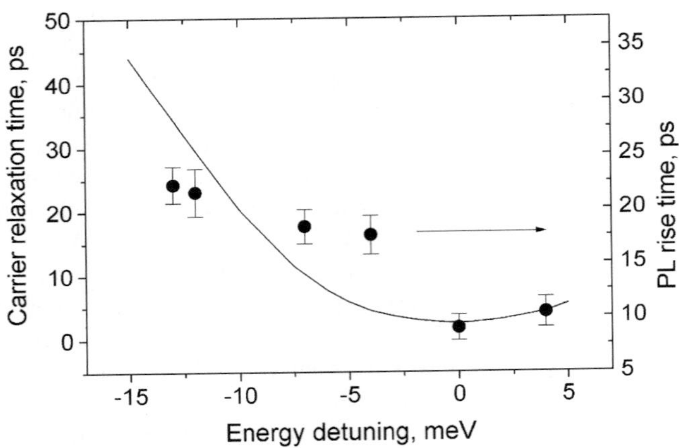

Fig. 5.18 Electron relaxation time from the first excited state to the ground state at 300 K as a function of detuning of interlevel spacing from the LO phonon energy after Ref. [28] (full line). Dots represent ground state PL rise times for the intermixed samples at low excitation intensity. Zero detuning is assumed for 30 meV interlevel energy separation (From MarcinkeviCius et al. [82])

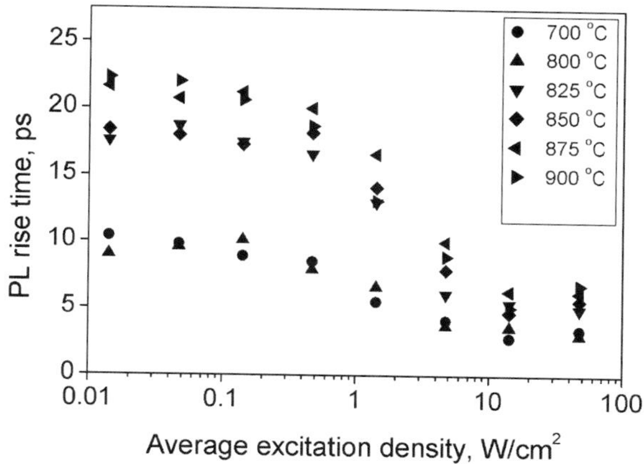

Fig. 5.19 PL rise times as a function of average photoexcitation density for QD samples annealed at different temperatures (From MarcinkeviCius et al. [82])

QDs the interlevel energies were larger, and multiphonon relaxation had to be considered.

With increased photoexcited carrier density, the PL rise times decrease and attain comparable values for all samples (Fig. 5.19, see also Fig. 5.12(a)). Similar excitation intensity dependence has been reported by several groups [22, 84–86] and has been attributed to an increased rate of carrier relaxation by the Auger-type process. High relaxation efficiency by carrier-carrier scattering at high carrier densities shadows the importance of the LO phonon-assisted process in the high photoexcitation density regime.

5.4.5 *Relaxation Through Continuum States*

In a number of works, efficient carrier relaxation through continuum states in the barrier and WL band tail states has been proposed [62, 87–91]. This suggestion has been grounded on PL excitation measurements that showed considerable PL excitation signal at wavelengths below the WL band gap. In other measurements [92], however, the PL excitation spectrum showed no such tails, leaving the problem somewhat open.

Toda *et al.* were the first to report WL continuum states in InGaAs/GaAs QD structures protruding about 80 meV into the WL band gap (can be compared to ~100 meV energy difference between the QD ground state and the WL transitions) and merging with the zero-dimensional QD states [87]. These continuum states were observed in single dot and ensemble PL excitation measurements and were suggested to have a universal character for self-assembled QDs. Even though the origin of these states was not proposed, they were suggested to play a major role in the carrier relaxation. According to this model, electrons first relax through the continuum with emission of LA phonons and then take the final leap into the

QD ground state with the aid of an LO phonon. WL continuum states in the QD transition region have also been reported in Ref. [88]. In an attempt to explain why in some cases the PL excitation continuum signal has and in others has not been observed, Kammerer *at al.* attributed the origin of the continuum states to the WL roughness and, hence, to the band gap fluctuations [89]. In a theoretical study, Vasanelli with coworkers suggested that the continuum of a PL excitation spectrum stems from indirect transitions between the barrier/WL states and discrete QD states and, therefore, is an intrinsic property of QD structures [90]. This idea has been further developed by Oulton *et al.* using PL excitation and photoconductivity measurements [91]. It was suggested that the interaction between the continuum states and the QDs could be controlled by, for instance, using QDs with a deeper confining potential or increasing WL potential by adding Al to the QD alloy.

Bogaart and coworkers explored the importance of such spatially-indirect transitions for the carrier relaxation [62] by performing two color pump-probe measurements on InAs/GaAs QDs with the pump wavelength tuned to transitions in the barrier and the probe wavelength corresponding to the ground and excited QD state transitions. The experimental data were interpreted by the following model: An electron first relaxes through the continuum states in the barriers and then, by emitting a single LO phonon, is captured into the QD (Fig. 5.20). The slower rise of the differential reflectivity signal at the excited state transition as compared to the ground state (25 vs. 12 ps) was presented as the main argument for such a model, besides, the temperature dependence for the ground state rise time was found to fit the model of relaxation via single LO phonon emission. It was suggested that efficiency of this relaxation channel directly depends on the coupling strength between

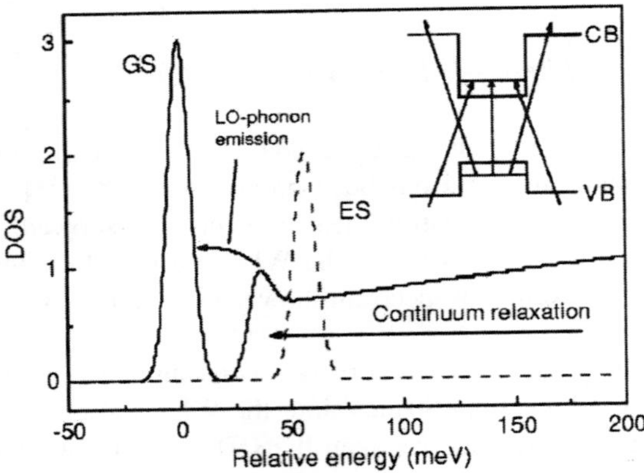

Fig. 5.20 Schematic representation of the QD density of states, showing the continuum background, which gives rise to a pathway for carrier relaxation. The LO-phonon replica is located near the onset of the continuum background indicating that the carrier relaxation through the continuum is followed by a single LO phonon emission. The inset depicts the indirect in real space transitions, which give rise to the continuum background (Reprinted figure with permission from [62]. Copyright (2005) by the American Physical Society)

the continuum and states and the discrete QD levels. This coupling is governed by Coulomb interaction and acoustic phonons and is stronger for the ground state than the first excited state because of the symmetry considerations.

One should note, however, that there is an alternative explanation to the slower signal rise for the excited state compared to the ground state transitions. Previously, such an observation has been interpreted by faster carrier down-scattering from the excited state as compared to the carrier supply by capture from the WL [85]. Carrier relaxation faster than capture has also been observed in QD optical amplifiers, as discussed below.

5.4.6 Ultrafast Carrier Relaxation in Quantum Dot Devices

Carrier relaxation has also been studied in QD devices, more precisely, QD optical amplifiers. Such investigations, often performed under operating conditions, are important for understanding intrinsic limitations of the device high-speed performance. In an amplifier under bias, the carrier dynamics can be tested under gain, as opposed to absorptive, conditions. Carrier relaxation in QD amplifiers is evaluated from measurements of the gain dynamics using pump-probe and four-wave-mixing techniques.

Borri *et al.* explored spectral hole burning in an InAs/InGaAs/GaAs QD amplifier using heterodyne pump-probe experiment with carrier excitation in excited and the ground QD states [93]. In the case of pump and probe wavelengths tuned to the excited state transition, the transmission change showed a double-exponential decay allowing to extract electron and hole interlevel relaxation times equal to 1.6 and 0.3 ps, respectively. In Ref. [94] it was shown that under electrical pumping and gain conditions the spectral hole at the ground and excited states recovers in less than 200 fs, showing a dramatic reduction of the interlevel carrier relaxation under the high QD population conditions. The carrier relaxation was found to be faster than the carrier capture [95, 96]; 90 fs relaxation and 2 ps capture times were measured by pump-probe and four-wave-mixing spectroscopy in InAs QD amplifiers [96]. Investigations of 1.3 μm QD amplifiers showed that even LO phonon-assisted relaxation under gain conditions occurs on a subpicosecond scale [97]. A complicated level population dynamics in InAs QD amplifiers has been explored both theoretically and experimentally by van der Poel and coworkers [98]. It has been found that level population is determined by multiple processes involving emission and absorption of phonons, multiphonon interactions and interaction with the carriers and phonons of the WL.

Summarizing one can state that carrier relaxation in QDs is a fast process with the phonon bottleneck effect manifesting only in special cases. Carrier relaxation can proceed via several different mechanisms. Interaction with spectrally-broadened LO phonons leads to a fast relaxation, which is clearly evidenced in numerous experiments performed at low carrier densities. At high carrier densities or in doped dots, the carrier-carrier scattering typically takes over as the most efficient process with relaxation times approaching subpicosecond durations. Carrier relaxation through the barrier or wetting layer states and spatially-indirect capture directly into the ground QD state has been suggested as an alternative to the intradot relaxation.

5.5 Conclusion

In this chapter, ultrafast dynamics of carrier transfer into InAs/GaAs and InGaAs/GaAs QDs have been discussed. The transfer process includes carrier transport in the barriers, capture into the dots and relaxation in the dots. Considered capture mechanisms include capture with LO phonon emission and via carrier-carrier scattering, and capture through defect states. The described results provide a rather coherent picture for the carrier capture in the QDs. For high density dots, the capture occurs in the time range from a subpicosecond to a few picoseconds. In low QD density structures, the carrier capture is slower, $\sim$10 ps, which is attributed to the presence of potential barriers at the QD and WL interfaces. Relatively weak dependence of the capture time on temperature and carrier density suggests LO phonon-assisted capture as a rather universal capture mechanism. Hot electron distribution in the barriers and the WL relaxes the energy conservation requirements for such a phonon-assisted process. Carrier relaxation to the ground state of the dots may also proceed via several mechanisms, such as carrier-carrier and carrier-phonon scattering as well as relaxation through the barrier/wetting layer continuum states. In doped QDs and/or at high carrier densities, the carrier-carrier scattering-assisted relaxation dominates. In undoped dots and low excitation levels, relaxation through LO phonon emission is the relevant process. Participation of energetically-different barrier- and QD-related phonons as well as spectral broadening of the phonon energies ease the strict energy conservation requirements for the scattering process assuring rapid, 1 to 10 ps, relaxation. Whatever the mechanisms, the carrier transfer in QD structures is, as a rule, fast, occurring in 1 to 20 ps time range.

Acknowledgments I would like to thank all my collaborators who have contributed to the results presented in this overview, and especially Rosa Leon, Jörg Siegert and Qing Xiang Zhao. Financial support from Swedish Foundation for International Cooperation in Research and Higher Education (STINT) and Swedish Foundation for Strategic Research (SSF) is gratefully acknowledged.

References

1. Shah J (ed) (1992) Hot carriers in semiconductor nanostructures: Physics and applications. Academic Press, Boston
2. Shah J (1999) Ultrafast spectroscopy of semiconductors and semiconductor nanostructures. Springer, Berlin
3. Dekel E, Regelman DV, Gershoni D, Ehrenfreund E, Schoenfeld WV Petroff PM (2000) Cascade evolution and radiative recombination of quantum dot multiexcitons studied by time-resolved spectroscopy. Phys Rev B 62:11038–11045

4. Finley JJ, Lemaître A, Ashmore AD, Mowbray DJ, Skolnick MS, Hopkinson M, Krauss TF (2001) Excitation and relaxation mechanisms in single In(Ga)As quantum dots. Phys Stat Sol (b) 224:373–378
5. Kono S, Kirihara A, Tomita A, Nakamura K, Fujikata J, Ohashi K, Saito H, Nishii K (2005) Excitonic molecule in a quantum dot: Photoluminescence lifetime of a single InAs/GaAs quantum dot. Phys Rev B 72:155307
6. Moskalenko ES, Larsson S, Schoenfeld WV, Petroff PM, Holtz PO (2006) Carrier transport in self-organized InAs/GaAs quantum-dot structures studied by single-dot spectroscopy. Phys Rev B 73:155336
7. Unold T, Mueller K, Lienau C, Elsaesser T (2004) Space and time resolved coherent optical spectroscopy of single quantum dots. Semicond Sci Technol 19:S260–S263
8. Li X, Wu Y, Xu X, Steel DG, Gammon D (2006) Transient nonlinear optical spectroscopy studies involving biexciton coherence in single quantum dots. Phys Rev B 73:153304
9. Jacak L, Hawrylak P, Wójs A (1998) Quantum dots. Springer, Berlin
10. Bimberg D, Grundmann M, Ledentsov NN (1999) Quantum dot heterostructures. John Wiley and Sons, Chichester
11. Mukai K, Sugawara M (1999) The phonon bottleneck effect in quantum dots. In: Sugawara M (ed) Self-assembled InGaAs/GaAs quantum dots (Semiconductors and Semimetals vol 60) Academic Press, Boston, pp 209–240
12. Borri P (2002) Ultrafast optical properties of quantum dot amplifiers. In: Grundmann M (ed) Nano-optoelectronics: concepts, physics and devices. Springer, Berlin, pp 411–430
13. Ignatiev IV, Kozin IE (2002) Dynamics of carrier relaxation in self-assembled quantum dots. In: Masumoto Y, Takagahara T (eds) Semiconductor quantum dots: physics, spectroscopy and applications. Springer, Berlin, pp 245–293
14. Bockelmann U (2005) Carrier relaxation in nanostructures and its implication for optical properties. In: Bryant G, Solomon GS (eds) Optics of quantum dots and wires. Artech House, Boston, pp 35–55
15. Lipsanen H, Sopanen M, Tullki J (2005) Optical properties of quantum dots induced by self-assembled stressors. In: Bryant G, Solomon GS (eds) Optics of quantum dots and wires. Artech House, Boston, pp 97–131
16. Pan JL (1994) Intraband Auger processes and simple models of the ionization balance in semiconductor quantum-dot lasers. Phys Rev B 49:11272–11287
17. Ferreira R, Bastard G (1999) Phonon-assisted capture and intradot Auger relaxation in quantum dots. Appl Phys Lett 74:2818–2820
18. Magnusdottir I, Bischoff S, Uskov AV, Mørk J (2002) Geometry dependence of Auger carrier capture rates into cone-shaped self-assembled quantum dots. Phys Rev B 67:205326
19. Uskov AV, McInerney J, Adler F, Schweizer H, Pilkuhn MH (1998) Auger carrier capture kinetics in self-assembled quantum dot structures. Appl Phys Lett 72:58–60
20. Zhang J-Z, Galbraith I (2006) Rapid hot-electron capture in self-assembled quantum dots via phonon processes. Appl Phys Lett 89:153119
21. Magnusdottir I, Uskov AV, Bischoff S, Tromborg B, Mørk J (2002) One- and two-phonon capture processes in quantum dots. J Appl Phys 92:5982–5990
22. Ohnesorge B, Albrecht M, Oshinowo J, Forchel A, Arakawa Y (1996) Rapid carrier relaxation in self-assembled $In_xGa_{1-x}As$/GaAs quantum dots. Phys Rev B 54:11532–11538
23. Ignatiev IV, Kozin IE, Davydov VG, Nair SV, Lee J-S, Ren H-W, Sugou S, Masumoto Y (2001) Phonon resonances in photoluminescence spectra of self-assembled quantum dots in an electric field. Phys Rev B 63:075316
24. Bockelmann U, Bastard G (1990) Phonon scattering and energy relaxation in two-, one-, and zero-dimensional electron gases. Phys Rev B 42:8947–8951
25. Benisty H, Sotomayor-Torrès CM, Weisbuch C (1991) Intrinsic mechanism for the poor luminescence of quantum-box systems. Phys Rev B 44:10945–10948
26. Uskov AV, Adler F, Schweizer H, Pilkuhn MH (1997) Auger carrier relaxation in self-assembled quantum dots by collisions with two-dimensional carriers. J Appl Phys 81:7895–7899

27. Nielsen TR, Gartner P, Jahnke F (2004) Many-body theory of carrier capture and relaxation in semiconductor quantum-dot lasers. Phys Rev B 69:235314
28. Narvaez GA, Bester G, Zunger A (2006) Carrier relaxation mechanisms in self-assembled (In,Ga)As/GaAs quantum dots: Efficient $P \rightarrow S$ Auger relaxation of electrons. Phys. Rev. B 74:075403
29. Li X-Q, Nakayama H, Arakawa Y (1999) Phonon bottleneck in quantum dots: Role of lifetime of the confined optical phonons. Phys Rev B 59:506915
30. Verzelen O, Bastard G, Ferreira R (2002) Energy relaxation in quantum dots. Phys Rev B 66:081308(R)
31. Sercel PC (1995) Multiphonon-assisted tunneling through deep levels: A rapid energy-relaxation mechanism in nonideal quantum-dot heterostructures. Phys Rev B 51:14532–14541
32. Schroeter DF, Griffits DJ, Sercel PC (1996) Defect-assisted relaxation in quantum dots at low temperature. Phys Rev B 54:1486:1489
33. Sercel P, Efros AL, Rosen M (1999) Intrinsic gap states in semiconductor nanocrystals. Phys Rev Lett 83:2394–2397
34. Siegert J, Marcinkevičius S, Zhao QX (2005) Carrier dynamics in modulation-doped InAs/GaAs quantum dots. Phys. Rev. B 72:085316
35. Leon R, Kim Y, Jagadish C, Gal M, Zou J, Cockayne DJH (1996) Effects of interdiffusion on the luminescence of InGaAs/GaAs quantum dots. Appl Phys Lett 69:1888–1890
36. Leon R, Lobo C, Zou J, Romeo T, Cockayne DJH (1998) Stable and metastable InGaAs/GaAs island shapes and surfactantlike suppression of the wetting transformation. Phys Rev Lett 81:2486–2489
37. Leon R, Lobo C, Clark A, Bozek R, Wysmolek A, Kurpiewski A, Kaminska M (1998) Different paths to tunability in III–V quantum dots. J Appl Phys 84:248–254
38. Itskevich IE, Skolnick MS, Mowbray DJ, Trojan IA, Lyapin SG, Wilson LR, Steer MJ, Hopkinson M, Eaves L, Main PC (1999) Excited states and selection rules in self-assembled InAs/GaAs quantum dots. Phys Rev B 60:R2185–R2188
39. Williamson AJ, Wang LW, Zunger A (2000) Theoretical interpretation of the experimental electronic structure of lens-shaped self-assembled InAs/GaAs quantum dots. Phys Rev B 62:12963–12977
40. Schmidt KH, Medeiros-Ribeiro G, Oestreich M, Petroff PM, Döhler GH (1996) Carrier relaxation and electronic structure in InAs self-assembled quantum dots. Phys Rev B 54:11346–11353
41. Warburton RJ, Dürr CS, Karrai K, Kotthaus JP, Medeiros-Ribeiro G, Petroff PM (1997) Charged excitons in self-assembled semiconductor quantum dots. Phys Rev Lett 79:5282–5285
42. Chang W-H, Hsu TM, Yeh NT, Chyi J-I (2000) Electron distribution and level occupation in an ensemble of $In_xGa_{1-x}As$/GaAs self-assembled quantum dots. Phys Rev B 62: 13040–13047
43. Chang W-H, Chen WY, Hsu TM, Yeah N-T, Chyi J-I (2002) Hole emission processes in InAs/GaAs self-assembled quantum dots. Phys Rev B 66:195337
44. Chu L, Zrenner A, Böhm G, Abstreiter G (2000) Lateral intersubband photocurrent spectroscopy on InAs/GaAs quantum dots. Appl Phys Lett 76:1944–1946
45. Raymond S, Fafard S, Poole PJ, Wojs A, Hawrylak P, Charbonneau S, Leonard D, Leon R, Petroff PM, Merz JL (1996) State filling and time-resolved photoluminescence of excited states in $In_xGa_{1-x}As$/GaAs quantum dots. Phys Rev B 54:11548–11554
46. Müller T, Schrey FF, Strasser G, Unterrainer K (2003) Ultrafast intraband spectroscopy of electron capture and relaxation in InAs/GaAs quantum dots. Appl Phys Lett 83: 3572–3574
47. Haacke S, Taylor RA, Bar-Joseph I, Brasil MJSP, Hartig M, Deveaud B (1998) Improving the signal-to-noise ratio of femtosecond luminescence upconversion by multichannel detection. J Opt Soc Am B 15:1410–1417
48. Oh I-K, Singh J, Thilagam A, Vengurlekar AS (2000) Exciton formation assisted by LO phonons in quantum wells. Phys Rev B 62:2045–2050 and references therein

49. Dawson P, Rubel O, Baranovskii SD, Pierz K, Thomas P, Göbel EO (2005) Temperature-dependent optical properties of InAs/GaAs quantum dots: Independent carrier versus exciton relaxation. Phys Rev B 72:235301
50. Quochi F, Dinu M, Bonadeo NH, Shah J, Pfeiffer LN, West KW, Platzman PM (2002) Ultrafast carrier dynamics in resonantly excited 1.3-μm InAs/GaAs self-assembled quantum dots. Physica B 314:263–267
51. Le Ru EC, Fack J, Murray R (2003) Temperature and excitation density dependence of the photoluminescence from annealed InAs/GaAs quantum dots. Phys Rev B 67: 245318
52. Sosnovski TS, Norris TB, Jiang H, Singh J, Kamath K, Bhattacharya P (1998) Rapid carrier relaxation in $In_{0.4}Ga_{0.6}As$/GaAs quantum dots characterized by differential transmission spectroscopy. Phys Rev B 57:R9423–R9426
53. Shchekin OB, Deppe DG, Lu D (2001) Fermi-level effect on the interdiffusion of InAs and InGaAs quantum dots. Appl Phys Lett 78:3115–3117
54. Heitz R, Veit M, Ledentsov NN, Hoffmann A, Bimberg D, Ustinov VM, Kop'ev PS, Alferov ZhI (1997) Energy relaxation by multiphonon processes in InAs/GaAs quantum dots. Phys Rev B 56:10435–10445
55. Deveaud B, Shah J, Damen TC, Lambert B, Chomette A, Regreny A (1988) Optical studies of perpendicular transport in semiconductor superlattices. IEEE J. Quantum Electron 24:1641–1651
56. Marcinkevičius S, Olin U, Wallin J, Streubel K, Landgren G (1994) Photoexcited carrier transport in InGaAsP/InP quantum well laser structure. Appl Phys Lett 65:2057–2059
57. Raymond S, Hinzer K, Fafard S Merz JL (2000) Experimental determination of Auger capture coefficients in self-assembled quantum dots. Phys Rev B 61:R16331–R16334
58. Sun KW, Chen JW, Lee BC, Lee CP, Kechiantz AM (2005) Carrier capture and relaxation in InAs quantum dots. Nanotechnology 16:1530–1535
59. Zhang L, Boggess TF, Gundogdu K, Flatté ME, Deppe DG, Cao C, Shchekin OB (2001) Excited-state dynamics and carrier capture in InGaAs/GaAs quantum dots. Appl Phys Lett 79:3320–3322
60. Urayama J, Norris TB, Jiang H, Singh J, Bhattacharya P (2002) Temperature-dependent carrier dynamics in self-assembled InGaAs quantum dots. Appl Phys Lett 80:2162–2164
61. Wesseli M, Ruppert C, Trumm S, Krenner HJ, Finley JJ, Betz M (2006) Nonequilibrium carrier dynamics in self-assembled InGaAs quantum dots. Phys Stat Sol (b) 243:2217–2233
62. Bogaart EW, Haverkort JEM, Mano T, van Lippen T, Nötzel R, Wolter JH (2005) Role of continuum background for carrier relaxation in InAs quantum dots. Phys Rev B 72:195301
63. Feldmann J, Cundiff ST, Arzberger M, Böhm G, Abstreiter G (2001) Carrier capture into InAs/GaAs quantum dots via multiple optical phonon emission. J Appl Phys 89:1180–1183
64. Lobo C, Leon R, Marcinkevičius S, Yang W, Sercel PC, Liao XZ, Zou J, Cockayne DJH (1999) Inhibited carrier transfer in ensembles of isolated quantum dots. Phys Rev B 60:16647–16651
65. Leon R, Marcinkevicius S, Liao XZ, Zou J, Cockayne DJH, Fafard S (1999) Ensemble interactions in strained semiconductor quantum dots. Phys Rev B 60:R8517–R8520
66. Marcinkevičius S, Leon R (2000) Photoexcited carrier transfer in InGaAs quantum dot structures: Dependence on the dot density. Appl Phys Lett 76:2406–2408
67. Grundmann M, Ledentsov NN, Heitz R, Eckey L, Christen J, Böhrer J, Bimberg D, Ruvimov SS, Werner P, Richter U, Heydenreich J, Ustinov VM, Egorov AYu, Zhukov AE, Kopev PS, Alferov ZhI (1995) InAs/GaAs quantum dots radiative recombination from zero-dimensional states. Phys Stat Sol (b) 188:249–258
68. Williamson AJ, Zunger A (1999) InAs quantum dots: Predicted electronic structure of free-standing versus GaAs-embedded structues. Phys Rev B 59:15819–15824
69. Wang HL, Yang FH, Feng SL, Zhu HJ, Ning D, Wang H, Wang XD, (2000) Experimental determination of local strain effect on InAs/GaAs self-organized quantum dots. Phys Rev B 61:5530–5534
70. Popescu DP, Eliseev PG, Stintz A, Malloy KJ (2004) Temperature dependence of the photoluminescence emission from InAs quantum dots in a strained $Ga_{0.85}$ $In_{0.15}$ As quantum well. Semicond Sci Technol 19:33–38

71. Leon R, Marcinkevicius S, Siegert J, CechaviCius B, Magness B, Taylor WA, Lobo C (2002) Effects of proton irradiation on luminescence emission and carrier dynamics of self-assembled III-V quantum dots. IEEE Trans Nuclear Sci 49:2844–2851
72. Marcinkevičius S, Siegert J, Leon R, CechaviCius B, Magness B, Taylor W, Lobo C (2002) Changes in luminescence intensities and carrier dynamics induced by proton irradiation in InGaAs/GaAs quantum dots. Phys Rev B 66:235314
73. Ji Y, Chen G, Tang N, Wang Q, Wang XG, Shao J, Chen XS, Lu W (2003) Proton-implantation-induced photoluminescence enhancement in self-assembled InAs/GaAs quantum dots. Appl Phys Lett 62:2802–2804
74. Ferrini R, Galli M, Guizzetti G, Patrini M, Nava F, Canali C, Vanni P (1997) Optical evaluation of the ionized EL2 fraction in proton (24 GeV) irradiated semi-insulating GaAs. Appl Phys Lett 71:3084–3086
75. Inoshita T, Sakaki H (1992) Electron relaxation in a quantum dot: Significance of multiphonon processes. Phys Rev B 46:7260–7263
76. Marcinkevičius S, Leon R (1997) Carrier dynamics in InGaAs/GaAs quantum dots. Phys Stat Sol (b) 204:290–292
77. Gündoğdu K, Hall KC, Boggess TF, Deppe DG, Shchekin OB (2004) Ultrafast electron capture into p-modulation-doped quantum dots. Appl Phys Lett 85:4570–4572
78. Bockelmann U, Egeler T (1992) Electron relaxation in quantum dots by means of Auger scattering. Phys Rev B 46:15574–15577
79. Urayama J, Norris TB, Singh J, Bhattacharya P (2001) Observation of phonon bottleneck in quantum dot electronic relaxation. Phys Rev Lett 86:4930–4933
80. Sauvage S, Boucaud P, Glotin F, Prazeres R, Ortega J-M, Lemaître A, Gérard J-M, Thierry-Flieg V (1998) Saturation of intraband absorption and electron relaxation time in n-doped InAs/GaAs self-assembled quantum dots. Appl Phys Lett 73:3818–3820
81. Steer MJ, Mowbray DJ, Tribe WR, Skolnick MS, Sturge MD, Hopkinson M, Cullis AG, Whitehouse CR, Murray R (1996) Electronic energy levels and energy relaxation mechanisms in self-organized InAs/GaAs quantum dots. Phys Rev B 54:17738–17744
82. MarcinkeviCius S, Gaarder A, Leon R (2001) Rapid carrier relaxation by phonon emission in InGaAs/GaAs quantum dots. Phys Rev B 64:115307
83. Malik S, Le Ru EC, Childs D, Murray R (2001) Time-resolved studies of annealed InAs/GaAs self-assembled quantum dots. Phys Rev B 63:155313
84. MarcinkeviCius S and Leon R (1999) Carrier capture and escape in $In_xGa_{1-x}As$/GaAs quantum dots: Effects of intermixing. Phys Rev B 59:4630–4633
85. Yuan ZL, Foo ERAD, Ryan JF, Mowbray DJ, Skolnick MS, Hopkinson M (1999) Many-body effects in carrier capture and energy relaxation in self-organized InAs/GaAs quantum dots. Physica B 272:12–14
86. Morris D, Perret N, Fafard S (1999) Carrier energy relaxation by means of Auger processes in InAs/GaAs self-assembled quantum dots. Appl Phys Lett 75:3593–3595
87. Toda Y, Moriwaki O, Nishioka M, Arakawa Y (1999) Efficient carrier relaxation mechanism in InGaAs/GaAs self-assembled quantum dots based on the existence of continuum states. Phys Rev Lett 82:4114–4117
88. Finley JJ, Ashmore AD, Lemaître A, Mowbray DJ, Skolnick MS, Itskevich IE, Maksym PA, Hopkinson M, Krauss TF (2005) Charged and neutral exciton complexes in individual self-assembled In(Ga)As quantum dots. Phys Rev B 63:073307
89. Kammerer C, Cassabois G, Voisin C, Delalande C, Roussignol Ph, Gérard J-M (2001) Photoluminescence up-conversion in single self-assembled InAs/GaAs quantum dots. Phys Rev Lett 87:207401
90. Vasanelli A, Ferreira R, Bastard G (2002) Continuous absorption background and decoherence in quantum dots. Phys Rev Lett 89:216804
91. Oulton R, Finley JJ, Tartakovskii AI, Mowbray DJ, Skolnick MS, Hopkinson M, Vasanelli A, Ferreira R, Bastard G (2003) Continuum transitions and phonon coupling in single self-assembled Stranski-Krastanow quantum dots. Phys Rev B 68:235301

92. Monte AFG, Finley JJ, Ashmore AD, Fox AM, Mowbray DJ, Skolnick MS, Hopkinson M (2003) Carrier dynamics in short wavelength self-assembled InAs/$Al_{0.6}Ga_{0.4}As$ quantum dots with indirect barriers. J Appl Phys 93:3524–3528
93. Borri P, Langbein W, Hwam JM, Heinrichsdorff F, Mao M-H, Bimberg D (2000) Spectral hole-burning and carrier-heating dynamics in InGaAs quantum-dot amplifiers. IEEE J. Selected Topics Quantum Electron. 6:544–551
94. Schneider S, Borri P, Langbein W, Woggon U, Sellin RL, Ouyang D, Bimberg D (2005) Excited-State Gain Dynamics in InGaAsQuantum-Dot Amplifiers. IEEE Photon Technology Lett 17:2014–2016
95. Akiyama T, Kuwatsuka H, Simoyama T, Nakata Y, Mukai K, Ishikawa H (2001) Ultrafast nonlinear processes in quantum-dot optical amplifiers. Opt. Quantum Electron. 33:927–938
96. Berg TW, Bishoff S, Magnusdottir I, Mørk J (2001) Ultrafast gain recovery and modulation limitations in self-assembled quantum-dot devices. IEEE Photon Technology Lett 13:541–543
97. Van der Poel M, Gehrig E, Hess O, Birkedal D, Hvam JM (2005) Ultrafast gain dynamics in quantum-dot amplifiers: theoretical analysis and experimental investigations. IEEE J. Quantum Electron. 41:1115–1123
98. Van der Poel M, Mørk J, Somers A, Forchel A, Reithmaier JP, Eisenstein G (2006) Ultrafast gain and index dynamics of quantum dash structures emitting at 1.55 μm. Appl. Phys. Lett. 89:81102

Chapter 6
Spin Phenomena in Self-assembled Quantum Dots

Alexander Tartakovskii

6.1 Introduction

6.1.1 Spin in Quantum Dots: A Building Block for New Generation Electronics

The precise coherent manipulation of quantum mechanical degrees of freedom in nanostructured solids is currently one of the most fascinating and highly sought after goals in condensed matter science. This ability is not only of strong fundamental interest but is likely to lead to a breakthrough in information technologies that exploit the quantum mechanical nature of nanostructured materials. In particular, the spin of isolated charges (electrons or holes) trapped in nanometer sized semiconductor quantum dots (QDs) has recently emerged as one of the most promising solid-state systems for the implementation of a quantum bit (qubit) [1–4].

The major advantages of implementing the spin-based systems are: (a) spin couples much more weakly to the solid-state environment when compared to charge and (b) localization of spins in quantum dot nanostructures is known to dramatically prolong their quantum phase coherence time when compared with nanostructures with higher dimensionality. This arises due to the full motional quantization in QDs, which effectively decouples spin from the orbital motion. This property in turn suppresses the spin decoherence rate due to scattering processes (phonons, charge fluctuations) which couple to the spin via the spin-orbit interaction. The spin flip time in self-assembled QDs was measured to lie in the millisecond range [5], and coherence time is expected to be similarly long. However, in order to reach this theoretical limit other possible decoherence mechanisms need to be controlled. Most important amongst these is the hyperfine coupling between the confined electron spin and the ensemble of nuclear spins in the QD.

Combination of robustness of spin-based approach with newly developed methods for the controlled fabrication of few-dot nanostructures that can be optically addressed over ultrafast timescales, opens a new window of opportunity to develop a class of fully coherent, spin based opto-electronic devices that may facilitate future information technologies.

Z. M. Wang, *Self-Assembled Quantum Dots.*

Over the past years several research groups around the world have developed the capabilities to perform experiments on the spin of individual charge carriers (electrons and holes) localized in semiconductor quantum dot (QD) nano-structures. The selective and controlled generation of charges in individual QDs and the orientation and readout of their spin has become possible, both for electrically defined and optically active dots [6–9]. The ultimate goal will remain to realize a well defined spin-qubit system that is optically addressable, electrically controllable and can be scaled-up beyond one qubit with future materials optimization.

In this chapter we will discuss the prerequisite for the exciting prospects to exploit single spins trapped in nanometer-sized quantum dots for the quantum information processing. We will consider polarization properties of electron-hole complexes on which we will have to rely (at least initially) for ultra-fast control of the spin states by optical pulses. We will discuss what consequences electron-hole interactions in self-assembled InGaAs quantum dots will have on the ability to excite and manipulate single spin in this class of nanostructures. Most of the analysis will be performed on the data collected on large ensembles of quantum dots where technological challenges such as inhomogeneous broadening due to statistical fluctuations in the properties of the dots become apparent. In the three last sections of this chapter we will also discuss experiments on single quantum dots, where we realize the necessary conditions for spin "writing" in a single quantum dot and encounter new challenges in manipulation of the electron spin on the nano-scale due to its coupling to large spin reservoir of lattice nuclei.

6.1.2 Self-assembled In(Ga)Aa Quantum Dots

We will present results obtained on In(Ga)As dots grown by molecular beam epitaxy on GaAs substrates. Due to a large lattice mismatch ($\approx 7\%$) two-dimensional growth of InAs (or InGaAs) on GaAs is only possible for extremely thin films. The strain accumulating in the initial InGaAs layer ("wetting" layer) is released when instead of a layer by layer coverage three-dimensional In-rich islands form (see Fig. 6.1). Note, that even if pure InAs is deposited on GaAs, all three elements will contribute to the composition of the dots. This occurs due to a high temperature growth mode (>500 C), where high mobility of atoms lead to strong intermixing. Inclusion of Ga atoms inside the dots is also energetically preferable due to reduction of the strain.

Such so-called Stranski-Krastanow growth mode described above and schematically shown in Fig. 6.1 has attracted increasing interest in recent years and has allowed an efficient means to fabricate arrays of virtually defect-free uniform QDs with shape, density and emission wavelength which are reasonably controllable during the growth by the temperature of the substrate, deposition rate and composition of the material deposited [10]

Figure 6.2 shows a transmission electron microscopy image of a typical InGaAs quantum dot. As seen from the image the dot is about 50 nm at the base and <10 nm

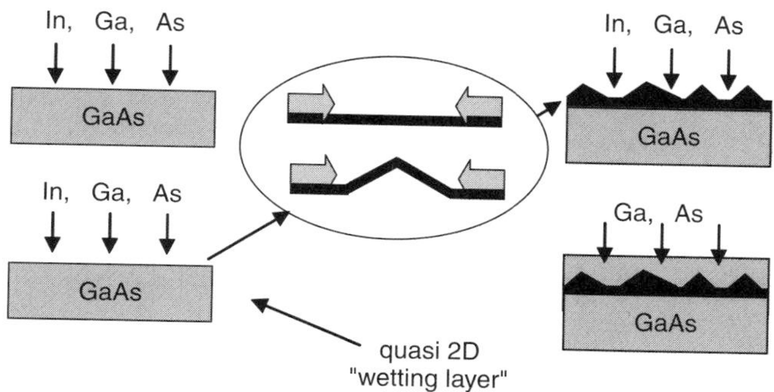

Fig. 6.1 Schematic diagram of Stranski-Krastanow growth mode. First, In,Ga and As are deposited on a GaAs substrate and form a thin strained "wetting layer". The compressive strain lead to the island formation. The atoms forming the islands can effectively have larger lattice constant, since they do not occupy the sites prepared by the underlying GaAs lattice. The dots are then covered with GaAs

tall. It is superimposed on top of a very thin wetting layer, which in effect is a two-dimensional quasi-continuous layer of InGaAs. Typical dimensions of InGaAs quantum dots are 20–50 nm at the base and 3–9 nm in height. They normally have truncated pyramid or lens shapes. The wetting layer is on average $\approx$1 mono-layer thick and can be considered as a very poor quality quantum well.

InGaAs dots are optically active and composed of a direct band-gap semiconductor. This makes them very efficient light emitters, which leads to potential for many applications in lasers [10] and other type of light sources (such as, for example, single-photon emitters [11, 12]). In the context of the research presented in this chapter, this property importantly enables optical addressing of the spin states by circularly polarized light as well as coherent manipulation of the exciton states by precisely tailored ultra-fast pulses.

The material choice allows the growth of very versatile quantum dot structures with emission wavelength ranging from 900 to 1400 nm even when using InGaAs alloy only. The wavelength range can be extended into visible if the dots are grown on AlGaAs. The dots provide strong confinement for trapped charge carriers,

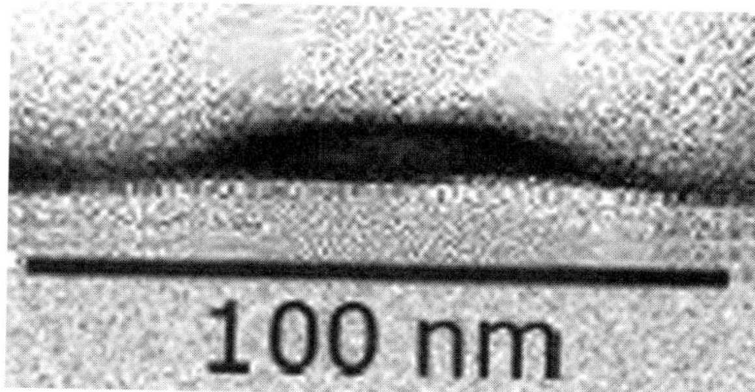

Fig. 6.2 Transmission electron microscopy image of a single InGaAs/GaAs quantum dot

especially for electrons for which confinement energies of 100–150 meV are typical. This property underlies the robustness of the quantum dot carrier population against thermal activation processes at elevated temperatures. In what follows we will also address the robustness of coherent properties of electrons and holes in quantum dots at elevated temperatures.

6.2 Polarization Properties of Quantum Dot Excitons

6.2.1 Exciton Fine Structure Splitting

At low temperatures electron and hole occupy lowest energy levels available in a quantum dot, however the exact energy of the electron-hole pair will depend on the spin configuration of the charge carriers. Strong confinement results in a tight overlap of the electron and hole wave-functions and enhancement of their mutual interactions. This includes the electron-hole exchange interaction leading to a fine structure splitting of the multiplet of the exciton [1] states with this effect mainly determined by the symmetry of the system. Usually these effects are still small in the energy domain, which makes their detection rather difficult due to various inhomogeneous broadenings present in quantum dot systems. However, as we will see in the following Sections, the fine structure splitting will have a very pronounced effect on the spin dynamics. It can also be detected in single dot spectroscopy, where individual dots are probed and small energies typical for the fine structure are not masked by the inhomogeneity.

The general form of the spin Hamiltonian for the short-range electron-hole exchange interaction of an exciton formed by a hole with spin J_h and by an electron with spin S_e is given by [13]:

$$H_{ex} = \sum_{i=x,y,z} (a_i J_{h,i} S_{e,i} + b_i J_{h,i}^3 S_{e,i}) \tag{6.1}$$

Here the z direction points along the sample growth direction. Neglecting the heavy- and light-hole mixing, the single-particle basis from which the excitons are constructed consists of a heavy hole with $J_h = 3/2$, $J_{h,z} = \pm 3/2$ and an electron $S_e = 1/2$, $S_{e,z} = \pm 1/2$. From these states four excitons are formed, which are degenerate when the spin H_{ex} is neglected. These states are characterized by their angular momentum projections $M = S_{e,z} + J_{h,z}$. States with $|M| = 2$ do not couple to the photons, and are therefore dark excitons, whereas states with $|M| = 1$ are optically active bright excitons.

With these angular-momentum eigenstates the matrix representation of H_{ex} can be constructed. It turns out that the dark and bright excitons are not mixed by H_{ex}, but within the pairs the states with opposite angular momenta are in general

[1] We will use this notation to describe electron-hole complexes in quantum dots

hybridized. The off-diagonal elements describing the mixing of the bright and dark states are $\delta_1 = 0.75(b_x - b_y)$ and $\delta_2 = 0.75(b_x + b_y)$, respectively. It thus follows that in the system with rotational symmetry where $b_x = b_y$ the bright exciton states are degenerate and not mixed, whereas when the rotational symmetry is broken, i.e. $b_x \neq b_y$, the bright excitons mix and their degeneracy is lifted. The dark exciton states with $|M| = 2$ are always mixed by the exchange interaction. There is also a splitting between the states with $|M| = 2$ and $|M| = 1$ given by $\delta_0 = 1.5(a_z + 2.25b_z)$ [14].

Figure 6.3 shows a diagram of the exciton states described above. Note, that in the case of lowered symmetry the split exciton state pairs are linear combinations of the $M = \pm 2$ or $M = \pm 1$ states. In the high symmetry case radiative recombination of the e-h pair with $M = \pm 1$ will produce a photon which will be detected as circularly polarized light. The mixing of the exciton states in the case of the lowered symmetry means that the emission from the exciton state will be linearly polarized. The same transformation applies to the photon absorption into the ground exciton state. The mixed bright exciton states form symmetric and anti-symmetric linear combinations of the original spin states in GaAs-based nano-structures [grown on (100) substrates] emit orthogonally polarized photons with linear polarization along [110] and [–110] crystallographic directions.

Note, that the situation changes radically, when additional charge, an electron or a hole, is introduced into the ground state of the dot as shown in Fig. 6.4. According to the Pauli exclusion principle the spins of the two electrons or holes have to be anti-parallel. As follows from Eq. 6.1, the exchange interaction in this case is canceled since either electrons in X^- or holes in X^+ form a spin singlet. Indeed, the spin of

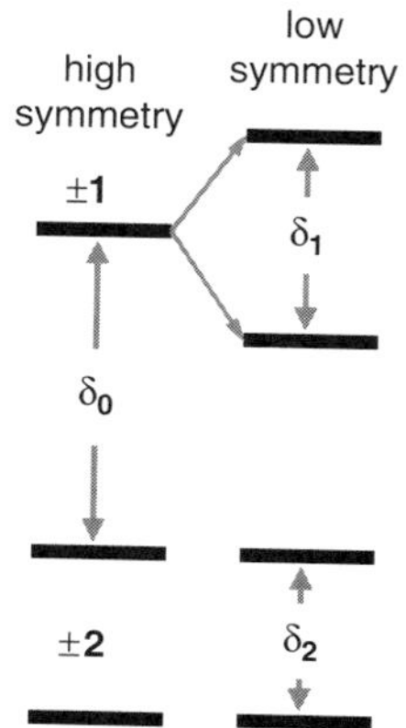

Fig. 6.3 Schematic diagram of the exciton states using the bright ($|M| = 1$) and dark ($|M| = 2$) state representation. Electron-hole exchange interaction results in splitting between the $|M| = 1$ and $|M| = 2$ states in both high and low symmetry confining potentials (denoted δ_0). $|M| = 2$ states are split (δ_2) and hybridized regardless the symmetry, whereas $|M| = 1$ states only mix and split in a low symmetry ($< D_{2d}$) potential

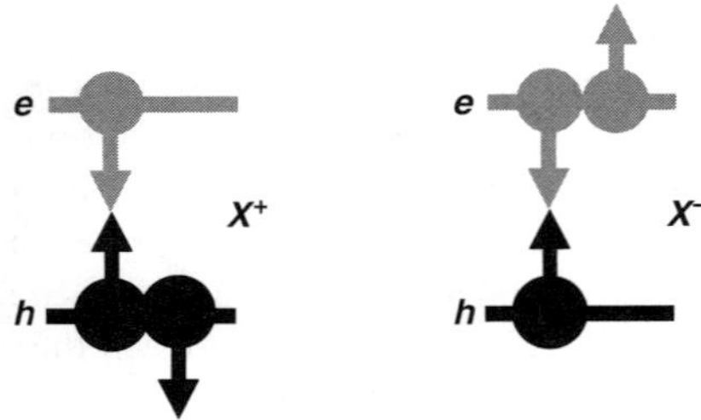

Fig. 6.4 Schematic diagram of the charged exciton states with the ground state populated with two holes (X^+, left) or two electrons (X^-, right) with anti-parallel spins forming spin singlets

a positive and negative trion is well defined by the uncoupled spin of the electron or hole, respectively. Note also, that there is no dark state for a charged exciton.

In the following Sections we will review the strong effects which the polarization properties arising from the electron-hole exchange interaction have on the spin dynamics and spin properties in general in InGaAs quantum dots.

6.2.2 *Ultra-fast Quantum Beat Measurements on Quantum Dot Ensembles*

In this Section we will review polarization-resolved pump-probe experiments on ensembles of InGaAs QDs with a focus on coherence properties of the optically induced exciton polarization. From differential transmission data we deduce the dynamics of coherent spin polarization manifested in quantum beats [15]. Our main aim in this Section will be to consider the effect of the coupling of QDs to their environment [16–19] on the dynamics of coherent spin polarizations. We show the robustness of the spin coherence compared to other optical coherence phenomena in the QD system: this is deduced from the order of magnitude longer spin coherence decay (or slower spin dephasing) time we find at high T, as compared to the exciton dephasing measured in four-wave-mixing experiments [16].

The sample investigated consists of 16 InGaAs QD layers (dot density $\approx 5 \cdot 10^{10}\,\text{cm}^{-2}$) grown using molecular beam epitaxy. Each layer is grown in the middle of a 20 nm GaAs quantum well clad by 10 nm thick AlGaAs barriers. The sample was thermally annealed at 700° C for 12 min [2]. Differential pump-probe transmission $\Delta P/P$ (P - laser power transmitted through the sample) was measured using lock-in techniques schematically shown in Fig. 6.5a. As shown in Fig. 5b, pulses of a femto-second Tisapphire laser were tuned into resonance with the peak of the QD exciton ground. In a relatively large range of energies ($\pm$15meV) around the GS peak the dynamics are similar to those reported here. A notable modification to the dynamics is observed when the laser is tuned >30meV above the GS peak [20], close to the first excited state. Thus the contribution of the excited state absorption to the data measured at the GS peak is very weak. We use 3(0.5) mW pump (probe)

[2] The effect of annealing leading to an increase in the quantum beat period will be discussed in Section 6.2.3.

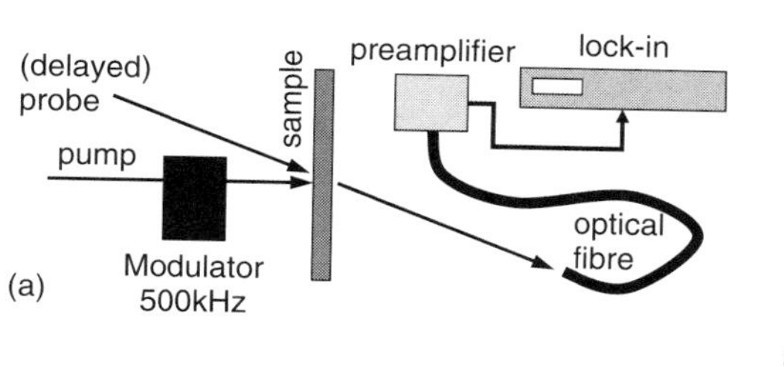

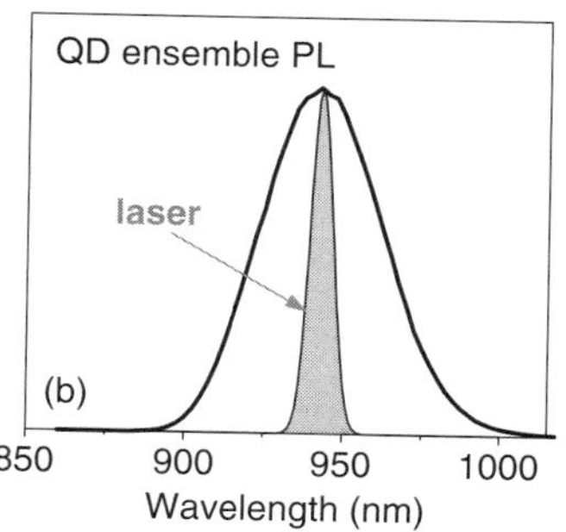

Fig. 6.5 (a) Schematic diagram of the ultra-fast pump-probe set-up employed in the experiments presented. Both pump and probe beams can be individually polarized using various wave-plates. High frequency modulation of the pump beam is achieved by using acousto-optical modulator. (b) Normalized low temperature (T=7K) quantum dot photoluminescence and femto-second laser spectra (black and gray, respectively). The laser is tuned in resonance with the peak of the dot ground state energy distribution

power focused to a 100 μm spot, corresponding to $\approx 2(0.3) \cdot 10^{12}$ photons/cm^2 per pulse. From our experiments we estimate an absorption of $< 10^{-4}$ per dot layer, which leads to a conclusion that either pump or probe pulses excite < 0.1 excitons per dot, excluding contributions from biexciton effects.

Figure 6.6a shows $\Delta P/P$ pump-probe traces measured at $T = 7$K with co- and cross-circularly polarized pulses. In the $\sigma^+\sigma^+$ configuration the $\Delta P/P$ signal is maximum at zero pump-probe delay. The signal reaches a local minimum at a delay time $\tau_d \approx 100$ ps, a maximum at $\tau_d \approx 170$ ps and subsequently decays with a time constant of ≈ 550 ps. The behavior for the $\sigma^+\sigma^-$ configuration is in antiphase to that of the $\sigma^+\sigma^+$ trace for $\tau_d < 170$ ps, with a negligible signal at short delays $\tau_d \approx 0$ and decay with a time constant ≈ 420 ps at long delays.

At long delays the $\sigma^+\sigma^-$ signal is markedly weaker than that in the $\sigma^+\sigma^+$ configuration. This difference results in a "residual" circular polarization or exciton spin-memory in the sample. This is seen clearly in Fig. 6.6b where we plot the degree of circular polarization $\Pi = (I^+ - I^-)/(I^+ + I^-)$. Here I^+ (I^-) is the signal intensity in the $\sigma^+\sigma^+$ ($\sigma^+\sigma^-$) configuration. $\Pi \approx 1$ at zero delay and then reaches a minimum of $\Pi = -0.147$ at $\tau_d \approx 90$ ps. At long delays ($\tau_d > 200$ ps) $\Pi > 0.2$ and increases gradually with time.

The oscillatory behavior observed at short delays originates from the fine structure splitting (E_{FS}) of the two linearly polarized QD exciton modes [15, 21, 22] and can be described within the framework of the quantum beat theory of Ref. [23]. We consider the populations of the two linearly polarized exciton modes split by E_{FS}, which are excited simultaneously with an initial relative phase $\Delta\Theta_0$ by the circularly polarized pump pulse. Following the excitation, $\Delta\Theta$ then oscillates in time with period $2\pi/\omega = h/E_{FS}$. This leads to a periodic variation of the coherent polarization in the sample. In a large ensemble of dots a slight dot-to-dot variation of E_{FS} is present (described in our model by a Gaussian of width δ) leading to a gradual loss of the amplitude in the collective beat pattern. In addition, the phase

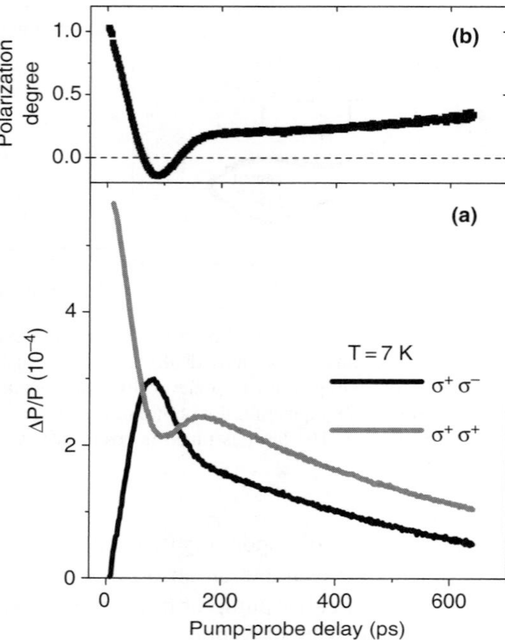

Fig. 6.6 (a) Dependence of the differential transmission on the pump-probe delay measured at $T = 7\text{K}$ for σ^+ polarized pump and σ^+ or σ^- polarized probe (gray and black line, respectively). Both $\approx$ 100 fs pulses are tuned to the QD ground state. (b) Temporal dependence of the degree of σ^+ circular polarization after excitation of the sample with a σ^+ polarized pulse at $T = 7\text{K}$

relation between different dots as well as the periodic evolution of $\Delta\Theta$ within a dot can be perturbed by scattering, such as interactions with phonons etc. These effects are introduced in the equation below via the polarization coherence time T_{coh}. The carrier population decay is given by τ_X, describing the incoherent contribution to the observed signal. We thus obtain:

$$I^{\pm} = e^{-t/\tau_X} \pm e^{-t/T_{coh}} e^{-\delta^2 t^2} cos(\omega t) \tag{6.2}$$

Equation 6.2 gives $I^+ \approx I^- \approx e^{-t/\tau_X}$ at long delays. However, the residual polarization observed in Fig. 6.22 indicates that in addition to the dots exhibiting the oscillating behavior there is a subset of the dot ensemble with no oscillatory component, i.e. for which $\omega \approx 0$. We attribute the non-oscillating contribution to dots containing a charge carrier prior to the pump excitation, arising from the carriers ionized from impurities in the AlGaAs barriers. This attribution is based in part on cw single-dot experiments, which showed the absence of the fine structure for charged excitons [14, 24, 25], for which the electron-hole exchange interaction is blocked by the presence of an additional carrier. This situation is illustrated in Fig. 6.7. We assume equal distribution of carriers captured in dots between spin up

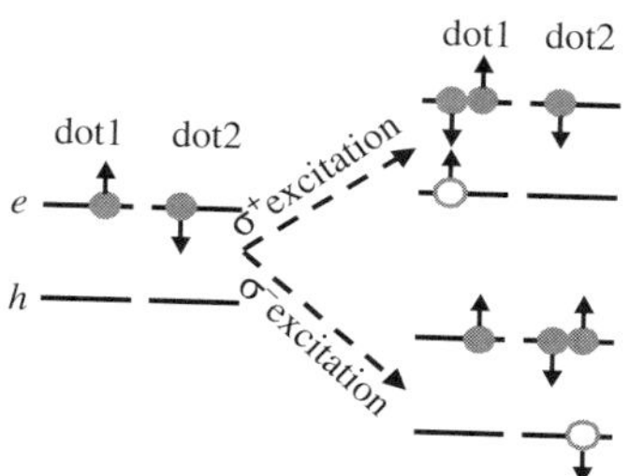

Fig. 6.7 Schematic diagram of excitation of a charged dot with circularly polarized light. Population of the dot with an electron introduces new selection rules for light absorption

and spin down populations. The sign of the excess carriers cannot be determined in the experiments presented. However, the arguments proceed in exactly the same way for either sign of the charge of the excess carrier. As seen from the figure for dot 1(2) the $\sigma^+(\sigma^-)$-polarized excitation creates a charged exciton, where due to the Pauli principle the electrons form a spin singlet state and hence the exchange interaction in such e-h complex is suppressed [14, 24, 25] (see also Section 6.2.1 and Fig. 6.4).

It is notable in Fig. 6.7 that only dot 1(2) can be excited with a $\sigma^+(\sigma^-)$-polarized pulse. At the same time the creation of a charged exciton in dot 1 by a σ^+-polarized pump pulse will not change the absorption of a σ^--polarized probe pulse, since the absorption of the latter pulse in dot 1 was blocked even before the pump excitation. Thus, the signal I^+ measured from the whole ensemble of dots has contributions from both neutral and charged dots, while I^- originates from neutral dots only. The contribution from the charged dots to I^+ is taken into account by adding a term $A_{chX}exp(-t/\tau_{chX})$ to Eq. 6.2. Here A_{chX} is half of the weight of charged dots in the ensemble and τ_{chX} is the charged exciton lifetime. It also follows that τ_X in Eq. 6.2 can be measured directly from the decay of I^- at long delays and corresponds to the neutral exciton life-time.

As seen from Eq. 6.2 the damping of the quantum beats is strongly dependent on the spin polarization coherence time T_{coh}, which implies that the measurement of QBs can serve as a method to study the spin coherence and, in particular for our purposes, its temperature dependence. An enhanced damping of the beat pattern with increasing T is observed in Fig. 6.8. Small change in the QB pattern is seen in the whole range of $7 < T < 55$ K, with the decay of the beats governed by the dot-to-dot variation of E_{FS}. However, at $T > 60 - 65$ K a progressively weaker oscillation pattern is seen. At $T \geq 90$K the beats are washed out and only a strong initial peak (minimum) is observed in I^+ (I^-).

The black curves in Fig. 6.8a,d show the results of fitting using a modified Eq. 6.2 with the contribution of charged excitons taken into account. From the low T data we determine δ and ω, which are then kept unchanged for the higher T fits. The change of the other parameters with T is deduced from the experimental results. For example, A_{chX} increases by $\approx 25\%$ from 10 to 130 K due to the extra charges released from the AlGaAs barriers.

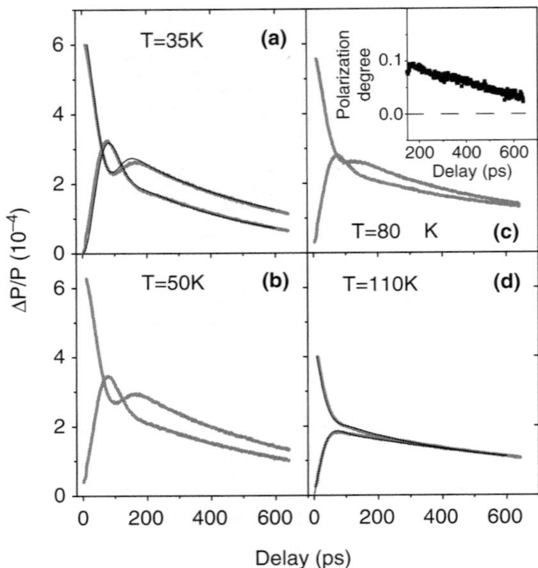

Fig. 6.8 Dependence of the differential transmission on the pump-probe delay measured at (a) $T = 35$K, (b) $T = 50$K, (c) $T = 80$K and (d) $T = 110$K for σ^+ polarized pump and $\sigma^\pm$ polarized probe. The inset in (c) shows the temporal dependence of the degree of circular polarization in the sample at $T = 80$K

Although the Gaussian distribution of E_{FS} used in Eq. 6.2 is a reliable starting point, the actual physical picture can be more complex, as has been discussed for quantum wells in Ref. [26]. From Eq. 6.2 it is notable that the form of the distribution of E_{FS} could affect mainly T_{coh}, the latter influencing the decay of the beat pattern. However, we find that even when using a Lorentz distribution, magnitudes of T_{coh} within 20% of those obtained with a Gaussian distribution are found. The weak sensitivity to the form of the distribution of E_{FS} arises from the relatively small width of the distribution in our case as can be concluded from the fact that unlike Nickolaus *et al.*[26] quantum beats are observed in our experiments.

Using a Gaussian distribution of E_{FS} we find that the oscillation period corresponds to E_{FS} ($\hbar\omega \pm \delta/2$) of $18 \pm 3\,\mu$eV (The splitting for the "as grown" sample was $\approx 30\,\mu$eV). At low temperature the damping of quantum beats is determined by the distribution of E_{FS} with the width of $\approx 6\,\mu$eV (see Fig. 6.9). Above 40K temperature-induced dephasing starts to contribute to the damping of the beat pattern. Eventually, at 130 K T_{coh} falls to 28 ps. At $T \geq 65$K the temperature dependence of T_{coh} can be fitted with an ionization-type exponent with activation energy $E_A \approx 22.5$ meV (see top panel in Fig. 6.9), similar to the 30 meV found in measurements of linear polarization decay due to population redistribution in QDs [17]. The magnitude of T_{coh} observed at $T = 130$ K is about 1 order of magnitude larger than that reported at high T for neutral exciton dephasing in four-wave-mixing

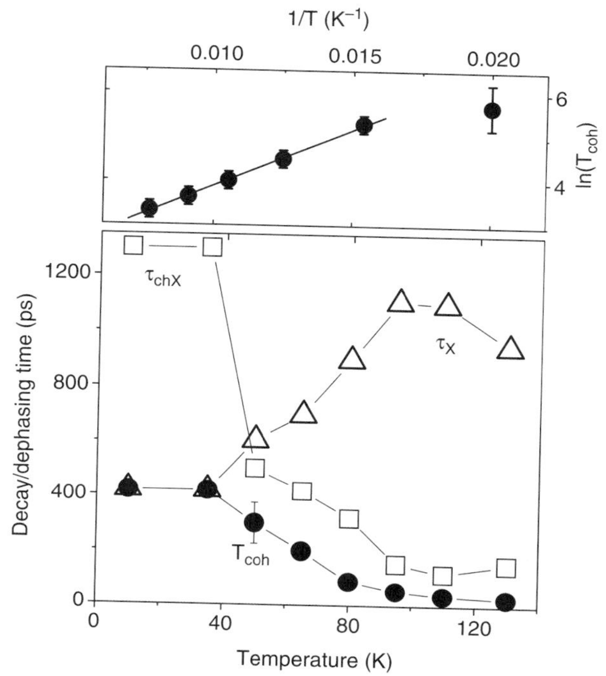

Fig. 6.9 Temperature dependence of T_{coh} (circles), τ_{chX} (squares) and τ_X (triangles). Inset shows logarithm of T_{coh} versus $1/T$. The fit is a linear function with slope of $\approx$ 260K (22.5 meV)

experiments [16]. This observation indicates the robustness of spin coherence in the QD system.

Equation 6.2 describes also the long-lived spin-memory related to incoherent population decay. The variation of τ_{chX} and τ_X with temperature is shown in Fig. 6.9. At $T \leq 35$ K $\tau_{chX} \approx 1300$ ps and is $\approx$ 3 times longer than $\tau_X = 420$ ps. This difference in incoherent population decay rates is responsible for the long-lived circular polarization (or exciton spin orientation) in the sample excited by the pump beam. As the temperature reaches 50 K a sharp drop in τ_{chX} to 500 ps is observed. In the whole range $T \geq 50$ K τ_{chX} is smaller than τ_X, with the latter growing with T at $T > 35$ K. This in turn leads to decay of the residual circular polarization with time. This is seen for example in the inset of Fig. 6.8c at $T = 80$ K, in marked contrast to the growing polarization observed at $T = 7$ K in Fig. 6.6b.

The observed temperature dependence arises from the onset at 35 K$< T <$50 K of the efficient thermal excitation of carriers from the QD ground state, which is dominated by the hole activation [18, 19, 27, 28]. Other mechanisms such as hole spin-flip have also been reported for n-type charged II-VI dots [29], which are, however, likely to be less significant in our case. In our samples the charged excitons are most likely p-type, due to the well known residual p-doping in MBE grown

GaAs structures. In addition, the thermal activation time at $T \approx 70\,\mathrm{K}$ reported from measurements of PL [18, 27, 28] and photocurrent [19] on individual dots is > 10 times faster than spin-flip reported in Ref. [29].

The decrease of τ_{chX} indicates the shortening of the time during which the ground state is occupied simultaneously by 3 carriers, which in turn leads to a faster decrease of the degree of circular polarization in the sample. The increase of τ_X with T is explained by the increased probability of excitation of carriers to excited states and the consequent relaxation to lower energy, thus leading to an increase in the exciton lifetime relative to the radiative recombination time at low T ($\tau_X = 420\,\mathrm{ps}$ at $T \leq 35\,\mathrm{K}$). The slight deviation from the trend at $T > 110\,\mathrm{K}$ may arise from irreversible excitation of carriers from the dot (leading to the fall in τ_X) and/or contribution from increasingly faster electron dynamics. Due to the complex population dynamics neither the τ_X nor τ_{chX} T-dependence can be fitted with a single activation energy. The magnitude of $\tau_{chX} \approx 1300\mathrm{ps}$, observed below the onset of the strong thermal activation, represents the radiative lifetime of charged excitons.

To summarize this Section, the dynamics of spin-coherence were studied in a wide range of temperatures in an ensemble of QDs by measuring quantum beats in pump-probe experiments. We find gradual shortening with T of T_{coh} (to 28ps at $T = 130\,\mathrm{K}$), which can be fitted by a single energy activation-type process. The same set of experimental data reveals a long-lived spin-memory in the sample arising from the slow incoherent decay of the exciton population in charged dots, where the spin oscillation is suppressed. We note that the spin polarization memory can be maximized by employing doped dots, and in addition can be controlled by using gated structures.

6.2.3 Effect of Post-growth High Temperature Annealing on Polarization Properties of Quantum Dots

As was shown by Bester and co-workers [30], the reduced atomistic symmetry of zinc-blende semiconductors (C_{2v} symmetry) lead to the fine structure splitting even in self-assembled dots with cylindrical symmetry about the growth axis. However, in this case splittings of only up to $\approx 9\,\mu\mathrm{eV}$ are predicted [30], considerably smaller than those found in most as-grown dot samples. The splitting will then be enhanced by anisotropy in dot shape occuring during crystal growth.

In a similar way it is known that the strain field is anisotropic for self-assmbled dots, distinguishing the [110] and [–110] directions, even for dots of spherical shape [31–34]. The strain leads to piezoelectric fields in the samples, and when included in **k.p** theory treatments of electronic band structure, leads to markedly enhanced p-state splittings [33, 34]; the s-like ground state is expected to show similarly enhanced splitting when strain and piezoelectric fields are enhanced. In accord with the above arguments, in unstrained dots in a glass matrix, the fine structure is zero [35]. Further support to the above reasoning is given by recent experiments where externally applied uniaxial strain was shown to produce a large splitting of bright excitons in quantum wells [36, 37].

In this Section we present systematic studies of the fine structure splitting in ensembles of InGaAs dots, providing experimental evidence for the factors determining the magnitude of the electron-hole exchange interaction. In particular, we show that E_{FS} can be accurately tuned in InGaAs dots by post-growth thermal annealing, enhancing the In/Ga intermixing and reducing strain [38, 39]. The notable In/Ga intermixing occurring during thermal annealing is shown to result in a smaller E_{FS}. These observations suggest that strain present in dots due to the lattice mismatch of In(Ga)As and GaAs plays an important role in controlling the fine structure splitting. By comparing different QDs we find that E_{FS} has a similar trend as the depth of the exciton confinement, which is less pronounced in annealed quantum dots.

We deduce E_{FS} from low T measurements of the period of quantum beats in polarization-resolved pump-probe experiments on dot ensembles as was described in Section 6.2.2. In the experiments a circularly polarized pump pulse excites the two linearly polarized exciton modes (split by the electron-hole exchange interaction) and their coherent superposition is then probed by a second circularly polarized probe pulse [15, 40].

As was shown in Section 6.2.2, by varying the pump-probe delay the time-dependent differential transmission signal $\Delta P/P$ can be measured (P laser power transmitted through the sample). $\Delta P/P$ has contributions from (i) the coherent polarization dynamics (responsible for quantum beats) and (ii) incoherent carrier population decay (leading to a slow exponential decay) and can be described by the Eq. 6.2. In what follows, we will focus on the deduction of $\omega = E_{FS}/\hbar$ from our experimental data, which we show can be achieved with accuracy better than $\pm 10\%$.

We studied dots grown on GaAs with low T photoluminescence (PL) of "as grown" QDs in the range 980–1035 nm. The first type of samples (A) comprises 16 QD layers grown at $T = 500^\circ$ C using molecular beam epitaxy (MBE). The dot density is $\approx 6 \cdot 10^{10}\,\mathrm{cm}^{-2}$. In the samples of type A with nominally InAs dots (considered in detail below) each dot layer (2.2 ML thick) is grown in the middle of a 20 nm GaAs quantum well clad by 10 nm thick AlGaAs barriers. Different Sections of the wafer were thermally annealed at 700° C for up to $t_{an} = 17.5$ minutes. In addition, we studied 7 ML thick $In_{0.5}Ga_{0.5}As$ dots (low T emission at 1035 nm) embedded in Al-free GaAs barriers (type B).

Finally, samples of type C, an edge-emitting laser structure comprising long wavelength quantum dots was also studied. The laser grown by MBE contained 3 layers of so-called "dot-in-a-well" (DWELL) structures [41–44]. These are nominally InAs dots grown between two $In_{0.15}Ga_{0.85}As$ strain-reducing layers (2 nm below and 6 nm above the dots). The low T PL peak at $\approx$ 1200 nm, which is a considerably longer wavelength than normally achievable with InAs dots grown in GaAs (normally in the range <1100 nm). The dot density is $3.6 \cdot 10^{10}\,\mathrm{cm}^{-2}$. Such laser structures contain AlGaAs claddings to provide confinement of the optical modes. In order to obtain high material quality of the claddings the substrate temperature of 620C was maintained for over an hour during the growth, so effectively

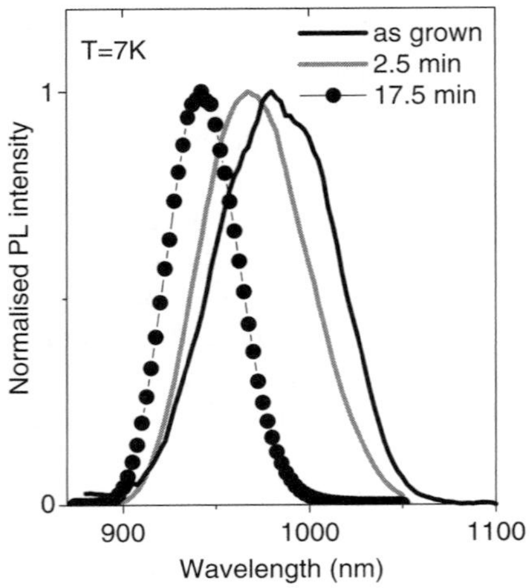

Fig. 6.10 Normalised PL spectra measured at $T = 7$K using low excitation intensity on the unannealed sample with nominally InAs dots (black curve), and samples annealed at 700°C 700°C for 2.5 and 17.5 min (gray curve and

the dots underwent thermal annealing in "soft" (moderate temperature) conditions. More results on DWELL samples are presented in Section 6.2.4.

Figure 6.10 shows the results of the PL characterization (at $T = 7$K) of the annealed pieces of samples of type A. The PL spectrum measured for the "as grown" sample exhibits a peak at 980 nm with a FWHM of 77 nm ($\approx$ 100 meV). After annealing for 2.5 min at 700°C the PL peak shifts to 967 nm and the width reduces by 10%. As the annealing time is further increased, the PL peak shifts to shorter wavelength and reaches 942 nm (FWHM 63 meV) for the sample annealed for 17.5 min. The observed behavior arises due to In/Ga intermixing [45, 46], which leads to a more Ga rich dot composition and hence weaker confinement potential.

Figure 6.11 shows $\Delta P/P$ pump-probe traces measured at $T = 7$K for the annealed samples with co- and cross-circularly polarized pulses. The dynamics measured for the as grown sample and samples annealed for $t_{an} = 7.5$ and 17.5 minutes are shown in Fig. 6.11a,b and c, respectively. In the $\sigma^+\sigma^+$ configuration the $\Delta P/P$ signal is maximum at zero delay between the pump and probe pulses. The signal exhibits a fast oscillation and then decays with a time constant of few hundred ps at delay times $\tau_d > 200$ ps (see the inset of Fig. 6.11a for pump-probe curves on a longer time scale). The behavior in the case of the $\sigma^+\sigma^-$ configuration is in antiphase to that of the $\sigma^+\sigma^+$ traces at $\tau_d < 200$ ps, with a negligible signal at $\tau_d \approx 0$ [15, 40].

As seen from Fig. 6.11 a systematic variation of the exciton dynamics is observed in the pump-probe curves measured from samples annealed for different times: As indicated by the vertical arrows, the oscillations in both $\sigma^+\sigma^+$ and $\sigma^+\sigma^-$ configurations occur on a progressively longer time scale as the annealing time is increased. The increase of the oscillation period corresponds to a reduction in E_{FS}. The oscillation period, t_{osc}, (and hence $E_{FS} = h/t_{osc}$) is deduced from fitting of the $\Delta P/P$ traces using Eq. 6.2. The dependences of t_{osc} and E_{FS} on t_{an} are plotted in Fig. 6.12. After annealing for 17.5 minutes t_{osc} reaches ≈ 260 ps, corresponding to $E_{FS} \approx 15\ \mu$eV. These data are summarized in Fig. 12.

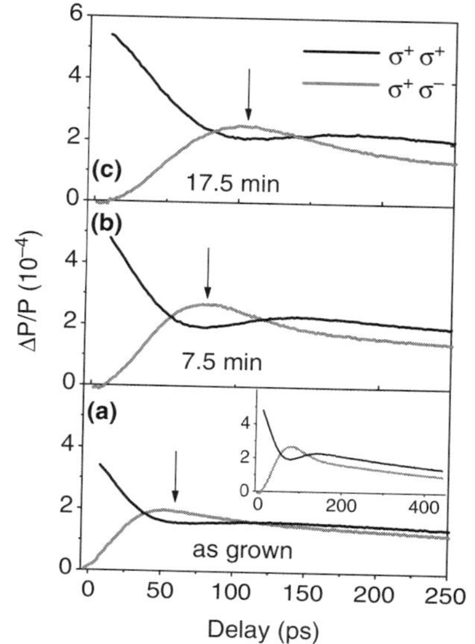

Fig. 6.11 Dependence of the differential transmission $\Delta P/P$ on the pump-probe delay measured at $T = 7$K for σ^+ polarized pump and σ^+ or σ^- polarized probe (black and gray lines, respectively). Both fs pulsed are tuned to the QD ground state. The curves are measured for as grown sample (a) and samples after 7.5 and 17.5 min annealing (figures b and c, respectively). The vertical arrows mark the shift in oscillation time. Inset in Fig. a shows the traces from Fig. b on a larger time-scale

We find that for the unannealed sample a satisfactory fit of the $\Delta P/P$ curves is only possible if the contributions of at least two dot distributions to the oscillation behaviour are taken into account. This is consistent with the observation of multi-modal structure in the PL spectrum of the dot ensemble in the as grown sample (Fig. 6.10). The annealing results in suppression of the multi-modal dot distribution in the PL spectra, which we are able to confirm also on a very fine energy scale in the pump-probe measurements.

Figure 6.13 compares the results presented in Fig. 6.12 with E_{FS} data measured on two other samples of type B. E_{FS} is plotted in Fig. 6.13 versus the PL peak wavelength, λ_{PL}, which decreases with annealing time for these samples. The data

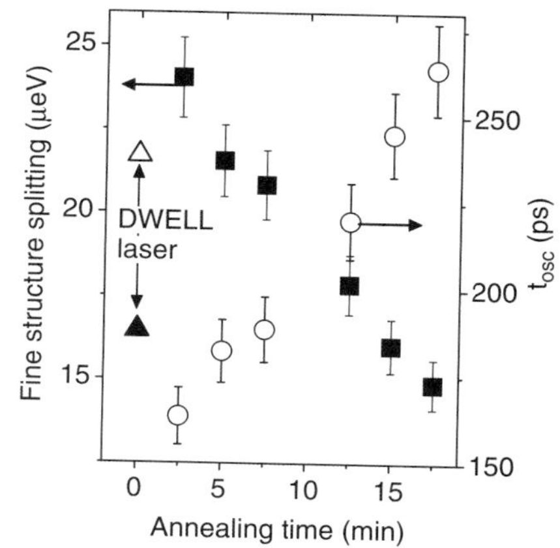

Fig. 6.12 Dependences on annealing time of the oscillation period t_{osc} [circles, obtained from fitting of data in Fig. 6.11 by Eq. (1)] and fine structure splitting E_{FS} (squares), obtained for the sample with nominally InAs dots grown on GaAs. Triangles show the results for a DWELL laser structure

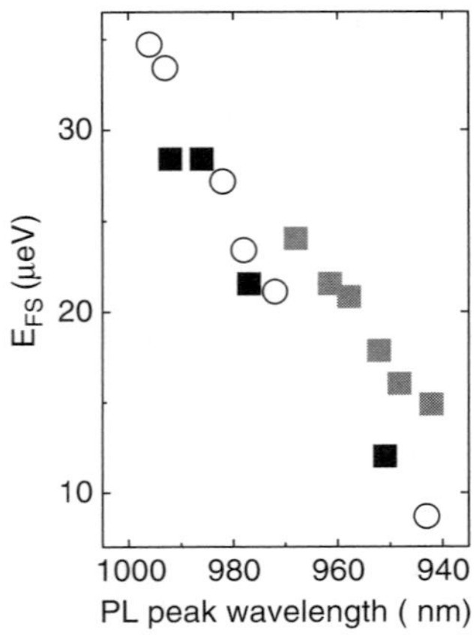

Fig. 6.13 E_{FS} versus PL peak wavelength for 3 In(Ga)As dot samples annealed at 700, 750 and 800 °C (see text for details)

from Fig. 6.12 (for the nominally InAs QD sample of type A) is shown by gray squares, while the results for two samples of type B are shown by open and black symbols, respectively. These samples were annealed for up to 20 minutes at 750° (points above 970 nm) and 800° C (points below 950 nm). As seen, E_{FS} can be tuned by annealing over a very large range from $\approx$ 36 to 8 μeV. The almost linear correlation between λ_{PL} and E_{FS} for all three samples is notable; however, a difference is observed in the dependences $E_{FS}(\lambda_{PL})$ for different types of samples. The sample with nominally InAs dots has larger E_{FS} in the range of λ_{PL}, where the dependences from the different samples overlap: for $\lambda_{PL} \approx$ 950nm E_{FS} measured for this sample is up to a factor of 2 higher than for the samples with nominally $In_{0.5}Ga_{0.5}As$ dots.

One of the important parameters which has an impact on E_{FS} and is modified by the In/Ga intermixing during annealing is strain. During the growth of quantum dots the strain due to the lattice mismatch with the underlying GaAs is relieved partially via the coarsening of the In-rich surface and eventual formation of dots. However, the consequent capping of In-rich dots with GaAs results in additional compressive strain in dots and tensile strain in GaAs [31–34]. These strains will be reduced during thermal annealing [46], producing more Ga-rich dots due to the In/Ga intermixing and thus reducing the lattice constant contrast in the material surrounding the dots. The reduction of strain (with the latter predicted to induce further anisotropy in dots [31–34]) will result in more s-like exciton wave-functions and hence weaker e-h exchange interaction [47].

Figure 6.12 also presents data for the dot-in-a-well laser structure (sample C). For this experiments the long wavelength pulses were supplied by the optical parametric oscillator (OPO) pumped synchronously with a femtosecond Ti-sapphire laser. The OPO generated pulses with repetition rate 76 MHz and emission FWHM of $\approx$ 10 meV ($\approx$ 200 fs pulse duration). The excitation pulses were tuned into resonance with the peak of the QD exciton ground state $\approx$ 1190 nm. Analysis of the results for the DWELL laser structure using Eq. 6.2 provides the data shown in Fig. 6.12 by triangles: $t_{osc} = 238$ ps and $E_{FS} = 16.5$ μeV.

The magnitude of E_{FS} is about 2 times smaller than that detected for unannealed dots grown on GaAs. This deviates notably from the trend in $E_{FS}(\lambda_{PL})$ observed in Fig. 6.13, showing that the magnitude of the exciton fine energy structure is not directly related to the magnitude of confinement of the electron and hole. As will be shown in the next session, as grown DWELL structures exhibit notably larger fine structure splittings roughly following the trend observed in Fig. 6.13. Thus, the effect of annealing on the fine structure splitting is rather pronounced in the DWELL laser structure investigated. This contradicts a rather weak effect on the peak wavelength which is approximately the same as for non-laser DWELL structures, which are all grown at temperatures below 600 C. Indeed, notably higher temperatures above 700 C and similar to those used for dots grown in GaAs are required to shift the emission wavelength of a DWELL structure. One possibility could be that In/Ga intermixing during annealing acting to blue-shift the emission also acts to reduce the strain in the dot, with the effect of the latter being a red-shift in emission. This two effects may cancel each other with the emission remaining still at long wavelengths but the dots becoming much less strained as seen from the notably reduced magnitude of the fine structure splitting.

To summarize this Section, experimental evidence is found for the effect of strain reduction on the magnitude of the fine structure splitting of quantum dot excitons. The splitting is shown to decrease in dots after annealing. During annealing In/Ga intermixing leads to smaller lattice constant contrast in the dot environment, leading to strain reduction. Very small magnitudes of the fine structure splitting is found in an edge emitting long wavelength laser based on dot-in-a-well structure. There the dots become annealed during the growth of the laser claddings at 620 C. Finally, thermal annealing is shown to be a suitable tool to produce small fine structure splittings required for sources of entangled photons based on individual dots.

6.2.4 Polarization Properties in Long-wavelength Dot-in-a-well Structures

In contrast to single-dot spectrally resolved spectroscopy [14, 24, 48], ultra-fast spectroscopy permits the measurement of very small fine structure splitting [15, 40, 49, 50]. In addition, measurements of dots emitting at $> 1\,\mu$m become possible, single dot spectroscopy in this spectral region being difficult due to a lack of sensitive detectors [51]. Hence ultra-fast spectroscopy allows a direct insight into the properties of excitons in technologically important dot structures used for lasers emitting at $\lambda \geq 1.3\,\mu$m[41–44]. Such structures are typically formed by growing InAs QDs in an InGaAs quantum well (a so-called dot-in-a-well, DWELL).

In this Section we present a systematic study of the fine structure splitting of the bright ground state excitons in a range of DWELL samples with different compositions and thicknesses of the quantum well. It is demonstrated that these parameters allow the gradual tuning of the fine structure splitting and hence the coherent spin dynamics of the QDs [15, 40, 49, 50].

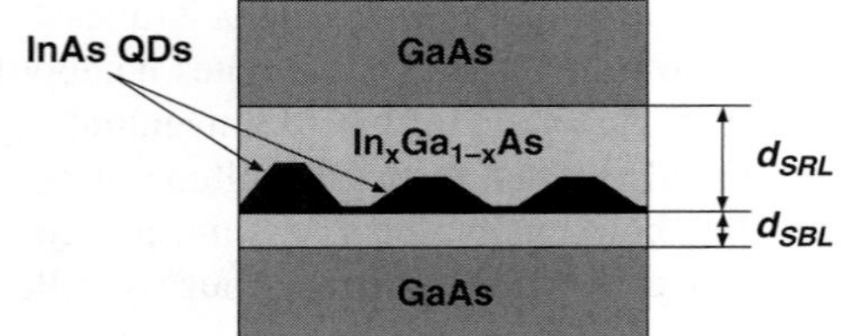

Fig. 6.14 Schematic diagram of a DWELL structure. The thicknesses of the strained buffer layer and strain reducing layer are denoted d_{SBL} and d_{SRL}, respectively

The schematic diagram of the sample structure is shown in Fig. 6.14 with the InAs QDs (2.9 ML of InAs) grown on a strained buffer layer (SBL) of thickness d_{SBL} and capped with a strain reducing layer (SRL) of thickness d_{SRL}. In samples of set A the In content of the SBL and SRL is 6, 12, 15 and 20% and $d_{SRL} = 6$ nm and $d_{SBL} = 2$ nm. Details of the growth and characterization of these samples, which contain a single DWELL, can be found in Ref. [44]. The samples of set B contain 10 DWELLs with a thickness of the SRL of 6, 8, 10, 12 and 16 nm, $d_{SBL} = 2$ nm and an In composition of 15%.

Figure 6.15a shows PL spectra (at $T = 7$K) for the samples of set A, measured using low power excitation with a HeNe laser. It is seen that as the In content, x, in the $In_xGa_{1-x}As$ well increases, the PL peak shifts to increasingly longer wavelength. This is consistent with an increasing size and In content of the dots [44] and also strain reduction inside the dots [42] when the nominal In composition in the well is increased.

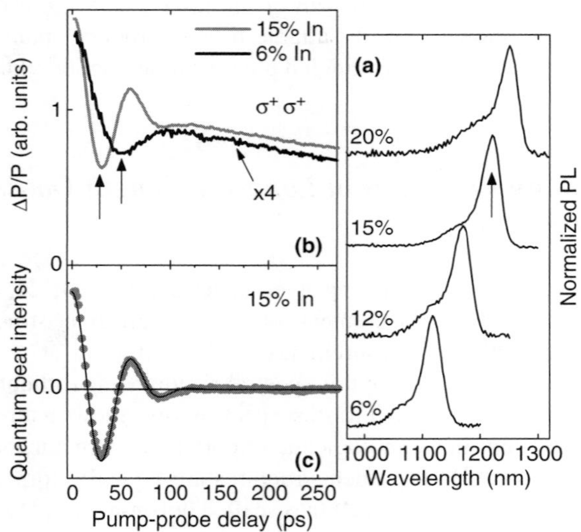

Fig. 6.15 (a) PL spectra of the DWELL samples of set A with $d_{SBL} = 2$ nm and $d_{SRL} = 6$ nm and differring In concentration in the well. (b) Differential transmission curves measured in the co-polarized pump-probe configuration for the samples of set A with 15% (gray line) and 6% (black line) In in the well. (c) The symbols show the data for the sample with 15% with the exponential decay (due to exciton recombination) subtracted. The solid line is the fit using Eq. 6.3

Figure 6.15b shows the results of pump-probe measurements (at $T = 7\text{K}$) with both circularly polarized pump and probe beams tuned to the QD ground state. Similarly to the experiments presented in Sections 6.2.2 and 6.2.3, the time-resolved differential transmission $\Delta P/P$ (P - laser power transmitted through the sample) transients were measured using lock-in detection [40, 50]. Femto-second pulses from an optical parametric oscillator pumped by a Ti-sapphire laser (Mira, Coherent) were used for excitation.

In Fig. 6.15b the transients are shown for co-polarized pump and probe beams for the samples of set A with an In content in the well of $x = 0.15$ (gray) and $x = 0.06$ (black). Both traces show similar features, with oscillations [quantum beats (QBs)] up to a delay of $\tau_d \approx 120\,\text{ps}$ followed by a slow mono-exponential decay. Minima corresponding to the first half-periods of the oscillations at $\tau_d < 120\,\text{ps}$ are indicated by vertical arrows. The period is significantly shorter for the sample with 15% In: $\approx 60\,\text{ps}$ compared to $\approx 100\,\text{ps}$ for the sample with 6% In.

The symbols in Fig. 6.15c show the data from Fig. 6.15b (for the 15% In sample) with the incoherent exciton population decay contribution to the signal subtracted [40, 50]. The remaining coherent polarization signal, I, can be fitted by the following expression [40, 50] deduced from Eq. 6.2:

$$I = e^{-t/T_{coh}} e^{-\delta^2 t^2} cos(\omega t) \tag{6.3}$$

Here $\omega = E_{FS}/\hbar$, δ describes a Gaussian distribution of the dot-to-dot variation of E_{FS}, and T_{coh} describes the spin decoherence due to scattering, the notations introduced in the previous two Sections. In what follows, we focus on the deduction of $E_{FS} = \hbar\omega$ from our experimental data. The curve obtained using the expression above (line in Fig. 6.15c) with $E_{FS}(\pm\delta/2) = 64(\pm 5)\mu\text{eV}$ provides an excellent fit to the experimental data [3].

The results of a similar fitting procedure to the data obtained on all samples from set A are shown in Fig. 6.16a (circles). The PL peak wavelength as a function of the In composition of the quantum well (from the data in Fig. 6.15a) is presented in Fig. 6.16b (circles). As can be seen from Fig. 6.16a the tuning of E_{FS} over a very wide range is possible by using different In compositions in the $In_xGa_{1-x}As$ well: E_{FS} increases from 27 to 64 μeV as x changes from 0.06 to 0.15. A notable decrease of E_{FS} to 32 μeV is observed when the In content is increased to 0.2. The triangles in Fig. 6.16a,b show results for a second sample with 15% In and $d_{SBL} = 2\text{nm}$, $d_{SRL} = 6\text{nm}$. The magnitude of E_{FS} for this sample is very similar to that determined for the corresponding sample from set A.

Similar measurements and analysis have been performed on the samples from set B, where the In content was kept constant at 15% and only the thickness of the top InGaAs layer was varied (Fig. 6.16c,d). As seen from Fig. 6.16c, for $d_{SRL} = 6$ and

[3] δ varies from sample to sample and can reach up to 30% of E_{FS}. Even in those cases such relatively low magnitudes of δ do not result in masking of quantum beats and thus have limited effect on the accuracy with which E_{FS} is measured.

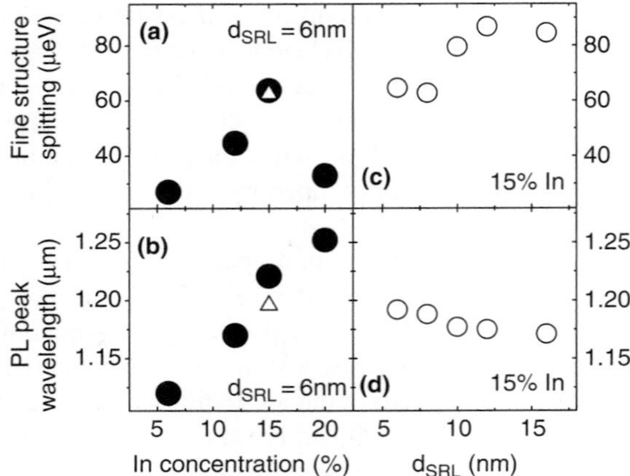

Fig. 6.16 Dependence of the fine structure splitting (a) and PL peak wavelength (b) on In concentration in the well for the samples of set A. Triangles show data for an additionally grown sample with 15% In in the well and $d_{SBL} = 2\,\text{nm}$ and $d_{SRL} = 6\,\text{nm}$. (c) and (d) show the dependences of the fine structure splitting and PL peak wavelength on d_{SRL} for the samples of set B

8 nm the magnitude of E_{FS} is consistent with the values in Fig. 6.16a for the two samples with 15% In and $d_{SRL} = 6\,\text{nm}$. As the thickness of the upper InGaAs layer is increased E_{FS} increases by $\approx$ 30%, saturating at $\approx$ 87μeV for $d_{SRL} \geq$ 10nm (Fig. 6.16c). In contrast the thickness of the upper InGaAs layer has a relatively weak effect on the emission wavelength, which decreases from 1192 to 1171 nm when d_{SRL} is increased from 6 to 12 nm.

Figure 6.17 shows the data from Fig. 6.16a-d plotted as the dependence of E_{FS} on emission wavelength, λ_{PL}. In addition, this figure includes data from our previous work [50] (squares), where a series of annealed samples with InGaAs dots grown in GaAs were studied. As seen from this figure most of the data points follow a trend of increasing E_{FS} with increasing λ_{PL} [52].

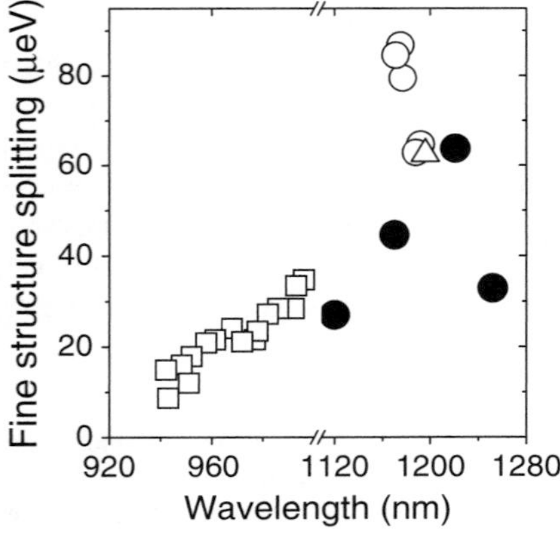

Fig. 6.17 Data from Fig. 6.16a-d plotted as E_{FS} versus the emission wavelength (the same symbols as in Fig. 6.16a-d are used). Squares show data from Section 6.2.3 (and Ref. [50]) for annealed InGaAs/GaAs dots

A larger magnitude of E_{FS} corresponds to stronger anisotropy of the electron and hole wave-functions. As shown in recent theoretical work by Bester and Zunger [30], even for idealized, structurally isotropic islands, symmetry lowering still arises from (i) interfacial symmetry, (ii) atomistic strain, and (iii) piezoelectricity [33]. In annealed samples where In/Ga intermixing occurs, the observed reduction of E_{FS} indicates that structural and electronic asymmetry can be reduced by annealing [49, 50], partly due to the blurring of the dot-matrix boundaries. In addition, In/Ga intermixing results in a weaker contrast between the barrier and the dot material, leading to decreased strain and hence reduced piezoelectric fields inside the dot. In/Ga intermixing also causes the PL to blueshift leading to the experimentally observed increase of E_{FS} with increasing λ_{PL} (Fig. 6.16c).

In contrast to InGaAs/GaAs dots, dots in DWELLs have a higher In content since i) the dots are surrounded by InGaAs, which makes In/Ga intermixing during the growth less significant and ii) a strain driven decomposition of the well occurs with additional In atoms transferred into the dots[42, 43]. The higher In content in the dot (increasing with In content in the InGaAs well) will result in a larger spread of the exciton wave-function inside the dot compared with InGaAs/GaAs dots, where In segregation leads to a strong localization of the hole at the dot apex [53, 54]. The larger spread of the wave-function in DWELLs is also favored by the smaller confinement potential than in InGaAs/GaAs dots, a result of the growth of the dots in the lower band gap InGaAs. Thus the piezoelectric field acting near the dot-barrier interfaces, which acts to increase the asymmetry of the exciton wave-function [30, 33], will have a stronger effect on the more extended exciton wave-function in DWELLs. This is consistent with the strong effect of the piezo-electric field on the asymmetry of the exciton wave-function found in Ref. [30], where uniform InAs dots were considered. The increasing trend in $E_{FS}(\lambda_{PL})$ is consistent with the increasing In content in the dot resulting in both a larger spatial extent of the exciton wave-function and a reduction of the emission energy.

A deviation from the increasing trend in $E_{FS}(\lambda_{PL})$ is observed for samples of set B with a thick (>8nm) top InGaAs layer and also for the sample of set A with 20% In content in the well (see Fig. 6.16). In the samples of set B a slight emission blue-shift is observed as the thickness of the top layer is increased. The behaviour of the set B samples may be caused by increased strain in the In-rich InGaAs well, with the strain acting on the exciton states in the dots. In support of this interpretation, dislocation formation has been reported for DWELL samples with thick InGaAs strain-reducing layers[43] consistent with a large amount of strain. This increased strain results in larger piezoelectric fields and hence increased anisotropy of the exciton wave-function. As in the case of annealed InGaAs/GaAs dots and the DWELLs of set A, a stronger relative change is observed in E_{FS} than in λ_{PL} since the former is sensitive to the shape of the wave-function whereas the latter depends on the effective volume occupied by the exciton.

In the sample of set A with 20% In, a considerably weaker room temperature PL intensity was found, indicating the presence of defects (such as dislocations) related to strain relaxation. Similar effects have been reported by other groups for samples of high In content [42, 43]. The strain relaxation in this sample results in a reduced piezoelectric field inside the dot and, as a consequence, a reduced E_{FS}.

In summary, in this Section measurements of the exciton fine structure splitting are presented for long-wavelength dot-in-a-well structures. A detailed study of 10 samples using ultra-fast pump-probe techniques reveals a strong effect of the well In content and, to a smaller degree, thickness of the InGaAs well on the properties of the dot exciton wave-function. A comparison with InGaAs/GaAs dots reveals a general trend of increasing E_{FS} with emission wavelength. The results suggest that E_{FS} can be controllably tuned by changing the structure parameters.

6.3 Spin Properties of Charged Excitons in Quantum Dots

In the following two Sections we will present the studies of the spin properties of quantum dots where in addition to the photo-generated electrons and holes either a resident carrier is present which relaxed to the dot from a closely located dopant or another carrier is excited electrically from a contact using gate voltage. The studies of the class of dots containing resident carriers is conducted on ensemble samples (Section 6.3.1), whereas individual quantum dots are probed in the experiments where electrical and optical excitation of carriers is combined (Section 6.3.2).

6.3.1 Spin Dynamics in Ensembles of Charged Quantum Dots

In this Section we demonstrate a method to measure accurately charging levels (i.e. fraction of charged dots, f_{ch}) in QD ensembles with <1 resident carrier per dot. The method is based on quantum beat phenomenon observed in QDs and described in Section 6.2. More precisely, the method relies on the fact that the electron-hole exchange interaction is suppressed for charged QD excitons with one excess hole or electron, since either two holes or two electrons form a spin-singlet state due to the requirements of the Pauli exclusion principle [14, 25] (see also Section 6.2.1 and Fig. 6.4). Therefore the decay of spin polarization in dots containing charged excitons will be limited by the exciton life-time (e.g. determined by radiative recombination at low T), with no oscillations in the dynamics[20]. Furthermore, a dot with a resident carrier will contribute to the signal in a polarization-resolved pump-probe experiment only if pump and probe pulses are co-polarized as described in Fig. 6.7 and Section 6.2.2. This arises due to the selection rules (a consequence of the Pauli exclusion principle), which apply to such experiments (see Fig. 6.7): absorption in the dot with a spin-up(down) resident electron is blocked for the $\sigma^{-(+)}$-polarized probe even before excitation with the pump.

We can thus separate the signal in polarization-resolved pump-probe experiments into two types: (i) an oscillating signal from neutral QDs, which decays due to dephasing processes and the inhomogeneous distribution of E_{FS} in a large ensemble of dots; and (ii) a non-oscillating signal decaying with time constant given by the

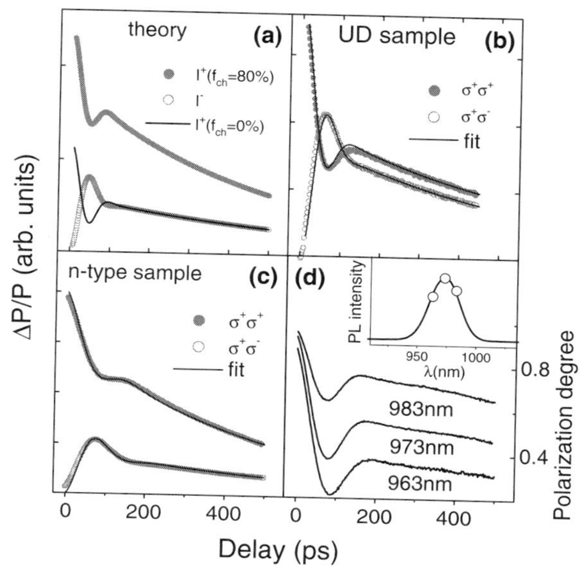

Fig. 6.18 (a-c) Dependence of differential transmission on pump-probe delay for σ^+ polarized pump and $\sigma^+(\sigma^-)$ polarized probe [solid(open) symbols, respectively]. Black curves in Figs b and c are fits using Eq. 6.2. (a) Calculations using Eq. 6.2. Solid symbols and the black curve show I^+ (see text) calculated for 80 and 0% of charged dots in the dot ensemble. (b) and (c) Measured pump-probe curves ($T = 7$K) for samples of type A (annealed for 2.5 min at $T = 700°$C, Fig. b) and B (annealed for 20 min at $T = 750°$C, Fig. c). (d) Dependences of the degree of circular polarization on the pump-probe delay measured for the sample of Fig. c at different spectral positions within the inhomogeneously broadened peak of the QD ground state (shown with circles on the corresponding PL spectrum in the inset)

exciton life-times of uncharged (τ_X) and charged (τ_{chX}) dots. The resulting signal can be described by Eq. 6.2.

An example of pump-probe curves calculated using Eq. 6.2 is shown in Fig. 6.18a. If no charged dots are present in the system ($A = 0$) then I^+ and I^- curves merge at long delays, where the signals decay as e^{-t/τ_X}. This occurs due to the damping of the quantum beats (second term in Eq. 6.2), which are only observed at short delays, where the I^+ and I^- signals are in antiphase to each other. However, if a large number of charged dots is present (80% of all dots in an ensemble in Fig. 6.18a), then I^+ is larger than I^- at all delays. Clearly, even a small number of charged dots ($<10\%$) can be detected as a non-zero difference of $I^+ - I^-$ or, in other words, a residual circular polarization at long delays. Below we will demonstrate this in the discussion of our experimental results.

The samples under investigation were grown by MBE on semi-insulating GaAs substrates. 2 types of samples each containing 16 layers of InGaAs dots were studied: nominally undoped (UD) samples with dots grown in the middle of 20 nm GaAs quantum wells, which in turn are clad by 10 nm AlGaAs barriers (type A); QD layers grown with 25 nm separation in GaAs with Si delta-doping 2 nm below each

QD layer and with nominally <1 electron per dot (type B). Samples of type A were annealed at 700°C for up to 17.5 min, while samples of type B were annealed at 750°C for up to 20 min.

A fs Ti-sapphire laser (Mira, Coherent) was used for pump-probe measurements, which were performed in a near normal incidence geometry. Both pump and time-delayed probe pulses were circularly polarised using $\lambda/4$ plates. The pump beam was modulated at 500 kHz. The probe beam transmitted through the sample was detected by an InGaAs photo-receiver and the ac component of the signal was measured by a lock-in amplifier, thus providing the time-resolved differential transmission (DT) signal. The DT results reported here were obtained with both pump and probe pulses exciting resonantly the QD ground state transition at $T = 7$K.

Typical DT pump-probe curves measured on samples of type A and B are shown in Fig. 6.18b and c, respectively. The curves are similar to those shown in Fig. 6.18a and represent a superposition of damped quantum beats and an exponential decay as given by Eq. 6.2. Notably, even in the nominally undoped sample (Fig. 6.18b) a residual polarization is observed at long delays ($I^+ - I^- > 0$) indicating a finite weight of charged dots in the dot ensemble. A much stronger signature of charged dots is seen in the data recorded on the n-type sample (Fig. 6.18c): although a beat pattern is still observable at short delays, the degree of circular polarization, $\Pi = (I^+ - I^-)/(I^+ + I^-)$, at long delays exceeds 0.5. Furthermore, we find that the degree of residual polarization depends on the excitation wavelength as shown in the inset of Fig. 6.18c for the n-type sample: it increases for the lower energy dots, which by referring to Eq. 6.2 is ascribed to their increased charging level. Note, that Eq. 6.2 predicts $\Pi = 0$ at long delays for neutral dots.

We now determine the fraction of charged dots in the dot ensemble by fitting the experimental data by Eq. 6.2. In Figs. 6.2b,c the fitting curves are shown by black lines. The fitting in Fig. 6.18b,c gives 22 and 83 % fraction of charged dots in samples A and B, respectively, as measured at the wavelength corresponding to the maximum of the ground state PL peak. In addition, Fig. 6.18d shows that in the n-type sample the degree of residual polarization at long delays becomes larger with increasing detection wavelength, corresponding to an increased charging level of the dot subset with decreasing energy within the ensemble. The wavelengths at which the data in Fig. 6.18d were measured are shown relative to the QD ground state PL spectrum in the inset of Fig. 6.18d. Fitting of the pump-probe curves corresponding to the polarization degree data in Fig. 6.18d yields 73%, 83% and 93% for the measurement at 963, 973 (PL peak maximum) and 983nm. Note, that the latter observation can be used to study charging profiles in large inhomogeneously broadened QD ensembles. In addition, the pump-probe data provide information on dynamics (including relaxation, radiative life-time etc) of charged and neutral excitons in QDs.

The sign of the charges in charged dots cannot be determined from these data. However, we suggest that in the UD sample the dots are positively charged due to the residual p-type doping known to occur in MBE grown GaAs structures. We also find that in samples of type A (undoped but with AlGaAs barriers) f_{ch} increases strongly after thermal annealing. This is clearly seen in the plot of the depen-

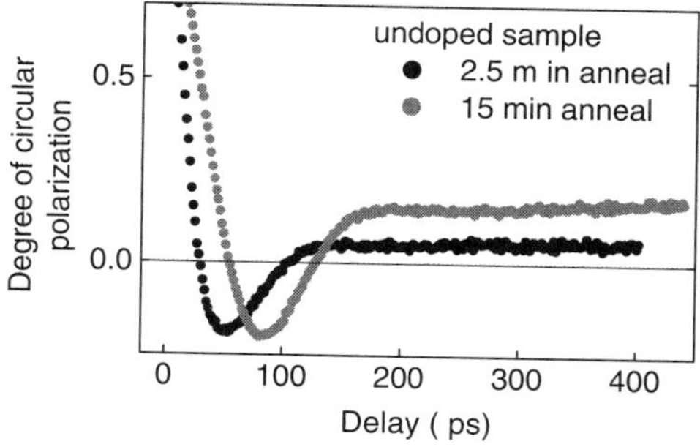

Fig. 6.19 Dependences of the degree of circular polarization on the pump-probe delay measured for two samples of type A annealed for 2.5 and 15 min (black and grey symbols respectively)

dence of the degree of polarization on pump-probe delay for two annealed samples originating from the same wafer (Fig. 6.19). Samples annealed for $t_{an} = 2.5$ and 15 min at $T_{an} = 700°C$ exhibit residual circular polarization of ≈ 0.05 and ≈ 0.17, respectively.

Fig. 6.20a shows the dependence of f_{ch} in dots in type A samples versus annealing time. A clear trend is observed with f_{ch} increasing with t_{an}: the fraction of charged dots increases from $\approx$ 20% to $\approx$ 40%. On the other hand, as seen in Fig. 6.20b, the n-type structures are much less affected by annealing: the fraction of negatively charged dots varies from $\approx$ 80% to $\approx$ 85%. We also observed a very weak dependence of f_{ch} on the annealing in UD samples with no Al containing layers in the structure. We suggest that the strong effect of annealing on the charging levels in the Al-containing sample is caused by charges released at high temperature from impurities in AlGaAs.

In conclusion, we have demonstrated a new method to measure the fraction of charged dots in dot ensembles and to study charged exciton dynamics. Previously charged excitons have been studied predominantly in single-dot spectroscopy [14, 55–57]. In many of those experiments some of the single-dot emission lines were tentatively ascribed to charged excitons, with firm evidence for their origin lacking. In our work, we demonstrate that even in nominally undoped QD structures a large fraction of the dot ensemble can be charged with an excess charge carrier. We also show that in structures where the dot layers are isolated by AlGaAs barriers, thermal

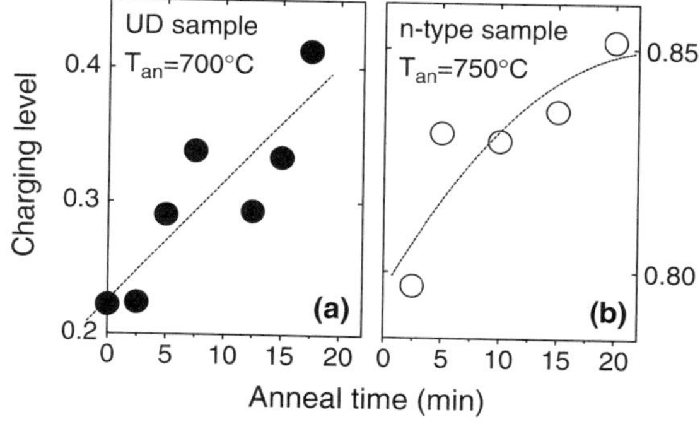

Fig. 6.20 Dependences of charging levels on the annealing time in samples of type A (Fig. a) and B (Fig. b). Dashed lines in (a) and (b) are guides for the eye

annealing results in increased charging of dots, which would have an immediate impact on spectra detected in single-dot experiments, where annealing is frequently employed to shift dot emission to shorter wavelength. We also find that annealing has little effect on charging levels in Al-free samples.

6.3.2 Electric Field Control of Spin Properties in Individual Quantum Dots

In this Section we present a study of the polarization properties of electronic states in individual InGaAs dots. Individual dots in our experiments are normally isolated by a metal shadow mask evaporated on the surface of the sample where sub-micron clear apertures are opened for optical access as shown in Fig. 6.21a. The dots are measured in a standard micro-PL set-up with the sample mounted on a cold finger inside a continuous flow He cryostat. The cold finger is placed inside a superconducting magnet with the fields up to 5 Tesla. The excitation as well as photoluminescence collection is performed by a long working distance (13mm) objective focusing the laser beam into $\approx 2\,\mu$m spot. The shadow mask deposited on the sample can also serve as a Schottky contact in so-called charge-tunable devices. Such devices are used to controllably generate charges in the dots by varying the bias applied across the intrinsic region where the dots are grown (see Fig. 6.21b). This procedure as well as the samples used in our experiments are described in more detail below.

The results presented in this Section are measured on an MBE grown sample with the dots embedded in a p-type Schottky diode. This allows controlled generation of positively charged excitons, X^+. At low temperature X^+ consists of two holes in the ground state with antiparallel spins (a consequence of the Pauli principle) and one electron with an unpaired spin (see the diagram in Fig. 6.22), the latter thus determining the X^+ spin [14, 25]. Thus excitation of a spin-polarized X^+ creates a spin-polarized electron, which relative to holes [29, 58] is expected to have an extended spin life-time [5, 59, 60].

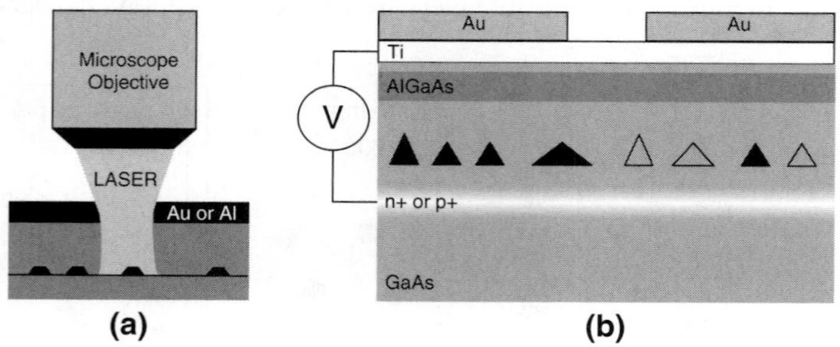

Fig. 6.21 (a) Schematic diagram of a micro-PL set-up and a sample covered with a metal shadow mask where a clear aperture is open using electron-beam lithography. (b) Schematic diagram of a Schottky diode device (charge-tunable device), where the shadow mask also plays a role of a top contact. Bias can be applied across the intrinsic region comprising quantum dots

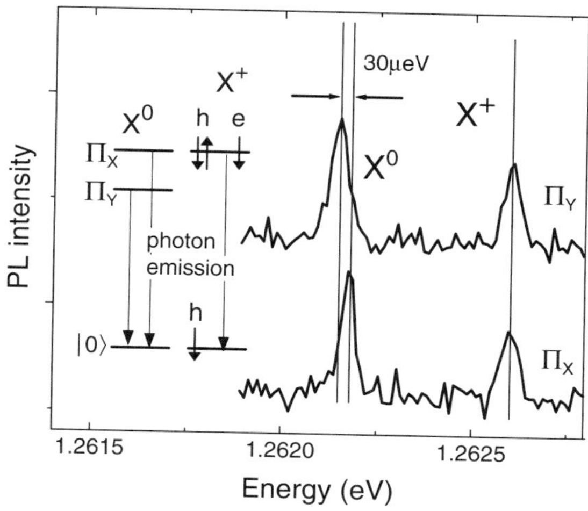

Fig. 6.22 PL spectra for dot B measured in two orthogonal linear polarizations. The fine structure splitting in X^0 emission is indicated. Diagram shows energy structure of the neutral (X^0) and charged (X^+) excitons. Short arrows show spin orientation of the holes (h) and electron (e). Long arrows show transitions from initial to final states accompanied by photon emission

In our experiments we perform optical generation ("writing") of spin-polarized individual electrons within the X^+ complex and demonstrate control of their degree of polarization by applying external electric and magnetic fields. The method could underlie the optical "writing" of coherent spin-polarized electrons required for recently proposed electron-spin-based quantum computation schemes [3, 4]. In our work we focus on X^+ since by contrast, at $B = 0$ spin states of a neutral exciton (X^0) are mixed by the electron-hole exchange interaction [14, 22, 25, 48] (Fig. 6.22).

We study low temperature ($T = 7$K) polarized emission from both neutral (X^0) and singly positively charged (X^+) QD excitons [61]. We find that at $B = 0$, excitation with circularly polarized light resonantly into the excited states leads to unpolarized X^0 emission and strongly (up to 70%) circularly polarized X^+ emission. Furthermore, we show that the degree of polarization of X^+ can be controlled by vertical electric field. The maximum of the degree of polarization (or spin-memory) corresponds to the bias regime with the maximum probability for the charging of the dot with a single hole which tunnels from the p^+ contact. When magnetic field is applied and the spin of the neutral exciton is well defined [25, 48] high spin-memory effects occur for both X^0 and X^+.

The sample investigated was grown using molecular beam epitaxy on an undoped GaAs substrate. The bottom contact was formed by a p^+ GaAs layer separated by a 25 nm undoped tunnel barrier from the InGaAs QDs. Optical access to individual QDs was provided by $\approx$ 800 nm apertures in an opaque Au mask (for more details of the sample see Ref. [62]). The diagram in Fig. 6.23b shows the band structure of the Schottky diode used in our work. Two bias regimes are shown corresponding to high probabilities of finding the dot either charged with holes tunneling from the p^+ region or empty. When the Fermi level of the structure is gradually tuned by the applied bias above the ground state (GS) in the dot, holes can tunnel to the excited states and eventually the wetting layer, thus populating the dot with multiple charge carriers. In the bias regime where tunneling occurs time-averaged PL spectra can exhibit peaks from both neutral and charged excitons, as a result of the finite probability of finding the dot empty prior to optical excitation [61, 62].

Figure 6.23a shows PL spectra recorded for Dot A in the two different bias regimes under resonant excitation at 1.274 eV with circularly (σ^+) polarized light and σ^+ or σ^- polarized detection. Fig. 6.23c shows the bias-dependence of $R = I(X^+)/I(X^0)$ for unpolarized emission (intensities in both polarizations are summed before the ratio is calculated). $R > 1$ (at $V < 0.75$V) corresponds to a high probability of hole tunneling into the dot and dominance of X^+, while at higher reverse biases the dot is increasingly unoccupied by holes and the X^0 line dominates in the PL ($R < 1$). As the reverse bias is further decreased towards the flat-band condition, more than one hole tunnels into the dot and wetting layer states, eventually leading to the occurrence of multi-charged exciton emission and quenching of X^+ and X^0. Thus in total in our PL measurements we can distinguish between 3 bias regimes (marked on Fig. 6.25) with differing emission spectra: 1) $V < 0.3$V, multiply charged exciton PL; 2) 0.3V$< V <$0.9V, singly charged and neutral exciton PL; 3) $0.9< V <1.4$, neutral exciton PL. Note that the transitions between the above three regimes are rather gradual for the case of hole charging.

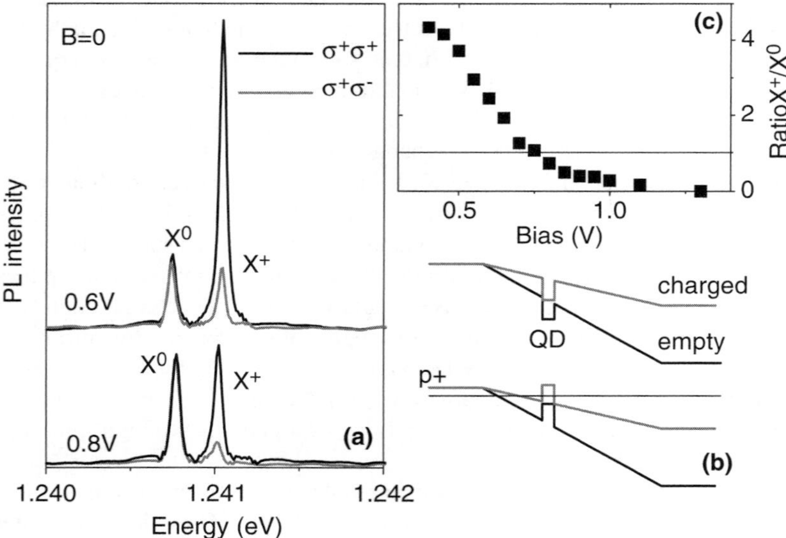

Fig. 6.23 (a) PL spectra for dot A excited with σ^+-polarized light at 1.274 eV. Spectra detected in σ^+ (σ^-) polarization are shown with black (grey) lines. Bottom (top) spectrum is measured at 0.8 (0.6) V. The diagram (b) shows the band structure of a Schottky diode in two bias regimes, when the dot (marked QD) is charged with extra holes or empty. (c) Bias-dependence of the intensity ratio of X^+ and X^0

Figure 6.23a shows that resonant σ^+ excitation into the excited states of the dot results in a high degree of circular polarization of the X^+ PL, $P_c = (I_{\sigma+} - I_{\sigma-})/(I_{\sigma+} + I_{\sigma-})$, ($I_{\sigma^\pm}$ is the PL intensity detected in $\sigma^\pm$ polarization). Such spin-memory is absent for X^0. Figure 6.22 shows PL spectra detected in crossed linear polarizations from another typical dot, Dot B, under linearly polarized excitation at 1.38 eV. X^0 emits in an orthogonally linearly polarized (Π_x, Π_y) doublet [14, 22, 25, 48] (split by 30 μeV), whereas the X^+ intensity and spectral position is the same for Π_x and Π_y polarizations. The combined evidence from the two polarization-resolved experiments shows the break-down of the spin nature of the X^0 neutral exciton states in InGaAs QDs, where instead two spin-mixed states form, split by the electron-hole exchange interaction. On the other hand, singly positively charged excitons are spin polarized with the spin of the e-h complex determined by the spin orientation of the uncoupled electron.

Figure 6.24 shows photoluminescence excitation (PLE) spectra obtained using σ^+ excitation in the two bias regimes when either X^+ or X^0 emission dominates (at 0.6 and 1V, respectively). For this measurement the excitation laser wavelength is scanned while the intensity of either X^0 or X^+ PL lines (PL emission from the dot ground state) is detected. PLE spectra at $V = 0.6$V for X^+ detection have much stronger intensity for co- relative to cross-polarized detection (black and grey lines, respectively) indicating efficient generation of spin-polarized X^+ in a wide range of excitation energies. We find that high efficiency selective pumping of spin-

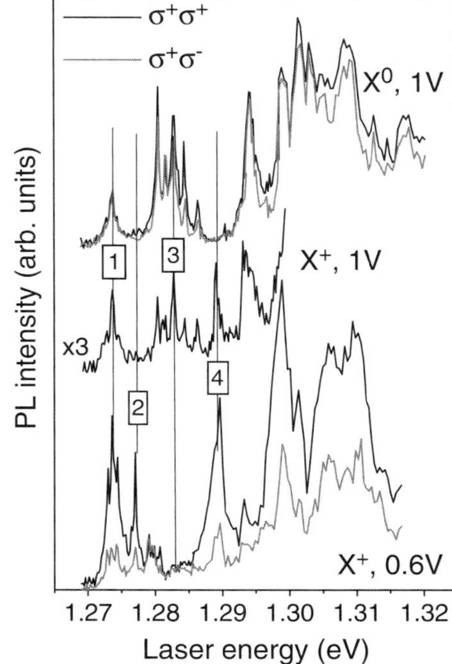

Fig. 6.24 Photoluminescence excitation (PLE) spectra measured for dot A for X^+ at V = 0.6 and 1 V and X^0 at V=1V. Spectra measured for σ^+ excitation and σ^+ (σ^-) detection are shown with black (grey) lines. The numbers mark the most pronounced features in the spectra (see text for details)

polarized X^+ states can be achieved at all excitation energies below the onset of the wetting layer at $\approx$ 1.42eV. In contrast, PLE spectra for X^0 detection in both σ^+ and σ^- polarizations (black and grey lines, respectively) have very similar intensities, providing evidence for negligible X^0 spin-memory.

Figure 6.24 also shows that the PLE spectra for X^0 detection have distinctly different spectral forms from those for X^+. The most notable features in the two types of spectra are marked 1-4 in Fig. 6.24. Feature 1 observed in both the X^0 and X^+ spectra was previously tentatively attributed to an exciton-InAs LO-phonon resonance[57]. By contrast with feature 1, features 2 and 4 (3) are observed for X^+ (X^0) only.

As seen from Fig. 6.23a, the relative intensity of X^+ for co- and cross-circularly polarized detection is very different at the two biases of 0.6 and 0.8 V. Fig. 6.25a shows the detailed bias dependence of the X^+ intensity for σ^+ and σ^- detection (referred to below as $I_{\sigma+}$ and $I_{\sigma-}$, respectively) and X^0 intensity detected in σ^+ polarization (I^0). The excitation is performed with σ^+ polarized light resonant with the maximum of feature 1 in Fig. 6.24. At low reverse bias ($V < 0.3$V) where the dot is filled with many holes the intensities are very weak. In this regime, P_c of X^+ (shown in Fig. 6.25b) is $\approx$ 0.35, indicating fast spin-depolarization following the optical excitation.

For 0.2V$< V < 0.4$V $I_{\sigma+}$ increases much faster with increasing bias than $I_{\sigma-}$ and I^0. The strong increase of the total X^+ intensity corresponds to the regime where the probability of single hole tunneling from the contact into the dot is high, whereas multiple charging is significantly less probable. Our data show that this regime also corresponds to a higher degree of circular polarization in X^+ PL. Indeed, P_c of X^+ reaches $\approx$ 0.65, indicating that spin-flip occurs on a time scale longer than the exciton life-time ($\approx$ 1ns).

For $V > 0.6$V $I_{\sigma+}$ decreases with bias and eventually becomes weaker than I^0 at $V > 0.8$, whereas I^0 grows monotonically in the whole range $V \leq 1.1$V. The behaviour at $V > 0.6$V reflects the gradual tuning of the Fermi level above the GS of the hole in the dot leading to suppression of hole tunneling and increasing probability of creation of X^0 in an empty dot. Fig. 6.25b shows that as $I_{\sigma+}$ decreases with V, the polarization degree of the X^+ emission also decreases: at 1V it has decreased to 0.45.

Fig. 6.25 demonstrates efficient control by applied electric field of the X^+ P_c following resonant optical excitation. The degree of spin-memory correlates well with the three regimes of hole tunneling from the contact discussed above, and is maximum when the probability is high for the dot to be charged with a single extra hole.

In order to investigate further the physics behind the control of the spin-memory we performed additional PLE measurements. Fig. 6.24 shows the PLE spectrum for X^+ at $V = 1$V, in the regime when X^0 dominates. The spectrum shows features from both the spectrum for X^+ at $V = 0.6$V and the spectrum for X^0 at $V = 1$V: all five lines of group 3 are present and peak 4 is very pronounced. The mixed X^+/X^0 character of the X^+ PLE spectra at $V = 1$V has a natural explanation if we assume that in this bias regime the X^+ emission can be excited via absorption in both charged and empty dots. The occurrence of the absorption lines of a neutral

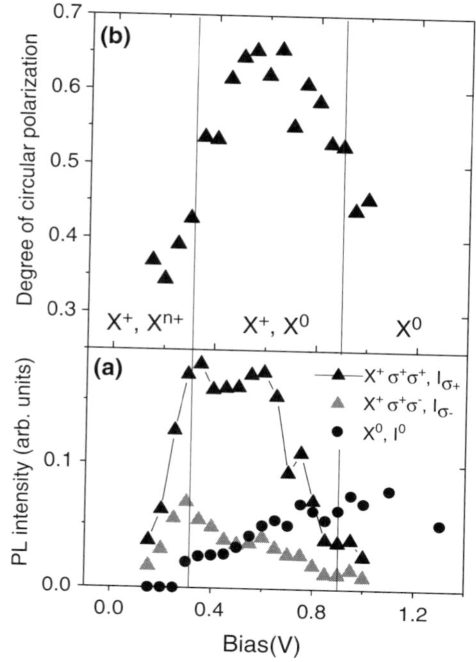

Fig. 6.25 Vertical lines and symbols in the top panel show the approximate boundaries of the three bias regimes where X^{n+} (multiply-charged exciton), X^+ and X^0 can be detected in PL. (a) Bias-dependence of X^+ and X^0 PL intensity for dot A excited at 1.274 eV with σ^+-polarized light. The intensity detected for X^+ in σ^+ (σ^-) polarization is shown with black (grey) triangles. Circles show the intensity of X^0 measured in σ^+ polarization. (b) Bias-dependence of the degree of circular polarization of X^+ calculated using the data in Fig. a

exciton in the X^+ PLE spectrum can only be explained by the tunneling of an extra hole after an e-h pair has been generated by the optical excitation. Relaxation of the e-h pair will generate a neutral exciton in the ground state, which as our experiments show does not possess spin information. Extra hole tunneling from the contact into the dot after the excitation will then create a charged exciton with random spin orientation, thus diminishing the degree of spin-memory in comparison with X^+ created in an occupied dot.

In the regime where the probability is high for a single hole to tunnel from the p^+ contact, a similar mechanism may be responsible for the P_c, although high, being less than 1 [4]. High X^+ P_c corresponds to efficient optical pumping of the electron spin (determining the spin of X^+), which as seen from the high P_c magnitude only loses its initial alignment within a charged exciton on a timescale longer than the

---

[4] Our experimental set-up allows excitation with $\approx$ 80% circularly polarized light. A further slight loss of polarization occurs on the detection side.

X^+ life-time. The observed effect can be used for the "write" algorithm in memories based on spin-polarized electrons in dots [59]: after optical excitation of X^+, the two holes can be removed from the dot by an electric field pulse [5] leaving an optically generated spin-polarized electron [5].

The already high P_c for X^+ at $B = 0$ in the intermediate bias regime can very likely be enhanced to close to $P_c = 1$ by use of devices with more abrupt charging thresholds. In this case at a given bias only one of either X^0 or X^+ will be present, as can be realized in structures with thinner tunnelling barriers between the dots and the p^+ region. On the other hand, strong depolarization still occurs in the low bias (or multiple-charging) regime, where the electron spin is flipped due to interaction with the randomly polarized holes tunneling in and out of the dot and the surrounding the dot wetting layer.

Further insight into the mechanism controlling the degree of exciton spin-memory is obtained from polarization-resolved PL measurements in magnetic field. Fig. 6.26 shows PL spectra for dot A measured for $B = 2$T in the bias regime where both X^0 and X^+ PL is strong. The spectra excited with linearly and circularly (σ^+) polarized light in resonance with feature 1 in Fig. 6.24 are shown with gray and black lines, respectively. Unpolarized PL is detected.

Under linearly polarized excitation both X^0 and X^+ PL exhibit Zeeman components of similar intensities (marked on the graph). The Zeeman splitting (the energy splitting between the PL peaks in σ^+ and σ^- polarization) for both X^+ and X^0 is $\approx 130\,\mu$eV ($\approx 70\mu$eV at $B = 1$T) and exceeds the fine-structure splitting of X^0. Thus at $B > 1$T the ground state of the neutral exciton is composed of pure spin eigenstates [25, 48]. When the excitation polarization is changed from linear to circular, the relative strength of the polarized PL components becomes strongly dependent on the polarization of the excitation. In marked contrast to the $B = 0$

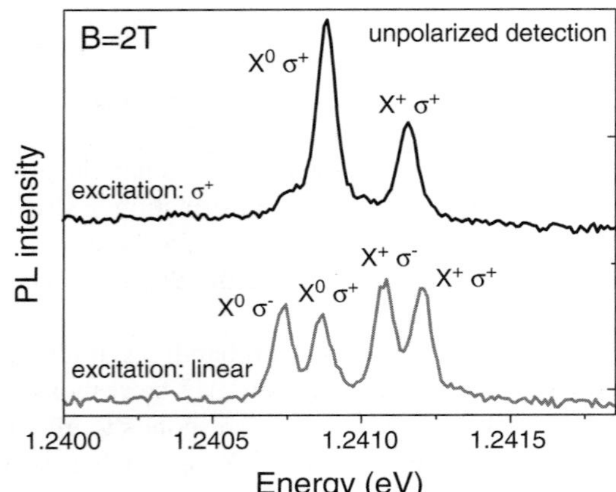

Fig. 6.26 Unpolarized PL spectra for dot A at $B = 2$T excited with linearly (gray) and circularly (black) polarized light resonant with feature 1 in Fig. 6.24

[5] Spin-memory effects will also permit optical pumping of spin-polarized holes since recombination of a spin-polarized X^+ will leave a hole with a well defined spin-orientation.

case in Fig. 6.22, and similarly to X^+, the X^0 emission becomes $\geq 80\%$ polarized at $B = 2$T, indicating efficient spin pumping of neutral excitons when their ground state has pure spin character in magnetic field.

In summary, in this Section we discussed optical orientation experiments on individual InGaAs QDs embedded in a Schottky diode using resonant circularly polarized excitation into the QD excited states. An optical "write" procedure for spin-polarized single electrons may be based on the controlled generation of spin-polarized singly-charged excitons achieved in our work. In particular, we demonstrated 1) a high degree of circular polarization of the emission of singly-charged excitons and its gate control, 2) absence of circular polarization for neutral excitons at $B = 0$, 3) a large degree of circular polarization for neutral excitons for $B > 1$T and finally 4) suppression of spin-memory effects in the regime of multiple-hole charging.

6.4 Electron-nuclear Spin Interaction in Quantum Dots

6.4.1 Hyper-fine Interaction

The hyperfine interaction in solids arises from the coupling between the magnetic dipole moments of nuclear and electron spins. Two major effects arising from the hyperfine interaction are (i) inelastic relaxation of electron spin via the "flip-flop" process (Fig. 6.27) and (ii) the Overhauser shift of the electron energy induced by the net magnetic field of the polarized nuclei [63, 64].

The dominant contribution to the coupling between the electron- and the nuclear-spin systems originates from the Fermi contact hyperfine interaction. It is notable that the nuclei interact only with electrons, the contact hyperfine interaction being negligible for holes due to the p-like nature of the Bloch functions [64]. In first order perturbation theory the electron-nuclear interaction in a dot can be expressed as:

$$\hat{H}_{hf} = \frac{v_0}{8} \sum_i A_i |\psi(\mathbf{R_i})|^2 \hat{\mathbf{S}} \cdot \hat{\mathbf{I}}^i \tag{6.4}$$

where v_0 is the volume of the InAs-crystal unit cell containing eight nuclei, $\hat{\mathbf{S}}$ is the dimensionless electron spin operator, $\psi(\mathbf{r})$ is the electron envelope wave function and $\hat{\mathbf{I}}^i$ and $\mathbf{R}_i$ are the spin and location of the i-th nucleus, respectively. $A_i = \frac{2}{3}\mu_0 g_0 \mu_B \mu_i I^i |u(\mathbf{R}_i)|^2$ is the hyperfine coupling constant, which depends on the nuclear magnetic moment μ_i, the nuclear spin I^i and on the value of the electron Bloch function $u(\mathbf{R}_i)$ at the nuclear site. μ_B is the Bohr magneton, g_0 is the free electron g-factor and μ_0 is the permeability of free space. A_i is of the order of 50 μeV for all nuclei in our InGaAs/GaAs dot system.

$\hat{\mathbf{S}}\cdot\hat{\mathbf{I}}^i$ can be rewritten as $1/2(I^i_+ S_- + I^i_- S_+) + I^i_z S_z$ where $I^i_\pm$ and $S_\pm$ are the nuclear and electron spin raising and lowering operators, respectively. Eq. 6.4 thus contains

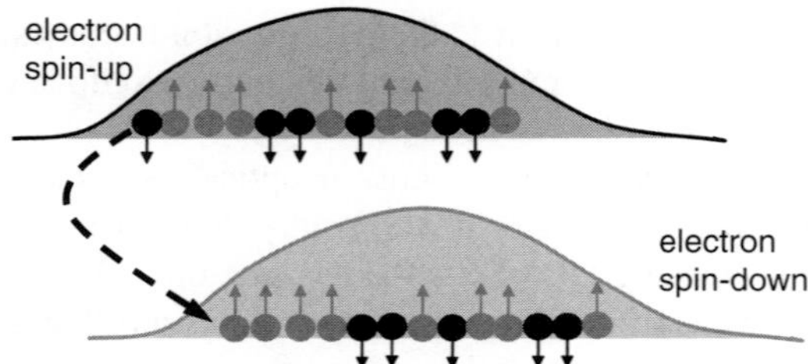

Fig. 6.27 Schematic diagram of the electron-to-nuclei spin-transfer mechanism often referred to as "flip-flop". The electron envelope wave-function "covers" $N = 10^4$ nuclei in a typical InGaAs dot. The electron-nuclear spin exchange process is described by the dynamical part in Eq. 6.4. The spin from one electron can be transferred to a single nucleus only

a dynamical part, $1/2(I_+^i S_- + I_-^i S_+)$ allowing for electron-nuclear spin exchange and a static part, $I_z^i S_z$, which simply acts to modify the energies of the two spin systems. The dynamical part is responsible for the transfer of angular momentum from spin-polarized electrons, which can be excited optically on the dot, to the nuclear spin reservoir of $N \approx 10^4$ nuclei in a typical InGaAs dot (see Fig. 6.27). The static part can be considered as an effective magnetic field acting either on electrons due to spin polarized nuclei (Overhauser field) [60, 65–69] or on the nuclei due to a spin polarized electron (Knight field [8]).

In the absence of electrons nuclei in the dot are unpolarized even at very low lattice temperatures and high magnetic fields. This is mainly caused by very small nuclear magnetic moments and hence Zeeman splittings, leading to negligible polarization due to the thermalization. Indeed, in magnetic field of 2 Tesla, the Zeeman splitting for Ga^{69} nucleus is 70 neV only. Such small splittings result also in suppression of the thermalization process due to a very small probability of phonon scattering. On the other hand, rather large fluctuations of the nuclear polarization are possible leading to the fluctuations of the nuclear magnetic field, B_N, of order of $A/(\sqrt(N) g_e \mu_B) = 30$mT [70]. Such fluctuations exist even at very low temperatures and their effect is further enhanced in a few nanometer dot due to a tight overlap of the electron wave-function with the underlying nuclei. The fluctuations lead to decoherence and relaxation of the electron spin [71, 72] and are detrimental for possible applications in quantum information devices.

One of the possibilities to overcome the fluctuations of the nuclear field is to artificially maintain very high degree of nuclear polarization [73], which can be achieved by optical pumping. Eq. 6.4 predicts that a single electron can transfer its spin to a single nucleus only (see Fig. 6.27), leading to a necessity to excite a large number of electrons on the dot. Populating the dot with a single electron at a time will allow flexibility and control of the electron polarization: multi-charge population may lead to additional polarization dissipation channels such as the formation of spin-singlets and spin relaxation without spin transfer to the nuclei. Thus the requirement is to provide a flow of spin polarized electrons through the dot, where the electrons are introduced into the dot one-by-one.

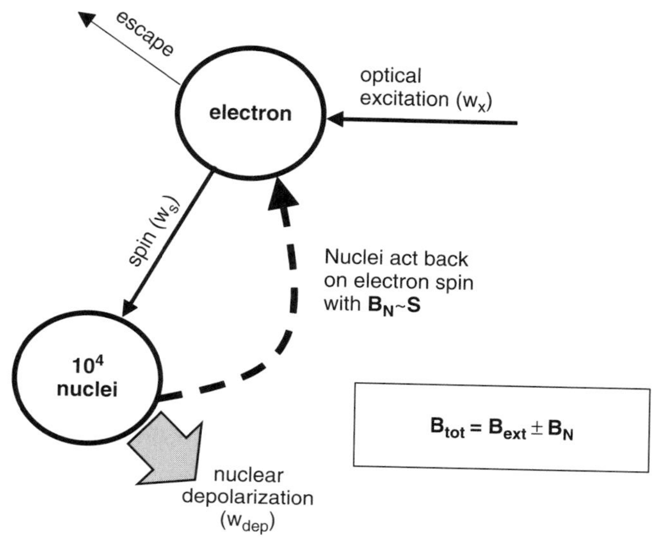

Fig. 6.28 Schematic diagram of the dynamical nuclear polarization in a quantum dot containing 10^4 nuclei. Spin polarized electrons are introduced to the dot by optical circularly polarized excitation with the rate w_x. They then escape from the dot either due to the radiative recombination with the holes or tunneling in the applied electric field. The electrons transfer their spin to the nuclear system with the rate w_s, which competes with the nuclear spin dissipation due to the diffusion into the bulk with the rate w_{dep}. The three rates define the magnitude of the effective nuclear magnetic field, B_N, acting back on electrons and proportional to the steady state nuclear polarization, S. The B_{tot} is the total magnetic field seen by the electron spin, which depends on the sign of the nuclear polarization

Fig. 6.28 shows the detailed procedure of the nuclear pumping in an InGaAs dot used in our experiments carried out at low sample temperature of 10-20K. The electrons and holes are excited using circularly polarized light resonant either with the wetting layer states (about 120 meV above the ground state of the dot) for photoluminescence measurements or with the ground state of the dot for photo-current experiments. After excitation at high energy the electrons relax to occupy the lowest energy state. Such energy relaxation is accompanied with partial loss of the spin polarization. Nevertheless, both excitation methods lead to a high degree of spin polarization of the electrons in the ground state of the dots and hence efficient nuclear pumping.

Most of the electrons excited in the dot will escape from the dot either via the radiative recombination or tunneling in applied electric field before they transfer spin to the nuclei. However, if a high enough pumping rate is established the nuclear polarization will build up. This is a dynamical process: the nuclear spin pumping competes with the spin diffusion into the bulk of the crystal. The electron pumping, spin transfer and spin diffusion rates introduced in Fig. 6.28 will be explained in more detail below in Sections 6.4.2 and 6.4.3 where we present the study of the electron-nuclear interactions in individual InGaAs dots.

6.4.2 Nuclear Spin Pumping in High Magnetic Fields

In this Section we study a p-i-n structure grown using molecular beam epitaxy on an n+ GaAs substrate. On top of n+ GaAs, undoped layers of AlGaAs (25nm) and GaAs (400nm) were deposited and a single dot layer (dot density of $5 \cdot 10^9 cm^{-2}$) was then grown. The structure was completed with undoped layers of GaAs (180nm) and AlGaAs (25nm) and the top p+ contact formed with 300 nm of Be-doped GaAs. The structure then was processed into diodes with the top diode surfaces covered with an opaque 80 nm thick Al mask where 400 nm clear apertures were opened for optical access to the individual QDs.

Photocurrent (PC) experiments were performed using a cw Ti-Sapphire laser ($<$ 1MHz linewidth) for excitation into the ground state of the dot. The PC signal arises from the tunneling to the contacts of electron-hole pairs excited resonantly into the dot. The PC measurements were performed by tuning the applied bias whilst fixing the laser excitation wavelength. This technique is based on the Stark shift (see the inset to Fig. 6.29a) exhibited by the QD exciton states under applied electric field, F. This allows a direct conversion of the voltage at which the PC signal is detected into the corresponding spectral position [62].

Figure 6.29a shows PC spectra for the exciton ground state with linearly polarized laser excitation at different magnetic fields. The PC spectrum peak at 1.2992 eV for $B_{ext} = 0T$ corresponds to the bias 3.94V. As the magnetic field is increased

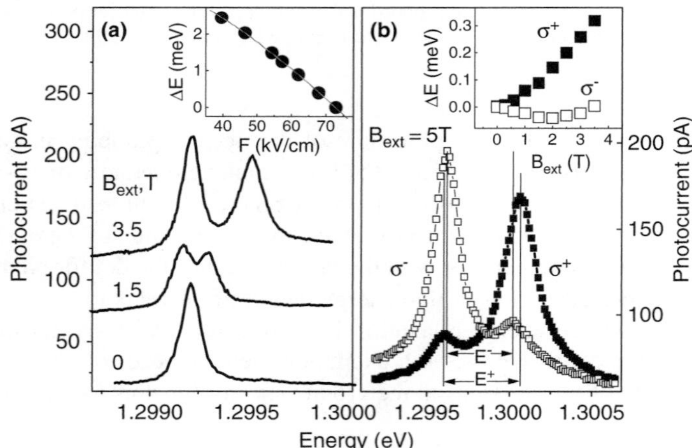

Fig. 6.29 (a) Photocurrent spectra measured with linearly polarized excitation into the ground exciton state of an individual dot. The single ground state peak at the external magnetic field $B_{ext} = 0T$ develops into the Zeeman doublet for $B_{ext} \neq 0$. Inset shows the Stark shift of the ground state exciton peak. (b) Photocurrent spectra measured with circularly polarized excitation into the ground exciton state of an individual dot at $B_{ext} = 5T$. Vertical lines indicate the peak maxima. The Zeeman doublet excited with $\sigma^+(\sigma^-)$ light is shown with solid (open) symbols and $E^+(E^-)$ denotes the Zeeman energy splitting. Inset shows the Zeeman peak energies as a function of B_{ext}

the ground state peak splits into a Zeeman doublet. In addition, as shown in the inset to Fig. 6.29b, the peaks exhibit the diamagnetic shift. As seen in Fig. 6.29b for spectra measured at $B_{ext} = 5$T the low (high) energy peak has the strongest intensity when excited with σ^- (σ^+) circularly polarized excitation. The finite intensity of the σ^- (σ^+) peak when exciting with σ^+ (σ^-) polarized light is related to imperfect polarization of the laser excitation of $\approx 80\%$.

As seen in Fig. 6.29b the splitting between the two Zeeman peaks is dependent on the sign of the circular (σ^+ or σ^-) polarization of excitation. At $B_{ext} = 5$T the splitting under σ^--polarized excitation, $E^- = 398\mu$eV, while for σ^+ polarization E^+ of 465μeV is measured. A dependence of the splitting on the polarization of excitation is a signature of the dynamical polarization of the nuclear spins in the dot [60, 65–69]. The sign of the effective nuclear magnetic field, B_N, observed in Fig. 6.29b is in agreement with that previously reported [60, 65–69]: B_N is parallel (anti-parallel) to B_{ext} when $\sigma^+(\sigma^-)$ excitation is used.

The difference of the Zeeman splittings for σ^- and σ^+ polarizations exhibits saturation at high excitation power. Fig. 6.30a shows the power-dependence of the shift, Δ^+, between the σ^+ (high energy) Zeeman peaks excited with σ^+ and σ^- polarized light for a dot at 1.275 eV. Here the high energy peak of the Zeeman doublet corresponds to the bias of 4.4V. Δ^+ is deduced from spectra where similarly to those in Fig. 6.29b both peaks of the Zeeman doublet are measured in each polarization. A clear saturation of Δ^+ is observed with increasing incident power (P), which reflects the saturation of B_N, reported previously for bulk, quantum wells and GaAs dots [64, 74–77]. In Fig. 6.30 the maximum $\Delta^+ = 47\mu$eV. The nuclear

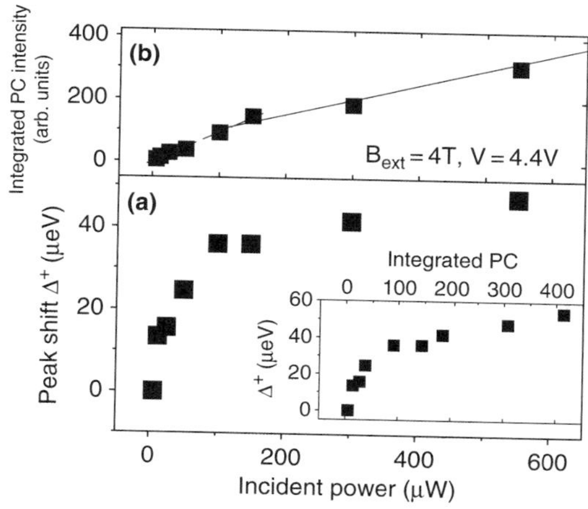

Fig. 6.30 (a) Excitation power dependence of the shift, Δ^+, between the σ^+ (high energy) Zeeman peaks excited with σ^+ and σ^- polarized light for a dot at 1.275 eV. (b) Power-dependence of the integrated photocurrent intensity, I_{PC}, of the high energy Zeeman peak excited with σ^+ light. Lines are guide for the eye only indicating a kink in power-dependence at $P \approx 150\mu$W. Inset in Fig. a shows Δ^+ from Fig. a plotted as a function of I_{PC} from Fig. b. The data in (a) and (b) is measured for $B_{ext} = 4$T and bias 4.4V

field can be estimated as $B_N \simeq \Delta^+/(g_e\mu_B)$, where g_e - electron g-factor, μ_B - Bohr magneton. We use $g_e = 0.55$ [78], which gives $B_N \approx 1.6$T for $\Delta^+ = 47\mu$eV.

The integrated PC, I_{PC}, the magnitude of which reflects the efficiency of the light absorption in the dot, is plotted as a function of P in Fig. 6.30b. For $P > 150\mu$W, although I_{PC} still increases with P, the rate of the increase is decreasing indicating that the ground state absorption is weakened by the increasing dot occupancy [62, 79]. However, the inset in Fig. 6.30a shows a similar saturating increase of Δ^+ when plotted as a function of I_{PC} instead of P, confirming the more significant effect on Δ^+ of B_N saturation rather than I_{PC}.

We also find that remarkably the degree of nuclear polarization and hence B_N can be controlled by applied bias, as demonstrated in Fig. 6.31. Fig. 6.31a(b) shows PC spectra of the high energy peak of the Zeeman doublet for the dot at ≈ 1.3eV (as in Fig. 6.29) measured at 2.9(4.6)V. The PC spectra in Fig. 6.31 excited with σ^+ or σ^- polarized light are shown with solid and open symbols, respectively. The spectra are measured at high power where a weak dependence of Δ^+ on power is observed. As seen in Fig. 6.31, at 4.6V Δ^+ considerably exceeds that at 2.9V: $\Delta^+(4.6V)$=40μV, $\Delta^+(2.9V)$=13μV.

The bias-dependent data are summarized in Fig. 6.32a. In the whole range of biases the measurements are done at a high $P = 500\mu$W. As seen from the figure the shift Δ^+ exhibits a practically linear dependence on the bias and varies from ≈ 6 to 40 μeV in the range 2.5–4.6V. Similar dependences were obtained on other dots studied. The peak shift is caused by the increase of B_N (acting on the electron spin only). B_N as high as 1.3T can be estimated from the figure. The data presented in Fig. 6.32b demonstrates an efficient control of the nuclear spin by applied electric field with B_N tuned in the range of $0 - 1.3$T.

The increase of B_N with bias anti-correlates with the electron life-time bias-dependence: Fig. 6.32b shows that the peak linewidth increases when the bias is changed from 2.9 to 4.6V. The change in the linewidth reflects the enhancement

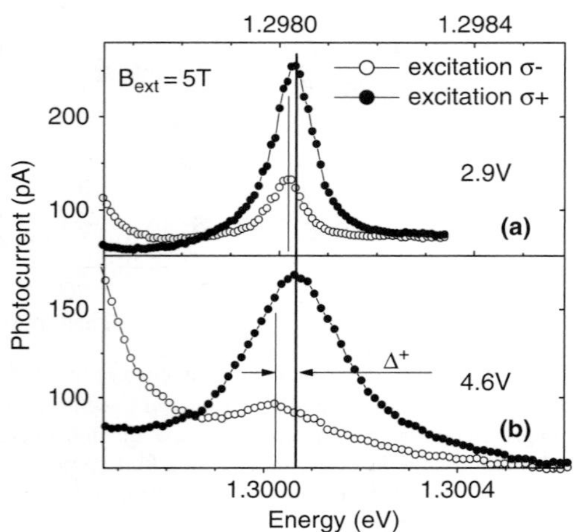

Fig. 6.31 Photocurrent spectra of the σ^+ (high energy) Zeeman peaks excited with σ^+ and σ^- polarized light (solid and open symbols, respectively) measured at the bias of 2.9V (a) and 4.6V (b). Vertical lines indicate peak maxima. The energy difference Δ^+ is shown with arrows in (b)

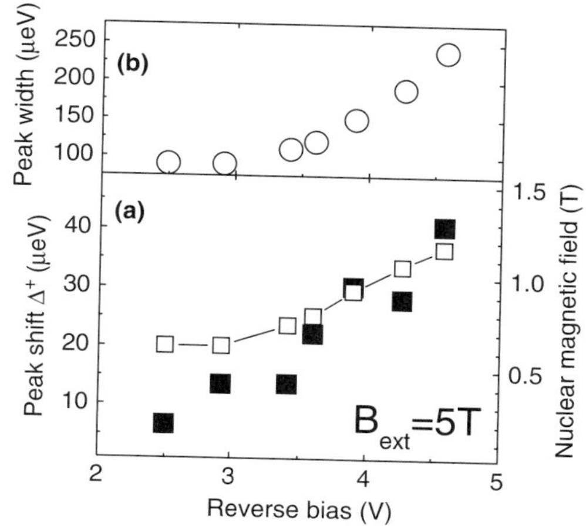

Fig. 6.32 (a) σ^+ peak shift Δ^+ measured at $B_{ext} = 5$T as a function of applied bias. Effective nuclear magnetic field is deduced as $B_N = \Delta^+/(g_e\mu_B)$ with $g_e = 0.55$. Squares show the theoretical fit obtained using the perturbation theory with the fitting constant $C = 6150$ (described in text). (b) Photocurrent peak line-width as a function of bias

of the tunneling rate (decrease of the carrier life-time) in the increased electric field [62, 79]. In the regime of efficient nuclear polarization at $V > 4$V the PC peak width up to 240μeV is observed. Taking into account the bias-independent contribution to the peak width (90μeV) we deduce that at 4.6V the exciton life-time, defined by the fastest tunneling process involving electron [62, 79], is $\approx$ 2ps.

The following important parameters will describe the main physics of the observed phenomenon: the strength of the hyperfine interaction h_{hf}, the electron Zeeman energy E_{eZ}, the electron tunneling rate w_e, the nuclear depolarization rate w_{dep}, and the pumping rate w_x.

The hyperfine interaction couples the two electron Zeeman split spin states and thus the probability of the spin "flip-flop" process is given by: $P_{hf} = |h_{hf}|^2/(E_{eZ}^2 + \hbar {w_e}^2/4)$. Since a real transition to the state with the opposite spin requires additional energy (E_{eZ}) a more probable scenario will be a virtual spin-flip, where the electron virtually occupies the state with the opposite spin, while remaining at the same energy. This is possible if the electron simultaneously leaves the dot: the electron can tunnel into a real state in the contact with the same energy, but opposite spin, thus leaving the spin behind in the nuclear system. The spin transfer rate will be obtained as $w_e P_{hf}$, as w_e describes the number of "attempts" an electron makes to transfer its spin to nuclei (i.e. to leave the dot). Neglecting the absorption saturation effects, the steady state nuclear polarization can be described as $S_N \propto w_e P_{hf} w_x / w_{dep}$. Thus $\Delta^+ \approx g_e \mu_B B_N = C\hbar w_e/(E_{eZ}^2 + \hbar {w_e}^2/4)$, where constant $C \propto w_x/w_{dep}$. At $B_{ext} = 5$T an estimate of E_{eZ} (for $g_e = 0.55$[78]) gives 160μeV. Using C as a fitting parameter and $\hbar w_e$ as measured in Fig. 6.32b we plot calculated Δ^+ in Fig. 6.32a (squares). Clearly a very similar trend to the experiment is observed with $C = 6150$.

The magnitude of C was chosen to fit the data at the high biases, where a more reliable measurement of the linewidth (w_e) is possible. Indeed, the deviation of the fitting at low electric field is most likely caused by the fact that the line-width measured in the experiment (90μeV) does not reflect the true life-time limited line-width of the exciton state, but is rather due to the fluctuations of the electric field [80]. Note also that $w_e/(E_{eZ}^2 + \hbar w_e{}^2/4)$ decreases at high w_e after reaching a maximum. This reflects a decreasing probability of the spin transfer when the electron life-time on the dot becomes very short. This situation is not observed in our experiments since even for the highest bias E_{eZ} is still 65% of the exciton linewidth.

In conclusion, optically induced nuclear spin polarization is achieved in individual InGaAs dots in a p-i-n diode allowing control of the spin state of $\approx 10^4$ nuclei. We employ optical "writing" of the electron spin directly into the ground state of a neutral exciton in high magnetic fields (4-5T). The nuclear magnetic field of up to 1.6T is achieved either by increasing the optical pumping rate or by tuning the vertical electric field. The latter enables control of the carrier tunneling rates with the carrier dynamics directly influencing the nuclear spin pumping rate.

6.4.3 Nuclear Spin Bi-stability in Moderate Magnetic Fields

In this Section we focus on fundamental properties of the coupling between the nuclear spin and the spin of a strongly localized electron. We show that in a few nanometre-scale QD a large local magnetic field (up to 3T) experienced by the spin of a single electron can be controllably switched on/off by manipulating the nuclear spin configuration. The origin of the nuclear spin switching is optically induced bistability of the nuclear spin polarization. We show that it arises due to feedback of the optically induced Overhauser field B_N on the dynamics of the spin transfer between the confined electron and nuclei.

In our experiments, samples containing InGaAs/GaAs self-assembled QDs are placed in an external magnetic field of $B_{ext} = 1 - 5$T and low electric field so that the non-radiative carrier escape through tunneling is negligible. Spin-polarized electrons are introduced one-by-one into a single 20nm InGaAs dot at a rate w_x (see Fig. 6.28) by the circularly polarized optical excitation of electron-hole pairs 120 meV above the lowest QD energy states. Due to the finite probability of hole spin-flip during energy relaxation, both bright and dark excitons can form in the dot ground state. The former will quickly recombine radiatively with a rate $w_{rec} \approx 10^9$ sec^{-1}, whereas the dark exciton can recombine with simultaneous spin transfer to a nucleus via a spin "flip-flop" process (as in Fig. 6.27) at the rate $w_{rec}P_{hf}$. Here P_{hf} is the probability of a "flip-flop" process, which from our perturbation theory treatment is given by:

$$P_{hf} = |h_{hf}|^2/(E_{eZ}^2 + \frac{1}{4}\gamma^2) \tag{6.5}$$

Here γ is the exciton life-time broadening, h_{hf} is the strength of the hyperfine interaction of the electron with a single nucleus and E_{eZ} is the electron Zeeman splitting. E_{eZ} is strongly dependent on the effective nuclear magnetic field B_N generated by the nuclei. This provides the feedback mechanism between the spin transfer rate and the degree of nuclear polarization ($S \propto B_N$) in the dot. The feedback gives rise to bistability in the nuclear polarization and threshold-like transitions between the spin configurations of 10^4 nuclei leading to abrupt changes of B_N by up to 3T in few nanometre sized QDs.

Such nuclear spin switch has been observed in several different structures containing self-assembled InGaAs/GaAs QD with ~3 × 20 × 20 nm size. Here we present results obtained at a temperature of 15K for two GaAs/AlGaAs Schottky diodes, where dots are grown in the intrinsic region of the device. In these structures a bias can be applied permitting control of the vertical electric field, F. Both n- and p-type Schottky devices were employed similar to those reported in Ref. [56] and Ref. [80]. For photoluminescence (PL) experiments, individual dots are isolated using 800 nm apertures in a gold shadow mask on the sample surface as was discussed in Section 6.3.2.

Figure 6.33 shows time-averaged (60s) PL spectra recorded for a neutral exciton in a single QD in an external magnetic field of 2T. Circularly polarized laser excitation in the low energy tail of the wetting layer (at 1.425eV) is employed and unpolarized PL from the dot is detected using a double spectrometer and a CCD. This excitation/detection configuration is used for all data presented. For each excitation polarization a spectrum consisting of an exciton Zeeman doublet is measured with the high (low) energy component dominating when σ^+ (σ^-) polarization is used.

A strong dependence of the exciton Zeeman splitting (E_{xZ}) on the polarization of the excitation is observed in Fig. 6.33: $E_{xZ}(\sigma^+) = 260\mu$eV and $E_{xZ}(\sigma^-) = 150\mu$eV. Such a dependence is a signature of dynamic nuclear polarization [8, 60, 65–68], which gives rise to a nuclear field B_N aligned parallel or anti-parallel to B_{ext} for σ^+ or σ^- excitation respectively. The difference between $E_{xZ}(\sigma^-)$

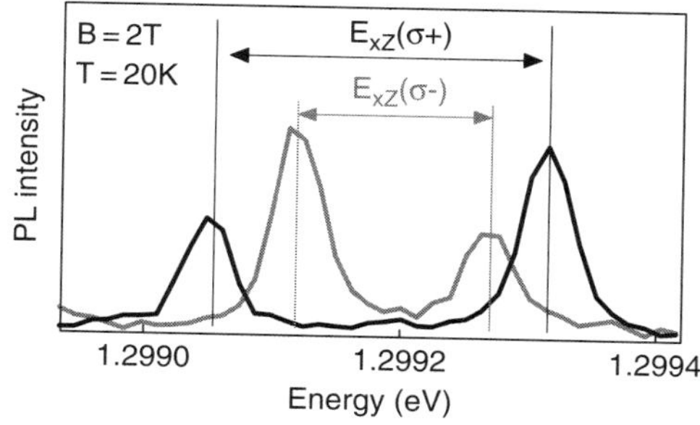

Fig. 6.33 X^0 photoluminescence spectra recorded for an individual InGaAs QD in an external magnetic field $B_{ext} = 2$T at $T = 20$K. The spectrum excited with σ^+ (σ^-) light resonant with the wetting layer is plotted in blue (red). The horizontal arrows show the corresponding exciton Zeeman splittings

and $E_{xZ}(\sigma^+)$ is usually termed the Overhauser shift and is given by $E_{Ovh} = |g_e|\mu_B(B_N(\sigma^+) + B_N(\sigma^-))$, where g_e is the electron g-factor, μ_B is the Bohr magneton, and $B_N(\sigma^+)$ and $B_N(\sigma^-)$ are the nuclear fields for the two excitation polarizations. It is notable that the nuclear field acts only on electrons, the contact hyperfine interaction being negligible for holes due to the p-like nature of the Bloch functions [64].

We now investigate the dependence of $B_N(\sigma^\pm)$ on the pumping power (or w_x). The grey-scale plot in Fig. 6.34a presents PL spectra at $B_{ext} = 2.5$T for a neutral exciton as a function of the power of σ^- excitation. At low power the Zeeman splitting $E_{xZ} = 310\mu$eV. As the power increases an abrupt decrease of the Zeeman splitting to $E_{xZ} = 225\mu$eV is observed, indicating the sudden appearance of a large nuclear field. Below we will refer to this effect as the nuclear spin switch.

E_{xZ} power dependences measured at $B_{ext} = 2$T for both σ^+ and σ^- excitation polarizations are shown in Fig. 6.34b. For σ^- excitation even below the switch,

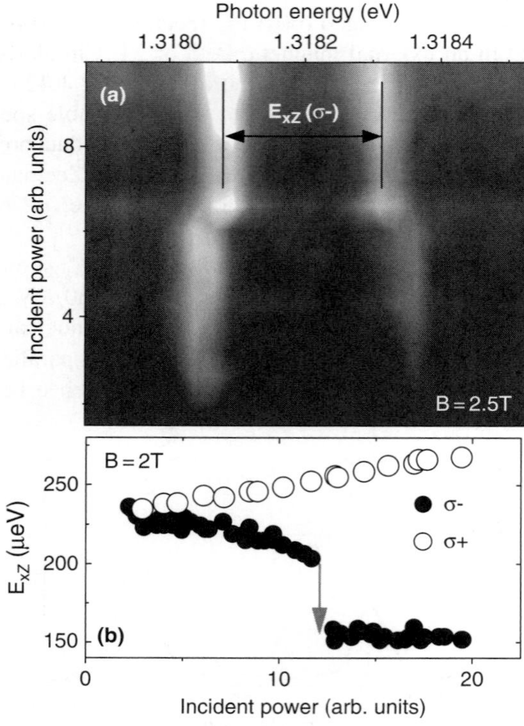

Fig. 6.34 (a) Grey-scale plot showing exciton PL spectra recorded for an individual InGaAs dot. The spectra are recorded at B_{ext}=2.5T using unpolarized detection and σ^- excitation into the wetting layer at 1.425 eV. The spectra are displaced along the vertical axis according to the excitation power at which they are measured. (b) E_{xZ} power dependences measured at $B_{ext} = 2$T for σ^+ and σ^- excitation polarizations

the magnitude of E_{XZ} decreases gradually, whereas above the threshold a very weak power dependence is observed. Very different behavior is observed for σ^+ excitation, with a weak monotonic increase of E_{XZ} found similar to that previously reported [77].

The Zeeman splitting dependence in Fig. 6.34 reflects the variation of the nuclear field B_N. At low excitation $B_N \approx 0$ and $E_{XZ} = |g_e + g_h|\mu_B B_{ext}$. As the excitation power is increased and non-zero B_N is generated, $E_{XZ}(\sigma^{\pm}) = |g_e + g_h|\mu_B B_{ext} \pm |g_e|\mu_B B_N(\sigma^{\pm})$. In Fig. 6.35 data of the type shown in Fig. 6.34 is presented as a function of B_{ext}. The E_{XZ} data are divided into two regimes, below threshold (low power) and above threshold (high power). At all fields in Fig. 6.35 the nuclear spin switch occurs, and there is a weak E_{XZ} power dependence above threshold (as in Fig. 6.34), motivating the division of the power dependence into two regimes with $B_N = 0$ at low power and high B_N above the threshold.

The nuclear field generated as a result of the spin switch is in its turn dependent on B_{ext}. The difference between the low and high power dependences $\Delta E_N = |g_e|\mu_B B_N(\sigma^-)$ is plotted as the triangles in Fig. 6.35. ΔE_N increases linearly with B_{ext} at low fields and then saturates at $B_{ext} \approx 2.5 - 3$T. The inset in Fig. 6.35 shows that the threshold-power for the switch increases nearly linearly with B_{ext}. No switch could be observed at $B_{ext} > 3$T in the range of powers employed in our studies.

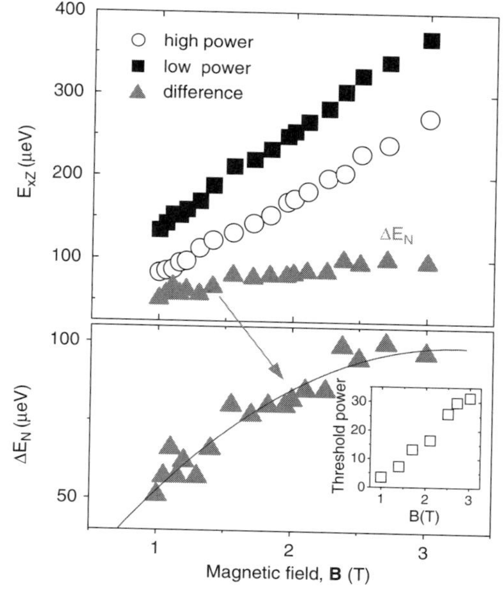

Fig. 6.35 Dependence of the QD exciton Zeeman splitting $E_{XZ}(\sigma^-)$ on the external magnetic field. Squares and circles show high and low power data, respectively, and triangles show their difference, ΔE_N, which is proportional to the nuclear magnetic field B_N. For all B_{ext} shown in the figure the nuclear switch threshold was observed with the threshold power shown in the inset as a function of B_{ext}

For $B_{ext} < 3T$, when the excitation power was gradually reduced from powers above the switching, E_{xZ} was found to vary weakly with power until another threshold was reached, where the magnitude of the exciton Zeeman splitting abruptly increased, as shown in Fig. 6.36. This increase of E_{xZ} corresponds to depolarization of the nuclei and hence reduction of B_N. The observed hysteresis of the nuclear polarization shows that two significantly different and stable nuclear spin configurations can exist for the same external parameters of magnetic field and excitation power. We find that high nuclear polarization persists at low excitation powers for more than 15 min, this time most likely being determined by the stability of the experimental set-up. In what follows we will distinguish two thresholds at which (i) B_N increases (E_{xZ} decreases) and (ii) B_N decreases (E_{xZ} increases) with threshold powers P_{up} and P_{down}, respectively.

We also show in Figs. 6.36a, b that the size of the hysteresis loop can be varied by changing either external magnetic or electric fields (the electric field is given by $F = (V_{rev} + 0.7V)/d$, where V_{rev} is the applied reverse bias and $d = 230\,nm$ is

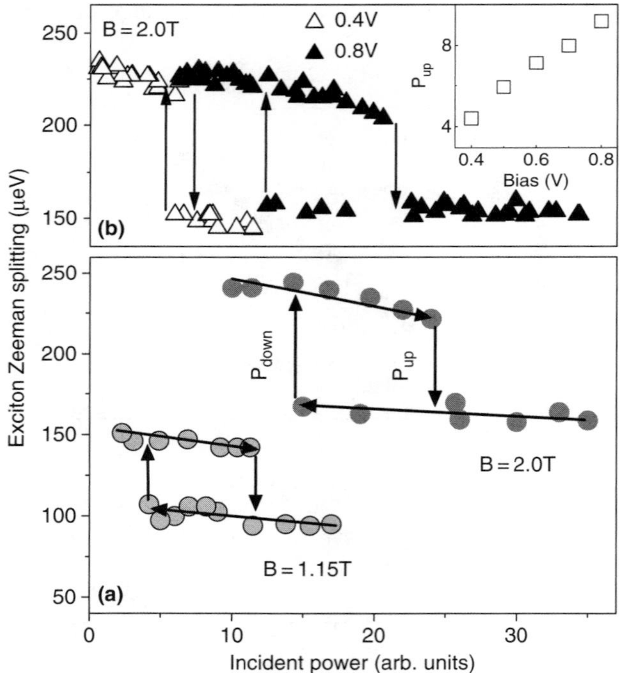

Fig. 6.36 (a) Power dependence of $E_{xZ}(\sigma^-)$ measured at $B_{ext} = 2T$ and 1.15T. The arrows show the direction in which the hysteresis loop is measured with two thresholds P_{up} and P_{down} at which $E_{xZ}(\sigma^-)$ abruptly decreases and increases, respectively. (b) $E_{xZ}(\sigma^-)$ power dependence measured at $B_{ext} = 2.0T$. The two hysteresis loops are measured at 0.4 and 0.8V applied bias. The inset shows the P_{up} dependence on the reverse bias applied to the diode

the width of the undoped region of the device). The inset in Fig. 6.36b shows the P_{up} reverse bias-dependence. In general, both P_{up} and P_{down} increase with B_{ext} and reverse bias, but also the difference between the two thresholds increases, leading to a greater range of incident powers in which the bistability is observed. The threshold bias dependence arises from the influence of the electric field on the charge state and carrier dynamics in the dot. The strong dependence of P_{up} and P_{down} on both B_{ext} and F and, importantly, the dependence of B_N on B_{ext} (as in Fig. 6.35) provides a flexible way to control precisely the degree of polarization of the nuclei in the dot.

In order to explain the nuclear switching and the bistability, we employ a model based on spin-flip assisted electron-hole radiative recombination. As briefly described above, circularly polarized excitation generates e-h pairs with a well defined spin. The spin of the hole is partially randomized during energy relaxation, whereas the electron has a high probability of retaining its initial spin orientation due to its weaker spin-orbit coupling to the lattice. Both dark and bright excitons can be formed on the dot with rates αw_x and $(1 - \alpha)w_x$, respectively. A bright exciton recombines at a rate w_{rec} without spin transfer to the nuclei. In contrast, a dark exciton can recombine with the electron simultaneously flipping its spin due to the hyperfine interaction: the electron virtually occupies an optically active state with the opposite spin and the same energy and then recombines with the hole. In this process, the probability of which (P_{hf}) is described by Eq. 6.5 [71], spin will be transferred to a nucleus.

The spin transfer rate $w_s \propto \alpha w_x P_{hf}$, is thus dependent on the dark exciton generation rate, and will compete with the nuclear depolarization rate $w_{dep} \approx 1 - 10$ sec^{-1} [81] due to spin diffusion away from the dot into the surrounding GaAs (see Fig. 6.28). At the same time, Eq. 6.5 shows that P_{hf} varies with the electron Zeeman splitting $E_{eZ} = |g_e|\mu_B[B_{ext} \pm B_N(\sigma^\pm)]$, a linear function of $B_N \propto S$, the degree of nuclear polarization. For the case of σ^- excitation, polarization of the nuclei leads to a decrease of E_{eZ}, and thus positive feedback and speeding up of the spin transfer process: the more spin is transferred to the nuclear system the faster becomes the spin transfer rate. By contrast for σ^+ excitation, spin transfer leads to an increase of E_{eZ}, leading to a saturation of S (and B_N) at high power.

For σ^- excitation the spin-pumping rate at high w_x may exceed the depolarization rate w_{dep}, and thus triggers a stimulated polarization process leading to an abrupt increase of the nuclear spin at the threshold P_{up}. The stimulation stops when either (i) E_{eZ} starts increasing again since $B_N > B_{ext}$, causing reduction of w_s or (ii) the maximum achievable B_N=B_N^{max} in the given dot is reached. This explains the dependence in Fig. 6.35, where ΔE_N, and hence the nuclear field, increases at low B_{ext} and saturates at high fields, from which B_N^{max} ≈2.5–3T can be estimated [6].

When reducing the power from beyond the P_{up} threshold, the spin transfer rate w_s first falls due to the decrease of the excitation rate, although P_{hf} remains high. When the condition $w_s < w_{dep}$ is reached at sufficiently low w_x, strong negative

[6] Note that a very similar magnitude of B_N is deduced from the dependence of $\Delta E_N(B_{ext})$ using $|g_e| \approx 0.5$.

feedback is expected: further nuclear depolarization will lead to even lower w_s due to the increase in the electron Zeeman energy E_{eZ}. Thus, an abrupt nuclear depolarization will take place (at the threshold P_{down}). This explains the observed hysteresis behavior in Fig. 6.36, and also accounts for the existence of a bistable state in the nuclear polarization at intermediate powers, $P_{down} < P < P_{up}$.

We have modelled the spin-transfer between the interacting electron-nuclear spin systems. Full details will be presented elesewhere. The degree of nuclear polarization S is obtained from rate equations describing the populations of bright and dark excitons and nuclei of both spin orientations. In the limit $\gamma \ll |g_e|\mu_B B_N^{max}$ we obtain the following equation for $\sigma^{\pm}$ excitation:

$$F(S,b) \equiv S\left[1 + b\frac{(x \pm S)^2}{1-S}\right] = a, \tag{6.6}$$

where, $x = B_{ext}/B_N^{max}$ and for $w_x \ll w_{rec}$ (low occupancy of the dot)

$$a = \frac{\alpha w_x}{N w_{dep}} \quad , \quad b = \frac{2\alpha N w_x}{w_{rec}}. \tag{6.7}$$

In Eq. 6.7, both a and b are proportional to the excitation power. For low excitation powers such that $b \ll 1$, for both σ^+ and σ^- excitation, Eq. 6.6 has a single solution for the degree of nuclear polarization, namely $S \approx a$. In the σ^+ excitation case, $F(S,b)$ is a monotonic function and for all a and b a single solution to Eq. 6.7 is obtained. On the other hand, for σ^- excitation, for higher powers such that b of order or larger 1, $F(S,b)$ acquires an N-shape, as illustrated in Fig. 6.37a. As shown in the diagram, an abrupt transition to $S > x$ ($S \approx a$) will be obtained when a_{max} (a_{min}) is reached at the local maximum (minimum) of $F(S,b)$. The transitions at a_{max} and a_{min} correspond to the P_{up} and P_{down} thresholds in Fig. 6.36, respectively, whereas for $a_{min} < a < a_{max}$, the polarization degree S enters a regime of bistability in which the cubic Eq. 6.6 has three solutions, two of which are stable with an unstable one in between.

We find that the occurrence of the switch to $S > x$ depends on the dimensionless ratio $\theta = a/b = w_{rec}/2N^2 w_{dep}$, since at small θ, a will grow more slowly with w_x than the magnitude of $F(S,b)$ at the local maximum. θ is determined by the dot parameters only, and can be estimated for the dots studied in our experiment: we obtain $\theta_{exp} \sim 1-10$ from $w_{rec} \sim 10^9$ sec^{-1}, $w_{dep} \sim 1-10$ sec^{-1} and $N \sim 10^4$. From Eq. 6.6 we calculate that for x<0.8 the spin switch is possible for a minimum $\theta_c = \frac{1}{16}(3-\sqrt{9-8x})(4x-3+\sqrt{9-8x})^2/(1+\sqrt{9-8x}) \approx 0.1 < \theta_{exp}$, consistent with our observations. A typical hysteresis loop calculated using Eq. 6.6 for $x = 0.7$ (with $\theta_c \approx 0.07$) and $\theta = 0.1$ is shown in Fig. 6.37b. Note that in GaAs interface dots $N \sim 10^5$ and $w_{rec} \approx 10^{10}$ sec^{-1} [82] leading to $\theta_{exp} \sim 0.1$, so that the bistability may also be expected although with less pronounced hysteresis loops than found in our experiments for InGaAs dots.

To summarize, in this Section we have demonstrated a strong optically induced bistability of the nuclear spin polarization in a few-nanometre size semiconductor

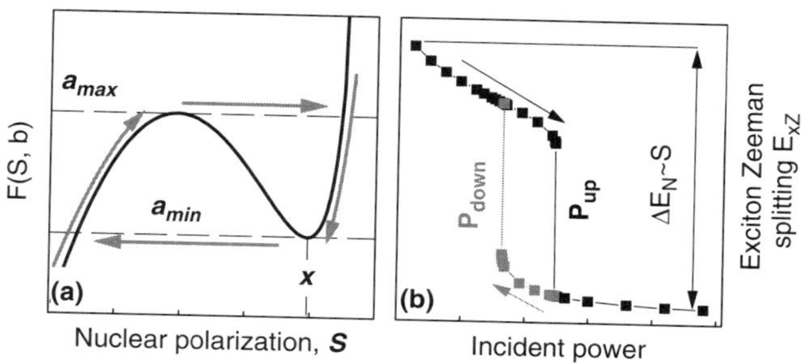

Fig. 6.37 (a) The full line shows the function $F(S, b)$ from Eq. 6.6. The red arrows show how the hysteresis loop is formed when the parameter a is varied ($\propto w_x$) for a fixed b. The horizontal arrows at a_{max} and a_{min} correspond to the abrupt spin polarization transitions, which are observed as the nuclear spin switch in the experiment. Note, however, that the graph is for illustrative purposes only, since for each value of a, $F(S, b)$ must be recalculated due to the interdependence between a and b (see Eq. 6.7). (b) Hysteresis loop of the exciton Zeeman splitting as a function of incident power calculated using Eq. 6.6 for $x \equiv B_{ext}/B_N^{max} = 0.7$ and $\theta \equiv a/b = 0.1$. As in the experimental data, E_{XZ} (B_N) decreases (increases) gradually below P_{up} threshold, followed by a weak power dependence for $P > P_{up}$. The vertical line shows the magnitude of ΔE_N, the experimental data for which is presented in Fig. 6.35

quantum dot. We show that nuclear magnetic fields up to 3 T can be switched on and off in individual InGaAs/GaAs dots by varying one of three external controlling parameters: electric and magnetic fields and intensity of circularly polarized excitation. We have found that the nuclear spin switch effect is a general phenomenon and has been observed in several different InGaAs/GaAs quantum dot samples at temperatures $T = 15 - 30$K and in the range of external magnetic fields $B_{ext} = 1 - 3$T. The effect arises due to the strong feedback of the nuclear spin polarization on the dynamics of the electron-nuclear spin interaction accompanying the radiative recombination process, which is enhanced when the Overhauser and external magnetic fields cancel each other. Since a major source of decoherence for electron spins in QDs arises from fluctuations of the local nuclear magnetic field, the ability to switch large fractions of the nuclei into a spin aligned state in a controlled fashion may point the way to a means to control or suppress such decoherence.

6.5 Conclusions

In conclusion, we have shown that optical measurements prove to be a powerful tool for investigating electron and nuclear spin dynamics in an important class of semiconductor nano-structures, self-assembled quantum dots. We have shown a great potential of these versatile nano-structures for the field of spintronics. The potential mainly lies on one hand in the robustness of the electron spin coherence and suppression of the quantum information dissipation channels in such structures and on the other hand in the tight connection between the properties of the spin, carrier confinement and photon polarization, enabling an efficient external control of the quantum states. We also outline the challenges on the path toward implementation of the quantum dot spin in future applications. Firstly it is the electron-hole exchange interaction, which mixes the spin states leading to fast oscillations of the optically imprinted polarization. Secondly it is the electron-nuclear spin interaction leading to the loss of the coherent spin polarization due to the fluctuating nuclear magnetic field. The ways to deal with these problems have been investigated, leading to striking observations such as nuclear spin bi-stability and discovery of long-lived spin effects in charged quantum dots. We have shown a variety of experimental techniques where ensembles as well as individual quantum dots have been studied, using both optical and electrical addressing of their quantum states. It is the possibility for such flexible approach, a combination of the electrical and optical excitation, and a full compatibility with the advanced GaAs-based technology which strengthens the potential of InGaAs quantum dots. In future a focus will probably be on use all-electrical excitation and manipulation of the spin states in individually addressable nano-structures.

Acknowledgments Many people directly contributed to this work including Maxim Makhonin, Maurice Skolnick, Dmitri Krizhanovskii, David Whittaker, Hui-Yun Liu, Andy Ebbens and Tim Wright. I am grateful for their enthusiasm and skill. I would also like to acknowledge the financial support from the EPSRC through the Advanced Research Fellowship No. EP/C54563X/1 and research Grant No. EP/C545648/1.

References

1. S. A. Wolf, D. D. Awschalom, R. A. Buhrman, J. M. Daughton, S. von Molnar, M. L. Roukes, A. Y. Chtchelkanova, and D. M. Treger, Science **294**, 1488 (2001).
2. E. Pazy, E. Biolatti, T. Calarco, I. DAmico, P. Zanardi, F. Rossi, P. Zoller, Europhys. Lett. **62**, 175 (2003).
3. Pochung Chen, C. Piermarocchi, L. J. Sham, D. Gammon, and D. G. Steel, Phys. Rev. B **69**, 075320 (2004).
4. A. Imamoglu, D. D. Awschalom, G. Burkard, D. P. DiVincenzo, D. Loss, and M. Sherwin, A. Small, Phys. Rev. Lett. **83**, 4204 (1999).
5. M. Krouvtar, Y. Ducommun, D. Heiss, M. Bichler, D. Schuh, and G. Abstreiter, J. J. Finley, Nature **432**, 81 (2004).
6. F. H. L. Koppens, J. A. Folk, J. M. Elzerman, R. Hanson, L. H. Willems van Beveren, I. T. Vink, H. P. Tranitz, W. Wegscheider, L. P. Kouwenhoven, and L. M. K. Vandersypen, Science **309**, 1346–1350 (2005).
7. J. R. Petta, A. C. Johnson, J. M. Taylor, E. A. Laird, A. Yacoby, M. D. Lukin, C. M. Marcus, M. P. Hanson, and A. C. Gossard, Science **309**, 2180–2184 (2005).
8. C. W. Lai, P. Maletinsky, A. Badolato, A. Imamoglu, Phys. Rev. Lett. **96**, 167403 (2006).
9. M. Atatre, J. Dreiser, A. Badolato, A. Hgele, K. Karrai, and A. Imamoglu, Science **312** 551 (2006).
10. D. Bimberg, M. Grundmann, and N. N. Ledentsov *Quantum Dot Heterostructures* (Wiley, Chichester, 1998).
11. Z. Yuan, B. E. Kardynal, R. M. Stevenson, A. J. Shields, C. J. Lobo, K. Cooper, N. S. Beattie, D. A. Ritchie, and M. Pepper, Science **295**, 102 (2002).
12. R. M. Stevenson, R. J. Young, P. Atkinson, K. Cooper, D. A. Ritchie, and A. J. Shields, Nature **439**, 179 (2006).

13. E. L. Ivchenko and G. E. Pikus, *Superlattices and other Heterostructures*, Springer Series in Solid-State Sciences, **110**, Springer-Verlag, Berlin (1997).
14. M. Bayer, G. Ortner, O. Stern, A. Kuther, A. A. Gorbunov, A. Forchel, P. Hawrylak, S. Fafard, K. Hinzer, T. L. Reinecke, S. N. Walck, J. P. Reithmaier, F. Klopf, and F. Schäfer, Phys. Rev. **B 65**, 195315 (2002).
15. A. S. Lenihan and M. V. Gurudev Dutt, D. G. Steel, S. Ghosh, P. K. Bhattacharya, Phys. Rev. Lett. **88**, 223601 (2002).
16. P. Borri, W. Langbein, S. Schneider, U. Woggon, R. L. Sellin, D. Ouyang, and D. Bimberg, Phys. Rev. Lett. **87**, 157401 (2001).
17. M. Paillard, X. Marie, P. Renucci, T. Amand, A. Jbeli, and J. M. Gérard, Phys. Rev. Lett. **B 86**, 1634 (2001).
18. M. Bayer and A. Forchel, Phys. Rev. **B 65**, 041308(R) (2002).
19. R. Oulton, A. I. Tartakovskii, A. Ebbens, J. Cahill, J. J. Finley, D. J. Mowbray, M. S. Skolnick and M. Hopkinson, Phys. Rev. **B 69**, 155323 (2004).
20. Importance of excited states in trion dynamics is discussed by I. E. Kozin, V. G. Davydov, I. V. Ignatiev, A. V. Kavokin, K. V. Kavokin, G. Malpuech, Hong-Wen Ren, M. Sugisaki, S. Sugou, and Y. Masumoto, Phys. Rev. **B 65**, 241312(R) (2002).
21. N. H. Bonadeo, J. Erland, D. Gammon, D. Park, D. S. Katzer, and D. G. Steel, Science **282**, 1473 (1998).
22. T. Flissikowski, A. Hundt, M. Lowisch, M. Rabe, and F. Henneberger, Phys. Rev. Lett. **86**, 3172 (2001).
23. M. Mitsunaga and C. L. Tang, Phys. Rev. **A 35**, 1720 (1987).
24. M. Bayer, A. Kuther, A. Forchel, A. Gorbunov, V. B. Timofeev, F. Schafer, J. P. Reithmaier, T. L. Reinecke, and S. N. Walck, Phys. Rev. Lett. **82**, 1748 (1999).
25. J. J. Finley, D. J. Mowbray, M. S. Skolnick, A. D. Ashmore, C. Baker, A. F. G. Monte, and M. Hopkinson, Phys. Rev. **B 66**, 153316 (2002).
26. H. Nickolaus, H.-J. Wünsche, and F. Henneberger, Phys. Rev. Lett. **81**, 2586 (1998).
27. L. Besombes, K. Kheng, L. Marsal, and H. Mariette, Phys. Rev. **B 63**, 155307 (2001)
28. C. Kammerer, G. Cassabois, C. Voisin, C. Delalande, Ph. Roussignol, A. Lemaître, and J. M. Gérard, Phys. Rev. **B 65**, 033313 (2001).
29. T. Flissikowski, I. A. Akimov, A. Hundt, and F. Henneberger, Phys. Rev. **B 68**, 161309(R) (2003).
30. G. Bester, S. Nair, and A. Zunger, Phys. Rev. **B 67**, 161306(R) (2003).
31. John H. Davies, J. Appl. Phys. **84**, 1358 (1998).
32. A. D. Andreev, J. R. Downes, D. A. Faux, and E. P. O'Reilly, J. Appl. Phys. **86**, 297 (1999).
33. M. Grundmann, O. Stier, and D. Bimberg, Phys. Rev. **B 52**, 11969 (1995).
34. O. Stier, M. Grundmann, and D. Bimberg, Phys. Rev. **B 59**, 5688 (1999).
35. A. Franceschetti, H. Fu, L. W. Wang, and A. Zunger, Phys. Rev. B 60, 1819 (1999).
36. G. Dasbach, A. A. Dremin, M. Bayer, V. D. Kulakovskii, N. A. Gippius, and A. Forchel, Phys. Rev. **B 65**, 245316 (2002).
37. G. Dasbach, A. A. Dremin, M. Bayer, V. D. Kulakovskii, N. A. Gippius, and A. Forchel, Physica **E 13**, 394 (2002).
38. M. O. Lipinski, H. Schuler, O. G. Schmidt, K. Eberl, and N. Y. Jin-Phillipp, Appl. Phys. Lett. **77**, 1789 (2000)
39. O. G. Schmidt, and K. Eberl, Phys. Rev. **B 61**, 13 721 (2000).
40. A. I. Tartakovskii, J. Cahill, M. N. Makhonin, D. M. Whittaker, J-P. R. Wells, A. M. Fox, D. J. Mowbray, M. S. Skolnick, K. M. Groom, M. J. Steer, and M. Hopkinson, Phys. Rev. Lett. **93**, 057401 (2004).
41. K. Nishi, H. Saito, S. Sugou, and J.-S. Lee, Appl. Phys. Lett. **74**, 1111 (1998).
42. F. Guffarth, R. Heitz, A. Schliwa, O. Stier, N. N. Ledentsov, A. R. Kovsh, V. M. Ustinov, and D. Bimberg, Phys. Rev. **B 64**, 085305 (2001) and references therein.
43. M. V. Maximov, A. F. Tsatsulnikov, B. V. Volovik, D. S. Sizov, Yu. M. Shernyakov, I. N. Kaiander, A. E. Zhukov, A. R. Kovsh, S. S. Mikhrin, V. M. Ustinov, Zh. I. Alferov,

R. Heitz, V. A. Shchukin, N. N. Ledentsov, D. Bimberg, Yu. G. Musikhin, and W. Neumann, Phys. Rev. **B 62**, 16 671 (2000).
44. H. Y. Liu, M. Hopkinson, C. N. Harrison, M. J. Steer, R. Frith, I. R. Sellers, D. J. Mowbray, and M. S. Skolnick, J. Appl. Phys. **93**, 2931 (2003).
45. S. Fafard and C. Ni. Allen, Appl. Phys. Lett. **75**, 2374 (1999).
46. C. Lobo, R. Leon, S. Fafard, and P. G. Piva, Appl. Phys. Lett. **72**, 2850 (1998).
47. T. Takagahara, Phys. Rev. **B 47**, 4569 (1993); T. Takagahara, Phys. Rev. **B 62**, 16 840 (2000).
48. V. D. Kulakovskii, G. Bacher, R. Weigand, T. Kümmell, A. Forchel, E. Borovitskaya, K. Leonardi, and D. Hommel, Phys. Rev. Lett. **82**, 1780 (1999).
49. W. Langbein, P. Borri, U. Woggon, V. Stavarache, D. Reuter, and A. D. Wieck, Phys. Rev. **B 69**, 161301(R) (2004).
50. A. I. Tartakovskii, M. N. Makhonin, I. R. Sellers, J. Cahill, A. D. Andreev, D. M. Whittaker, J-P. R. Wells, A. M. Fox, D. J. Mowbray, M. S. Skolnick, K. M. Groom, M. J. Steer, H. Y. Liu, and M. Hopkinson, Phys. Rev. **B 70**, 193303 (2004).
51. S. Kaiser, T. Mensing, L. Worschech, F. Klopf, J. P. Reithmaier, and A. Forchel, Appl. Phys. Lett. **81**, 4898 (2002).
52. Similar behaviour was also observed for individual InGaAs quntum dots in R. J. Young, R. M. Stevenson, A. J. Shields, P. Atkinson, K. Cooper, D. A. Ritchie, K. M. Groom, A. I. Tartakovskii, and M. S. Skolnick, Phys. Rev. **B 72**, 113305 (2005).
53. J. A. Barker, E. P. O'Reilly, Phys. Rev. **B 61** 13840 (2000).
54. T. Walther, A. G. Cullis, D. J. Norris, and M. Hopkinson, Phys. Rev. Lett. **86** 2381 (2001).
55. J. J. Finley, A. D. Ashmore, A. Lemaitre, D. J. Mowbray, M. S. Skolnick, I. E. Itskevich, P. A. Maksym, M. Hopkinson, and T. F. Krauss, Phys. Rev. **B 63** 073307 (2001).
56. R. J. Warburton, C. Schäflein, D. Haft, F. Bickel, A. Lorke, K. Karrai, J. M. Garcia, W. Schoenfeld, P. M. Petroff, Nature **405**, 926 (2000).
57. A. Lemaitre, A. D. Ashmore, J. J. Finley, D. J. Mowbray, M. S. Skolnick, M. Hopkinson, and T.F. Krauss, Phys. Rev. **B 63**, 161309(R) (2002).
58. S. Laurent, B. Eble, O. Krebs, A. Lemaitre, B. Urbaszek, X. Marie, T. Amand, and P. Voisin, Phys. Rev. Lett. **94**, 147401 (2005).
59. S. Cortez, O. Krebs, S. Laurent, M. Senes, X. Marie, P. Voisin, R. Ferreira, G. Bastard, J-M. Gérard, and T. Amand, Phys. Rev. Lett. **89**, 207401 (2002).
60. A. S. Bracker, E. A. Stinaff, D. Gammon, M. E. Ware, J. G. Tischler, A. Shabaev, Al. L. Efros, D. Park, D. Gershoni, V. L. Korenev, and I. A. Merkulov, Phys. Rev. Lett. **94**, 047402 (2005).
61. J. J. Finley, M. Sabathil, P. Vogl, G. Abstreiter, R. Oulton, A. I. Tartakovskii, D. J. Mowbray, M. S. Skolnick, S. L. Liew, A. G. Cullis, and M. Hopkinson, Phys. Rev. B 70, 201308(R) (2004).
62. R. Oulton, A. I. Tartakovskii, A. Ebbens, J. Cahill, J. J. Finley, D. J. Mowbray, M. S. Skolnick, and M. Hopkinson, Phys. Rev. **B 69**, 155323 (2004).
63. A. W. Overhauser, Phys. Rev. **92**, 411 (1953).
64. F. Meier and B. P. Zakarchenya, *Optical Orientation.* (Elsevier, New York, 1984).
65. D. Gammon, S. W. Brown, E. S. Snow, T. A. Kennedy, D. S. Katzer, and D. Park, Science **277**, 85 (1997).
66. T. Yokoi, S. Adachi, H. Sasakura, S. Muto, H. Z. Song, T. Usuki, and S. Hirose, Phys. Rev. **B 71**, 041307(R) (2005).
67. P.-F. Braun, X. Marie, L. Lombez, B. Urbaszek, T. Amand, P. Renucci, V. K. Kalevich, K.V. Kavokin, O. Krebs, P. Voisin, and Y. Masumoto, Phys. Rev. Lett. **94**, 116601 (2005).
68. B. Eble, O. Krebs, A. Lemaitre, K. Kowalik, A. Kudelski, P. Voisin, B. Urbaszek, X. Marie, and T. Amand, Phys. Rev. **B 74**, 081306(R) (2006).
69. A. I. Tartakovskii, T. Wright, A. Russell, V. I. Falko, A. B. Vankov, J. Skiba-Szymanska, I. Drouzas, R. S. Kolodka, M. S. Skolnick, P. W. Fry, A. Tahraoui, H.-Y. Liu, and M. Hopkinson, Phys. Rev. Lett. **B 98**, 026806 (2006).
70. I. A. Merkulov, A. L. Efros, and M. Rosen, Phys. Rev. **B 65**, 205309 (2002).
71. S. I. Erlingsson, Y. V. Nazarov, and V. I. Fal'ko, Phys. Rev. **B 64**, 195306 (2001).

72. A. V. Khaetskii, D. Loss, and L. Glazman, Phys. Rev. Lett. **88**, 186802 (2002).
73. A. Imamoglu, E. Knill, L. Tian, and P. Zoller, Phys. Rev. Lett. **91**, 017402 (2003).
74. S. E. Barrett, R. Tycko, L. N. Pfeiffer, and K. W. West, Phys. Rev. Lett. **72**, 1386 (1994).
75. H. Sanada, S. Matsuzaka, K. Morita, C. Y. Hu, Y. Ohno, and H. Ohno, Phys. Rev. Lett. **94**, 097601 (2005).
76. G. Salis, D. T. Fuchs, J. M. Kikkawa, D. D. Awschalom, Y. Ohno, and H. Ohno, Phys. Rev. Lett. **86**, 2677 (2001).
77. S. W. Brown, T. A. Kennedy, D. Gammon, and E. S. Snow, Phys. Rev. **B 54**, 17339 (1996).
78. D. N. Krizhanovskii, A. Ebbens, A. I. Tartakovskii, F. Pulizzi, T. Wright, M. S. Skolnick, and M. Hopkinson, Phys. Rev. **B 72**, 161312(R) (2005).
79. E. Beham, A. Zrenner, F. Findeis, M. Bichler, and G. Abstreiter, Appl. Phys. Lett. **79**, 2808 (2001);
80. R. Oulton, J. J. Finley, A. D. Ashmore, I. S. Gregory, D. J. Mowbray, M. S. Skolnick, M. J. Steer, San-Lin Liew, M. A. Migliorato, and A. J. Cullis Phys. Rev. **B 66**, 045313 (2002).
81. D. Paget, Phys. Rev. **B 25**, 4444 (1982).
82. D. Gammon, A. L. Efros, T. A. Kennedy, M. Rosen, D. S. Katzer, D. Park, S. W. Brown, V. L. Korenev, I. A. Merkulov, Phys. Rev. Lett. **86**, 5176 (2001).
83. Y. Toda, T. Sugimoto, M. Nishioka, and Y. Arakawa, Appl. Phys. Lett. **76**, 3887 (2000).
84. L. Besombes, J. J. Baumberg, and J. Motohisa, Phys. Phys. Lett. **90**, 257402 (2003).
85. D. Birkedal, K. Leosson, and J. M. Hvam, Phys. Rev. Lett. **B 87**, 227401 (2001).
86. L. Sham, Science **277**, 1258 (1997).
87. J. M. Kikkawa, I. P. Smorchkova, N. Samarth, and D. D. Awschalom, Science **277**, 1284 (1997).
88. J. M. Kikkawa, and D. D. Awschalom, Phys. Rev. Lett. **80**, 4313 (1998).

Chapter 7
Excitons and Spins in Quantum Dots Coupled to a Continuum of States

Alexander O. Govorov

7.1 Introduction

One of the most interesting topics in solid-state physics concerns a localized electron coupled to the continuum of extended states. When the extended state continuum is not filled with electrons, the problem is described by the Fano model (Fano 1961). A localized state in the presence of filled continuum corresponds to the model based on the Anderson Hamiltonian that involves explicitly the electron spin (Anderson 1961). One of the well-known consequences of this model is the Kondo effect (Kondo 1964, Hewson 1993). The Kondo effect was mostly introduced to understand electrical transport phenomena in solids and nanostructures with localized spins (Hewson 1993; Glazman and Raikh 1988; Goldhaber-Gordon et al. 1998). The Fano model was first introduced to describe optical and energy-loss spectra of atoms in which electrons couple with delocalized states via Coulomb-induced auto-ionization (Fano 1961).

Quantum dots are a natural extension of solid-state physics and the physics of atoms. This chapter describes optical properties of self-assembled InAs/GaAs quantum dots (QDs) interacting with the continuum. The mechanisms of interaction come from the tunneling (Govorov et al. 2003; Smith et al. 2005), Auger (auto-ionization) processes (Ferreira and Bastard 2000; Karrai et al. 2004; Govorov et al. 2004; Ferreira and Bastard 2006), phonon-assisted transitions (Kammerer et al. 2001), and shake-up spin-dependent effects (Kikoin and Avishai 2000). The tunnel coupling with filled continuum leads to the formation of Kondo excitons in the low-temperature regime. At slightly elevated temperatures, the tunnel coupling with the Fermi sea permits voltage-control of spin flips of electrons in a QD (Smith et al. 2005). This method to control the spin can be used in the research related to quantum information. Auger processes and auto-ionization are typical for systems with few interacting electrons. There effects represent characteristic properties of a collection of charged particles that interact and scatter due to the Coulomb force. In QDs, the auto-ionization of excitons leads to the peculiar shapes of optical lines (Karrai et al. 2004; Govorov et al. 2004). In the presence of magnetic field, the auto-ionization process creates anti-crossings between a localized state and Landau levels (Karrai et al. 2004; Govorov et al. 2004). Although the physics of QDs inherits much from the atomic physics, the difference between atomic systems (like the auto-ionized level of He) and QDs is quite remarkable. This difference comes from: (7.1) the existence of Fermi sea in QD structures, (7.2) the ability to tune electron states by the voltage, (7.3) spin physics including the Kondo effect, and (7.4) strong magnetic-field effects due a relatively large size of QD and a small effective mass of carriers.

7.2 Models of Self-assembled Quantum Dots

Voltage-tunable quantum dots permit an electrical control of the number of electrons trapped in one nanocrystal (Drexler et al. 1994, Luyken et al. 1999; Warburton et al. 2000; Findeis et al., 2001; Finley et al. 2001). Figure 7.1 shows schematic of a transistor structure with voltage-tunable QDs. By applying the gate voltage, electrons tunnel to the QD from the back contact and the QD traps a well controlled number of electrons, typically few carriers. In the presence of illumination, incident photons generate additional electrons and holes in the system. When a QD traps a single hole, the electrons and hole form a charged exciton (Fig. 7.2). The formation of stable charged excitons in voltage-controlled QDs has been demonstrated in the experimental papers (Warburton et al. 2000; Findeis et al., 2001; Finley et al. 2001; Karrai et al. 2004). These structures demonstrate the regions of gate voltage with constant excitonic charge, with the excitonic charge changing abruptly at particular values of gate voltage U_{top}. The charged excitons, labeled X^{n-}, include $N_e = n + 1$ electrons and one hole. Excitons with n up to 3 have been observed in InAs/GaAs QDs (Warburton et al. 2000; Findeis et al., 2001; Finley et al. 2001; Karrai et al., 2004). We should note that, in principle, extra electrons can come from impurities. Such doped QDs have been discussed theoretically in the early paper by Wojs and Hawrylak (Wojs and Hawrylak 1997) . Figure 7.3 illustrated few first charged states of exciton.

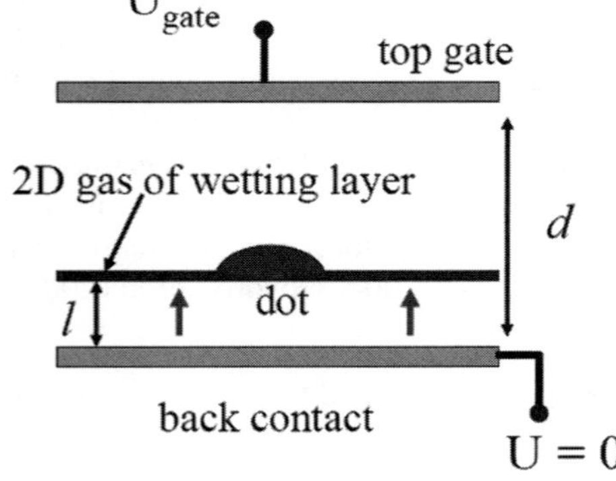

Fig. 7.1 Schematic of the transistor structure with a quantum dot embedded between the top gate and back contact. The top gate is kept at nonzero voltage. Typical dimensions of such structures relate as $l << d$. A quantum dot is formed on a wetting layer

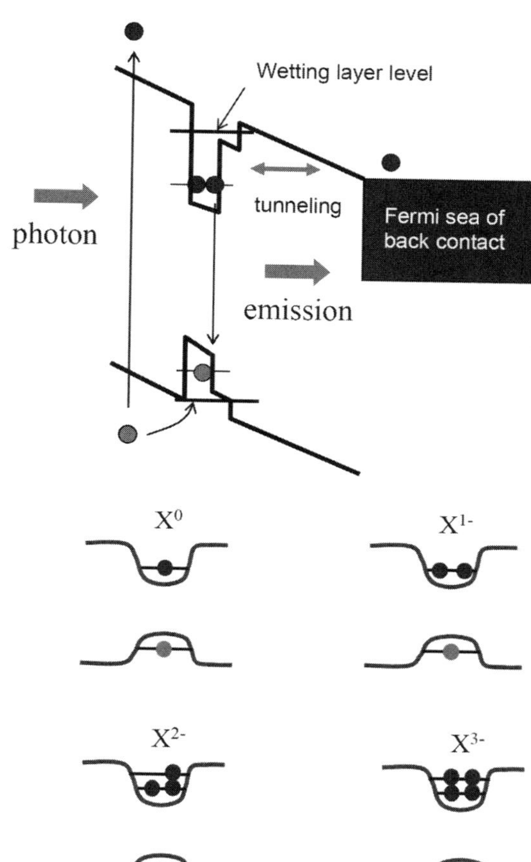

Fig. 7.2 Band diagram of a voltage tunable quantum dot. Quantum dot accommodates one optically-generated hole and one or few electrons coming from the back contact. The resultant state of quantum dot is a charged exciton

Fig. 7.3 Schematic of four first states of charged exciton

7.3 Quantum Dot Coupled to the Continuum

Now we consider explicit mechanisms of coupling between a changed exciton and extended states. At smaller voltages, electrons in the system shown in Fig. 7.1 are loaded into QDs. This can be monitored by looking at the capacitance spectrum (Drexler et al. 1994; Luyken et al. 1999); the capacitance spectra typically show peaks that reflect the tunneling of single electrons from a back contact to QDs. At higher voltages, the capacitance spectra indicate that electrons fill the wetting layer (Luyken et al. 1999). The wetting layer is a two-dimensional quantum well at the interface of two materials (AlGaAs and GaAs); self-assembled QDs are formed right on this layer (see Fig. 7.1). The importance of wetting layer for optical and electronic properties of QDs was noticed in several papers (Luyken et al. 1999; Kammerer et al. 2001; Alén et al. 2005; Mazur et al. 2007). It was suggested in paper by Govorov et al. (Govorov et al. 2003) that the wetting layer can be used as a simple method to generate a Fermi sea in close proximity of charged exciton trapped in a QD. When electrons start to fill the wetting layer, the Fermi energy of a two-dimensional (2D) electron gas in the wetting layer depends linearly on U_{top}. Using a simple flat capacitor model and the 2D density of states in the wetting layer, we obtain (Govorov et al., 2003):

$$E_F = \frac{a_0^*}{4d'}(U_{top} - U_{top}^0), \tag{7.1}$$

where a_0^* is the effective Bohr radius, d' is the distance between the top gate and wetting layer, and U_{top}^0 is the threshold gate voltage at which electrons start to fill the wetting layer (see Fig. 7.1). A charged exciton localized in a QD can interact with the continuum of states in a wetting layer through tunneling. This becomes possible for changed excitons with two or more extra electrons ($n \geq 2$). At the same time, the excitons with the smallest numbers n (X^0 and X^{1-}) exist only at gate voltages where the wetting layer is empty (Warburton et al. 2000). Also, the excitons $X^0$and X^{1-} are strongly bound to a QD. Therefore, we focus on the exciton X^{2-}. Self-assembled QDs are typically anisotropic and, therefore, have non-degenerate p-orbitals. In the ground state of exciton X^{2-}, two electrons occupy the s-orbital and one electron resides in the lower p-orbital (Fig. 7.4); a hole in the valence band occupies the lowest s-state. For the p-state electron, the potential consists of the short-range QD confinement and long-range Coulomb repulsion associated with the net bound charge in the s-states. The resultant potential for the p-electron will therefore have a barrier at the edge of the QD (Fig. 7.4). This barrier permits tunnel coupling with the continuum.

To describe the tunnel coupling to the continuum, we employ the Anderson Hamiltonian adopted for the case of exciton. The following description assumes that an exciton lives long enough to form a collective state with continuum. The Hamiltonian has the form:

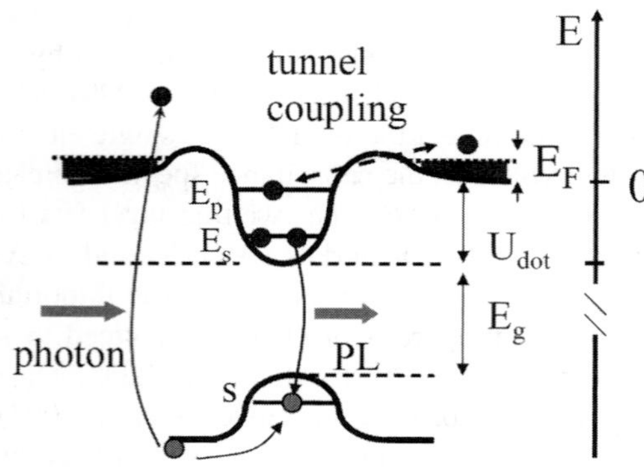

Fig. 7.4 The band diagram of a quantum dot and associated wetting layer. The exciton X^{2-} interacts with the wetting-layer continuum via tunneling

$$\hat{H} = \hat{H}_{sp} + \frac{1}{2} \sum_{\alpha 1,\alpha 2,\alpha 3,\alpha 4} U^{ee}_{\alpha 1,\alpha 2,\alpha 3,\alpha 4} \hat{a}^{+}_{\alpha 1} \hat{a}^{+}_{\alpha 2} \hat{a}_{\alpha 3} \hat{a}_{\alpha 4} - \sum_{\alpha 1,\alpha 2,\alpha 3,\alpha 4} U^{eh}_{\alpha 1,\alpha 2,\alpha 3,\alpha 4} \hat{a}^{+}_{\alpha 1} \hat{b}^{+}_{\alpha 2} \hat{b}_{\alpha 3} \hat{a}^{+}_{\alpha 4} \hat{H}_{tun}, \quad (7.2)$$

where $\hat{H}_{sp}$ is the single-particle intra-dot energy, $\hat{a}_{\alpha}(\hat{b}_{\alpha})$ are the intra-dot annihilation operators for electrons (holes); U^{ee} are U^{eh} are the matrix elements for electron-electron and electron-hole Coulomb interactions. The quantum state index α involves two sub-indexes: $\alpha = (\beta, \sigma)$, where β is the orbital state (s or p) and σ is the spin. For the case of electrons, $\sigma = \uparrow, \downarrow$ and, for holes, $\sigma = \pm 3/2$. The operator $\hat{H}_{tun}$ introduces the tunnel coupling between the exciton and Fermi sea:

$$\hat{H}_{tun} = \sum_{k,\sigma} V_k c^{+}_{k,\sigma} \hat{a}_{p,\sigma} + V^{*}_{k} \hat{a}^{+}_{p,\sigma} c_{k,\sigma}, \quad (7.3)$$

where $c^{+}_{k,\sigma}$ describes the delocalized states of electrons in the wetting layer. The tunneling operator (7.3) includes only the p-state of QD since the energy of s-states is far below the continuum. The intra-QD many-electron states can be calculated treating Coulomb interaction within the perturbation theory; this approach is valid for QDs with a strong confinement and reproduces well experimental data (Warburton et al. 2000; Schulhauser et al. 2002; Govorov et al. 2004).

Another physical situation with strong coupling between an exciton and Fermi sea appears in the same structures due to the tunneling between a QD and metal back contact (Fig. 7.2). This type of coupling exists already at smaller voltages when electrons do not fill the wetting layer. The tunnel Hamiltonian for the coupling to the Fermi sea in the back contact has the same form as Eq. 7.3. In this case, the operators $c^{+}_{k,\sigma}$ relate to the delocalized states in the back contact.

7.4 Spins and Kondo Excitons

7.4.1 Coupling to the Wetting Layer

First we discuss the case of exciton X^{2-} and filled wetting layer. This exciton is interesting because the net electron spin in a QD is nonzero (1/2) and therefore the ground state of this exciton can be Kondo-type. The Kondo state is characterized by zero total spin. In other words, in such a state, electrons of Fermi sea "screen" the spin of QD. Below we will discuss the conditions to observe such Kondo state. Regarding other charged states, the excitons studied to date in the regime of a filled wetting layer are X^{3-} and X^{4-}. The exciton X^{3-} has zero spin electron spin ($n = 4$) and the exciton X^{4-} suffers from strong Auger processes and have very broad emission line.

At zero temperature, the optical emission spectrum is given by the correlation function (Govorov et al. 2003):

$$I(\omega) = \mathrm{Re} \int_0^{+\infty} e^{-i\omega t} \langle i | \hat{V}_{opt}^{+}(t) \hat{V}_{opt}(0) | i \rangle, \tag{7.4}$$

where ω is the frequency of emitted light and $|i\rangle$ is the initial state of exciton. In the presence of tunneling, the wave function $|i\rangle$ is a collective state of exciton and Fermi sea. The optical emission operator

$$\hat{V}_{opt} = V_{opt}(\hat{b}_{s,-3/2}\hat{a}_{s,\uparrow} + \hat{b}_{s,+3/2}\hat{a}_{s,\downarrow}). \tag{7.5}$$

The above operator (7.5) describes the s-s inter-band transitions that correspond to the strongest line in the PL spectrum of X^{2-}.

It is instructional to start with a zero bandwidth model, in which the Fermi sea is replaced just by one "delocalized" state of energy E_F. In this model, the energy of delocalized state plays a role of the Fermi energy. Even though this model is greatly simplified, it allows us to predict the main features of the optical spectrum in the presence of the Kondo effect. We note that, in the QD systems, the Fermi energy is a linear function of the voltage. Therefore, the energy ε_F can be considered as a voltage tunable parameter. To describe the coupling with the "extended" state, we should involve two intra-dot excitonic states X^{2-} and X^{3-}. The exciton X^{3-} can decay to the state X^{2-} and one "delocalized" electron. Overall, we involve here four electrons in the conduction band (see Fig. 7.5). The hole in the valence band can be taken in the state $+3/2$; for the state $-3/2$, we can obtain the equivalent results.

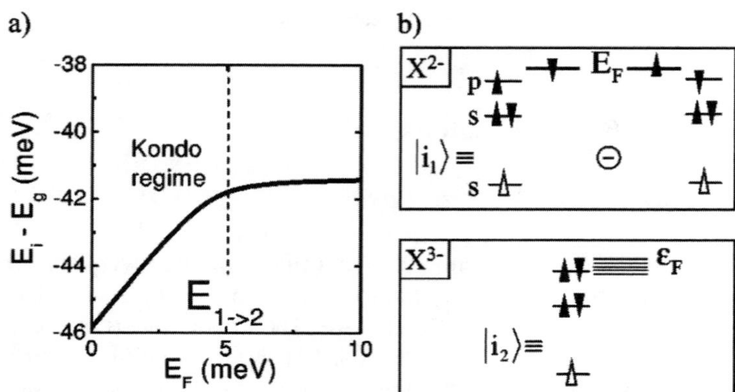

Fig. 7.5 (a) Calculated energy of the ground state of exciton X^{2-} coupled with one extended state; E_g is the band gap of the QD. (b) Electron configurations for two wave functions with zero total electron spin. These wave functions contribute to the ground state of exciton. Electrons and holes are represented by solid and open triangles, respectively. Reproduced from (Govorov et al. 2003)

At zero temperature, the initial state of exciton should be taken as a state with a minimum energy. The corresponding electron wave function should have zero total spin. We therefore express the initial state as a linear combination of two many-particle wave functions:

$$|i\rangle = A_1 |1\rangle + A_2 |2\rangle , \tag{7.6}$$

where

$$\begin{aligned} |1\rangle &= \frac{1}{\sqrt{2}} \left(\left| s_\uparrow, s_\downarrow, p_\uparrow, e_\downarrow \right\rangle - \left| s_\uparrow, s_\downarrow, p_\downarrow, e_\uparrow \right\rangle \right) \left| h_{s,+3/2} \right\rangle , \\ |2\rangle &= \frac{1}{\sqrt{2}} \left| s_\uparrow, s_\downarrow, p_\uparrow, p_\downarrow \right\rangle \left| h_{s,+3/2} \right\rangle . \end{aligned} \tag{7.7}$$

We note that both functions ($|1\rangle$ and $|2\rangle$) have zero electron spin; the states $|1\rangle$ and $|2\rangle$ describe the excitons X^{2-} and X^{3-}, respectively (Fig. 7.5). Their energies are $E_1 = E_{\text{intra}-dot,X2-} + E_F$ and $E_2 = E_{\text{intra}-dot,X3-}$, where the intra-dot contributions $E_{\text{intra}-dot,X3(2)-}$ include the band gap, quantization energy, and Coulomb terms. The function $|1\rangle$ is the Kondo state; in this state, the "delocalized" electron screens the spin of exciton. By diagonalising the Hamiltonian, we find the energy of the ground state $|i\rangle$:

$$E_i = \frac{1}{2} \left[E_1 + E_2 - \sqrt{(E_1 - E_2)^2 + 8V^2} \right] , \tag{7.7}$$

where V is the matrix element of coupling between the p-state and delocalized state. As the energy E_F increases, the ground states evolves from the exciton X^{2-} to exciton X^{3-}. The smooth transition occurs when $E_1 \approx E_2$ that corresponds to $E_{\text{intra}-dot,X3-} - E_{\text{intra}-dot,X2-} = E_{1\to 2} \approx E_F$. The region $E_{1\to 2} > E_F$ is the Kondo regime, the region $E_{1\to 2} \approx E_F$ is the co-called mixed-valence regime (Hewson 1993), and the interval $E_{1\to 2} < E_F$ corresponds to the non-Kondo state (see Fig. 7.5a). For numerical calculations, we now represent a QD as a 2D anisotropic harmonic oscillator with characteristic frequencies $\omega_{e,x(y)}$ and $\omega_{h,x(y)}$. For the oscillator frequencies, we take typical numbers: $\hbar\omega_{e,x(y)} = 25\,meV$ $(20\,meV)$ and $\hbar\omega_{h,x(y)} = 12.5\,meV$ $(10\,meV)$. For the depth of QD (see Fig. 7.4), we take $U_{dot} = 64\,meV$; the tunnel matrix element $V = 0.5\ meV$. For more sophisticated models of self-assembled QD involved in numerical simulations, one can see, for example, the paper (Narvaez et al. 2006).

To calculate the optical emission spectrum, we should now describe all possible final states. After emission of a photon, all final states should have the electron spin ½. Figure 7.6a shows three possible intra-dot final states (Warburton et al. 2000). These final states correspond to three lines in photoluminescence spectrum (Fig. 7.6b). The upper panel in Fig. 7.6b demonstrates the optical spectrum in the absence of the coupling with the extended state ($V = 0$). We see that, as the

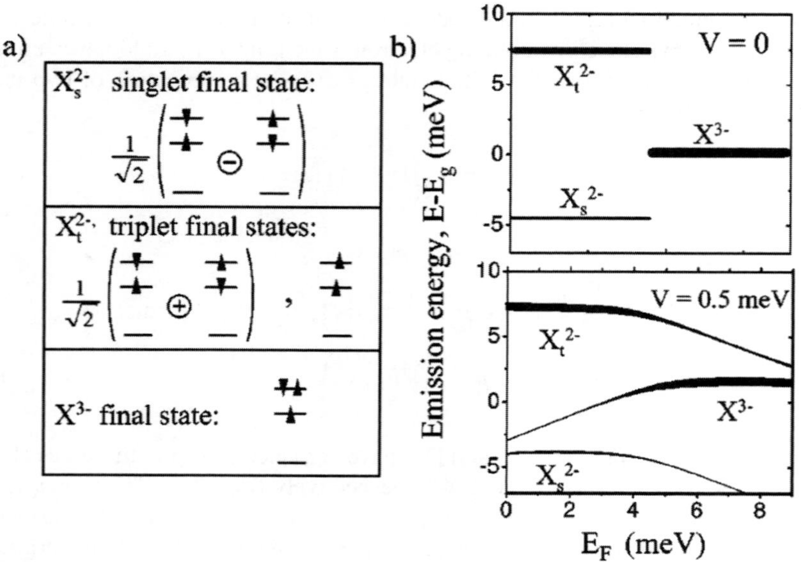

Fig. 7.6 a) Final intra-dot electron configurations for the excitons X^{2-} and X^{3-}. b) Calculated energies of the optical transitions as a function of the delocalized state energy without ($V = 0$) and with ($V = 0.5\, meV$) the tunnel coupling. The thickness of the line represents the intensity of the optical lines. Reproduced from (Govorov et al. 2003)

energy E_F increases, the spectrum changes abruptly at $E_{1\to 2} = E_F$. This is due to the transition between two ground states, X^{2-} and X^{3-}. In the presence of tunnel hybridization, this transition becomes smooth and the Kondo exciton (X^{2-}) evolves gradually to the non-Kondo state (X^{3-}). In the Kondo state (X^{2-}), the exciton wave function is strongly mixed with the delocalized electron, whereas, in the stateX^{3-}, the hybridization is not so strong.

In the next step, we focus on the photoluminescence line shapes of excitons in the presence of the tunnel coupling. While the zero bandwidth model is very useful for understanding of the general picture of the optical emission, we have to employ a finite bandwidth model in order to calculate the emission line shapes. This problem was considered analytically, using a trial functions, (Govorov et al. 2003) and numerically, using Wilson's renormalization group method (Helmes et al. 2005). Here we mention some of the results obtained using a trial function. One simple form of a trial function for the Kondo state is (Hewson 1993; Govorov et al. 2003):

$$|i\rangle = \left[A_0 \left|\phi_0\right\rangle + \sum_{k>k_F} A_k \left|\phi_k\right\rangle \right] \cdot \left|h_{s,+3/2}\right\rangle, \tag{7.8}$$

where $|\phi_k\rangle$ and $|\phi_0\rangle$ are the many-body states that involve the bound exciton and Fermi sea (Fig. 7.7a). The summation in the function (7.8) is performed over

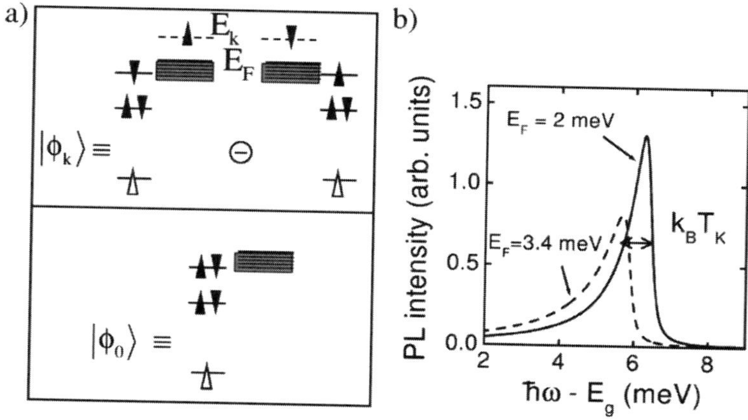

Fig. 7.7 a) Contributions to the initial state of exciton coupled with the Fermi sea. b) Calculated emission spectrum of the Kondo- exciton X^{2-} with the triplet final state configuration; Kondo effect manifests as an asymmetric peak. Reproduced from (Govorov et al. 2003)

all unoccupied states k above the Fermi see level. Figure 7.7a illustrates these states. The state $|\phi_k\rangle$ is the exciton X^{2-} mixed with the Femi sea electrons and the state $|\phi_0\rangle$ corresponds to X^{3-}. The strong mixing of the state $|\phi_k\rangle$ with the Fermi sea occurs due to the non-zero electron spin of exciton. The net spins for both states are zero. In the Kondo regime ($E_F < E_{1\to 2}$), the energy of the ground state ground state is given by $E_i = E_{\text{intra}-dot,X2-} + E_F - \delta$. Here $E_{1\to 2} = E_{\text{intra}-dot,X3-} - E_{\text{intra}-dot,X2-}$ and the parameter δ plays a role of Kondo temperature:

$$\delta = k_B T_K = (D - E_F)\cdot\exp\left(-\frac{\pi\left(E_{\text{intra}-dot,X3-} - E_{\text{intra}-dot,X2-} - E_F\right)}{2\cdot\Delta}\right), \quad (7.9)$$

where D is the cut-off parameter, i.e. $V_k = V$ for $0 < E_k < D$and $V_k = 0$ elsewhere. The matrix element V_k describes the tunnel coupling between the p-state and k-extended state. The corresponding tunnel broadening of the p-state $\Delta = \pi V^2 \rho$, where ρ is the density of states. More details of the derivation can be found in the paper (Govorov et al. 2003). The temperature T_K can be as high as 7–14 K for realistic parameters $\Delta = 1\,meV$ and $D \approx 30\,meV$.

The above estimate for the Kondo temperature suggests that the Kondo excitons can be observed experimentally. The photoluminescence spectrum of Kondo exciton X^{2-} calculated using the trial function (7.8) is shown in Fig. 7.7b. The Kondo temperature manifests itself in this spectrum as a broadening of asymmetric peak. Simultaneously, the peak asymmetry is also an indication of the Kondo state. Another example of the spectrum, where T_K is responsible for the broadening, was studied in the paper by (Gunnarsson and Schönhammer 1983).

The Kondo effect exists for a bound state with nonzero spin. Along with the Kondo coupling, there may be a strong effect of hybridization. This effect is not necessarily spin-dependent. Already a simple zero-band width model shows that the exciton lines strongly shift with the Fermi energy it the transition regime $E_{1\to 2} \approx E_F$ (Fig. 7.6b). Since the Fermi energy is controlled by the gate voltage, such exciton states should strongly depend on the voltage. The trial exciton functions used in the paper by Govorov et al. (Govorov et al. 2003) can be used to calculate the whole spectrum incorporating both X^{2-} and X^{3-}. Figure 7.8 shows a grey-scale plot of exciton emission. We can see that the exciton lines in the regime of the transition$X^{2-} \to X^{3-}$(i.e. when $E_F \approx E_{1\to 2}$) become strongly voltage-dependent. This is due to the hybridization of exciton and Fermi level. The voltage dispersion of the main exciton line in the regime $E_F \approx E_{1\to 2}$ obeys a simple formula:

$$\hbar\omega_{exc} = const - E_F = const - \frac{a_0^*}{4d}\Delta U,$$

where $\Delta U = U_{top} - U_{top}^0$ and $d \approx d'$. Basically, the exciton line follows the Fermi level that strongly depends on the applied voltage. The main optical line in the mixed-valence regime ($E_F \approx E_{1\to 2}$, $X^{2-} \to X^{3-}$) comes from a mixed state that involves the Fermi sea electrons and two excitons, X^{2-} and X^{3-}.

It is interesting to compare the effect of voltage-tunable emission due to the mixed $X^{2-} - X^{3-}$ state and the usual Stark effect in self-assembled QDs. Due to the strong localization in the growth direction (z-axis), self-assembled QDs have a

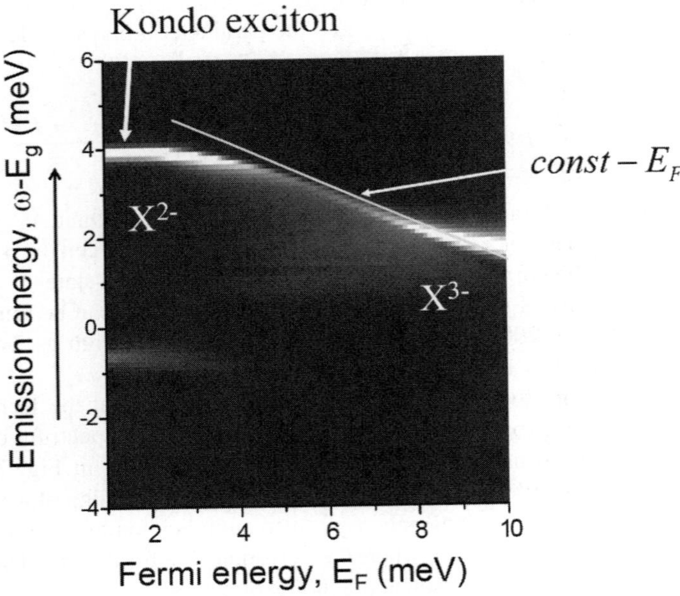

Fig. 7.8 Grey-scale plot of exciton emission as a function of the gate voltage and emission energy. The parameters of quantum dot are given in the text

relatively weak exciton dipole. Typically, the exciton dipole, d_{exc}, in the "vertical" direction is about few $\dot{A}$. This means that the exciton line shifts with applied electric field as $\hbar\omega_{exc} = const - \frac{d_{exc}}{d}\Delta U$. The hybridization effect described here is several times stronger! The "effective dipole" of the hybrid exciton is about $a_0^*/4 \approx 25\,\dot{A}$. This strong electric-field dependence comes from the coupling between a localized exciton and Fermi sea of charged carriers. The experimental spectra reported in papers (Karrai et al. 2000; Schulhauser et al. 2004)) showed excitonic lines with a very strong voltage dispersion which is consistent with our prediction $\hbar\omega_{exc} = const - \frac{a_0^*}{4d}\Delta U$.

7.4.2 Coupling to the Back Contact: Voltage Control of Spin Flips

In the regime of empty wetting layer, excitons can couple with the Fermi sea in the back contact (see Fig. 9). This type of coupling was recently studied by Smith et al. (Smith et al. 2005). Smith et al. (Smith et al. 2005) reported time-resolved photoluminescence spectra of single QDs. The temporal relaxation of photoluminescence of single QDs showed a bi-exponential behavior (Fig. 10a). The appearance of two exponents was explained in the following way: the first voltage-independent exponent is due to the radiative recombination and the second exponent is due to spin-flip transitions in the exciton state. Interestingly, the second exponent was strongly voltage-dependent. Figure 7.9 illustrates the processes involved in the experiment (Smith et al. 2005).

The fast exponent in the temporal dynamics is the usual radiative rate (Fig. 7.9). The radiative transition occurs only for bright excitons with the total spin $J_z = \pm 1$. These bright excitons are composed of electrons with spin $+1/2$ $(-1/2)$ and holes with angular momentum $-3/2$ $(+3/2)$. Along with bright excitons, the QD is populated with dark excitons ($J_z = \pm 2$). Because the system is excited above the barrier by a non-polarized light, initial populations of dark and bright excitons are similar.

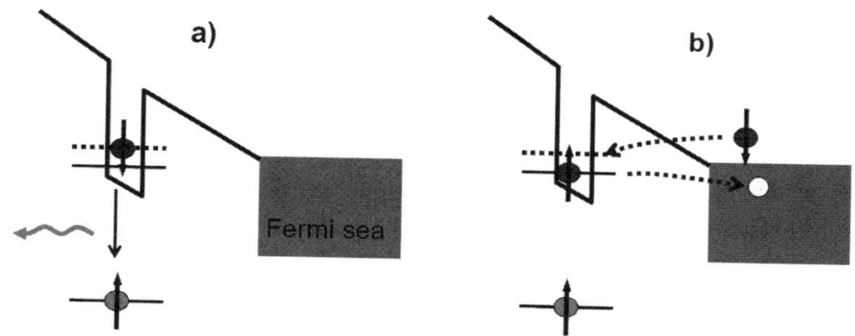

Fig. 7.9 Two processes contributing to the time-resolved optical emission from quantum dots located in the vicinity of a metal contact. a) Radiative recombination of the bright exciton. b) Spin-flip transition from dark to bright state assisted by the Fermi sea

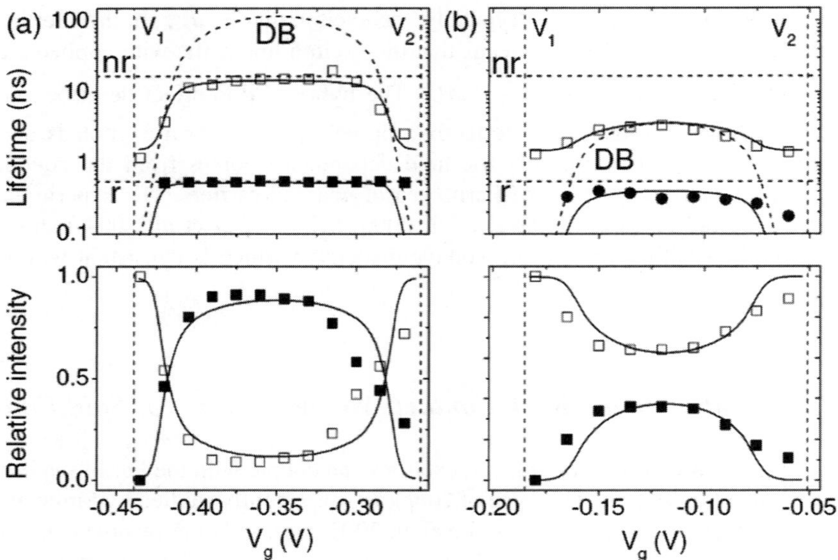

Fig. 7.10 (a) Lifetimes and relative intensities of the primary (filled symbols) and secondary (open symbols) components of the emission decay vs the gate voltage for a single quantum dot. (b) As for (a) but for a different dot. The solid lines are the results of the calculations taking radiative and no-radiative recombination rates as $\gamma_r = 1.8\,ns^{-1}$ and $\gamma_{nr} = 0.06\,ns^{-1}$. The tunneling parameters: $\Delta_t = 0.07\,meV$ and $\Delta_t = 0.3\,meV$ for the dot (a) and (b), respectively. The individual components γ_r, γ_{nr}, and $\gamma_{Dark\to Bright}$ are shown by the dashed lines. Reproduced from (Smith et al. 2005)

Then, bright excitons are able to recombine fast due to their coupling with photons. But, dark excitons can not recombine radiatively and may have much longer lifetimes. In the presence of the Fermi sea, an electron from a dark exciton can tunnel to the back contact and another electron with opposite spin can come back to the QD (Fig. 7.9b). This spin exchange mechanism leads to the conversion of a dark exciton into a bright excitonic state, i.e. $J_z = \pm 2 \to J_z = \pm 1$. Since the resultant exciton is optically active, it contributes to the photoluminescence. To calculate the conversion rate for the processes $J = 2 \to J = 1$, we again employ the Anderson Hamiltonian. In fact, we deal here with the high-temperature limit of the Kondo effect. In other words, it is assumed here that the coupling between the exciton and Fermi sea is not strong enough and is broken by thermal fluctuations. This assumption is confirmed by the excellent agreement between theory and experiment. Within the second-order perturbation theory, the spin-flip rate for the process $J = 2 \to J = 1$ is give by:

$$\gamma_{Dark\to Bright} = \\ = \frac{\Delta_t^2}{h} \int \left| \frac{1}{\varepsilon + e\left(U_g - U_1\right)/\lambda + i\Gamma} + \frac{1}{-\varepsilon + e\left(U_2 - U_g\right)/\lambda + i\Gamma} \right| \\ f_F(\varepsilon)\left[1 - f_F(\varepsilon - \Delta_{BD})\right] d\varepsilon, \tag{7.10}$$

where $f_F(\varepsilon)$ is the Fermi distribution at non-zero temperature T, $\Delta_t = \pi V^2 \rho$ is the tunnel broadening, the tunnel element V_k is taken as a constant V; the parameter Δ_{BD} is the exchange splitting between the bright and dark excitons. The factor $f_F(\varepsilon)\,[1 - f_F(\varepsilon - \Delta_{BD})]$ implies that only the energy interval $k_B T$ in the vicinity of the Fermi level ($\varepsilon \approx \varepsilon_F$) is important for the spin flips. It is also seen clearly from the diagram of Fig. 7.9b: the electron, which tunnels into the QD, comes from the above of the Fermi level and the QD electron should tunnel to an empty state below the Fermi level. Therefore, the important energy interval, where all events happen, is in the vicinity of ε_F. In addition, the two terms in Eq. 7.10 comes from two simple diagrams in which either the QD electron or Fermi-sea carrier tunnels first. Another interesting property of Eq. 7.10 is that it diverges in the limit $\Gamma \to 0$. Therefore, to obtain a meaningful result, one should include a final broadening of the states; this broadening may come from the radiative lifetime or from phonon-induced or electron-electron scatterings.

It is important to note that, in this scheme of spin manipulation, the spin of hole is not changed. In other words, we assume a sufficiently long lifetime for the spin of hole. Simultaneously, the spin lifetime of electron can be quite short because of the presence of Fermi sea. The Fermi sea is composed of nearly infinite number of electrons and can create spin flips in the localized exciton. The spin-flip rate of exciton has two important properties:

$$\gamma_{Dark \to Bright} = f\,(\varepsilon_{QD} - \varepsilon_F) = F(U_g), \quad \gamma_{Dark \to Bright} = f\left(\frac{\Delta_{BD}}{k_B T}\right).$$

First, the spin-flip rate depends on the deference between the electron energy in the QD and the Fermi energy and, therefore, on the voltage. Second, it strongly depends on the temperature and, more precisely, on the ratio $\dfrac{\Delta_{BD}}{k_B T}$. In the limit $T \to 0$, $\gamma_{Dark \to Bright} \to 0$. Generally speaking, in the limit $T \to 0$, we can not use the perturbation theory and should consider the Kondo state.

For this type of this type of electrical manipulation of spin is that a dark exciton with $J_z = +2$ can be converted into the bright state $J_z = +1$. Then, the bright state $J_z = +1$ will create a photon with the momentum$+1$. In the case of initial dark state $J_z = -2$, a photon emitted after such manipulation should be with the momentum -1. We now assume that the information is stored in the state of long-lived dark exciton. Then, application of voltage pulse, that brings the dark exciton into resonance with the Fermi level, can allow us to read the quantum state of dark exciton by looking at the momentum of emitted photon. A recently published paper by Atatüre et al. (Atatüre et al. 2006) reports the ability to prepare a spin state of electron with a high fidelity in voltage-tunable QDs in a magnetic field. The spin preparation process involves interplay between spontaneous spin-flip Raman scattering, hyperfine-induced electron spin-flips, and electronic co-tunneling (Kondo-type) processes. Another recent paper discussed the QD with a single magnetic impurity (Govorov and Kalameitsev 2005). The spin state of magnetic impurity (Mn) can be controlled optically via creation and annihilation of excitons in a QD.

Certain schemes of Mn-spin manipulation in a QD require fast relaxation of the electron spin (Govorov and Kalameitsev 2005) and the co-tunneling mechanism shown in Fig. 7.9 can be used for this purpose.

7.5 Coupling to the Continuum via Coulomb Interaction: Auto-ionization and Auger Processes

7.5.1 Fano Resonance

It is known that He atom in its doubly exited state undergoes auto-ionization induced by the Coulomb interaction. In Fig. 7.11, we show a schematic of electron transitions in He atom. Classical experiments on inelastic scattering of electrons by Silverman and Lassettre involve 2s2p ^{1}P electronic levels of He [See (Fano 1961) and references therein]. Under electron excitation, He atom makes transition from the ground state 1s1s ^{1}S to the state 2s2p ^{1}P. This excited state has one electron in the 2s state and another in the 2p orbital. The final state can decay since its energy is degenerate with the ionized states. The coupling between the auto-ionized 2s2p state and the continuum leads to peculiar line shapes in the scattering spectra. This line shape is known as Fano resonance. Interestingly, a variety of spectra obtained with different types of spectroscopic methods can be described by the Fano resonance formula:

$$I(\omega) \propto \frac{(\hbar\omega - E_0 + q \cdot \Delta)^2}{(\hbar\omega - E_0)^2 + \Delta^2}, \tag{7.11}$$

where $\hbar\omega$ is the energy loss in a spectroscopic experiment, E_0 is the energy of transition between the ground state and quasi-stationary excited state, Δ is the auto-ionization broadening on the resonance, and q is the Fano factor. The broadening and Fano factor are given by $\Delta = \pi \cdot W^2 \rho$ and

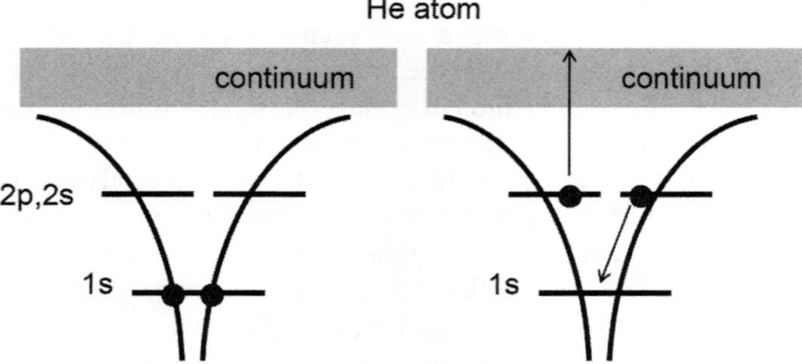

Fig. 7.11 Ground (left) and doubly excited states (right) of He atom. The exciton state is not stable and can decay due to the intra-atomic Coulomb interaction

$$q = \frac{V_0 \cdot W}{\Delta \cdot \upsilon}, \tag{7.12}$$

respectively. The parameters V_0, W, and υ are the amplitudes of transitions in a scattering process involving an auto-ionized state (Fig. 7.12).

The asymmetric shape in the energy loss spectra is well described by the Fano-resonance equation (7.11) and comes from the interference. The system can be excited to the continuum over two paths: ground state → continuum and ground state → quasi-stationary state → continuum. On one side of the absorption resonance, the interference coming from these two paths is constructive. On the other side of the resonance, the interference is destructive. The asymmetry is controlled by the Fano factor q. When $q \to \infty$ or $\upsilon \to 0$, the Fano resonance becomes symmetric

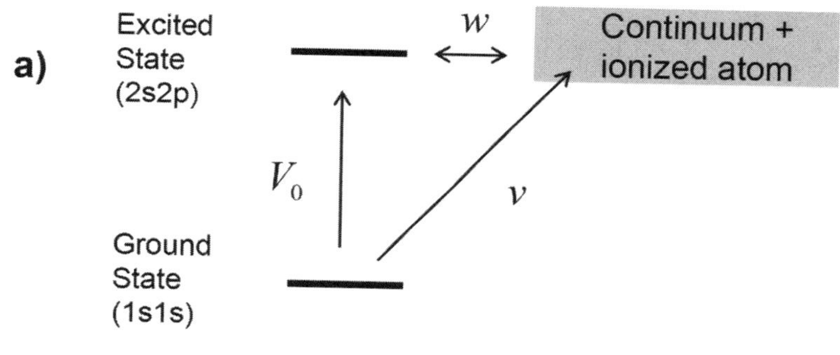

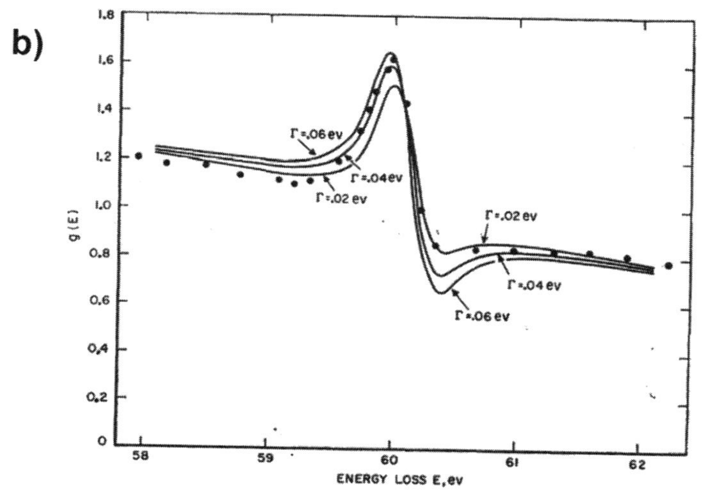

Fig. 7.12 a) Diagram of transitions for an auto-ionized atom. b) Theoretical shapes for inelastic scattering of electrons by He atom (solid lines) and the experimental data by Silverman and Lassettre (dots). The broadening parameter $\Gamma = 2\pi \cdot W^2\rho$. The graph b) is reproduced from (Fano 1961)

and Lorentzian. In the next subsection, we compare the Fano model of He atom with the models of QDs developed in recent papers.

7.5.2 Coulomb-induced Processes, Intra-band Auger Processes, and Auto-Ionization in Quantum Dots

The importance of Coulomb-induced transitions in the emission of heavily-charged excitons in QDs was noticed in the paper (Wojs and Hawrylak 1997) . Figure 7.13 illustrates the Coulomb scattering in the state remaining after emission of exciton X^{5-}. The immediate final state in the optical emission process can couple with two other configurations in which one of the electrons is promoted to the upper d-levels.

Now we look at more shallow QDs that bound only s- and p-states. In other words, we assume that d-states are in the continuum. This situation corresponds to the QDs studied experimentally in the papers (Warburton et al. 2000; Karrai et al. 2004). It turns out that the final states of excitons become unstable and auto-ionized starting from the exciton X^{3-}. The exciton X^{1-}in its final state has a single electron in the s-level; such a state can not decay (Fig. 7.14). The final state of X^{2-} has one s-electron and one p-electron (Fig. 7.14). This state can not decay through the Coulomb-induced processes since there is not enough energy for such a process in the system. However, the final state of X^{2-} can decay to the ground state due to the phonon emission; for the triplet configuration shown in Fig. 7.14, the relaxation requires also spin flip. Overall, the final state of X^{2-} is pretty stable. The situation with the exciton X^{3-} is totally different. The final state of X^{3-} has two p-electrons and an empty state in the s-shell. Conservation of energy, angular

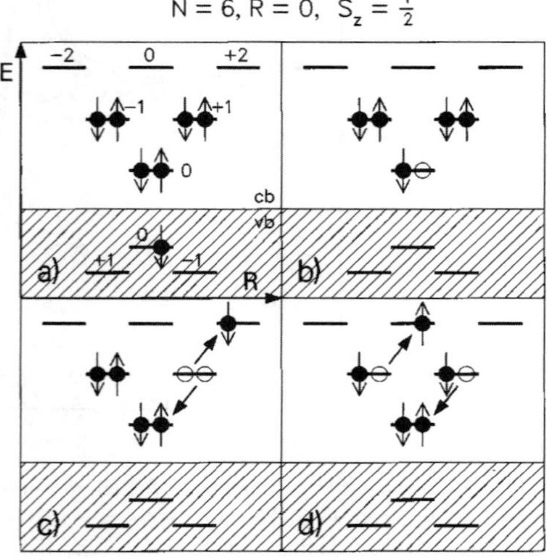

Fig. 7.13 Electronic configurations for the exciton X^{5-} with six electrons ($N = 6$). a) Initial ground state of exciton; b) Final state; c) and d) Diagrams of Coulomb-induced transitions in the final state. Open circles indicate holes in shells s and p. Reproduced from (Wojs and Hawrylak 1997)

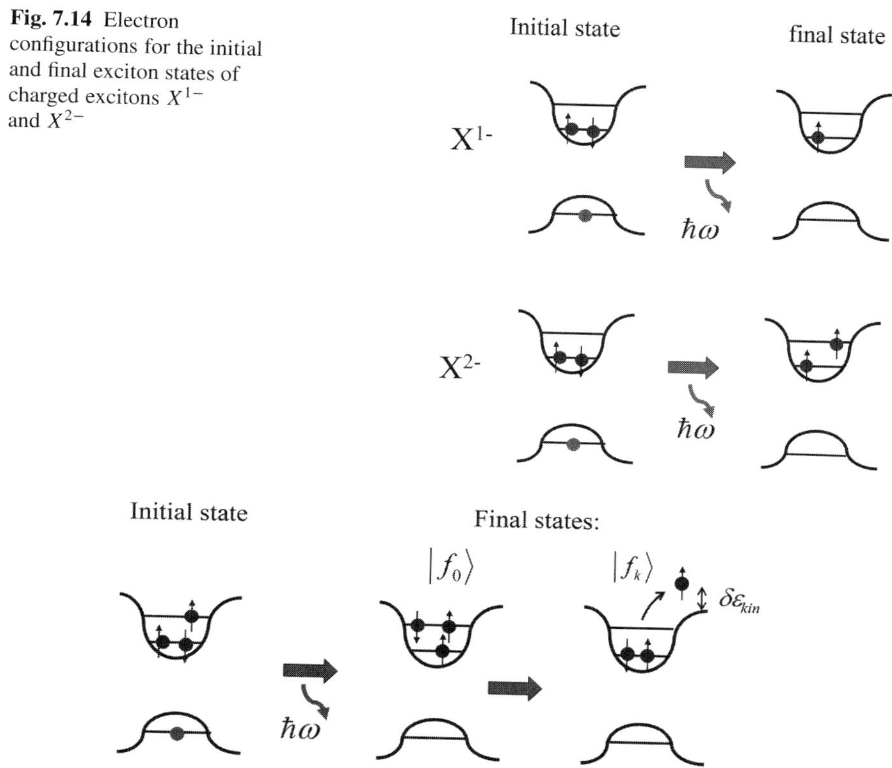

Fig. 7.14 Electron configurations for the initial and final exciton states of charged excitons X^{1-} and X^{2-}

Fig. 7.15 Electron configurations for the initial and final exciton states of X^{3-}. The final state undergoes ionization

momentum, and spin permit this state to be auto-ionized (Fig. 7.15). The Hamiltonian employed to describe this exciton has a form (Govorov et al. 2004):

$$\hat{H}_{X^{3-}} = E_0 \left| f_0 \right\rangle \left\langle f_0 \right| + \sum_k E_k \left| f_k \right\rangle \left\langle f_k \right| + \sum_k W_k \left| f_k \right\rangle \left\langle f_0 \right| + W_k^* \left| f_0 \right\rangle \left\langle f_k \right|. \quad (7.13)$$

Here $|f_0\rangle$ and $|f_k\rangle$ are the final states of X^{3-} ; the state$|f_k\rangle$ has two localized s-electrons and one electron excited to the continuum. The energies of final states are denoted as E_0 and E_k; $W_k = \langle f_0| \hat{U}_{Coul} |f_{1k}\rangle$ is the amplitude of Coulomb-induced auto-ionization. The continuum of states in this case corresponds to the 2D wetting layer discussed above in detail. The spectrum of emission should be calculated with Eq. 7.4 in which $|i\rangle = |X^{3-}\rangle$. In the absence of Fermi-sea electrons in the wetting layer, this problem can be solved exactly. The emission spectrum is given by:

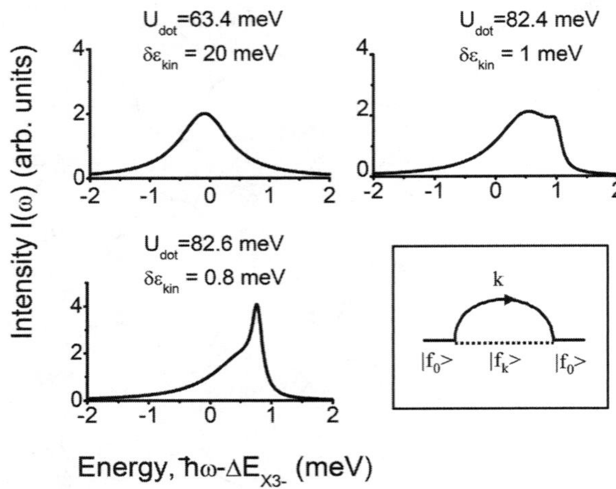

Fig. 7.16 Emission lines for several QD depths and the diagram for the self-energy in Eq. 7.14

$$I(\omega) = V_{opt}^2 \mathrm{Re}\frac{-i}{\tilde{\omega} - \Sigma_1 - i\gamma_{ph}}, \quad \Sigma_1 = \int_0^{\infty} \frac{|W_\varepsilon|^2 \rho(\varepsilon) d\varepsilon}{\tilde{\omega} - \delta\varepsilon_{kin} + \varepsilon - i\gamma_{ph}}, \quad (7.14)$$

where $\tilde{\omega} = \omega - \Delta E_{X3-}$ and $\Delta E_{X3-} = E_{X3-}^{initial} - E_o$ is the PL energy of the exciton X^{3-} in the absence of Auger coupling and E_{X3-}^{i} is the energy of exciton X^{3-},$\delta\varepsilon_{kin}$ is the excess kinetic energy in the Auger process (see Fig. 7.15) and $\rho(\varepsilon)$ is the 2D density of states of the wetting layer; γ_{ph} is a phonon-induced broadening. The self-energy Σ_1 in Eq. (7.14) was calculated according to the diagram shown in Fig. 7.16 (lower right panel). For the matrix element W_ε we assume $W_\varepsilon = W_a$ in the interval $0 < \varepsilon < D$ and $W_\varepsilon = 0$ elsewhere. In our model, the excess kinetic energy $\delta\varepsilon_{kin}$ depends on the QD depth U_{dot}. Figure 7.16 shows the calculated emission spectrum for $\Delta_a = \pi W_a^2 \rho_{2D} = 0.4\,meV$, $D \approx 30\,meV$ and several values of the parameter U_{dot}. The energy Δ_a is the broadening of the emission peak due to the Auger processes. Note that in the experiments (Karrai et al. 2004), $\Delta_a \approx 0.4\,meV$ and $\delta\varepsilon_{kin} \approx 20\,meV$. If $\delta\varepsilon_{kin} >> \Delta_a$, the spectrum is close to a Lorentzian line. Interesting shapes of the emission line appear when a QD is sufficiently deep and therefore $\delta\varepsilon_{kin} \approx \Delta_a$. In this case, the emission line shows quantum non-Lorentzian features due to the density of final states. In the case $\delta\varepsilon_{kin} \approx \Delta_a$, the Auger-coupling becomes partially coherent since the line is clearly non-Lorentzian. The reason for this is that the electron is excited to the states with $E_k \approx 0$ where the 2D density of states $\rho(E_k)$ has a discontinuity: $\rho(E_k) = \rho_{2D}$ for $\varepsilon > 0$ and $\rho(E_k) = 0$ for $\varepsilon < 0$.

Most interesting behavior for the auto-ionization process occurs in the presence of magnetic field (Karrai et al. 2004; Govorov et al. 2004). In a magnetic field, the continuum of 2D states turns into Landau levels. To describe

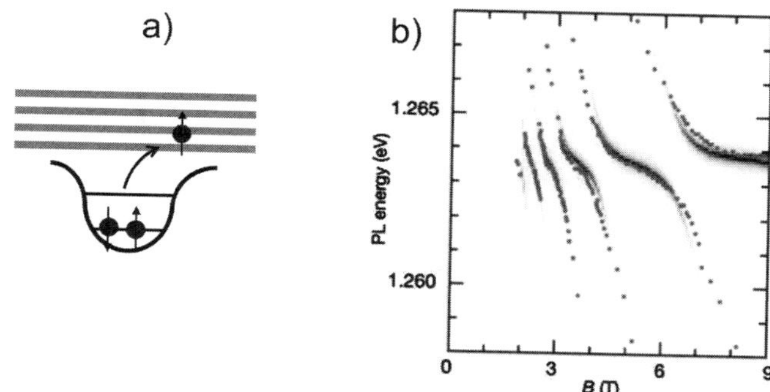

Fig. 7.17 a) Schematic of auto-ionization in the final state of X^{3-} in the magnetic field. One of the QD electrons is excited to a particular Landau level. b) A comparison of the results from the Anderson-Fano Hamiltonian model with the experiment for the X^{3-} exciton. The grey-scale shows the results of the calculation with the intensity representing the oscillator strength; the points represent the results of the experiment. Graph (b) is reproduced from (Karrai et al. 2004) . Reprinted by permission from Macmillan Publishers Ltd: Nature (Nature 427: 135–138, 2004), copyright (2004)

the Landau ladder in our model, we will use the following density of states: $\rho(\varepsilon) = \rho_0 \hbar\omega_c / \sqrt{\pi}\gamma_{LL} \sum_{n=0,1,2,..} \exp(-(\varepsilon - \varepsilon_n)^2/\gamma_{LL}^2)$, where $\varepsilon_n = \hbar\omega_c(n + 1/2)$ is the Landau-level energy, γ_{LL} is the Landau-level broadening, ω_c is the cyclotron frequency, and $n = 0, 1, 2, \ldots$. The emission spectrum as a function of the magnetic field demonstrates anti-crossing (Fig. 7.17b). The anti-crossings occur under the condition: $\delta\varepsilon_{kin} = \hbar\omega_c$. Generally speaking, an anti-crossing occurs when the kinetic energy of ionized electron is equal to the Landau-level energy. The reason is that, in quantum mechanics, two discrete levels always "repel" each other. Since the Coulomb-induced ionization process does not induce any additional broadening of exciton lines, we can conclude that this process is coherent. This is in contrast to the usual Auger processes in zero magnetic field.

Finally, we compare the model of auto-ionization in the QDs with the Fano model applied to He atom. Auto-ionization processes studied for QDs can be represented by the diagram shown in Fig. 7.18. This diagram has some elements of the Fano model, but it also has new features. First of all, for the case of QDs, we consider here the emission processes, whereas the Fano model was originally applied to the absorption (energy-loss) spectra. Another new feature is that, in the case of QDs, the amplitude υ (see Fig. 7.12a) can be neglected and the interference effects come mostly from the tunnel coupling between the QD discrete state and continuum. Under realistic conditions, the amplitude υ is much smaller than all other parameters. In other words, the effects in QD correspond to the Fano factor $q \to \infty(\upsilon \to 0)$. It can be easily seen that the Fano resonance (Eq. 7.12) turns into Lorentzian line in this limit ($q \to \infty$) and quantum interference vanishes. At the same time, for QDs under certain conditions, the strong interference effect exist even in the limit $q \to \infty$. Experimentally, the interference effects in QDs become

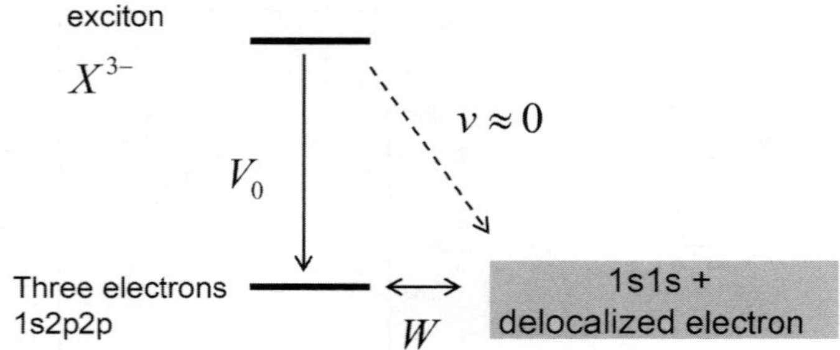

Fig. 7.18 Diagram of transitions for an auto-ionized charged exciton in a shallow QD. Detailed electron configurations of the states in this diagram are given in Fig. 7.15

accessible because the gate voltage and magnetic field are tunable parameters. For the case of atoms, there are not convenient tunable parameters.

We also should note that the papers (Ferreira and Bastard 2000; 2006) introduced the intra-dot Auger scattering for two p-electrons in the way similar to that shown in Fig. 7.15. These papers discussed incoherent dynamics and relaxation of electrons in QDs. The papers (Karrai et al. 2004; Govorov et al. 2004) treaded these processes differently involving both coherent and incoherent components of interaction. In particular, the coherent component of interaction leads to anti-crossings of terms in the magnetic field, non-Lorentzian lines in the emission spectrum, and emission-energy shifts.

7.6 Conclusions

To conclude, the chapter has reviewed recent studies on excitons in voltage-tunable quantum dots. In such systems, a new type of excitations, Kondo exciton, can exist. Kondo exciton can be formed when a charged electron-hole complex with nonzero spin interacts with the Fermi sea of electrons. The effect of hybridization between an excitons and Fermi sea leads to the optical lines which are strongly dependent on the electric field. Along with the tunnel coupling, Coulomb-induces auto-ionized states of excitons are described and observed. In the presence of magnetic field and under certain conditions, the auto-ionization (Auger) processes become totally or partially coherent. The optical features of the Coulomb-induced coupling for QDs and atomic systems look very different. This is due the different scales and confining potentials associated with QDs and atoms. In addition, effective masses of carriers involved in QD and atomic systems are very different.

Acknowledgments The Author likes to thank Richard Warburton and Khaled Karrai for many helpful discussions and joint work on several papers.

References

Alén B, Martínez-Pastor J, Granados D, García JM (2005) Continuum and discrete excitation spectrum of single quantum rings. Phys. Rev. B 72: 155331

Anderson PW (1961) Localized Magnetic States in Metals. Phys. Rev. 124: 41–53

Atatüre M, Dreiser J, Badolato A, Högele A, Karrai K, Imamoglu A (2006) Quantum-Dot Spin-State Preparation with Near-Unity Fidelity. Science 312: 551–553

Drexler H, Leonard D, Hansen W, Kotthaus JP, Petroff PM (1994) Spectroscopy of Quantum Levels in Charge-Tunable InGaAs Quantum Dots. Phys. Rev. Lett. 73, 2252–2255

Fano U (1961) Effects of Configuration Interaction on Intensities and Phase Shifts. Phys. Rev. 124: 1866–1878

Ferreira R, Bastard G (2000) Carrier Capture and Intra-Dot Auger Relaxation in Quantum Dots. *physica status solidi (a)*:178: 327–330

Ferreira R, Bastard G (2006) Unbound states in quantum heterostructures, Nanoscale Res. Lett.1: 120–136

Findeis F, Baier M, Zrenner A, Bichler M, Abstreiter G, Hohenester U, and Molinari E (2001) Optical excitations of a self-assembled artificial ion. Phys. Rev. B 63: 121309

Finley JJ, Ashmore AD, Lemaı tre A, Mowbray DJ, Skolnick MS, Itskevich IE, Maksym PA, Hopkinson M, Krauss TF (2001) Charged and neutral exciton complexes in individual self-assembled In(Ga)As quantum dots. Phys. Rev. B 63: 073307

Glazman LI, Raikh ME (1988) Resonant Kondo transparency of a barrier with quasilocal impurity states. JETP Lett. 47: 452–455

Goldhaber-Gordon D, H. Shtrikman, D. Mahalu, D. Abusch-Magder, U. Meirav, and M. A. Kastner (1998) Kondo effect in a single-electron transistor. Nature (London) 391: 156–160

Govorov AO, Karrai K, Warburton RJ (2003) Kondo excitons in self-assembled quantum dots. Phys. Rev. B 67 (RC): 241307

Govorov AO, Kalameitsev AV (2005) Optical properties of a semiconductor quantum dot with a single magnetic impurity: photo-induced spin orientation. Phys. Rev. B 71: 035338

Govorov AO, Karrai K, Warburton RJ, Kalameitsev AV (2004) Charged excitons in quantum dots: novel magnetic behavior and Auger processes. Physica E 20: 295–299

Gunnarsson O. and K. Schönhammer (1983) Electron spectroscopies for Ce compounds in the impurity model .Phys. Rev. B 28: 4315–4341

Helmes RW, Sindel M, Borda L, Van Delft J (2005) Absorption and emission in quantum dots: Fermi surface effects of Anderson excitons. Phys. Rev. B72: 125301

Hewson AC (1993) *The Kondo Problem to Heavy Fermions.* Cambridge University Press, Cambridge

Kammerer C, Cassabois G, Voisin C, Delalande C, Roussignol Ph., Gérard JM (2001) Photoluminescence Up-Conversion in Single Self-Assembled *InAs/GaAs* Quantum Dots. Phys. Rev. Lett. 87: 207401

Karrai K, Warburton RJ, Schulhauser C, Högele A, Urbaszek B, McGhee EJ, Govorov AO, Garcia JM, Gerardot BD, Petroff PM (2004) Hybridization of electronic states in quantum dots through photon emission. Nature 427: 135–138

Kikoin K, Avishai Y (2000) Many-particle resonances in excited states of semiconductor quantum dots. Phys. Rev. B 62: 4647–4655

Kondo J (1964) Resistance minimum in dilute magnetic alloys. Prog. Theor. Phys. 32: 37–49

Luyken RJ, Lorke A, Govorov AO, Kotthaus JP, Medeiros-Ribeiro G, Petroff PM (1999) The dynamics of tunneling into self-assembled InAs dots. Appl. Phys. Lett. 74: 2486–2488

Mazur YI, Liang BL, Wang ZM, Tarasov GG, Guzun D, Alamo GJ (2007) Development of continuum states in photoluminescence of self-assembled InGaAs/GaAs quantum dots. J. Appl. Phys. 101: 014301

Narvaez GA, Bester G, Franceschetti A, Zunger A (2006) Excitonic exchange effects on the radiative decay time of monoexcitons and biexcitons in quantum dots. Phys. Rev. B **74**: 205422

Schulhauser C., Haft D., Warburton RJ, Karrai K, Govorov AO, Kalameitsev AV, Chaplik A, Schoenfeld W, Garcia JM, Petroff PM (2002) Magneto-optical properties of charged excitons in quantum dots. Phys. Rev. B 66, 193303

Schulhauser C, Warburton RJ, Högele A, Karrai K, Govorov AO, Garcia JM, Gerardot BD, Petroff PM (2004) Emissison from neutral and charged excitons in a single quantum dot in a magnetic field. Physica E 21: 184–190

Smith JM, Dalgarno PA, Warburton RJ, Govorov AO, Karrai K, Gerardot BD, Petroff PM (2005). Voltage-control of the spin flip rate of an exciton in a semiconductor quantum dot, Physical Review Letters 94: 197402

Warburton RJ, Schäflein C, Haft D, Bickel F, Lorke A, Karrai K, Garcia JM, Schoenfeld W, Petroff PM (2000) Optical emission from a charge-tunable quantum ring. Nature 405: 926–930

Wojs A, Hawrylak P (1997) Theory of photoluminescence from modulation-doped self-assembled quantum dots in a magnetic field. Phys. Rev. B 55: 13066–13071

Chapter 8
Quantum Coupling in Quantum Dot Molecules

Xiulai Xu, Aleksey Andreev and David A. Williams

8.1 Introduction

Quantum information processing, including quantum computing and quantum communication (also called quantum cryptography), has attracted much interest both in theory and in experiment. In principle, a quantum computer is potentially more powerful than a classical computer for certain problems and quantum cryptography is believed to give perfect security for signal transmission [1–3]. Many systems have been proposed to implement quantum information processing. In this contribution, we will be reviewing solid-state based systems, in particular, semiconductor quantum dots.

Divincenzo *et al.* [3] pointed out the basic criteria for building a quantum computer: a collection of well characterized quantum two-level systems (qubits), which can be completely decoupled from each other and be initialised; quantum logic operations should be possible using external manipulations within the coherence time. The final result of computation can be measured at high fidelity. Solid-state systems are very promising to realise these points. Firstly, solid-state physics is a versatile branch of physics, which provides many potential candidates for realizing qubits. Solid-state qubits can be easily manipulated using external electric, optical and magnetic fields. Although coherence times are limited due to interaction with the environment, advanced technologies such as ultrafast optical pulses have enabled many quantum logic operations with a time less than the coherence time. Low noise measurement methods are now sufficiently developed to probe qubits. For instance, single electron transistors can be used to measure a total charge of a few electrons and highly sensitive photon detectors have been commercialised for single photon counting.

In the past twenty years, many candidates have been proposed and investigated for solid-state qubits. Nakamura, Pashkin and Tsai [4] have demonstrated single Cooper-pair qubits with observing Rabi oscillations. Mooij *et al.* [5] realized a single qubit using superconduction flux. Single ^{31}P-doped in Silicon for nuclear spin qubits has been proposed [6]. Spin-based qubits in quantum dots manipulating with cavity QED have been proposed by Imamoglu *et al* in 1999 [7] and realized afterwards [8–10]. Single nitrogen vacancy in diamond has

Z. M. Wang, *Self-Assembled Quantum Dots.*

been demonstrated to host a spin qubit [11]. Single ions feed into C_{60} carbon cages in single wall carbon nanotubes are very promising to scale the spin-based qubits [12].

Further, semiconductor quantum dots have attracted much interest in quantum information and quantum computation because of their quantized confined nature, also called 'artificial atoms' [13]. Single-qubit or two-qubit operation has been demonstrated in a single quantum dot [14, 15]. Naturally, scalability is the next issue to be solved for a quantum dot based quantum computer. Loss and DiVincenzo [16] proposed one or two qubit gates using coupled quantum dots (so called quantum dot molecules, QDMs), which have been investigated in several groups [17] using double quantum dots confined with negatively biased surface gates on the top of a 2DEG layer.

Besides spin-based qubits in QDMs, charge-based qubits have been demonstrated with a coherence time around 1 ns [18]. Recently, Gorman *et al.* [19] demonstrated charge qubits in isolated silicon QDMs. Figure 8.1 (a) shows the scanning electron micrograph of the single qubit device structure. All operations including initialisation, manipulation and measurement were demonstrated in a single device. Because the QDM is isolated from leads, the coherence time is about 200 ns using Ramsey-interference experiment (as shown in Fig. 8.1(b)), which is two orders longer than what was demonstrated by Hayashi *et al.* [18]. What is more fascinating is that the device shows the capability to scale up [19].

The qubit devices based on single electron charging and detection require diluted refrigerators with a temperature around tens of mK and they are difficult to manipulate or measure with optical methods. Direct band-gap semiconductor quantum dots grown by epitaxial self-assembly methods are optically active and can be operating around 4.2 K. Rabi oscillations of a single qubit based on III-V quantum dots (GaAs or InGaAs quantum dots) have been demonstrated using photoluminescence spectroscopy [14] or photocurrent [15] at 4.2 K. To scale up this quantum-dot based system, investigations are needed to observe and manipulate the coupling between two quantum dots.

Quantum cryptography (also called quantum key distribution) can be implemented with BB84 protocol with single-photon pulses [2]. In 1991, Ekert invented

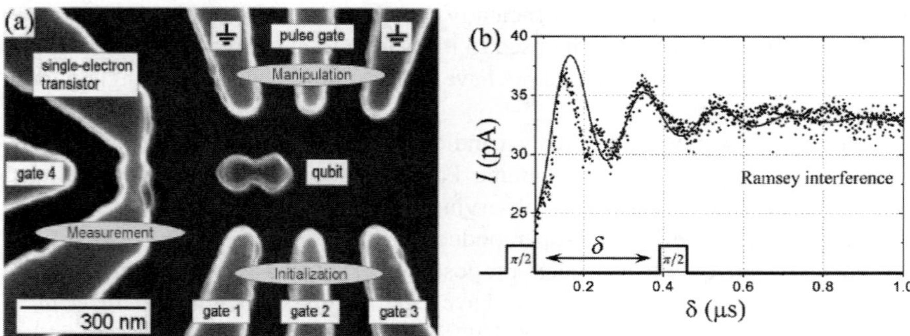

Fig. 8.1 (a) Scanning electron micrograph of a single Silicon QDM qubit device. (b) The free-evolution dephasing of the qubit using Ramsey-interference experiment. [from [19]]

an entanglement-based quantum key distribution [20], also called the Einstein-Podolsky-Rosen (EPR) [21] protocol. In this protocol, key distribution is not sent directly between sender and receiver but helped by the third party, a pair of polarization-entangled photons. In addition, the entangled photon sources are also potentially applicable for teleportation, quantum repeater and linear optical quantum computation [22]. To realize entangled photon sources, one usually uses parametric down-conversion [23]. However the photons generated by this method are random in time. Recently, entangled photons sources have been achieved in a single InAs quantum dot with diminishing fine splitting, which is induced by the asymmetry of the quantum dots [24]. However, the yield of the single-dot based entangled photon emission is very low due to the difficulties to find a quantum dot with small fine splitting. Gywat *et al.* [25] proposed the entangled generation from biexcitons in coupled quantum dots controlled with electric and magnetic fields, which has not been realized in experiment so far.

To achieve the scalable hardware for quantum information processing, quantum dot molecules are very promising to perform quantum operations because the excitons in a QDM are radiatively limited dephasing [26]. In this contribution, we concentrate on III-V QDMs, which can be controlled and detected optically. The quantum dots described here are InGaAs quantum dots using self-assembling methods. By controlling the growth conditions, the QDMs can be achieved both in lateral and vertical directions.

The chapter is arranged as follows: In Section 8.2, we discuss the electronic states of QDMs using $\mathbf{k}\cdot\mathbf{p}$ and plain-wave expansion methods including the strain. Then the lateral coupling of QDMs will be discussed using micro-photoluminescence spectroscopy in a high dot density sample in Section 8.3, followed by the investigations on the coupling of stacked QDMs. Lastly, a summary for this chapter will be given.

8.2 Theoretical Background of Quantum Dot Molecules

8.2.1 Strain Calculations

The calculation of the intrinsic strain field requires solving a 3D problem in elasticity theory for the non-trivial geometry of the QD shape. Commonly-used methods for the solution of this problem are finite-difference methods [27, 28] and atomistic techniques [29]. These methods require much computation time and computer memory. A simple and elegant method for calculating strain fields due to a single *isotropic* QD of arbitrary shape was presented by Downes *et al.* [30] as a simplification of the work of Eshelby [31]. This method provides *analytic* solutions for simple geometries such as cubic dots or pyramidal dots, and numerical solutions for more complex geometries, but neglects anisotropy.

Semiconductor compounds used in devices are cubic crystals with an anisotropy coefficient, defined in terms of elastic constants as $(C_{11}-C_{12})/2C_{44}$, typically equal to 0.5 compared to the isotropic value of 1. The sensitivity of some physical properties to strain suggests that anisotropic effects could be important in semiconductor materials and that an isotropic approximation may be poor, particularly in certain crystallographic directions.

In this chapter, we present an original method based on the Green's function technique for the calculation of strain distributions in QD structures [32] and demonstrate how this method can be applied to quantum dot molecules. Our method takes account of the anisotropy of the elastic properties in cubic crystals and produces a nearly analytical solution for the strain field due to QDs of arbitrary shape. The strain has a crucial effect on the electronic structure and optical properties of the quantum dot molecules.

The Green's tensor $G_{ln}(\mathbf{r})$ gives the displacement at $\mathbf{r}$ in the direction l due to a unit point force in direction n placed at the origin. The Green's tensor for infinite anisotropic elastic media [33] is the solution of the equation,

$$\lambda_{iklm}\frac{\partial G_{ln}(\mathbf{r})}{\partial x_k \partial x_m} = -\delta(\mathbf{r})\delta_{in} \tag{8.1}$$

with the boundary condition $G_{ln}(\mathbf{r}) \rightarrow 0$ as $|\mathbf{r}| \rightarrow \infty$. In equation (8.1), $\mathbf{r} = (x_1, x_2, x_3)$ is the space coordinate and λ_{iklm} is the tensor of elastic moduli. Here and below we use the usual rule for summation over 1,2,3 for repeating indices unless the sum is indicated explicitly. In this chapter, as a first approximation, we assume that the Green's tensor is the same for the matrix and QD material. If necessary, the different elastic moduli can be considered as a perturbation.

To solve equation (8.1) and find $G_{ln}(\mathbf{r})$ we use a Fourier transform technique. For the Fourier transform of the Green's tensor, $\widetilde{G}_{ln}(\xi)$, we obtain from equation (8.1) the following linear equation,

$$\lambda_{iklm}\xi_k\xi_m\,\widetilde{G}_{ln}(\xi) = \frac{\delta_{in}}{(2\pi)^3}. \tag{8.2}$$

The method of inclusions as proposed by Eshelby [31] is used to find the strain distribution in the QD structure. The displacement in a structure with a single QD can be expressed as the convolution of the Green's tensor and the forces spread over the QD surface,

$$u_i^s(\mathbf{r}) = u_i^T\chi_{QD}(\mathbf{r}) + \int G_{in}(\mathbf{r}-\mathbf{r}')\sigma_{nk}^T\, d\,S'_k \tag{8.3}$$

where $\chi_{QD}(\mathbf{r})$ is the characteristic function of the QD, equal to unity within the QD and zero outside; $\sigma_{nk}^T = \lambda_{nkpr}e_{pr}^T$ and σ_{nk}^T, e_{pr}^T and u_i^T are the components of the stress and strain tensors and the displacement caused by the "initial" strain due to the lattice mismatch. The superscript "s" indicates that this expression refers to a single QD.

The integration in equation (8.3) is carried out over the QD surface. Using Gauss's theorem, the strain tensor in a single QD structure is given by

$$e_{ij}^{s}(\mathbf{r}) = e_{ij}^{T}\chi_{QD}(\mathbf{r}) + \frac{1}{2}\int_{QD}\left[\frac{\partial G_{in}(\mathbf{r}-\mathbf{r}')}{\partial x_j \partial x_k} + \frac{\partial G_{jn}(\mathbf{r}-\mathbf{r}')}{\partial x_i \partial x_k}\right]\lambda_{nkpr}e_{pr}^{T}dV', \quad (8.4)$$

where integration is carried out over the QD volume. Using the convolution theorem and then taking the Fourier transform gives

$$\widetilde{e}_{ij}^{s} = e_{ij}^{T}\widetilde{\chi}_{QD}(\xi) - \frac{(2\pi)^3}{2}\left\{\xi_i \widetilde{G}_{jn}(\xi) + \xi_j \widetilde{G}_{in}(\xi)\right\}\lambda_{nkpr}\xi_k e_{pr}^{T}\widetilde{\chi}_{QD}(\xi) \quad (8.5)$$

where $\widetilde{\chi}_{QD}(\xi)$ is the Fourier transform of the QD characteristic function. Equation (8.5) gives the general expression for the Fourier transform of the strain tensor in a structure containing a single QD of arbitrary shape. This is a general formula valid for crystals of cubic or any other symmetry. Note that the QD shape enters only as the Fourier transform of the QD characteristic function.

The elastic problem is a linear one and so the solution for a QD array is obtained as a superposition of the elastic fields for single QDs, namely,

$$e_{ij} = \sum_{n_1,n_2,n_3} e_{ij}^{s}(x_1 - n_1 d_1, x_2 - n_2 d_2, x_3 - n_3 d_3) \quad (8.6)$$

where d_1, d_2, d_3 are the periods in the x, y and z directions respectively. An additional condition for e_{ij} arises from the requirement of minimum elastic energy for the periodic QD array. Equivalently, the strain tensor averaged over the elementary 3D superlattice unit cell is zero ($\overline{e_{ij}} = 0$). From equation (8.6) it follows that the coefficients for the Fourier series expansion of e_{ij} are equal to $\left[(2\pi)^3 / (d_1 d_2 d_3)\right]\widetilde{e}_{ij}^{s}(\xi_n)$, where $\xi_n = 2\pi(n_1/d_1, n_2/d_2, n_3/d_3)$. Finally, for the strain tensor in a QD array we obtain

$$e_{ij} = \frac{(2\pi)^3}{d_1 d_2 d_3}\sum_{n_1,n_2,n_3}\widetilde{e}_{ij}^{s}(\xi_n)\exp(i\xi_n \cdot \mathbf{r}), \quad (8.7)$$

where the summation is carried out over all values of n_1, n_2, n_3, except the case when $n_1 = n_2 = n_3 = 0$. For cubic crystals it can be shown [32] that the Green's tensor has the form:

$$\widetilde{G}_{in}(\xi) = \frac{1}{(2\pi)^3}\frac{\delta_{in}}{C_{44}\xi^2 + C_{an}\xi_i^2} - \frac{1}{(2\pi)^3}\frac{(C_{12}+C_{44})\xi_i\xi_n}{(C_{44}\xi^2 + C_{an}\xi_i^2)(C_{44}\xi^2 + C_{an}\xi_n^2)} \times\left\{1 + (C_{12}+C_{44})\sum_{p=1}^{3}\frac{\xi_p^2}{C_{44}\xi^2 + C_{an}\xi_p^2}\right\}^{-1}. \quad (8.8)$$

The Green's function tensor can be found, in principle, by performing the inverse Fourier transform with the corresponding integral evaluated using the spherical coordinate system and the residue theorem [33]. In the case $C_{an} = 0$, equation (8.8) reduces to the well-known isotropic result [31].

The strain tensor may be obtained by substituting equation (8.8) into equation (8.5). For cubic crystals the initial strain is

$$e_{ij}^{T} = \frac{a_M - a_{QD}}{a_{QD}}\delta_{ij} \equiv \varepsilon_0 \delta_{ij} \tag{8.9}$$

where a_M and a_{QD} are the lattice constants of the matrix and the QD materials, respectively. Combining this with the explicit expression for the elastic tensor we find $\lambda_{nkpr} e_{pr}^{T} = \varepsilon_0 (C_{11} + 2C_{12})\delta_{nk}$ and equation (8.5) is simplified for cubic crystals to,

$$\widetilde{e}_{ij}^{s} = \varepsilon_0 \widetilde{\chi}_{QD}(\xi) \left\{ \delta_{ij} - \frac{(2\pi)^3}{2} (C_{11} + 2C_{12}) \left[\xi_i (\xi \widetilde{G})_j + \xi_j (\xi \widetilde{G})_i \right] \right\}. \tag{8.10}$$

Using the explicit expression for $(\xi \widetilde{G})_i$ the final formula for the Fourier transform of the strain tensor for QDs with cubic symmetry is obtained,

$$\widetilde{e}_{ij}^{s}(\xi) = \varepsilon_0 \widetilde{\chi}_{QD}(\xi) \left\{ \delta_{ij} - \frac{(C_{11} + 2C_1)\xi_i \xi_j / \xi^2}{1 + (C_{12} + C_{44}) \sum_{p=1}^{3} \frac{\xi_p^2}{C_{44}\xi^2 + C_{an}\xi_p^2}} \right.$$
$$\left. \times \frac{1}{2} \left[\frac{1}{C_{44} + C_{an}\xi_i^2/\xi^2} + \frac{1}{C_{44} + C_{an}\xi_j^2/\xi^2} \right] \right\}. \tag{8.11}$$

In the form of the Fourier series, this formula and equation (8.7) give analytical expressions for the strain distribution in structures containing QDs of arbitrary shape. The QD shape enters in equation (8.11) only in the form of the Fourier transform $\widetilde{\chi}_{QD}(\xi)$ of the QD characteristic function. Analytical expressions for $\widetilde{\chi}_{QD}(\xi)$ for different shapes (cube, cylinder, cone, pyramid, truncated pyramid) are given below. Note that equation (8.11) simplifies considerably in the isotropic approximation. Inserting $C_{an} = 0$ we obtain,

$$\widetilde{e}_{ij}^{\text{iso}}(\xi) = \varepsilon_0 \widetilde{\chi}_{QD}(\xi) \left\{ \delta_{ij} - \frac{3\lambda + 2\mu}{\lambda + 2\mu} \frac{\xi_i \xi_j}{\xi^2} \right\}, \tag{8.12}$$

where $\lambda = C_{12}$ and $\mu = C_{44}$ are the Lamé constants for an isotropic elastic medium. From this equation we immediately find that the hydrostatic component of the strain tensor, $e_{ii}^{\text{iso}} \equiv e_h$, is constant inside the QD and zero outside in the isotropic approximation and given by

$$e_{ii}^{\text{iso}} = \varepsilon_0 \frac{4\mu}{\lambda + 2\mu} \chi_{QD}(\mathbf{r}). \tag{8.13}$$

Thus, the deviation of the hydrostatic strain from this constant value is characteristic of the influence of elastic anisotropy on the strain distribution in QD structures.

The Fourier transform of the characteristic function is a 3D integral

$$\widetilde{\chi}_{QD}(\xi) = \int_{QD} e^{-i\xi\cdot\mathbf{r}} \frac{dV}{(2\pi)^3} \tag{8.14}$$

where the integration is carried out over the QD volume. For most QD shapes the function $\widetilde{\chi}_{QD}(\xi)$ can be found analytically. In this section we present formulae for the sphere, cube, pyramid, cylinder, hemisphere, cone and truncated pyramid.

For the cuboid,

$$\widetilde{\chi}_{QD}(\xi_n) = \frac{8}{\xi_1\xi_2\xi_3} \sin(\xi_1 a_1/2) \sin(\xi_2 a_2/2) \sin(\xi_3 a_3/2) \tag{8.15}$$

where a_1, a_2, a_3 are the cuboid dimensions and the origin of the coordinates is at the center of the QD.

For the pyramid,

$$\begin{aligned}\widetilde{\chi}_{QD}(\xi_n) &= \chi_1(\xi_1, \xi_2, \xi_3, L_x, L_y) + \chi_1(\xi_2, \xi_1, \xi_3, L_y, L_x)\\ &\quad + \chi_1(-\xi_1, \xi_2, \xi_3, L_x, L_y) + \chi_1(-\xi_2, \xi_1, \xi_3, L_y, L_x),\end{aligned} \tag{8.16}$$

with

$$\begin{aligned}\chi_1(\xi_1, \xi_2, \xi_3, L_x, L_y) = \frac{1}{\xi_2\xi_3} \Bigg\{ & e^{-i\xi_3 h} \left[I_{e0}\left(\frac{L_x}{2}, -\xi_1 + \xi_3\frac{L_y}{L_x} + \xi_2\frac{2h}{L_x}\right)\right.\\ & \left. - I_{e0}\left(\frac{L_x}{2}, -\xi_1 - \xi_3\frac{L_y}{L_x} + \xi_2\frac{2h}{L_x}\right)\right]\\ & - I_{e0}\left(\frac{L_x}{2}, -\xi_1 + \xi_2\frac{L_y}{L_x}\right)\\ & + I_{e0}\left(\frac{L_x}{2}, -\xi_1 - \xi_2\frac{L_y}{L_x}\right)\Bigg\}\end{aligned} \tag{8.17}$$

where L_x and L_y are the pyramid base dimensions (the base is assumed to form a rectangle), h is the pyramid height, $I_{e0}(a, \xi) = \left[e^{i\xi a} - 1\right]/(i\xi)$, the origin of the coordinate system is at the center of the base and the x and y axes are parallel to the base sides.

For the cylinder,

$$\widetilde{\chi}_{QD}(\xi_n) = \frac{2\pi D}{\xi_{||}\xi_3} \sin(\xi_3 h/2)\, J_1\left(\frac{D\xi_{||}}{2}\right) \tag{8.18}$$

where J_1 is a Bessel function, h is the cylinder height, d is the diameter and the origin of the coordinate system is at the center of the cylinder.

For the hemisphere,

$$\widetilde{\chi}_{QD}(\xi) = \frac{1}{2}\widetilde{\chi}_{QD}^{sphere}(\xi) + \frac{2\pi i}{\xi_3}\left\{\frac{R}{\xi_{||}}J_1(R\xi_{||}) - R^2 I_{j0}(R\xi_3, R\xi_{||})\right\} \tag{8.19}$$

where $\widetilde{\chi}_{QD}^{sphere}$ is the Fourier transform for the sphere, R is hemisphere radius, $\xi_{||} = \sqrt{\xi_1^2 + \xi_2^2}$ and I_{j0} denotes the integral

$$I_{j0}(\alpha, \beta) = \int_0^1 x\cos\left(\alpha\sqrt{1-x^2}\right) J_0(\beta x)dx, \tag{8.20}$$

where J_0 is a Bessel function. The integral in equation (8.20) is best calculated numerically.

For the cone,

$$\widetilde{\chi}_{QD}(\xi) = \frac{2\pi i}{\xi_z}\left\{e^{-i\xi_z h} I_2(\xi_{||}R, \xi_z h) - \frac{R}{\xi_{||}}J_1(\xi_{||}R)\right\} \tag{8.21}$$

where R is the radius of the cone base, h is the cone height and I_2 denotes the integral

$$I_2(\alpha, \beta) = \int_0^1 xJ_0(\alpha x)e^{i3x}\,dx. \tag{8.22}$$

This integral can be expressed as a power series or calculated numerically.

For the truncated pyramid,

$$\widetilde{\chi}_{QD}(\xi) = \widetilde{\chi}_{QD}^{pyr}(\xi, L_x^b, L_y^b, h_l) - e^{-i\xi_z h_t}\,\widetilde{\chi}_{QD}^{pyr}(\xi, L_x^t, L_y^t, h_s) \tag{8.23}$$

where L_x^b and L_y^b are the base lengths, L_x^t and L_y^t are the dimensions of the truncated face, h_t is the height, $\widetilde{\chi}_{QD}^{pyr}(\xi, L_1, L_2, h)$ is the Fourier transform for a pyramid with base lengths L_1 and L_2 and height h; $h_s = L_x^t h_t/(L_x^b - L_x^t)$ and $h_l = L_x^b h_t/(L_x^b - L_x^t)$. In the above formula the origin of the coordinate system is at the center of the QD base.

Finally, we will show how to derive the Fourier transform of the characteristic function of a quantum dot molecule. From Eq. 8.14 we find for the quantum dot molecule (QDM):

$$\widetilde{\chi}_{QDM}(\xi) = \widetilde{\chi}_{QD1}(\xi) + e^{-i\xi_3 d_c}\,\widetilde{\chi}_{QD2}(\xi) \tag{8.24}$$

For the simple case, when both QDs forming the quantum dot molecule are identical, we have:

$$\widetilde{\chi}_{QDM}(\xi) = \widetilde{\chi}_{QD}(\xi)\left(1 + e^{-i\xi_3 d_c}\right) \tag{8.25}$$

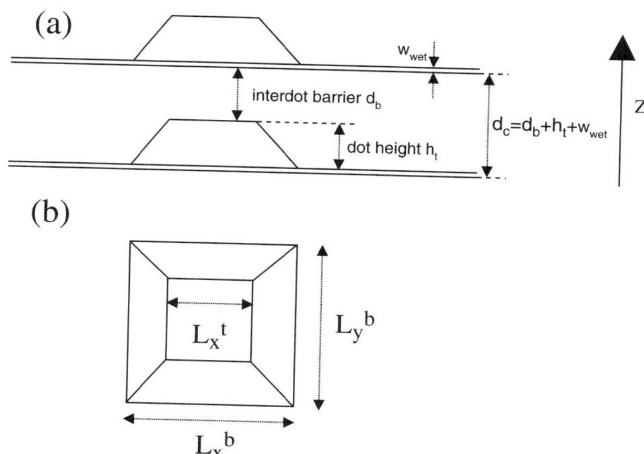

Fig. 8.2 (a) A schematic cross-section view of the quantum dot molecule. The upper dot is obtained by shifting the lower dot in z direction by distance d_c, equal to the sum of the wetting layer width, w_{wet}, interdot barrier, d_b, and the QD height, h_t: $d_c = w_{wet} + d_b + h_t$. (b) The top view of the truncated pyramid quantum dot showing the bottom and top sizes

In the above formulas we assumed that d_c is the distance in z direction between the centers of the quantum dots forming the QD molecule (i.e. the second dot is obtained by shifting the first dot by distance d_c in z direction), see Fig. 8.2 for an illustration.

8.2.2 Determination of Piezoelectric Field and Potential

It is by now well established that built-in piezoelectric fields modify the electronic structure of GaAs-based quantum dots. Typically, the potential amplitude of the piezoelectric potential is around tens of meV, or less. The effects are considerably weaker than those observed in wurtize GaN-based structures.

We start with the Maxwell equation $\text{div}\mathbf{D} = 0$ for the displacement field $\mathbf{D}$, which is defined here by

$$\mathbf{D} = \varepsilon_r \mathbf{E} + 4\pi \mathbf{P}_{strain} \tag{8.26}$$

where the strain-induced polarization, P_{strain} depends on the strain tensor e_{kl} as

$$P_i^{strain}(\mathbf{r}) = \varepsilon_{i,kl} e_{kl} = \varepsilon_{i,kl}^M e_{kl} + \delta\varepsilon_{i,kl} \chi_{QD} e_{kl}, \tag{8.27}$$

where $\varepsilon_{i,kl}^M$ and $\varepsilon_{i,kl}^{QD}$ are the piezoelectric constants for the matrix and dot materials respectively, and $\delta\varepsilon_{i,kl} \equiv \varepsilon_{i,kl}^{QD} - \varepsilon_{i,kl}^M$. It should be noted that for cubic crystals there is only one independent piezoelectric constant, ε_{14}.

By taking the Fourier transform of the Maxwell equation divD $= 0$ with **D** defined by Eq. (8.1), we find the Fourier transform of the built-in piezoelectric field, **E**,

$$\widetilde{E}_i = -\frac{4\pi\,\xi_i\xi_l}{\varepsilon_r\xi^2}\,\widetilde{P}_l^{strain}, \tag{8.28}$$

where tilde denotes the Fourier transform. When using the Fourier transform technique to find the built-in electric field for a free-standing array of quantum dots or quantum molecules, we include the additional condition that the electric field averaged over the unit cell of the QD superlattice is zero: $\overline{E}_i = 0$ (this follows from the requirement that the electric field energy is minimised). This is achieved by requiring that $\widetilde{E}_i$ is zero at ξ equals zero. The Fourier transform of the electrostatic potential, φ, is related to the built-in electric field by $\widetilde{\varphi} = -i\,\widetilde{E}_i\,/\xi_i$. The Fourier transform of the piezoelectric potential is then made up of two contributions:

$$\widetilde{\varphi} = \widetilde{\varphi}^{strain} + \widetilde{\varphi}^{\delta\varepsilon} \tag{8.29}$$

$$\widetilde{\varphi}^{str} = -i\frac{8\pi}{\varepsilon_r\xi^2}\varepsilon_{14}\left[\xi_1\,\widetilde{e}_{23} + \xi_2\,\widetilde{e}_{13} + \xi_3\,\widetilde{e}_{12}\right] \tag{8.30}$$

$$\widetilde{\varphi}^{\delta\varepsilon} = -i\frac{8\pi}{\varepsilon_r\xi^2}(\delta\varepsilon)_{14}\left[\xi_1\,\widetilde{e}^{\chi}_{23} + \xi_2\,\widetilde{e}^{\chi}_{13} + \xi_3\,\widetilde{e}^{\chi}_{12}\right], \tag{8.31}$$

where $\widetilde{e}^{\chi}_{ij}$ denotes the Fourier transform of the product $\chi_{QD}e_{ij}$ of the QD characteristic function and the elastic strain in the structure. The Fourier transform of the product is the convolution of the Fourier transforms of the individual terms, with $\widetilde{e}^{\chi}_{ij}$ therefore given by

$$\widetilde{e}^{\chi}_{ij}(\xi) = \sum_{\xi}\widetilde{\chi}_{QD}(\xi - \xi')\,\widetilde{e}_{ij}(\xi') \tag{8.32}$$

8.2.3 Plane-wave Expansion Method for Band Structure Calculations

We describe in this section our method to calculate the carrier spectrum and wave functions of QD structures, using the analytical expressions for the built-in strain and electric fields derived in the two previous sections.

The multiband envelope function approximation is widely used to calculate carrier spectra in semiconductor quantum structures. It has proved to be a convenient and reliable tool, describing well e.g. the experimentally observed variation in interband transition energies due to quantum size effects in quantum well (QW) and

QD structures [34, 35]. In the envelope function method, the carrier states in a quantum structure are calculated by solving a Schroedinger-like equation with an effective multi-band Hamiltonian, $\widehat{\mathbf{H}}\Psi = E\Psi$. The number of bands included, N_H and the form of $\widehat{\mathbf{H}}$ differ depending on the particular multi-band Hamiltonian which is chosen. From a mathematical point of view the effective Hamiltonian equation describes a system of coupled differential equations. By using a plane-wave method, we can solve this system of equations using a Fourier transform technique. From a physical point of view, this corresponds to describing the carrier states in terms of a linear combination of suitably chosen bulk states [36], associated with a periodic array of bulk wavevectors. The effective Hamiltonian can then be naturally represented in the form

$$\widehat{\mathbf{H}} = \widehat{\mathbf{H}}_0 + \widehat{\mathbf{V}} \tag{8.33}$$

where the "perturbation" $\widehat{\mathrm{V}}$ describes the difference between the potential in the quantum structure considered and the potential in the bulk Hamiltonian $\widehat{\mathrm{H}}_0$ used for the basis states. The eigenstates of the bulk Hamiltonian are chosen as the basis states for the plane-wave expansion; each of these eigenstates then has the form:

$$\Psi_{\mathbf{p},\mathbf{n},S}(\mathbf{r}) = \frac{1}{\sqrt{d_1 d_2 d_3}} \sum_{\alpha=1}^{N_H} B_\alpha^S(\mathbf{p},\mathbf{n}) u_\alpha(\mathbf{r}) \exp\{i[(\mathbf{p} - \mathbf{p_n})\mathbf{r}]\} \tag{8.34}$$

where $\alpha = 1, \ldots N_H$ labels the basic Bloch functions $u_\alpha(\mathbf{r})$, p is the "quasimometum" label for the 3D superlattice of quantums dots, $(p_\mathbf{n})_i = 2\pi n_i/d_i$, **n** is the plane wave number, and S denotes the different types of state included, (e.g. doubly degenerate electron, heavy-, light-hole and spin-split-off bands for $N_H = 8$). The operator matrix $\widehat{\mathbf{V}}$ in Eq. (8.8) is obtained from the bulk-like Hamiltonian by making the substitution $k_i \rightarrow -i\partial/\partial_i$, to take account of the spatial dependence of the band parameters. Details of the interface boundary conditions are included by an appropriate application of the differential operators at each interface. Each wavefunction of the effective Hamiltonian $\widehat{\mathbf{H}}$ is then found as a series expansion with respect to the plane wave states of Eq. (8.9):

$$\Psi_{\mathbf{p}}(\mathbf{r}) = \sum_S \sum_{\mathbf{n}} C_{\mathbf{p},\mathbf{n},S} \Psi_{\mathbf{p},\mathbf{n},S}(\mathbf{r}) \tag{8.35}$$

where the summation over S takes into account such effects as light- and heavy-hole mixing in heterostructures. The next step is to obtain the matrix **A**, whose eigenvectors and eigenvalues are the coefficients $C_{\mathrm{p,n},S}$ and the energy spectrum of the QD. This matrix has the form:

$$A_{i'i} = E_S(\mathbf{p}-\mathbf{n})\delta_{S'S}\delta_{\mathbf{n}',\mathbf{n}} + \sum_{\alpha'=1}^{N_H} \sum_{\alpha=1}^{N_H} \left[B_{\alpha'}^{i'}\right]^* B_\alpha^i \, \widetilde{V}_{\alpha',\alpha}^{i'i}(\mathbf{n}, \mathbf{n}') \tag{8.36}$$

where the numbers i' and i denote the set of quantum numbers $(\mathbf{p}, \mathbf{n}, S)$, $E_S(\mathbf{k})$ is the energy dispersion of the bulk state of type S; and $\widetilde{V}_{\alpha',\alpha}(\mathbf{n}, \mathbf{n'})$ is the Fourier transform of $V_{\alpha',\alpha}$:

$$\widetilde{V}_{\alpha',\alpha}(\mathbf{n}, \mathbf{n}') = \frac{1}{d_1 d_2 d_3} \int_{\Omega_0} e^{i\mathbf{n}'\mathbf{r}} V_{\alpha'\alpha} e^{-1\mathbf{n}\mathbf{r}} d^3 r \tag{8.37}$$

The matrix elements $V_{\alpha',\alpha}$ depend *linearly* on the strain tensor components and on the built-in electric potential. Therefore the Fourier transform of $V_{\alpha',\alpha}$ is expressed through the Fourier transforms of the strain tensor, the built-in electrostatic potential and the QD characteristic function χ_{QD}, introduced in the previous sections. Using a plane-wave expansion method in conjunction with the techniques introduced in previous sections, there is then no need to calculate the full spatial distribution of the strain and of the built-in electric field, unlike in other methods. This simple trick considerably reduces the computation time to set up calculations and makes the plane-wave method very effective for the further study of QD optical properties and modeling QD devices. We also note that the number of bulk states ("plane waves") which must be included to obtain a given level of accuracy is reduced in periodic structures with partly coupled QDs. In addition, the maximum number of bulk plane wave states, N_i^{max}, which can be included along any direction, i, in an envelope function calculation, is less than the number of atomic layers in one period of the QD superlattice along that direction: $N_i^{max} < d_i/a_i$. The envelope-function approximation is generally valid only when the envelope function varies smoothly over distances of the order of the lattice constant a_i. As a result, all terms with large wavevectors in the Fourier series, $(k_i > 2\pi/a_i)$, should be neglected, since they must be negligibly small in the envelope function approximation. This therefore provides an estimate for the maximum number of plane waves which should be included in the matrix expansion of Eq. (8.11). It also provides a means of testing the applicability of the plane wave expansion method to calculate the carrier spectrum and wave functions in any given QD structure. If the number of plane waves required along a particular direction to calculate the wave function and carrier energy of a given level is less than the maximum number N_i^{max}, then the plane wave expansion method should be valid. This is because the terms in Eq. (8.11) which have large wavevectors $(k_i > 2\pi/a_i)$ and which should therefore be thrown away do indeed make a negligibly small contribution to the solution of the Schroedinger-like equation $\widehat{\mathbf{H}}\Psi = E\Psi$. We find for all the QD structures considered in this chapter that the number of plane waves required along any direction is always much less than N_i^{max} and it is therefore appropriate to use the plane wave expansion method.

8.2.4 Coupling Dependence on QD Separation

In this chapter we have used the above-described method in the framework of the 8×8 $\mathbf{k} \cdot \mathbf{p}$ Hamiltonian. The explicit form of the Hamiltonian can be found, for example, in [37]. For modelling the properties of the QDM structures we considered the structures schematically shown in Fig. 8.2. The quantum dot shape is a truncated pyramid and the interdot barrier is defined as the distance from the top of the lower dot to the wetting layer of the upper dot in the quantum dot molecule.

Figure 8.3 shows the dependence of splitting of the electron levels in InGaAs quantum dots in GaAs matrix with 50% In content in the dot. For the quantum dot size considered we found only four localised levels for small interdot separation (not counting spin degeneracy; with spin degeneracy there are totally 8 levels or 4 degenerate pairs of levels.). For the dot parameters considered in this example we see that 6 nm interdot spacing is enough to nearly vanish the splitting of the ground state. As expected, the splitting of the ground state, shown in Fig. 8.4, exponentially depends on the interdot barrier width. For the interdot barrier equal around 1 nm the calculated splitting is in good agreement with experimental results by Vorobjev et.al. [38]. The splitting of the excited levels remains finite even at large interdot barrier widths, see Fig. 8.3. We note that some levels shown in Fig. 8.3 are not fully localised in the QDM. The surfaces of the constant probability density of electrons and hole for the first several energy levels are shown in Figs. 8.5 and 8.6. It is clear the electron states are much stronger coupled compared to the holes.

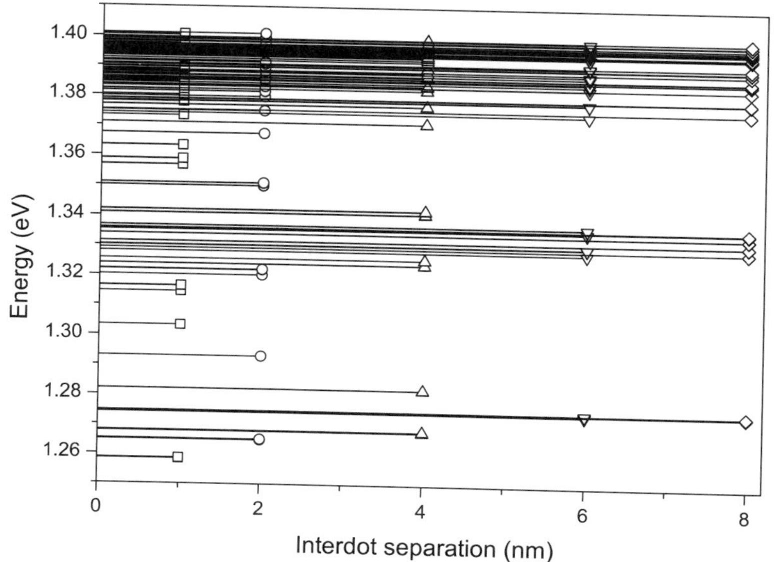

Fig. 8.3 The electron energy in coupled QD molecules vs interdot barrier d_b at 10K. Parameters of the $In_{0.5}Ga_{0.5}As$ dots: L_b=9nm, L_t=4nm and h_t=4 nm

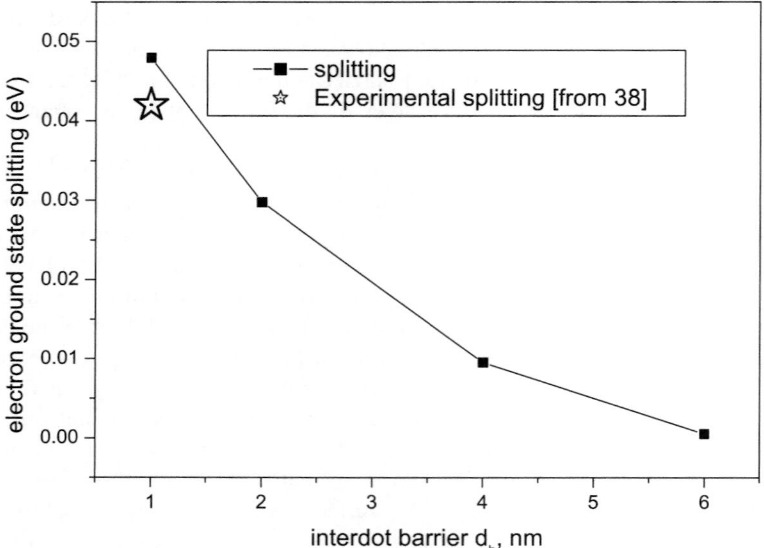

Fig. 8.4 The splitting of the ground state vs interdot barrier width for the same system as in Fig. 8.3

8.2.5 Coupling Variation with an Applied Electric Field

The presence of the electric field can be taken into account using the plane wave expansion method. The homogeneous electric field E corresponds to the following electric potential:

$$\phi = -\mathbf{Er} \tag{8.38}$$

If the electric field is directed along z, then we have:

$$\phi = -\mathrm{Ez}. \tag{8.39}$$

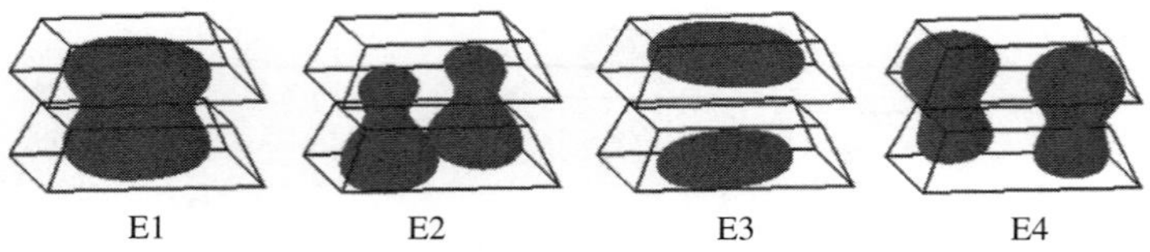

Fig. 8.5 Surface of constant electron probability density, $|\Psi|^2$, equal to 45% of its maximum value for each state for the first four energy levels (not counting spin degeneracy), E1, E2, E3 and E4. The parameters of the QDs are Lb=11.3 nm; Lt=8.5 nm, ht=3 nm, dc=4.5 nm; $In_{0.5}Ga_{0.5}As$ QD in GaAs matrix (Note there are just four localised states in this QDM)

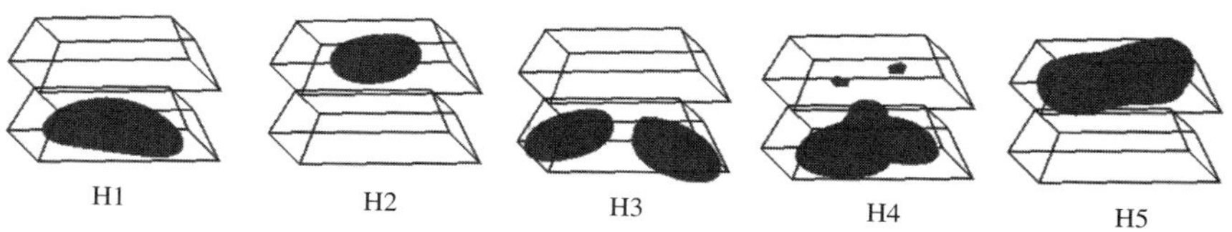

Fig. 8.6 Surface of constant hole probability density, $|\Psi|^2$, equal to 45% of its maximum value for each state for the first five energy levels (not counting spin degeneracy), H1, H2, H3, H4 and H5. The parameters of the QDs are the same as in Fig. 8.5

The next step is to take the Fourier transform of the potential ϕ and use it directly in the planewave expansion method. The Fourier transform in this case can be easily calculated analytically.

To characterise the coupling between the two dots in the QDM, we consider the coupling degree ν defined as

$$\nu = 4P_1 P_2, \tag{8.40}$$

Where P1 and P2 are the probabilities to find the electrons in the lower (QD1) and upper (QD2) dots, respectively:

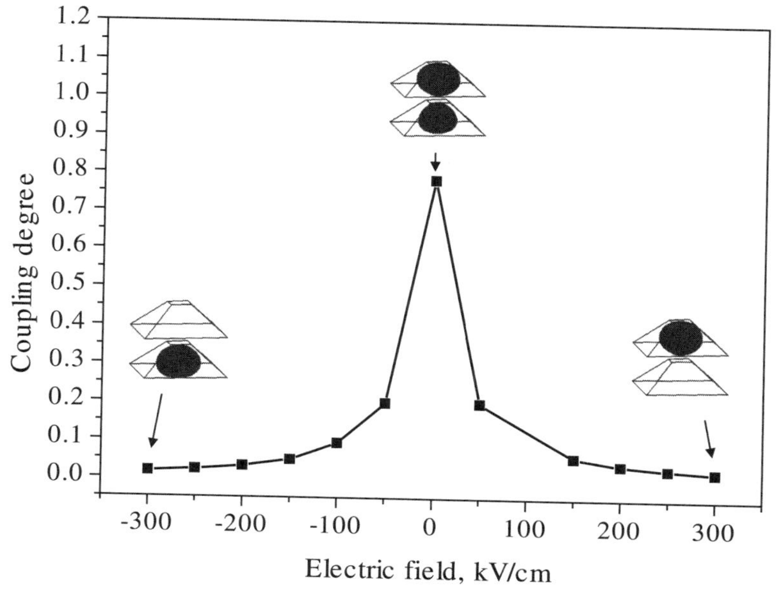

Fig. 8.7 Coupling strength as a function of electric field. Inset: electron wavefuntions corresponding to the electric fields

$$P_1 = \int_{QD1} |\Psi|^2 d^3 r \tag{8.41}$$

$$P_2 = \int_{QD2} |\Psi|^2 d^3 r \tag{8.42}$$

where Ψ(r) is the wave function of the electron on the ground level and the QDM is formed from two dots, QD1 and QD2.

The degree of coupling is zero, if the electron probability P in one of the dots is zero. When the probability to find the electron in both dots is equal to ½, the degree of coupling is 1. Figure 8.7 illustrates that by applying the electric file in growth direction it is possible to control the coupling of the electron states in the QDM. When the field is zero, the electron spends nearly equal time in both dots and the coupling is degree has a sharp maximum. When the electric field is applied, the electron is forced to occupy mainly only one dot, depending on the direction of the electric field.

8.3 Coupling in Lateral Quantum Dot Molecules

Since coupling in QDMs provide the potential application for building blocks of quantum computers and entanglement photon sources, experimental investigations have been pursued intensively to obtain it. In the theory, strong evidence that two quantum dots in a single QDM are coupled is to observe the bonding and antibonding states with a certain separation distance. The splitting between bonding and antibonding states decreases with increasing distance between the two dots, that are coupled quantum mechanically. Spatially, two categories of QDMs are normally investigated: lateral QDMs, and vertical (stacked) QDMs. For lateral QDMs, most work has focused on micro-fabricated isolated QDMs [19] or negative bias confined quantum dots using surface metal on the top of two dimentional electron gases [17] for charge- or spin- based quantum bits. For III-V quantum dots, Schedelbeck *et al.* [39] demonstrated bonding and antibonding states of two quantum dots with different separations using microscopic photoluminescence spectroscopy. The QDMs are fabricated using cleaved edge overgrowth that requires three MBE growth steps. Although the separation between two quantum dots can be precisely controlled with different barrier thickness between two quantum wells, the rest growth steps especially the edge epitaxy are very difficult. In the rest of this section, we discuss the observation of lateral coupling between two quantum dots with a high dot density wafer [40].

8.3.1 Experimental Details

The photoluminescence measurement is one of the basic tools that allows for the investigation of discrete energy levels in quantum dots. A laser beam usually with energy larger than recombination energy excites the electrons from the valence band to the conduction band and creates electron-hole pairs. The electron-hole pairs (also called excitons) recombine following the selection rule to ensure momentum conservation. Recently, microphotoluminescence spectroscopy techniques probing the optical properties of single quantum dots have been well established, in which the size of the laser spot can be focused up to a diameter around 1 μm. Usually in a confocal microphotoluminescence system, the quantum dot sample is mounted in a He-flow cryostat and cooled to 5K. The QDs are excited with a laser, the excitation power density being adjusted with neutral density filters. The laser spot is focused with a long working distance objective, which has a large numerical aperture. A high precision XYZ stage with high resolution is used to align the laser spot to the sample. The emission light is collected by the same objective, dispersed through a spectrometer and then detected with a cooled charge-coupled device (CCD) camera.

As discussed in the chapter of quantum dot growth, the dot density varies from 10^7 to $10^{10}/\text{cm}^2$ using different growth conditions. One easy way to realize different dot densities is to stop rotating the substrate during growth. With a non-rotating substrate, the Indium component will be different across the wafer because of the asymmetry of the Indium source with respect to the wafer, resulting in variation of the InAs dot density [41]. Using this method, the dot density varies from negligible at the low In flux edge of the wafer, to $10^{10}\,\text{cm}^{-2}$ on the opposite side. Figure 8.8 shows AFM images of samples with different dot densities across the wafer.

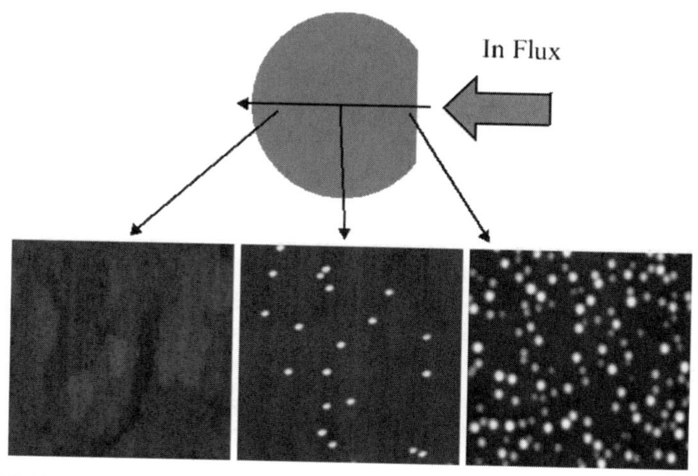

Fig. 8.8 AFM images (1 μm × 1 μm) of InAs quantum dots (Approximate coverage at the center of a 1.7 monolayer) without rotating substrate during InAs growth

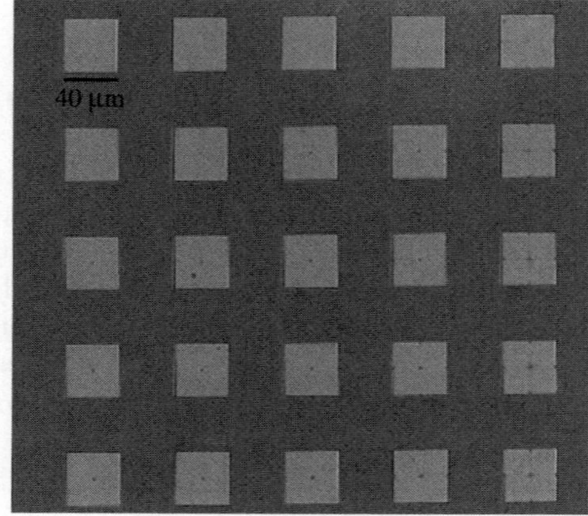

Fig. 8.9 Optical image of an array of metal apertures

Although the laser spot can be focused to 1 μm or even to subwavelengths, it is still challenging to observe single dot emission with bare wafers. Therefore, it is necessary to isolate a single dot. One way to isolate a single dot is to etch a pillar structure with a small enough cross-section so that only a single dot is present [42]. However, the depleted surface layer always quenches light emission via nonradiative recombination centres. To avoid this, metal masking can isolate a single dot effectively. One method to make metal masks is known as the 'lift-off' technique. The idea of this technique uses an electron beam to expose a large square in PMMA with a small hole left at the centre of the square. After metal deposition and lift-off, metal squares with small holes in the centre are achieved. Figure 8.9 shows an optical image of an array of aluminum masks with different apertures for PL measurements.

8.3.2 Coupling of Lateral QDMs

When aluminium mask arrays are deposited on high density regions, it is possible to have two quantum dots within one metal aperture, which are coupled when they are in close proximity. Figure 8.10 (a) shows a wide scan of PL emission from the dots. The observed emission lines fall into three distinct groups at 1.271 eV (P0), 1.298 eV (P1) and 1.361 eV (P2), respectively. At low excitation intensity, only P1 can be observed (around 1.298 eV). A low-energy peak (P0) appears when the excitation is increased to 1.67 μW and a high-energy peak appears with increasing excitation power. This peak grouping has been observed at different lithographically-defined sites on the wafer. P0 and P1 were attributed respectively to the transitions of electron to heavy-hole and electron to light-hole in the s-shell [43]. P2 is due to excited state emission (p-shell); the p-shell emission is around 60 meV higher than the s-shell emission, as seen elsewhere [44–47].

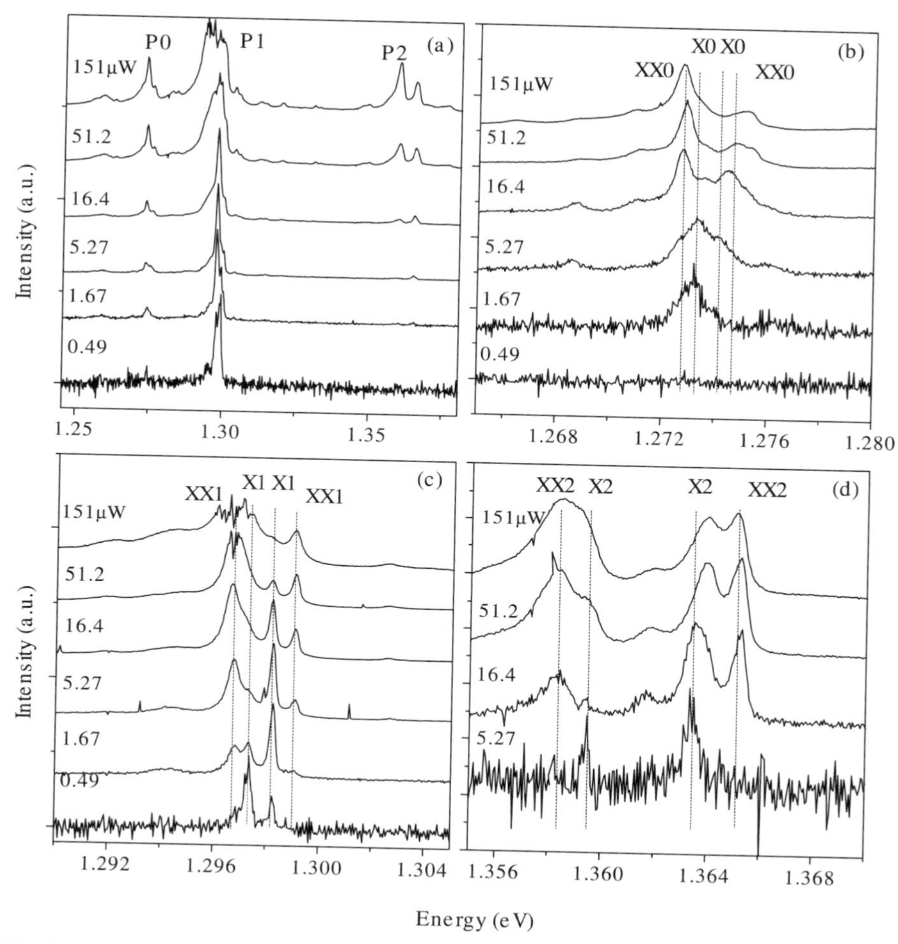

Fig. 8.10 Photoluminescence spectra of InAs QDMs for different excitation power intensity. (a) low-resolution scan spectrum peaks P0, P1 and P2. (b)–(d) details of splitting for P0, P1 and P2 respectively. In each case, the spectra are normalized to the highest peak value and dashed lines are used to guide the eye

The fine structures of these peaks were investigated by high-resolution scans, and are shown in Figs. 8.10 (b)–(d). For P1, s-shell emission is bunched into a four-peak structure. At very low excitation intensities, two peaks (X1) separated by 926 μeV are observed. Such a two-peak structure cannot be explained by exciton and biexciton transitions, or exciton and charged-exciton transitions with studying power dependences and temperature-assisted charging effects (for more details please refer to [40]). Therefore, both X1 were due to exciton emission. As the excitation is increased, two sidebands (XX1) can be seen which grow superlinearly by factors approximately equal to the square of the excitation intensity, indicating biexciton emission, respectively [48].

To confirm the origin of this double peak structure, a reference sample with a low dot density was measured to exclude the single quantum dot signature. This comparison also excludes the possibility that the double peaks are due to the ground state splitting of one dot. If the four peaks were due to the exciton and biexciton of two different dots without coupling, the two exciton peaks at the higher energy side would be expected to decrease and the two biexciton peaks increase on the lower side with increasing excitation intensity, which does not correspond with the spectra here. From all these considerations, we concluded that there are two dots within the aperture and these are coupled to each other.

Without coulomb interaction, the coupling of two identical quantum dots leads to a splitting of the single particle state into a bonding and an antibonding state. The magnitude of the splitting depends on the height and width of the barrier. In stacked double dots, the coupling energy can be up to several tens of meVs because a very short separation of several nanometers can be achieved [45, 49]. On the other hand, the splitting of coupled dots made by cleaved edge overgrowth [39] is much weaker, of the order of several hundred μeV. The splitting of X from P1 peak is around 1 meV, which is a little larger than that observed in the Schedelbeck [39] structure, which has dots of comparable spacing to those reported here. The lateral distance estimated from AFM images is approximately 20–100 nm. The larger coupling in this structure may come from the low GaAs barrier instead of AlGaAs in their structure. The biexciton splitting energy is around 2.35 meV.

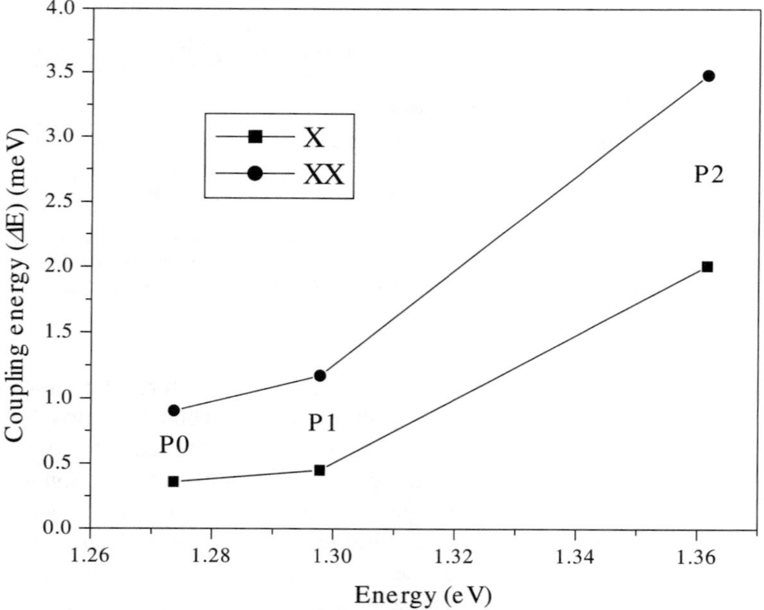

Fig. 8.11 Coupling energy of exciton and biexciton at different energy levels, corresponding to P0, P1 and P2 in Fig. 8.10

It can be seen that the four-peak structure is resolvable. From first-order perturbation theory, the energy levels can be written as $E = E_0 \pm \Delta E$, where E_0 is the energy level of an individual dot and ΔE is the coupling energy. $2\Delta E$ is the energy splitting corresponding to the energy between bonding and antibonding states [39]. For the higher energy states, the splitting energies between antibonding and bonding states of exciton and biexciton is higher than for ground state for a fixed barrier width, because these states have a larger penetration into the GaAs barrier resulting in a larger tunnel matrix element [13]. Figure 8.11 shows the coupling energies of exciton and biexciton at different energy levels corresponding to Fig. 8.10. The coupling energy increases with energy levels, as expected.

In this section, we discussed the coupling between laterally coupled quantum dot molecules at different energy levels when different apertures were selected for micro-photoluminescence measurements. Splitting of the exciton and biexciton increases at high energy levels because of the larger interaction between the dots. Although the observation of coupling can be obtained by chance when two randomly self-assembled dots in one aperture have been measured, the laterally coupled QDMs produce much interest because this alignment may provide a good way of generating maximum entanglement [49] and they also scale up easily. Recently experiments [50, 51] have successfully obtained laterally coupled QDMs using self-assembled nanoholes with MBE, which are very promising to have spatial ordering on the long-range scale. The QDMs in large scale ordering provides potential application to scale up quantum bits in quantum computation.

8.4 Stacked Quantum Dot Molecules

To implement a qubit or a quantum gate, several requirements are necessary [16]: reliable state preparation, low decoherence, accurate quantum gate operations, and strong quantum measurement. Although bonding and antibonding states can be observed in lateral QDMs (they were observed by chance), it is not easy to control the coupling strength in this system. External controls such as magnetic fields [16], electric fields [17] and optical fields [50, 52] have been proposed to manipulate the entanglement. For this purpose, vertically stacked QDMs have been suggested to host a single quantum bit, or double quantum bits; these can be controlled by optical pulses, by an electric field, or by a magnetic field [53, 54]. In this chapter, stacked QDMs will be discussed under electric fields, especially in electroluminescence.

To confine the InAs quantum dots in three dimensions, a cover layer of GaAs will be deposited after the growth of InAs quantum dots. Additionally, it is also interesting to see what happens if a second layer of InAs is deposited on top of the cover layer. Xie *et al.* [55, 56] observed that a second layer of InAs quantum dots positions itself on top of the quantum dots in the first layer. This is because the first InAs quantum dot layer induces strain in the thin GaAs spacer layer (up to a few nanometers) and this strain drives the adatom migration of the second layer. This kind of stacked quantum dot structure has also been realized in several other systems, such as InP [57], SiGe/Si [58] and $PbSe/Pb_{(1-x)}Eu_xTe$ [59]. It provides a nice method of growing stacked double quantum dots, which are potential candidates to embody qubits for quantum information processing [39, 60].

Coupling between stacked quantum dots was first observed in multi-layer quantum dots. The luminescence spectra have a larger spectral linewidth than single layer quantum dots because of the coupling between different layers. Recently, investigations on the coupling are focusing on single QDMs manipulated with electric and magnetic fields. Ortner *et al.* [61] demonstrated that the mixing of exciton states on the two quantum dots, with different separation distances can be controlled with an electric field. One strong evidence for the coupling has been shown with an anticrossing of direct and indirect excitons with changing electric field across the stacked QDMs [62]. Direct and indirect excitons are defined where electron and hole wave functions are spatially occupied in the same quantum dot, or different quantum dots in a single QDM. Usually holes are located in one of the single dots because of their larger effective mass relative to the electrons. When the distance between two quantum dots decreases, the electron component of exciton wave functions hybridizes into bonding and antibonding states. If the distance is larger enough, the electron wave functions are localized forming direct or indirect excitons, in which electrons and holes are occupying in the same dot or different dots, respectively. Because of the different spatial electron-hole configurations, the quantum confined Stark effect is different for various applied electric fields. The indirect exciton energy changes more dramatically than direct exciton energy with electric field because of the larger excitonic dipole moment. By applying an electric field, it is possible to bring direct excitons and indirect excitons into resonance; electron wavefunctions are delocalized over two dots resulting in symmetric and antisymmetric molecular states, an anticrossing can be observed as a function of electric field. Stinaff *et al.* [63] demonstrated by applying an electric field that a single hole can form a coherent molecular wave funtion in QDMs using optical spectroscopy. The resonance between charge exciton and single hole state with certain bias provides a way that the coupling can be potentially manipulated with optical pulses.

All these works are investigated using electric-field induced photoluminescence spectroscopy. From an application point of veiw, it is more interesting to investigate the coupling with electroluminescence spectroscopy, for example, electrically pumped entangled photon sources. In the rest of this section, we focus on the observation of coupling in QDMs using electroluminescence spectroscopy.

8.4.1 Lateral p-i-n Junctions

To measure the optical spectroscopy of single quantum dots or single quantum dots molecules, metal apertures may be made on the surface to select the emission site. The low coincidence between quantum dots and metal apertures produces a low yield of functional sites because of the random distribution of the self-assembled quantum dots. To avoid this problem, lateral *p-i-n* junctions are designed for electroluminescence measurement in this section. The lateral *p-n* junction has been proposed and is investigated by several groups [64, 65], since it would be more easily incorporated with other lateral devices than the conventional *p-n* junction [66]. Kaestner *et al.* [67] proposed a new method to realize a lateral *p-n* junction. The structure is designed as follows: A δ doped *n*-type layer substitutes for the *n*-type layer in conventional *p-i-n* structures. The n-type layer is fully depleted when the top *p*-type layer is in place and it becomes conductive when the top layer is etched away. Therefore, the electrons and holes will accumulate at the etched edge when a forward bias is applied, resulting in electroluminescence [67]. If quantum dots are inserted into the intrinsic region of this structure, the dots at the etched edge will be excited first by using this confinement of electron and hole injection at the etched edge.

Figure 8.12(a) shows a schematic vertical section of a lateral *p-i-n* structure. Using standard microfabrication processing, lateral *p-i-n* devices have been achieved, as shown in Figure 8.12 (b). Typically $1 \times 10\,\mu m^2$ in area and with the top *p*-type layer removed where the *n*-type layer is to conduct; this material was contacted with a AuGeNi annealed contact, while the *p*-type mesa was contacted with Cr/Au which did not have to be annealed because of the heavily doped surface layer. Ideal diode *I-V* characteristics have been observed from this structure. When a wire-shaped device is formed, the dots at the etched edge in the active channel are excited at low voltages whilst those far away from the edge can be excited by increasing the voltage when there is no dot near the edge. By controlling the voltages, single-dot emission can be obtained from most devices because of the long active channel. The structure is then self-selecting for emission from a single dot, with no need for metal apertures. Using this structure, electrically-pumped single-photon sources have been demonstrated with a repetition up to 100 MHz [68].

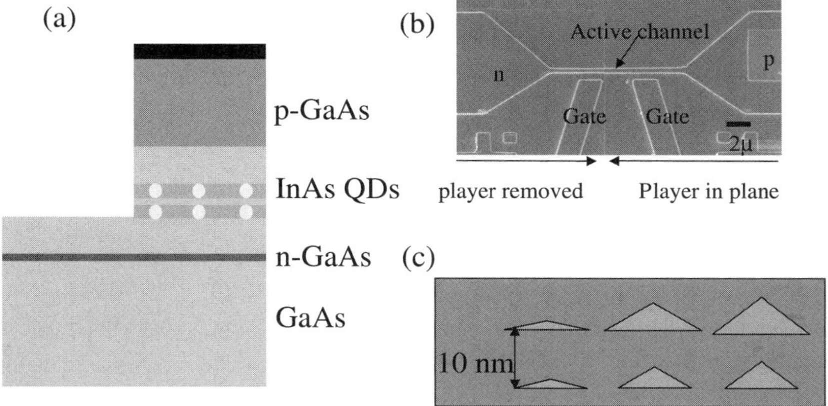

Fig. 8.12 (a) Schematic vertical section of lateral *p-i-n* structure. (b) SEM image of the device, the bar is $2\,\mu m$. (c) Schematic sketch of QDMs with different dots sizes grown with non-rotating substrate

To investigate the optical properties of QDMs, double layer quantum dots are incorporated into lateral *p-i-n* junctions. The double quantum dots were grown by molecular beam epitaxy on a nonrotating GaAs substrate; the asymmetry of the In source with respect to the wafer induces a graded In flux, resulting in variation of the InAs dot density and dot size [69]. The different dot sizes induce the different dot-dot separations on the same wafer [as sketched in Figure 8.12(c)], resulting in different coupling strength, which can be observed both in photoluminescence and electroluminescence.

8.4.2 Photoluminescence from QDMs

For a wafer with graded QDMs sandwiched in lateral *p-i-n* junction, the photoluminescence measurement was performed using a micro-photoluminescence system with metal apertures patterned on the wafer surface. Figure 8.13(a) shows the PL spectra with different excitation intensities at the sites with low In flux (with dot density less than $10^9\,\mathrm{cm}^{-2}$). Only two peaks can be observed in addition to the wetting layer peak (around 1425 meV), one at 1368.93 meV (P1) and the other at 1378.97 meV (P2). Such a two-peak structure has been observed at several sites on the wafer. P1 appears first at low excitation intensity, saturates very quickly and decreases with further increasing excitation. P2 appears with increasing excitation intensity and saturates at very high excitation intensity. The fact that the low energy peak comes first at low excitation power may result from the upper dots in the stacked pair being larger than the lower dots [56], with larger electron and hole wave-function overlaps. Both P1 and P2 are red-shifted with increasing excitation power.

PL spectra at a site with dot density around $1 \sim 3 \times 10^9\,\mathrm{cm}^{-2}$ are shown in Fig. 8.13(b). At very low excitation intensity 0.49 μW, only one peak can be

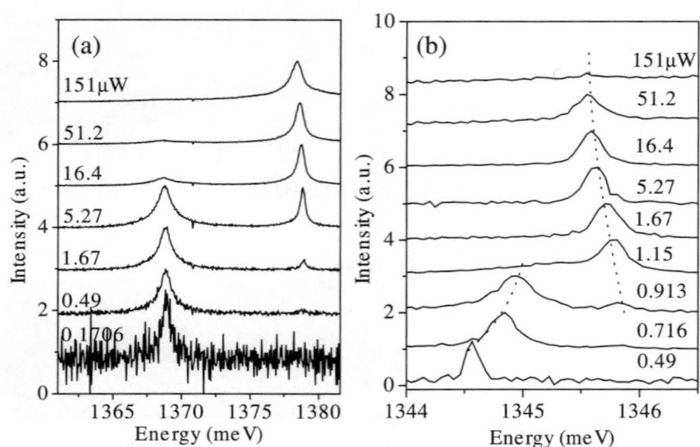

Fig. 8.13 Photoluminescence spectra of QDMs as a function of excitation power on low-density sites (a) and high-density sites (b) on the wafer

observed at 1344.56 meV. With slightly increasing excitation power up to 0.716 μW, a high-energy peak (1345.81 meV) appears. With increasing excitation power, the first peak is blue-shifted and is quenched with excitation power at 1.15 μW. By contrast, the peak at 1345.81 meV is red-shifted. An anticrossing can be observed at 0.913 μW, when the second peak emerges. From the power dependences, these two peaks cannot be the exciton and biexciton emission from a single quantum dot. We ascribe these peaks at 1344.56 and 1345.86 meV to indirect exciton and direct exciton recombinations respectively.

With increasing excitation power, optically generated carriers are redistributing because of the built-in electric field in p-i-n junction, resulting in decreasing the effective electric field across the quantum dots. The red shifts of the two peaks in Fig. 8.13(a) are the Stark shift induced by this electric field change. At low-density sites, the similar Stark shift being observed for two dots indicates that these two dots are not coupled or weakly coupled. At high-density sites, the anticrossing of direct and indirect exciton emission shows that the two dots are coupled and can be controlled by excitation power. The indirect exciton with a low energy around 1344.56 meV appears first at low excitation power and can be explained as follows. At very low power under nonresonant condition, electrons and holes generated around QDMs will separate by the built-in electric field and occupy the different dots, resulting in larger populations of indirect exciton emission, although the oscillation strength of direct excitons is larger than that of indirect excitons. With increasing excitation power, the built-in electric field is reduced where direct exciton recombination is dominant. No further emission peaks can be resolved at high power excitation because the luminescence from quantum dots is quenched.

8.4.3 Electroluminesce from QDMs

Using the same graded QDMs sandwiched wafer, electroluminescence devices are fabricated with different sites of the wafer. Figure 8.14(a) shows typical electroluminescence spectra from a device at low-density site for different current injections [70]. For low currents (5 μA), only one peak can be observed at 1.3371 eV. With increasing currents, a second peak around 1.3387 eV occurs; the low-energy peaks become saturated and the high-energy peaks become dominant with increasing injection current. For a single dot, the double peaks can be due to the increases in exciton and biexciton emission (or exciton and charge exciton emisssion). Both cases can be excluded by carefully studying power dependent and temperature dependent electroluminescence. Therefore, it is believed that the double peaks are due to the exciton emission in two uncoupled (or weakly coupled) quantum dots, similar to photoluminescence results.

Figure 8.14(b) shows the electroluminescence spectra as a function of injection current for another typical device at high density sites on the wafer. At low-current injection, two peaks around 1.3496 and 1.3523 eV (A and B) are observed. With increasing current, peak A is strongly blue-shifted and its intensity decreases, while

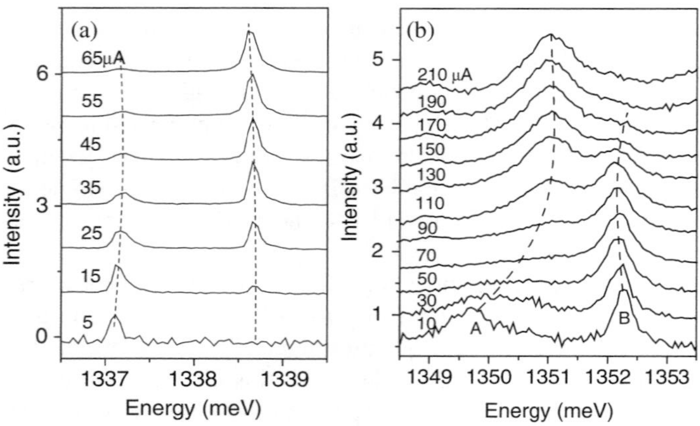

Fig. 8.14 Electroluminescence spectra of QDMs with different injection currents from devices fabricated from low-density sites (a) and high-density sites (b) on the wafer. Dashed lines through the peaks are used to guide the eyes

peak B is slightly red-shifted with increasing intensity. By increasing the current above 90 μA, peak A appears again with increasing intensity while peak B quenches rapidly and shifts to the higher-energy side. The minimum energy splitting between A and B is about 1 meV with a current of 90 μA at 2.9 V. It is difficult to calculate the electric field because the devices are electroluminescence devices, and the electric fields are different when QDMs are located in different positions in the active channel (Fig. 8.12(a)). Clear anticrossing is a signature of coupling of the two quantum dots in electroluminescence.

The coupling strength depends strongly on the separation between the two stacked quantum dots [45]. In this structure, the quantum dots have different sizes because the substrate is stationary during growth, resulting in different dot-dot separations. Because of the different dot-dot separation, the coupling strength might be different, with some dots strongly coupled and some not. In this structure for a dot-dot separation around 10 nm, two kinds of exciton can be observed, direct and indirect. The direct exciton with the electron and hole wavefunctions localized within one dot (small electronic dipole moment) is less sensitive than indirect exciton with electron and hole in different dots (large electronic dipole moment). With certain dot-dot separations, it is possible to tune the indirect exciton to the resonance of the direct exciton, and to observe bonding and antibonding states in QDMs [71]. With these considerations, the peak A and peak B in Fig. 8.14(b) are ascribed to indirect exciton and direct exciton emission in a coupled QDM. The coupling can be controlled with applied forward bias. The intensity of peak B is higher than peak A before the anticrossing because of the higher oscillation strength for direct excitons than indirect excitons [61, 62]. For the device of Fig. 8.14(a) the two peaks show very small quantum-confined Stark effects, indicating both of them are direct-exciton effects in two separate dots. If we look carefully, some anticrossing can be resolved, which might be due to weak coupling. The different results for the devices in Fig. 8.14(a) and (b) are due to the different dots sizes because of the non-rotating substrate during the wafer growth, as expected. A splitting at the anticorssing point around 1 meV indicates a dot separation around 6 nm according to the theory, although the distance between the wetting layer to the wetting layer is 10 nm.

8.5 Conclutions

In this contribution, we discussed the coupling in semciconductor QDMs both in theory and in experiment. The engrgy splitting has been calculated with plane wave expension method as a funtion of dot separation and an applied electric field. Experimently, we demonstrated coupling both in lateral QDMs and stacked QDMs using photoluminescence and electroluminescence spetroscopy. For lateral QDMs, Splitting of the exciton and biexciton have been observed in different energy levels. These laterally coupled QDMs produce much interest because this alignment may provide a good way of generating maximum entanglement and they also scale up easily. For stacked QDMs, the separation distances between two quantum dots has been controlled using a non-rotating substrate in MBE growth, resulting in different coupling strengths. With a certain separation of a single QDM, the indirect exciton and direct exciton can be tuned into resonance, resulting in the observation of anti-crossings, when optical excitation power or forward bias is changed. Controlling the coupling in QDMs provides promising applications in manipulating scalable qubits optically or electrically and entangled-photon generation for quantum information processing.

References

1. M. A. Nielsen and I. L. Chuang, Quantum Computation and Quantum Information, Cambridge University Press, Cambridge (2002)
2. D. Bouwmeester, A. K. Ekert, and A. Zeilinger, The Physics of Quantum Information: Quantum Cryptography, Quantum Teleportation, Quantum Computation, Springer, Berlin, Heidelberg, New York, Barcelona, Hongkong, London, Milan, Paris, Singapore, Tokyo, (2000)
3. D. P. Divicenzo et al., ArXiv: cond-ma/9911245
4. Y. Nakamura, Y. A. Pashkin and J. S. Tsai, Nature, **398,** 786 (1999)
5. J. E. Mooij et al., Science **285**, 1036 (1999)
6. B. Kane, Nature, **393,** 133 (1998)
7. A. Imamoglu et al., Phys. Rev. Lett. **83**, 4204 (1999)
8. K. Hennessy et al., Nature, **445**, 896 (2007)
9. J. P. Reithmaier et al., Nature **432**, 197–200 (2004)
10. T. Yoshie et al., Nature **432**, 200–203 (2004)
11. L. Childress et al., Science, **314**, 281 (2006)
12. John J. L. Morton et al., Nature Physics **2**, 40 (2006)
13. P. Michler, Single Quantum Dots, Springer, Berlin, Heidelberg, New York, Barcelona, Hongkong, London, Milan, Paris, Singapore, Tokyo, 2003
14. X. Q. Li et al., Science **301**, 809 (2003)
15. A. Zrenner et al., Nature **418**, 612 (2002)
16. D. Loss and D. P. DiVincenzo, Phys, Rev. A **57,** 120 (1998)

17. W. G. van der Wiel et al., Rev. Mod. Phys. **75**, 1 (2002)
18. T. Hayashi et al., Phys. Rev. Lett. **91**, 226804 (2003)
19. J. Gorman, D. G. Hasko and D. A. Williams, Phys. Rev. Lett. **95**, 090502 (2005)
20. A. Ekert, Phys. Rev. Lett. **67**, 661 (1991)
21. A. Einstein, B. Podolsky, and N. Rosen, Phys. Rev. **47**, 777 (1935)
22. P. Kok et al., Rev. Mod. Phys. **79**, 135–174, (2007)
23. Paul G. Kwiat et al., Phys. Rev. Lett. **75**, 4337 (1995)
24. R. M. Stevenson et al., Nature **439**, 179–182 (2006)
25. O. Gywat, G. Burkard and D. Loss, Superlattices and Microstructures **31**, 127 (2002)
26. P. Borri et al., Phys. Rev. Lett. **91**, 267401 (2003)
27. M. Grundmann, O. Stier and D. Bimberg, Phys. Rev. B, **52**, 11969 (1995)
28. T. Benabbas, P. François, Y. Androussi and A. Lefebvre, J. Appl. Phys., **80 (5)**, 2763 (1996)
29. M. A. Cusack, P. R. Briddon and M. Jaros, Phys. Rev. B, **54**, R2300 (1996)
30. J. R. Downes, D. A. Faux and E. P. O'Reilly, J. Appl. Phys. **81(10)**, 6700 (1997)
31. J. D. Eshelby, Proc. R. Soc. London, Ser. A, **241**, 376 (1957)
32. A. D. Andreev, J. R. Downes, D. A. Faux and E. P. O'Reilly, J. of Appl. Phys., **84**, 297 (1999)
33. I. M. Lifshits and L. N. Rosentsverg, Zhurnal Exper. and Teor. Phiziki, **17**, 9 (1947)
34. A. L. Efros and M. Rosen, Phys. Rev. B **58**, 7120 (1998)
35. A. D. Andreev, and A. A. Lipovskii, Phys. Rev. B **59**, 15402 (1999)
36. A. D. Andreev, and R. A. Suris, Semiconductors, **30**, 285 (1996)
37. T. B. Bahder, Phys. Rev. B, **41**, 11992 (1990)
38. L. E. Vorobev et al., Semiconductors, **39 (1)**, 50 (2005)
39. G. Schedelbeck et al., Science **278**, 1792 (1997)
40. X. L. Xu, D. A. Williams and J. R. A. Cleaver, Appl. Phys. Lett. **86**, 012103 (2005)
41. D. Leonard, K. Pond, and P. M. Petroff, Phys. Rev. B **50**, 11687 (1994)
42. E. Moreau. et al, Phy. Rev. Lett. **87**, 183601 (2001)
43. O. Brandt. et al., Phys. Rev. B **41**, 12599 (1990)
44. L. Landin et al., Phys. Rev. B **60**, 16640 (1999)
45. M. Bayer et al., Science **291**, 451 (2001)
46. L. Landin et al., Science **280**, 262 (1998)
47. M. Bayer, O. Stern, S. Fafard and A. Forchel, Nature (London) **405**, 923 (2000)
48. H. Kamada, A. Ando, J. Temmyo and T. Tamamura, Phys. Rev. B **58**, 16243 (1998)
49. G. Ortner et al., Phys. Rev. Lett. **90**, 086404 (2003)
50. O. Gywat, G. Burkard and D. Loss, Superlattices and Microstructures **31**, 127 (2002)
51. R. Songmuang, S. Kiravittaya and O. G. Schmidt, Appl. Phys. Lett. **82,** 2892 (2003)
52. T. V. Lippen, R. NÄotzel, G. J. Hamhuis and J. H. Wolter, Appl. Phys. Lett. **85**, 118 (2004)
53. X. Q. Li and Y. Arakawa, Phys. Rev. A **63**, 012302 (2000)
54. J. M. Villas-Boas, A. O. Govorov and S. E. Ulloa, Phys. Rev. B **69**, 125342 (2004)
55. Q. Xie, P. Chen and A. Madhukar, Appl. Phys. Lett. **65**, 2051 (1994)
56. Q. Xie, A. Madhukar, P. Chen and N. P. Kobayashi, Phys. Rev. Lett. **75**, 2542 (1995)
57. M. Hayne, et al., Phys. Rev. B **62**, 10324 (2000)
58. J. Tersoff, C. Teichert and M. G. Lagally, Phys. Rev. Lett. **76**, 1675 (1996)
59. G. Springholz, V. Holy, M. Pinczolits and G. Bauer, Science **282,** 734 (1998)
60. K. Ono, D. G. Austing, Y. Tokura and S. Tarucha, Science **297**, 1313 (2002)
61. G. Ortner et al., Phys. Rev. Lett. **94**, 157401 (2005)
62. H. J. Krenner et al., Phys. Rev. Lett. **94**, 057402 (2005)
63. E. A. Stinaff et al., Science **311**, 636 (2006)
64. N. Saito, et al., Jpn. J. Appl. Phys. **36**, L896 (1997)
65. M. Inai,et al., Jpn. J. Appl.Phys. **32**, L1718 (1993)
66. B. Kaestner, D. H. Hasko and D. A. Williams, Jpn. J. Appl. Phys. **41**, 2513 (2002)
67. B. Kaestner, D. A. Williams and D. G. Hasko, Microelectronic Eng. **67–68**, 797 (2003)
68. X. L. Xu, D. A. Williams and J. R. A. Cleaver, Appl. Phys. Lett. **85**, 3238 (2004)
69. D. Leonard, K. Pond and P. M. Petroff, Phys. Rev. B **50**, 11687 (1994)
70. X. L. Xu, A. Andreev, D. A. Williams and J. R. A. Cleaver, Appl. Phys. Lett. **89**, 91120 (2006)
71. W. Sheng and J.-P. Leburton, Phys. Rev. Lett. **88**, 167401 (2002)

Chapter 9
Studies of Semiconductor Quantum Dots for Quantum Information Processing

H. Z. Song and T. Usuki

9.1 Introduction

In recent years, quantum information processing is of great theoretical and experimental interest. This new scientific field appears for people to devise and implement quantum-coherent strategies for computation and communication etc. A practical realization of a quantum computer would be quite significant, since there exist theoretical quantum algorithms which would make some classically hard computational problems tractable. Quantum communication is technologically very prospective because it offers the promise of private communication with unbreakable security assured by the laws of quantum mechanics.

Quantum computation encompasses almost every possible quantum phenomenon in nature, so that it can be physically realized in many systems such as optical cavity, ion trap, superconductor, nuclear spin and electron spin [1]. However, serious conditions such as scalable quantum bit (qubit) system, initialization ability, long coherence time, universal gates and quantum measurement capability must be simultaneously satisfied for quantum computation [2]. Semiconductor quantum

---

dots (QDs) have discrete electronic states, which can be considered as localized two level systems to construct the fundamental element, qubits of quantum logic gates. As another result of the fully quantized electronic structure, semiconductor QDs exhibit long coherence time of carriers. Owing to the well-established semiconductor manufacturing capabilities, scalable quantum computing system can be well constructed in solid state and, more favorably, be easily integrated with existing microelectronic technology. Satisfying all the necessary requirements of solid-state quantum gates, QDs have been investigated for building quantum computers in many theoretical proposals [3–9] and some experimental studies [10, 11]. In various QDs, self-assembled ones are more attractive due to their stronger carrier confinement, smaller size, longer coherence time and good homogeneity. For the purpose to apply self-assembled semiconductor QDs in quantum computers, one should be successful in coherent manipulation of QD qubits, which depends on fine control of QDs spatially and electronically. In the next section, we will show some of our studies on semiconductor QDs for the applications in quantum computation.

In the field of quantum communication, photons are mostly taken as the flying qubits to distantly transmit quantum information. In practical quantum communication, *e.g.* quantum key distribution, single photons are usually needed so that exploration of single photon emitters (SPEs) has been a very active research topic. Semiconductor self-assembled QDs have excellent potential in the realization of SPEs [12, 13]. Consider a single QD as a simple two-level system, an electron-hole pair created in its first excited state cannot be excited again until it returns to its ground state. If the recombination process is a radiative one, this leads to the emission of a single photon. As a SPE candidate, semiconductor QDs are more advantageous than other choices such as molecules [14], nitrogen-vacancy center [15], and impurity centers in semiconductors [16] because QDs have a narrow homogeneous width, optical stability, and no photobleaching effect. Semiconductor-QD-based SPEs had been demonstrated emitting at wavelengths mostly shorter than 1.2 μm [12, 13, 17, 18]. Quantum communication over long distances requires operations at longer wavelengths in the so-called telecommunication bands (1.26–1.57 μm), which are playing a great role in current optical networks running via silica-glass-based single-mode fibers. Therefore, QDs emitting at telecommunication wavelengths are of particular necessity [19]. In section 9.3, we will describe the current status of our experimental efforts towards the QD-based SPEs in telecommunication bands.

9.2 Quantum Dots for Quantum Computation

9.2.1 Individual Access to a Single Embedded Quantum Dot

In quantum computers built with self-assembled semiconductor QDs, simultaneous access of spatial and electronic information of single QDs is crucially required.

On QDs exposed to air or vacuum, the precision of this requirement may be not difficult to reach just by conventional tools of nano-scaled morphological study, *e.g.* atomic force microscope (AFM) or scanning tunneling microscope (STM) [20–23]. In the sense of application in quantum computing, however, the QDs should be embedded in a semiconductor matrix for effective interdot coherent coupling, which is essential for quantum gates. A real-time microscopic analysis of embedded QDs seems beyond the abilities of conventional AFM and STM. There were extensive transport studies on embedded QDs but scarcely to a precision of individual dots. It is thus necessary to find out a way of operating exactly a specific embedded QD. Here, it is seen that embedded QDs can be accessed individually by using suitably set STM [24].

The samples were fabricated using hydrogen-assisted molecular beam epitaxy (MBE) on a n^+-GaAs (311)B substrate. $In_{1-x}Ga_xAs$ QDs were self-assembled in Stranski-Krastanov (S-K) mode on 10 nm of non-doped GaAs and immediately capped by a thin GaAs layer. These thinly capped QDs were confirmed not so different in optical quality from conventional thickly capped QDs. One of the samples is of $In_{0.45}Ga_{0.55}As$ QDs, ~20 nm in lateral diameter and ~1.2 nm in height, buried under a GaAs capping layer as thin as 1.7 nm. The sampe surface appears featureless and smooth at atomic scale so that conventional STM brings about an image of no nanoscaled information, the same as observed by AFM. However, QD signals emerge as the measurement condition is suitably set in constant current mode. It exhibits a precise dot ensemble when measured at 4.5 K and under a sample bias of -1.5 V, as shown in Fig. 9.1(a). Actually, clear STM images of embedded QDs are observable in a particular bias range, *e.g.* -1.5 ± 0.2 V for the present sample, similar to the STM image of local charges in InAs/GaAs two-dimensional (2D) heterostructure [25]. Figure 9.1(b) shows an image obtained with sample bias of -1.8 V. The contrast comes down, and some of the dots are invisible. In deed, this is the transition to the featureless images indicative of a flat surface under biases well out of the above mentioned range. The image contrast also degrades with increasing temperature, but even at room temperature the QDs can still be distinguishable, as seen in the image of Fig. 9.1(c). Furthermore, positive bias hardly gives any informative images related to QDs, which is similar to the reported case for uncapped QDs [21]. With the fine resolution of STM, these results make it easy to access an individual embedded QD.

We shall now try to understand the accessibility of individual embedded QDs. On semiconductors, the effects of the electronic structures are so strong that they are essential for understanding the STM images [26]. To examine the electronic states in a thinly capped QD, scanning tunneling spectroscopy (STS) was measured on individual dots. The inset of Fig. 9.2 shows typical current-voltage (I-V) curves taken at the top point of a QD (on-QD) and on the wetting layer near this dot (off-QD) at 4.5 K on the same thinly capped sample as aforementioned. The on-QD current mostly iterates with the off-QD one, but is remarkably stronger (by the order of 1 nA) at sample bias around -1.5 V. This feature is suggestive of QD electronic information, which is clearer in normalized STS, *i.e.* $\overline{(dI/dV)/(I/V)}$

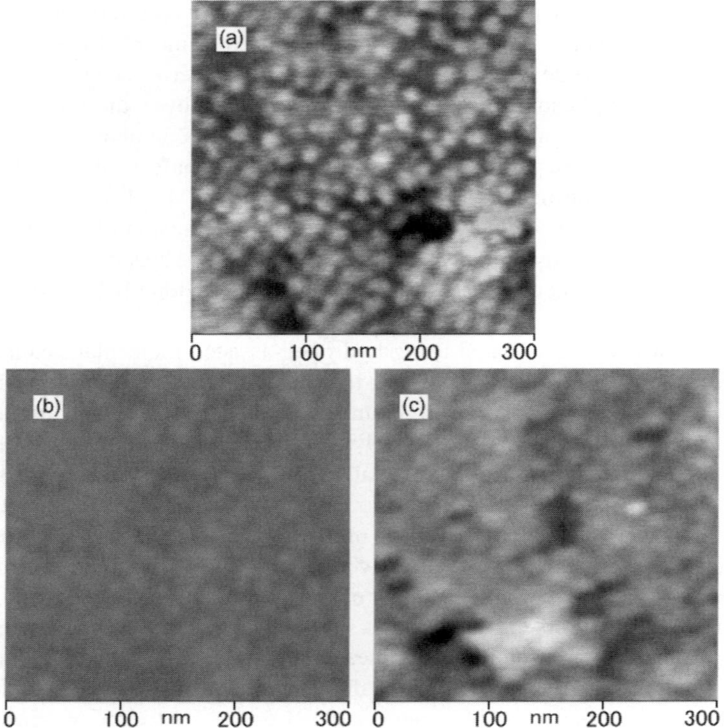

Fig. 9.1 STM images for the sample of thinly capped $In_{0.45}Ga_{0.55}As$ QDs taken with a constant tunneling current of 0.75 nA. The sample biases and measuring temperatures are (a) -1.5 V, 4.5 K; (b) -1.8 V, 4.5 K; and (c) -1.5 V, 300 K

versus V [27]. In Fig. 9.2, a current peak about 0.5 V wide at sample bias around -1.5 V is seen for every QD. As guided by a thick arrow, the dot-dependent peak positions distribute from -1.2 to -1.7 V, which is consistent with the restricting bias regime for the STM image. Since $\mathrm{d}I/\mathrm{d}V$ is proportional to the tunneling density of states [22], this peak can be ascribed to the electronic level of the single QD. It is most probably the ground state, because there appears nothing under negative sample bias less than 1.0 V. In addition, there also exists a feature at about -2.2 V of sample bias, which is guided by a thin arrow in Fig. 9.2, although it is weak compared with that around -1.5 V. Similarly, it can be attributed to the excited state of the measured QD. These two levels are also detectable when the tip resides at the side, *i.e.* somewhere in between the top point and the edge, of a QD. As presented by the offset curve in Fig. 9.2, the peak heights become lower than that at the top of the QD. With positive sample bias, such peaks are hardly observable, which reminds of the vanishing STM images under the same condition. At off-QD positions, the normalized STS curves are featureless in our measured range from -3 to 3 V. As the

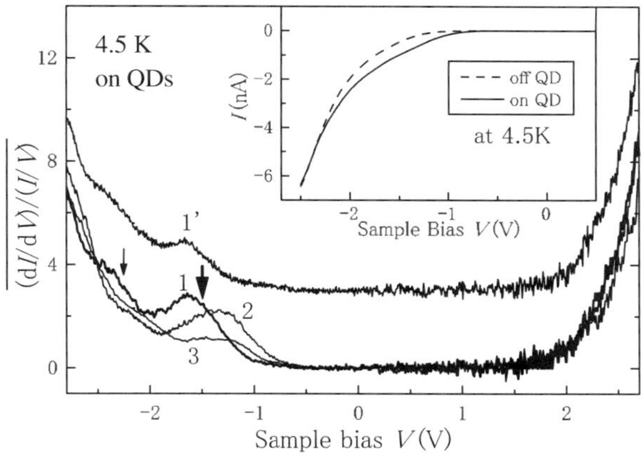

Fig. 9.2 Representing normalized STS curves measured on the top points of different single QDs 1, 2 and 3. The offset curve labeled 1' is taken for the same dot as QD 1 but with the tip positioning somewhere in between the top point and the edge. The inset shows the I-V curve measured at the top point of a thinly capped QD and that on the wetting layer nearby

temperature is raised, the current peaks of QDs are weakened, which is in agreement with the smeared STM image as shown in Fig. 9.1(c).

The above observations unambiguously demonstrate that the STM image of the capped QDs arises from the tunneling event associated with the capped QDs. At zero bias, the conduction band bending due to the surface potential lifts the QD levels well above the Fermi level E_f of the semiconductor. A suitable negative sample bias contributes a small part to the depletion region ($\sim$10%) so as to align the QD level with the filled conduction band states of the substrate. The probability of QD electron interacting with the tip is considerable as we learn from stacked InAs/GaAs QDs that strong coupling occurs even with spacer more than 5 nm [28]. For our InGaAs QDs, lower barrier gives even longer decay length, in the order of 10 nm [29]. Noting further that the on-QD current of $\sim$ 1 nA higher than the off-QD one is the same as that for naked InGaAs QDs [21], the 1.7 nm cap layer here does not prevent the QD from contributing remarkably to the tunneling current. At a certain bias for resonant tunneling with the QD, the current can be described as [30]

$$I \propto \left| \int \left(\Psi_t^* \frac{d}{dz} \Psi_d - \Psi_d \frac{d}{dz} \Psi_t^* \right) dr \right|^2, \tag{9.1}$$

where Ψ_t and Ψ_d are the electronic wavefunctions of the STM tip and the QD, respectively. With approximations of factorizing the wavefunctions into parallel and vertical (to the sample surface) parts, *i.e.* $\Psi_{t(d)}(r) = \Psi_{t(d)}^{//}(x, y)\Psi_{t(d)}^{\perp}(z)$, and $\left|\Psi_t^{//}(x, y)\right|^2 = \delta(x_t, y_t)$ as a result of much smaller tip than QD, we have

$$I \propto \left|\Psi_{\mathrm{d}}^{//}(x_{\mathrm{t}}, y_{\mathrm{t}})\right|^2 \left|\int \left(\Psi_{\mathrm{t}}^{\perp}(z)^* \frac{\mathrm{d}}{\mathrm{d}z} \Psi_{\mathrm{d}}^{\perp}(z) - \Psi_{\mathrm{d}}^{\perp}(z) \frac{\mathrm{d}}{\mathrm{d}z} \Psi_{\mathrm{t}}^{\perp}(z)^*\right) \mathrm{d}z\right|^2, \quad (9.2)$$

where x_{t}, y_{t} denote the present position of the tip. The z-related (vertical) part can be modeled to be proportional to $T = e^{-2\sqrt{2m\phi}z_{\mathrm{t}}/\hbar}$ [25], where ϕ is the effective vacuum barrier height with respect to E_{f} and z_{t} is the tip-sample distance. With sample bias of -1.5 ± 0.2 V, estimation by WKB approximation gives $\phi^{1/2}$ and T variations within $\pm 12\%$. Accordingly, the current straightly follows the in-plane QD wavefunctions to a high degree. In constant current mode as we presently use, expression (9.2) becomes:

$$z_t = z_0 + \frac{\hbar}{2\sqrt{2m\phi}} \ln \left|\Psi_d^{//}(x_t, y_t)\right|^2, \quad (9.3)$$

where z_0 is a constant. In this approximation, the in-plane wavefunctions of the capped QDs are visualized in quite a simple manner by the well detectable STM images. This is considered the mechanism of individual access to a single embedded QD, which is important in QD-based quantum computers. Accordingly, STM may further be useful to specify the lateral inter-dot coupling [29] and then help to manipulate quantum gates based on self-assembled semiconductor QDs.

9.2.2 Dilution of Quantum Dot Density

Research on self-assembled semiconductor QDs for quantum computation include clarifying the coherent electronic characteristics of single QDs. Single dot study in electrical and optical properties usually require QD density in the order of 10^9 cm^{-2} or lower. In conventional QD self-assembly in S-K mode, QD density increases abruptly from 0 to the order of 10^{10} cm^{-2} at the critical point for the transition from 2D to three-dimensional (3D) growth modes. It is rather difficult to prepare low density QDs, especially with desired QD size, due to this very narrow coverage range, so a method more controllable is then demanded. It is seen here that post-growth annealing of a wetting layer thinner than the critical thickness (subcritical) can controllably produce expected low density QD ensemble [31].

A systematic example is InAs/GaAs QDs fabricated by MBE on semi-insulating GaAs(001) substrates. Following the stop of InAs deposition, a period of post-growth annealing was performed without changing the substrate temperature and the arsenic pressure. To obtain morphology of the as-formed QDs, the sample holder is suddenly cooled down by switching off the substrate heater and then taken off from the heater within 20 s. In our experimental system, at 480 °C and with InAs growth rate of 0.031 ML/s, the nominal critical InAs wetting layer thickness for 2D–3D growth mode transition is 1.66 ML. Figure 9.3(a) shows the AFM image of a QD sample conventionally grown by depositing 1.69 ML of InAs followed by 60 s of post-growth annealing. It has a QD number density of 2.2×10^{10} cm^{-2}. If we stop depositing InAs before 2D–3D transition, post-growth annealing of such a subcrit-

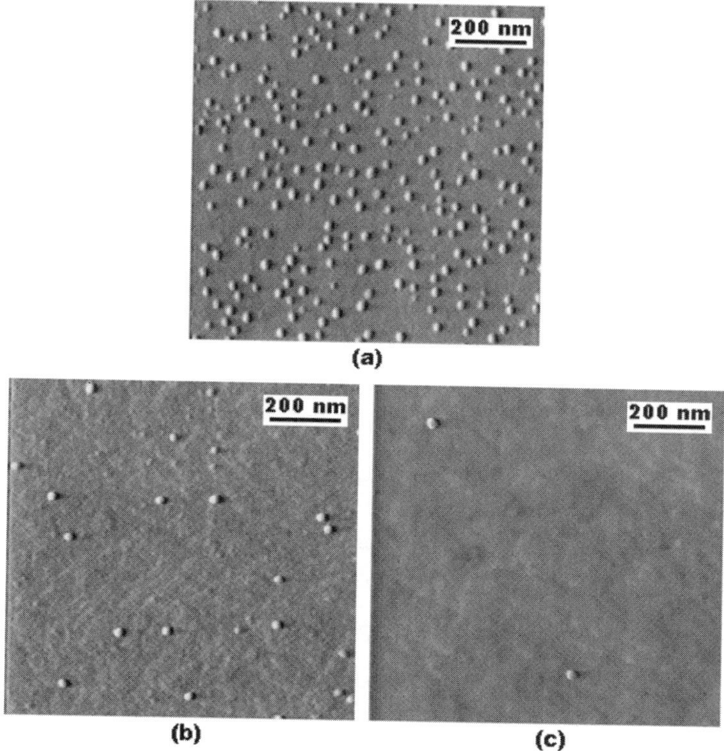

Fig. 9.3 AFM images of InAs/GaAs QDs formed by 60 s of post-growth annealing after depositing (a) 1.69, (b) 1.55, and (c) 1.49 ML of InAs at 480 °C with a growth rate of 0.031 ML/s

ical wetting layer also gives rise to finite density of QDs. As shown in Fig. 9.3(b) and (c) for 60 s of annealing, InAs coverage of 1.55 ML produces a QD ensemble with density of $2.0 \times 10^9\,\mathrm{cm}^{-2}$, and a decrease of coverage down to 1.49 ML leads to one order of magnitude lower QD density, $1.6 \times 10^8\,\mathrm{cm}^{-2}$. The condition for low density QDs is thus much relaxed compared with conventional self-assembly. These results also shows the fact that the size of the QDs, depending mainly on growth temperature and arsenic pressure etc., are almost unchanged with varying thickness of subcritical wetting layers. For other III-V semiconductor QDs, dilution of QD density can also be realized by the same way. We see that self-assembled low density QDs can be readily obtained with expected size.

To study quantitatively the controllability of QD density decreasing, real-time reflection high-energy-electron diffraction (RHEED) was used to monitor the QD formation process as has been reported [32–34]. During the growth and post-growth annealing, the substrate was fixed to observe the RHEED image along the [100] azimuth. Figure 9.4(a) shows typical RHEED images before, at the beginning of and after QD formation. The monitored area is selected to correspond to one of the brightest spots observed after QD formation, as indicated by the rectangular marks

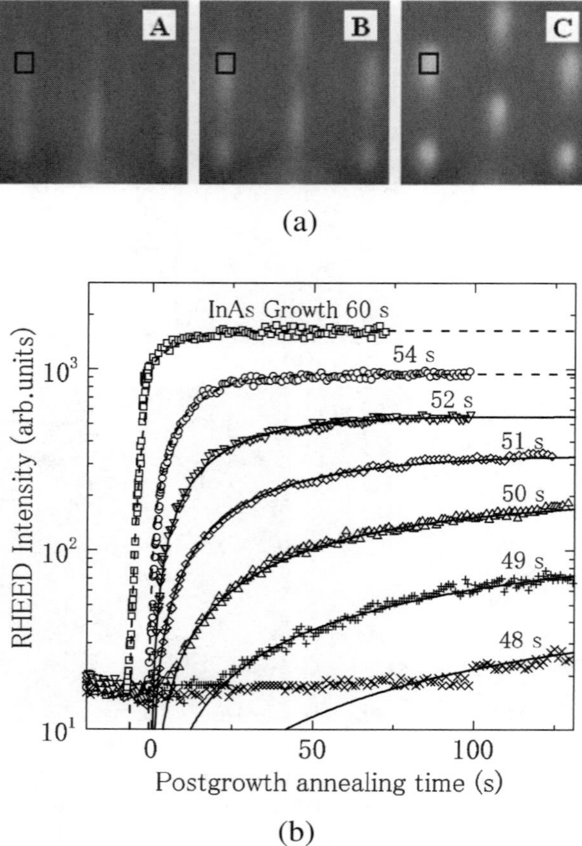

Fig. 9.4 (a) Typical RHEED images before (A), at the beginning of (B) and after (C) formation of QDs; (b) evolution of RHEED intensity integrated over marked area in (a) with QD formation process. The RHEED intensity is the measure of total QD volume. The curves are the simulated results according to Eq. (9.4) by setting the initial conditions: $t = t_0 = 43$ s, $n_1(t_0) = 0.03$ ML, $n_2(t_0) = 0.0006$ ML, $n_3(t_0) = 0$, $D\sigma_2 = 20$ and $D\sigma_3 = 40$

in Fig. 9.4(a). The integrated RHEED intensity in the marked area, is taken as a measure of total QD volume. In Fig. 9.4(b), the scattered symbols indicate a series of time-scanned RHEED intensity (total QD volume) for InAs/GaAs QD formation at 480 °C with different coverage of InAs. The InAs growth-stops before and after the 2D–3D growth mode transition are both included. For the sake of clarity, the zero point of the time axis is taken to be where InAs stops growing. The onsets of rapid increase for 60 and 54 s in InAs deposition, occur 7 and 1 s before the growth stop, but they are actually the same 53 s from the standpoint of InAs growth time. In such conventional QD growth cases, the initial increase in QD volume versus time is naturally the same, whereas the final saturation level decreases with decreasing InAs coverage. Using 53 s of InAs deposition, whose coverage equals the critical wetting layer thickness, QD volume increases immediately following the growth stop and as fast as the two above, but soon saturates at a lower level (not shown). When the InAs deposition time is shorter than 53 s, the apparent increase in QD volume is delayed, and this delay is prolonged as the coverage decreases. At the same time, the increase speed is reduced, and the final QD volume is further decreased. When the InAs coverage is less than 1.60 ML, the QD volume still increases after 2 min of post-growth annealing. In fact, there is certainly a saturation behavior even at low coverage as long as the post-growth annealing time is sufficiently long.

In the deposition of InAs wetting layer, RHEED does not show a change from 2D growth mode. However, it was found that the surface becomes much rougher even with a slight increase in the InAs coverage from 1.3 to 1.4 ML [35]. In some cases, small 3D islands can be formed from such an InAs wetting layer [36]. Our samples also exhibit that the apparently flat 2D wetting layer increases in roughness with its growth [31]. Theoretical studies have shown that due to roughness, a 2D strained film is always unstable [37, 38]. Nucleation can occur to form 3D islands as long as the initial surface is sufficiently rough [39]. On the other hand, self-assembled QDs are widely regarded as arising from precursors [35, 40–42], which are obviously some rough structures of the wetting layer. It is then reasonable to analyze the above observations by including QD nucleation from precursors on the wetting layer.

Considering mass transport between adatoms, 2D precursors and 3D QDs, Dobbs *et al.* proposed a mean-field theory on self-assembled QDs [43]. They started their calculation from the beginning of conventional QD formation. Here we extend the analysis to the stage well before the 2D–3D growth mode transition. Therefore, the pure 2D growth of the wetting layer has to be taken into account. Regarding the type of precursors, some floating features [35], 2D platelets [40, 41] and quasi-3D islands [42] are possible. We generally follow the mean-field theory of Dobbs *et al.* concerning the relation between the precursors and the adatoms/QDs whatever the exact precursor type is. Consequently, the rate equations for the amount of adatoms n_1, precursors n_2 and QDs n_3 are expressed as

$$\begin{aligned}\frac{dn_1}{dt} &= F - \beta n_1 - D(\sigma_2 n_2 + \sigma_3 n_3)n_1,\\ \frac{dn_2}{dt} &= D\sigma_2 n_1 n_2 - \gamma n_2,\\ \frac{dn_3}{dt} &= D\sigma_3 n_1 n_3 + \gamma n_2,\end{aligned} \tag{9.4}$$

where F is the InAs deposition rate, β is a coefficient for pure 2D growth of the wetting layer, D is the diffusion coefficient of adatoms, σ_2 and σ_3 are the so-called "capture number" for adatoms to be captured by the precursors and QDs, respectively, and γ is the rate of QD nucleation from the precursors.

In a growing 2D strained film such as a wetting layer, it has been proposed that some mechanisms such as long-range Van der Waals forces [44] and nonlinear elastic effects [39] act to stabilize the system, in other words, to suppress the formation of QDs as usually observed. These effects cause a negligible QD nucleation rate

γ on a growing 2D wetting layer. After the growth stop of InAs wetting layer, γ becomes finite so that QD nucleation occurs. This change can be described by the thermal activation behavior of γ [41] with activation energy (QD formation energy) exponentially depending on $(\theta–\theta_c)/\theta_c$ [45], where $\theta(\theta_c)$ is the thickness (critical thickness) of the wetting layer. In simulations, we take the activation energy of γ for the present growth rate F=0.031 ML/s 10 times that for F=0. As a good approximation, β and σ are reasonably taken to be constant. To set the initial conditions, we consider that the pure 2D wetting layer and adatoms are in dynamic equilibrium before the appearance of precursors, that is, $\mathrm{d}n_1/\mathrm{d}t=F-\beta n_1=0$ before n_2 starts to grow. The growth of pure 2D wetting layer usually proceeds in a way of forming large 2D islands. The precursors start to grow probably when the interactions of large 2D islands occur from about 1.4 ML of coverage [40]. Accordingly, we set the initial n_2 to be a finite small value around 1.4 ML. With RHEED intensity proportional to the total QD volume n_3, the simulation results of QD volume evolution are shown in Fig. 9.4(b) by the curves. It is clear that eq. (9.4) well describes the QD formation in the conventional way and by post-growth annealing of a subcritical wetting layer.

With γ neglected, the growth process before the appearance of QDs can be expressed as

$$\begin{aligned} \frac{dn_1}{dt} &= F - \beta n_1 - D\sigma_2 n_2 n_1 \\ \frac{dn_2}{dt} &= D\sigma_2 n_1 n_2 \end{aligned} \quad . \tag{9.5}$$

Simulation by Eq. (9.5) shows that the amount of the precursors n_2 increases fast with growth time and then slows down, as shown in Fig. 9.5. The adatom amount n_1 exhibits firstly a slow, but finally a fast degradation. As a result of growth stop, n_1 degrades steeply and n_2 saturates immediately. This result can simplify the analysis of QD formation after InAs growth stopped. It enables us to approximately take n_1=0 and $n_2 = n_{2s}$, the saturation value after growth stop, as the initial conditions for QD formation. Once InAs deposition is stopped, γ is suddenly changed to be finite so as to form QDs. Then we have $dn_3/dt=-dn_2/dt=\gamma n_2$, which gives

$$n_3 = n_{2s}(1 - \mathrm{e}^{-\gamma t}). \tag{9.6}$$

It is not surprising that the simulation results for the growth time of less than 53 s in Fig. 9.4(b) are almost the same as Eq. (9.6), whereas those for the longer deposition time do not match this relation. It means that, unlike the complicated process of conventional QD growth, the QD formation by post-growth annealing of a subcritical wetting layer is rather simply the nucleation from precursors appearing before growth stop. The observed saturation of QD volume corresponds to the stage when all the precursors (of amount n_{2s}) have changed into QDs by nucleation. This relation provides the way of controlling QD density by tuning the post-growth annealing time, especially in the cases of slow saturation of QD volume.

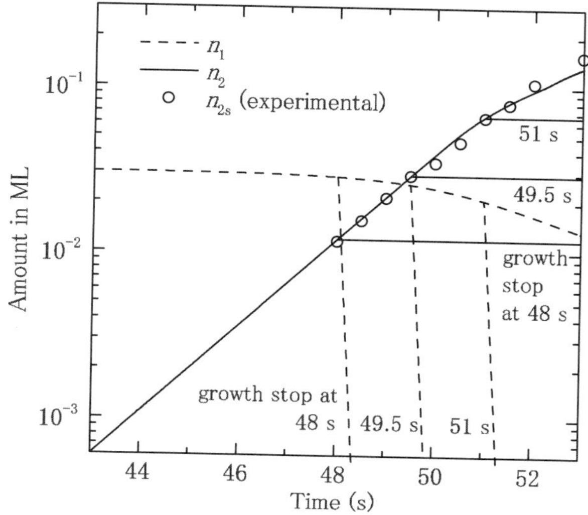

Fig. 9.5 Simulated results of the amount of adatoms n_1 and precursors n_2 before conventional 2D–3D growth mode transition using Eq. (9.5). The initial conditions are the same as those in Fig. 9.4. Scattered circles show the saturation QD volume (n_{2s}) for different InAs deposition time deduced from experimental data according to Eq. (9.6)

Fitting RHEED data to Eq. (9.6), we have the experimental saturation QD volume n_{2s} shown in Fig. 9.5 by scattered symbols. It well follows the change of n_2, which is nearly exponential with regard to growth time, until somewhere near the critical point. This is understandable noting that the little early change of n_1 and small initial value of n_2 in Eq. (9.5) lead to $\mathrm{d}n_2/\mathrm{d}t \approx D\sigma_2 F\beta^{-1} n_2$. Naturally, the saturation QD volume approximately follows $n_{2s} \propto \exp(D\sigma_2\beta^{-1}\theta)$. This relation establishes the controllability in prepareing low density QDs by changing the coverage of a subcritical wetting layer. We have used it to prepare samples for single dot and QD molecule studies [46, 47], which are fundamental for exploring QD-based quantum computers.

9.2.3 Site-control of Quantum Dots

In general, self-assembled semiconductor QDs grown in S-K mode are randomly distributed both in position and size. Quantum computers built by QDs require well defined dot-dot coupling [4], putting demands on precise control of QD sites. Efforts have ever been devoted to define the positions of self-assembled QDs by means of, *e.g.*, STM lithography [48], strain modulation [49] and nanotemplate [50], but they seem difficult in constructing qubits mainly due to dot separation still being out of noticeable inter-dot coupling. Here we introduce an AFM-assisted technique [51, 52], by which one can set QDs sufficiently close and prepare dot array sophisticated to fit the requirements of quantum computing.

The first stage of this technique is the fabrication of oxide dots, which is performed by AFM lithography at room temperature in a humid atmosphere. The substrate can be many semiconductors such as GaAs [51–53], InP [54] and Si [55] of any conduction type (n, p and intrinsic). As shown schematically in Fig. 9.6, when a negatively biased AFM tip approaches the flat surface of a semiconductor substrate, the electric field decomposes water molecules in the small region around the AFM tip into H^+ and OH^-. Then the OH^- ions locally oxidize the surface. In the case of GaAs substrate, this reaction is as follows [56]

$$2GaAs + 12OH^- \rightarrow Ga_2O_3 + As_2O_3 + 6H_2O + 12e^-. \tag{9.7}$$

Oxygen incorporation expands the volume, contributes a part above the original surface, and then forms a nanoscaled oxide dot outstanding beyond the surface as shown in Fig. 9.6(a). The oxidation rate depends on the electric field or current. Due to the pin-shape of a tip, the electric field/current decreases along the radial directions from the tip center. It thus gives rise to a lens-shaped oxide dot. The oxide dot size can be controlled by suitably tuning the applied voltage and reaction time.

The oxidized region is not limited above the original surface. Similar to the conventional oxidation of semiconductor surface, nearly half of the oxidized region lies below the original surface level, as can also be seen in Fig. 9.6(a). If the oxide

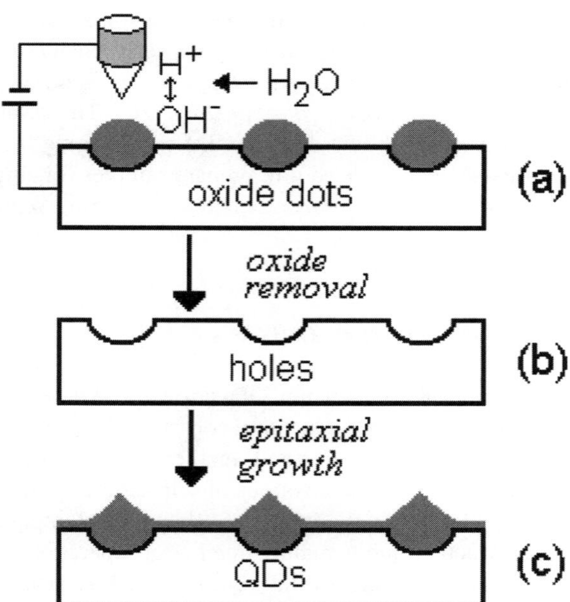

Fig. 9.6 Process of AFM-assisted control of semiconductor QDs

is removed, the space released from the oxide dot will give a hole. To remove the oxide dots, one can use chemical etching. The usually used solution is HCl : H_2O = 1:20 ∼ 1:100 at least for GaAs and InP. The etching time can be from 30 seconds to a few minutes. After etching, site-controlled holes are obtained, as shown schematically in Fig. 9.6(b). Immediately, the hole-patterned substrate is rinsed in flowing de-ionized water for enough time so that the surface is of little residual solution.

As the final process, QDs are epitaxially grown by methods such as MBE and metalorganic chemical vapor deposition (MOCVD) on the hole-patterned surface. Before overgrowing, the thin oxide layer, which is formed in the short time of mounting the sample into the growth chamber, has to be cleaned away from the surface. It is unsuitable to carry out thermal cleaning because the hole pattern may be smeared out or even destroyed at temperature as high as 600 °C. We can use irradiation of atomic hydrogen at temperature below 550 °C for a few minutes. Hereafter, heteroepitaxy is performed to grow QDs at the same temperatures as normally used for QD growth in S-K mode. The coverage is limited below the point of transition from 2D to 3D growth modes, which is a fundamental factor in S-K growth of self-assembled QDs. With growth condition well controlled, QDs are formed on the sites of holes as shown in Fig. 9.6(c). This site-selective growth of QDs may be understood by more strain-relaxation at the hole sites [57], although other explanations such as concentration of atomic steps [48] can not be completely excluded.

Next, we demonstrate how well the QDs can be controlled by this AFM-assisted technique. In general, the QDs can be well organized into an arbitrarily designed ordered array, with site precision as good as ±1.5 nm and the QD size fluctuation of about ±5%, much better than usual S-K growth and comparable to very recently achieved homogeneity of S-K QDs [58]. With the well achievable lateral size, 20 nm, the controllable inter-dot distance can be down to 25 nm, meaning QDs nearly touching their neighbors. Detailed studies show that the lateral size is determined by the hole diameter without changing with coverage, whereas the QD height increases with coverage at a speed related to the hole size [57]. This character enables observable coherent lateral interaction between neighboring QDs [59, 60]. The simultaneous availability of various QD sizes provides a way to controllably construct asymmetric QD molecules, which are often required in quantum computation based on semiconductor QDs. In a proposed model of quantum computer [6], one qubit consists of a big QD as the main dot and a few small QDs as the operation dots, as is schematically illustrated in Fig. 9.7(a). The inter-qubit interaction is controlled by pushing an electron into the main dots (weakly or not coupled) or neighboring operation dots (strongly coupled). In the stand-by state, the electron with spin up or down stays in the main dot. Applying a suitable π-pulse to any qubit, the electron will transfer to an operation dot. The quantum gate operation is implemented via swapping between the electron spins in operation dots belonging to neighboring qubits. Figure 9.7(b) shows such a QDs structure fabricated by the present technique. The big and small dots are ∼30 and ∼20 nm in diameter and 1.5 and 1.2 nm in height, respec-

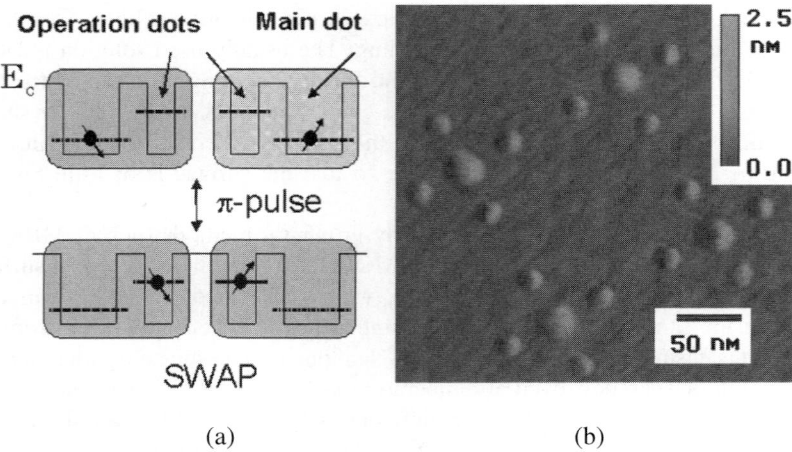

(a) (b)

Fig. 9.7 (a) Qubit structure and operation of an all-optical quantum computer using electron spins in asymmetrically coupled QDs; (b) AFM image of qubit-structured QDs array, which fits the model in (a), fabricated using the AFM-assisted technique

tively, and the center-center distance between big and small sites is 40 nm. This sort of QDs structure is now under test for manipulation as qubits in quantum computation.

9.3 Quantum Dots for Quantum Communication

9.3.1 Single Photon Emitting QDs

The requirement of telecommunication SPEs for quantum communication may be somehow fit by GaAs-based semiconductor QDs [61], but it is difficult to cover the whole telecommunication bands, 1.26–1.57 μm. InP-based QD is a better choice [62] in spite of the difficulty in forming well-shaped QDs by MBE. InAs/InP QDs self-assembled by MOCVD normally emit light at wavelengths longer than the telecommunication bands. Fortunately, there is a growth technique called 'double-cap' method [63–66], with which one can decrease the height of the dots so as to shorten the light emission wavelength. Here, we present the results of telecommunication single photon emission from double-capped InAs/InP QDs grown by MOCVD [67, 68].

On Fe-doped semi-insulating InP (001) substrates, QDs were formed by depositing 2.8 ML of InAs followed by 15 s of annealing in an AsH_3 ambient. On these QDs, a 2-nm thick InP layer, much thinner than dot height, was deposited followed by a 120-s growth interruption with PH_3 exposure. At this first-cap stage the InAs QDs were shortened to have no part beyond the capping layer through As/P exchange reactions. The subsequent second cap process is to grow an 18 nm InP

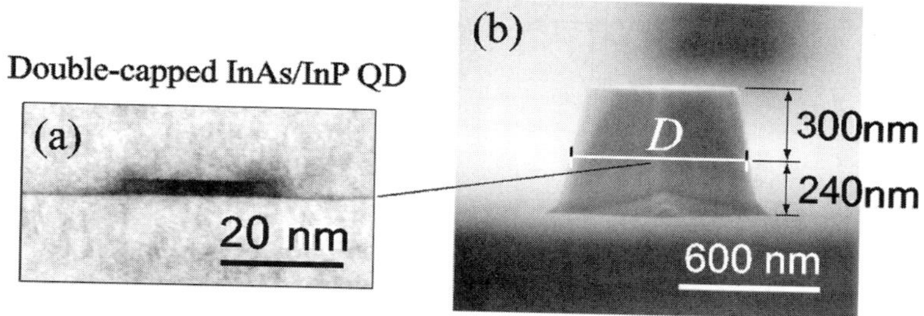

Fig. 9.8 (a) Cross-section TEM image of an InAs QD buried in InP; (b) SEM image of a mesa structure which is a tapered rectangular column. The QD in (a) clearly has a flat top due to the double-cap process. The QD layer is placed 300 nm below the top of the mesa to improve optical efficiency. The diagonal width D of the fabricated structures varies from 320 to 740 nm

layer on the fist one to fully embed the InAs QDs. A cross-section of the InAs QD layer is shown in Fig. 9.8(a) by a dark field transmission electron microscope (TEM) image. It shows the QD of about 30 nm in diameter and about 2 nm in height. Small mesa structures, as shown in the scanning electron microscope (SEM) image in Fig. 9.8(b), are prepared using photolithography and wet chemical etching to isolate the QDs with density of $1.8 \times 10^{10}\,\mathrm{cm}^{-2}$. Instead of a mountain-like shape [68], we obtained a tapered rectangular column as the mesa structure to improve optical efficiency. The mesas are 540 nm in height and 320–740 nm in diagonal width. The InAs QD layer is located 300 nm below the top mesa surface.

As a result of double-cap process, the InAs/InP QDs show emission wavelength covering the whole telecommunication bands [68]. The optical character of a single QD is examined by micro-photoluminescence (μ-PL) measurements at 10 K. The sample was excited by a mode-locked (80 MHz) Ti: sapphire laser at 780 nm. The laser beam was focused on a 2 μm spot at the top of the mesa structure through a microscope objective lens with a numerical aperture of 0.42. As shown in the inset of Fig. 9.9, a selected QD had a sharp main line at 1277.1 nm. The peak intensity of the main line shows single exciton behavior with linear dependence on the pump pulse energy in the low excitation region.

To test the possibility of using such QDs as single SPEs, photon correlation experiments were performed, in which the pump energy was adjusted to 6.25 fJ, slightly below the exciton saturation level presented in Fig. 9.9. Figure 9.10 shows a Hanbury-Brown and Twiss setup [69] for photon correlation analysis. In the transmitter shown in the upper half of Fig. 9.10, the pulse-repetition rate was reduced from 80 MHz to 2 MHz by a pulse-picker system, meaning the excitation pulse period t of 0.5 μs. The collected luminescence was coupled into a single-mode fiber after the scattered excitation light was eliminated by a silicon filter. The exciton emission was isolated using a tunable highly efficient bandpass filter with full width at half maximum of 1 nm. In the receiver shown in the lower half of Fig. 9.10, the signal was split by a 50:50 fiber coupler with each arm connected to a single photon detection module. These modules were Peltier-cooled InGaAs avalanche photodiodes (APDs) operated in a gated mode [69]. The gate timing was well synchronized

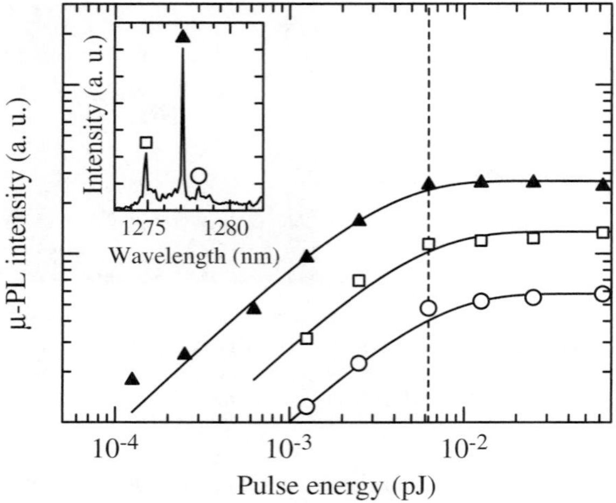

Fig. 9.9 Peak intensities of three μ-PL lines as a function of the pump pulse energy. Solid lines show linear dependence on the power density in the weakly excited region. Photon correlation measurements were performed for the main peak at 1277.1 nm (closed triangles), with a pulse energy of 6.25 fJ (dashed line). The inset shows the corresponding μ-PL spectrum of single InAs/InP QDs. The open square and circle indicate satellite structures of the main peak

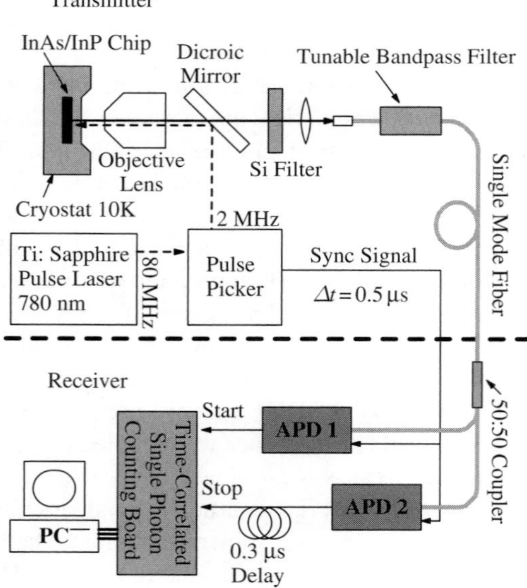

Fig. 9.10 Photon correlation measurement system. As a telecommunication medium, a single-mode fiber was used between the transmitter and the receiver

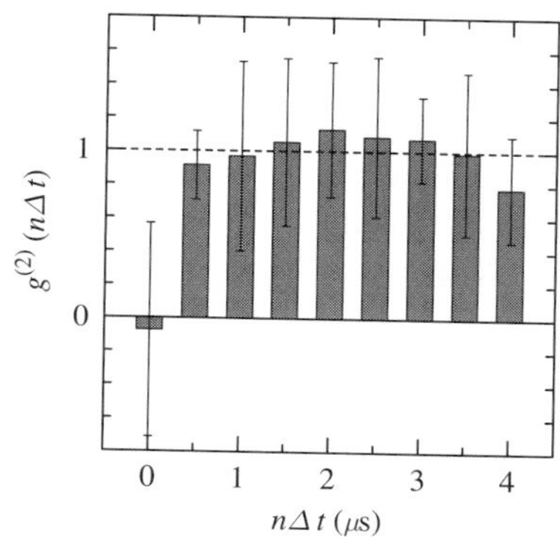

Fig. 9.11 Second-order correlation function $g^{(2)}(n\Delta t)$. Note that the gated operation of the APD makes the delay time discrete. At the time origin, $g^{(2)}{\sim}0$, so it can be concluded that the QD emits a single photon at each shot. The small negative value was caused by noise estimation error, and the error bars represent the variance of data from five trials and the noise estimation

with the 2 MHz trigger signal from the pulse-picker system in the transmitter. The start and stop signals produced by the two APDs were connected to a time-correlated single photon counting (TCSPC) board, which displays a histogram of the accumulated correlation events between start and stop timing as a function of the respective time delay. The stop signal was delayed by 0.3 μs to avoid an internal delay (~30 ns) of the TCSPC. The gate time and deadtime of the APDs were set to 5 ns and 10 μs, respectively. These values have been found to efficiently suppress dark counts and afterpulses of the detectors.

The photon correlation measurement by the above setup consists of five 2-hr trials with the total event number M of 7.2×10^{10}. The correlation count $C^{(2)}$(n) at delay time $n^{\Delta}t$ is expressed as

$$C^{(2)}(n) = M\langle C(m+n)C(m)\rangle \tag{9.8}$$

where C=1 (=0) when an APD is firing (not firing), and $\langle\cdots\rangle$ means the event average: $M^{-1}\sum_{m=1}^{M}\cdots$. Here, note that the gated operation of the APD makes the delay time of the correlation counts discrete. The result is that the correlation counts at a zero-time delay (n=0) were 251, lower than the $n{\neq}0$ counts, 2275 in average. These data obviously show an antibunching dip, although they include both the noise count C_{N}, which arises from APD misfiring (the dark count) and the signal count C_{S}. To cancel the effect of noise, we then measured the background contributions quantitatively by comparing the count rate with the central wavelength detuning of the bandpass filter from the exciton line. From the measurement, we obtained a ratio $\langle C_S\rangle/\langle C_N\rangle$ of 3/7. By subtracting the noise contribution from Eq. (9.8), we could estimate the second-order correlation function $g^{(2)}(n\Delta t) = \langle C_S(m+n)C_S(m)\rangle/\langle C_S(m)\rangle^2$. Figure 9.11 shows the estimated results for $g^{(2)}$, where the error bars indicate one-half of the 95% confidence interval given the variance of both the data from the five trials and the noise estimation. At the zero-time delay, $g^{(2)} \sim 0$, which for the first time demonstrated successfully single-photon emissions at a telecommunication wavelength, ~1.27 μm, from single InAs/InP QDs.

In principle, single-photon emission over the whole telecommunication bands can be realized on InAs/InP QDs prepared by double-cap process in MOCVD. By improving the efficiency, recently, we have also observed single photon emission at 1.55 μm from the same InAs/InP self-assembled QDs [70]. These results indicate the feasibility of using InAs/InP QDs to develop SPE devices for quantum key distribution systems in telecommunication bands.

9.3.2 Single Dot Emission Driven by Current Injection

In the view of application, a SPE must be integrated into small devices so as to make the quantum key distribution system compact and directly controllable by electrical signals. Electrically driven SPE is hence a very important but so far a challenging subject. There are some reports on the development of current injection SPEs with QDs [13, 71–73], but the emission wavelengths are mostly shorter than 1 μm, unavailable for quantum key distribution in telecommunication bands. Here, we describe the first observation of single QD electroluminescence (EL) at ~1.32 μm [74], which is approaching the realization of electrically driven SPEs in telecommunication bands.

On an n-type GaAs substrate, InAs-QDs-containing p–i–n diode structures, schematically shown in Fig. 9.12(a), were fabricated. An $Al_{0.1}Ga_{0.9}As$ barrier layer with doping concentration of $1 \times 10^{18}\,cm^{-3}$ is acting as a barrier for injected negative carriers. The i-type GaAs layer and the InAs QD layer serve as light-emitting matrices. A p^+-type GaAs layer with doping concentration of $1 \times 10^{19}\,cm^{-3}$ was grown to help making the top contact. The i- and p^+-GaAs layers above the QD layer were thin enough to increase the solid angle from the QD to the objective lens and to reduce free-carrier absorption.

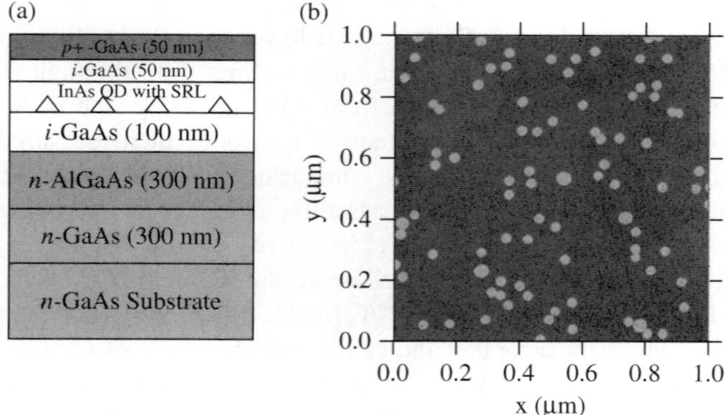

Fig. 9.12 (a) Schematic illustration of the epi-layer structure, and (b) AFM image of InAs QDs, for the sample to be fabricated into a light-emitting diode device

In the growth of the p-i-n diode structures, the most important is the formation of QDs. For the purpose of accessing a single QD optically and electrically, the QD density was firstly decreased, as the effort in section 9.2.2 aims at. Figure 9.12(b) shows a typical AFM image of the InAs QDs, with dot density decreased to be about $8.0 \times 10^9\,cm^{-2}$. Secondly, the InAs QDs were capped with an $In_xGa_{1-x}As$ strain-reducing layer (SRL) [75], to reduce the band gap of the QDs, originally larger than those corresponding to telecommunication bands, by relaxing the compressive strain in dots and suppressing the quantum confinement effect. As the example here, by setting the In content $x = 0.17$ in the SRL, 1.3 μm EL was obtained at 7 K.

A QD light-emitting-diode was made by putting the p–i–n structure between two electrodes. For observing macro-EL of QDs, the upper electrode has apertures of 20 μm in diameter. The I–V curve of this device clearly shows a diode behavior with a threshold voltage $V_{th} = 1.2$ V. Similarly, EL starts to appear above V_{th}. Hereafter, we term the applied voltage range below and above V_{th} "field effect region" and "current injection region", respectively. In order to access a single QD, the aperture diameter of the top electrode was decreased to typically 1.2 μm, as seen in Fig. 9.13(a). The photons emitted from QDs in the intrinsic region could pass through the aperture. To reduce the damage in devices, the fabrication processes of the p-i-n diodes were performed by wet etching techniques. There are seven main steps in a fabrication process: sputtering of SiO_2, patterning by photolithography, etching of SiO_2, etching of p^+-GaAs, evaporation of SiO_2, evaporation of Ti/Pt/Au

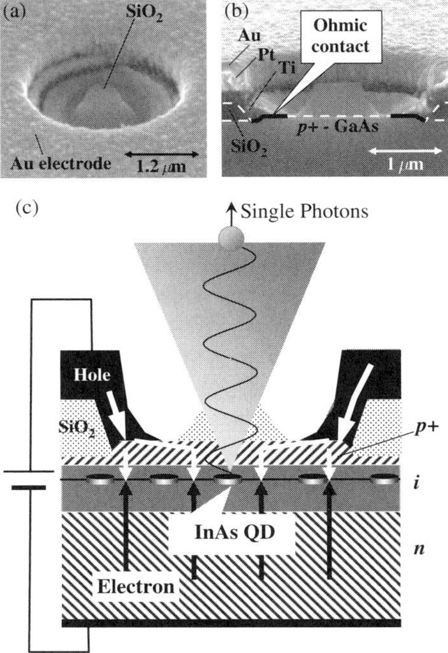

Fig. 9.13 (a) SEM image of the top electrode, (b) cross-sectional SEM image, and (c) schematic light emission process of the device used in μ-EL

electrodes and removal of resist. Some SiO_2 is left to reduce reflection at the boundary and increase the photon extraction efficiency. According to a finite-difference-time-domain simulation, the photon extraction efficiency of this device is expected to increase compared with the non-SiO_2 device. In addition, the extraction efficiency strongly depends on the diameter of the aperture. The improvement of the extraction efficiency is expected to be about 40% with the aperture of 1.2 μm in diameter. The ohmic contact was limited to only a small area around the center of the aperture as shown in Figs. 9.13(a) and (b), to constrain its negative effect on the device to be as small as possible. Figure 9.13(c) presents the EL process from a cross-sentional view of the device, where it is learned that electrons (holes) are injected from the bottom (top) electrode for recombining to emit photons in the QDs.

In the current injection region, micro-EL (μ-EL) spectra of a single QD was observed at 7K. The emitted photons from the QD were collected through an objective lens with numerical aperture of 0.5. The emitted photons were dispersed by a 0.6-m triple grating monochromator and detected with a liquid-nitrogen cooled InGaAs multi-channel detector. The energy resolution of this system was about 60 μeV in μ-EL measurements. Figure 9.14 shows typical μ-EL spectra for various injected currents. Sharp emission lines from 1319.9 to 1324.9 nm, corresponding to the ground state of excitons, are observed. Here we discuss the main two peaks at 1321.6 and 1324.9 nm by labeling them "X" and "XX", respectively. Their current dependence of the EL intensity is shown in the inset of Fig. 9.14. With increasing injection current, the peak X enhances linearly below and saturates above 55 nA. In contrast, quadratic dependence was clearly observed for the XX peak. It implies that X is an exciton peak whereas XX is the corresponding biexciton one. The energy difference between X and XX, 0.49 meV, is similar to that for the bi-exciton binding energy of an InAs/GaAs QD capped with an InGaAs layer [76], supporting the specification of the two peaks. What is more important, the emission wavelength here is longer than those of the single QDs attained before [61, 77]. This observation is a constructive step to electrically driven QD SPEs in telecommunication bands.

Wavelength tuning of QDs [78] is an important technique in traditional applications and probably also useful in quantum communication. On the present device,

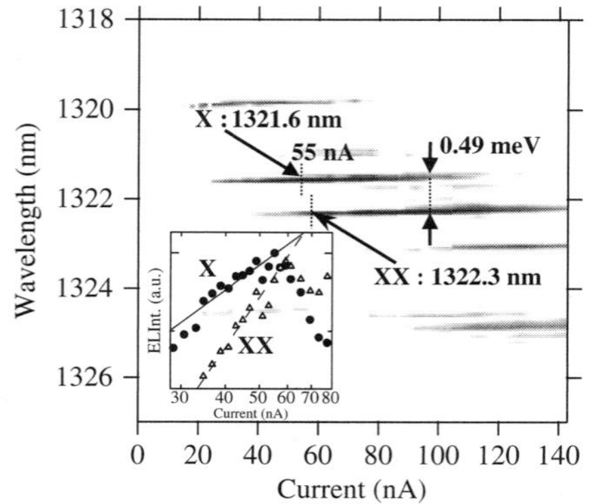

Fig. 9.14 Current dependence of the EL spectra of single QDs in the μ-EL device. Arrows indicate wavelengths for exciton (X) and biexciton (XX) and the maximum current for single-exciton emission. Inset shows the current dependences of the EL intensities of the exciton and biexciton. Lines are guides for the eye to linear and quadratic dependences on the injected current

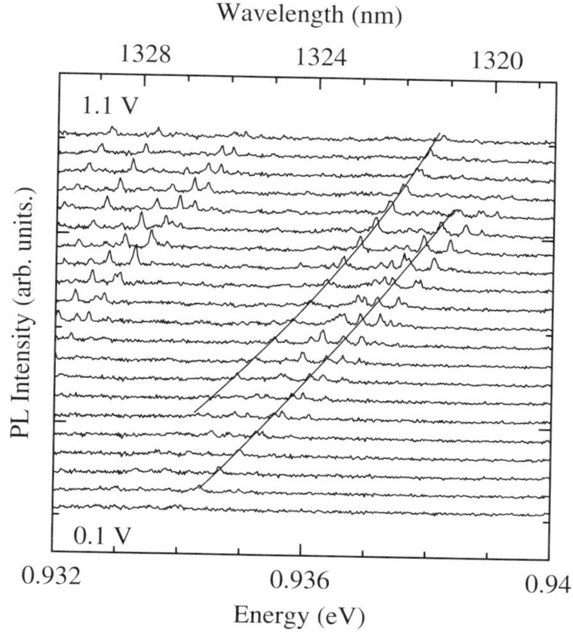

Fig. 9.15 Stark shift of the light emission from single QDs observed on the μ-EL device in the field effect region. Solid lines show quadratic dependence on the applied electric field

such an effect, Stark shift, was observed in the field effect region. In a typical diode, there is a built-in potential which originates from the dopant profile. With an externally applied voltage, the effective electric field around the QD is reduced. Then PL emission energy becomes higher because the hole in a QD shifts down and the electron in the same dot shifts up spatially. This type of Stark effect in QDs has been observed at shorter wavelengths [78–81]. Using here the same setup as that for μ-EL measurements, except for a CW Ti:Saphire laser at 780 nm as the excitation source, optical measurements demonstrate clear shift of single dot peaks versus applied voltage in the field effect region, as shown in Fig. 9.15. Each spectrum was taken by changing the applied voltage in the range 0.1–1.1 V by a step of 50 mV. Lines show the peak energy shifts of two excitons, both of which exhibit quadratic dependence on the applied voltage. With these facts, we estimated the dipole moment by differentiating the QD energy by the effective electric field, which is a combination of the build-in and external voltages. The dipole moment of QDs, ranging from 0.6 to 1.0 nm in the electric field range, agrees well with that in the previous report of InAs/GaAs QDs [78], confirming the applicability of the present QDs in quantum communication.

Further improvement of the photon extraction efficiency is needed to achieve the single-photon EL from InAs QDs on GaAs, for which we are now planning to optimize the material and device structures.

9.3.3 Towards Precisely Accessed Single Photon Emitters

Quantum communication based on self-assembled semiconductor QDs also requires site-controllability [50], with the same precision as for quantum computation since the ability to interconvert stationary and flying qubits is indispensable [2]. Using the technique described in section 9.2.3, it is possible to fabricate site-controlled single-photon emitting QDs in telecommunication bands. As an important progress towards precisely accessed SPEs, here the single-dot photon emission matching the telecommunication bands is observed from site-controlled InAs/InP QDs.

The sample preparation was performed on the surface of a flat semi-insulating InP(001) substrate. After formation of holes by chemical etching, MOCVD was carried out to form InAs QDs at the hole sites. In this case, the initial native oxides are removed by the reactive atomic hydrogen generated from decomposed PH_3. The InAs QDs, formed with 1.4 ML of InAs coverage, are 500 nm separated for the purpose of single dot PL measurements, as shown in Fig. 9.16(a). The cross section of such QDs is depicted schematically by the upper half in Fig. 9.16(b), where it is shown that they have a deep part, in fact $\sim$ 2 nm, below the wetting layer. Formation of site-controlled QDs are immediately followed by the "double-cap" growth mentioned in section 9.3.1, whose effect of cutting the QD part beyond the first cap layer is schematically shown by the lower diagram of Fig. 9.16(b).

The μ-PL was measured with the 532 nm line of a Nd:YAG laser on samples with the first cap layer thickness varying from 0.3 to 2 nm. At an arbitrary position, we first observe the strong wetting layer emission around 985 nm as shown in the inset of Fig. 9.16(c). In the region of site-controlled QDs, narrow PL peaks are observed. Taking the detector efficiency into account, the luminescence intensity is comparable to that of InAlAs/AlGaAs QDs self-assembled in conventional S-K mode. A typical single dot PL spectrum for each sample is presented in Fig. 9.16(c). For the sample with the first cap of 2 nm, the QDs emissions are estimated to be centered around 1.7 μm, which is beyond the instrument limit. As a result, most of the QDs are undetectable but few dots emitting at shorter wavelengths are observed, as exemplified by a peak at 1.59 μm in Fig. 9.16(c). The other three peaks in Fig. 9.16(c), at 1.33, 1.42 and 1.47 μm reflect approximately the average emission wavelength over all measured dots in one sample. The change of average position follows the quantum confinement effect and indicates that the emission wavelength can be tuned by the fabrication conditions to fit the application requirements.

Usually multi-peaks are observed in one measurement, but their relative strength can be changed by finely tuning the position of the laser spot, as exemplified in Fig. 9.17. On the sample with the first cap of 0.3 nm, we get the strongest single peak at 1448.9 nm (upper spectrum) by tuning the laser spot to a certain position. We reasonably consider that the laser spot of 1 μm wide straightly covers a single QD but neither of its neighbors 0.5 μm away, as is schemed above the upper spectrum. As we move the laser spot leftwards by 250 nm, there arises another peak at 1344.2 nm, where the excitation area also covers another QD, like the schematic illustration on the middle spectrum. As expected, with the laser spot further shifting leftwards by

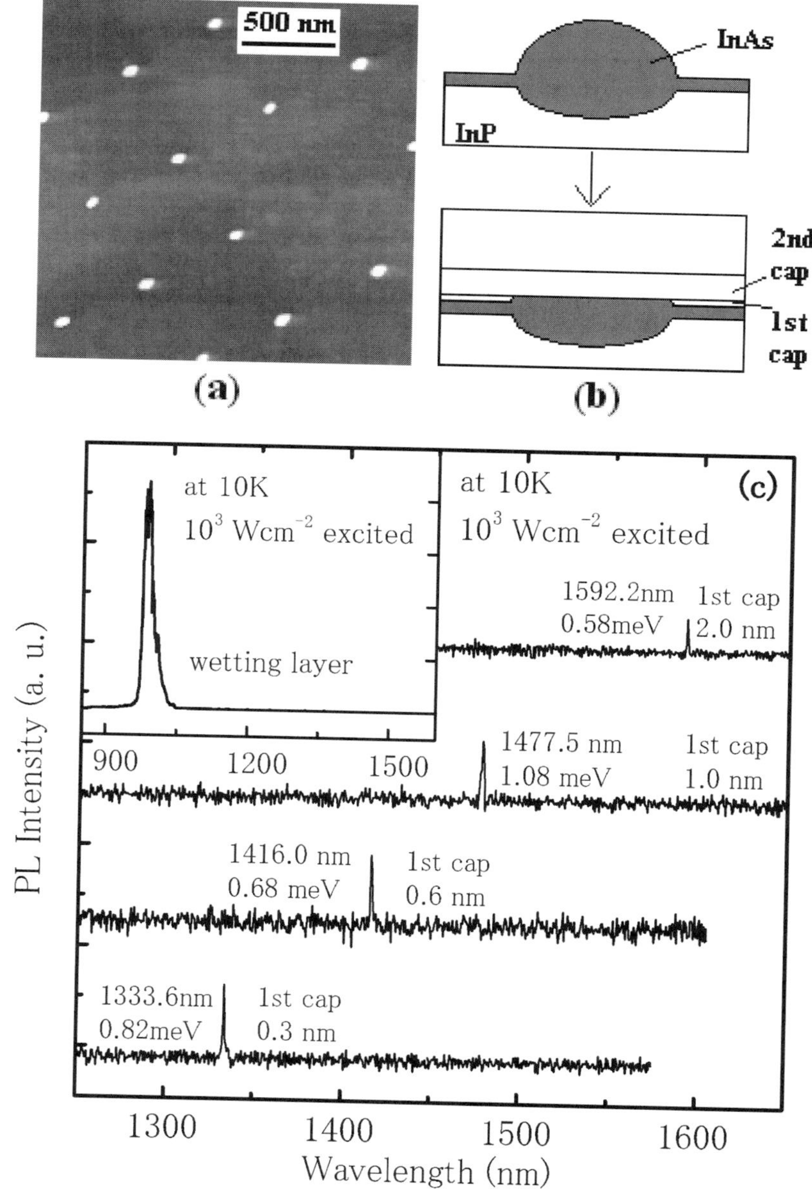

Fig. 9.16 (a) AFM image of site-controlled InAs/InP QDs after MOCVD regrowth of 1.4-ML InAs; (b) the schematic demonstration of the following "double-cap" growth; (c) typical single dot micro-PL peaks from site-controlled InAs/InP QDs with different thickness of the first cap layer. The inset in (c) shows the PL of the wetting layer

250 nm, the peak at 1448.9 nm disappears while that at 1344.2 nm remains, which corresponds to an exclusive pumping of the left dot, as shown by the lower spectrum and the drawing above it Fig. 9.17. The site-controllability of single-dot emission in telecommunication bands is thus revealed.

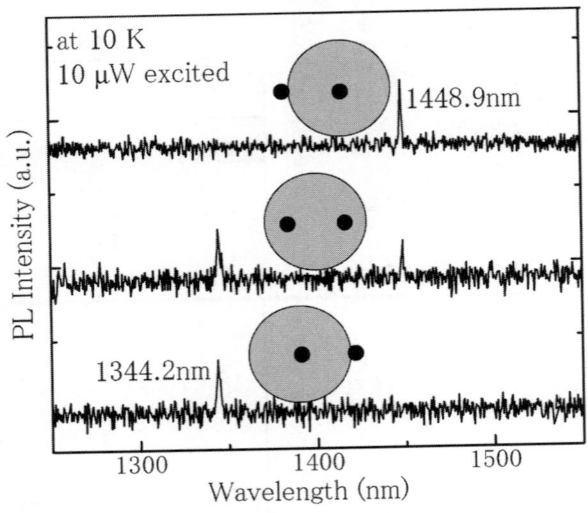

Fig. 9.17 Representative site dependence of the μ-PL peaks from site-controlled InAs/InP QDs. The schematic drawings above the spectra show the relative positions of the laser spot (gray circle, 1 μm in diameter) and the embedded QDs (solid circle, ∼100 nm in diameter and 500 nm separated)

To characterize the possibility of using such site-controlled QDs in quantum communication, their physical process of luminescence was studied. The single dot peaks look broader than conventionally grown self-assembled QDs, about 0.8 meV in full width at half maximum, but they are the same as that of InAs/InP QDs produced at nanotemplates [50], in which the formed QDs are a few hundred nm far away from the initially patterned surface processed by electron beam lithography. This suggests that regrowth of QDs directly on a chemically processed substrate may not be a severe factor to degrade the dot quality in our technique. To quantitatively characterize the possibly imperfect interface between the chemically processed substrate and the overgrown QDs, excitation density dependence of the single dot μ-PL intensity was investigated. It is found that the integrated intensities of single dot μ-PL peaks do not exhibit a universal excitation density dependence but vary from linear to quadratic for different QDs. Three examples, for dots a, b and c from the sample with first cap of 0.3 nm, shown in Fig. 9.18(a) by solid symbols follow the power law function with index of 1.0, 1.32 and 2.0, respectively. We may refer to a model in terms of nonradiative process as in quantum wells [82], in which the excitation density I_{ex} satisfies:

$$\alpha I_{\text{ex}} = \frac{n}{\tau_n} + \frac{np}{\tau_0} = \frac{p}{\tau_p} + \frac{np}{\tau_0} \tag{9.9}$$

where n (p) is the electron (hole) number in the QD, τ_n (τ_p) is the nonradiative decay time of electrons (holes), τ_0 is the radiative lifetime and α is an coefficient associated with the absorption. Here the radiative recombination takes the form of np/τ_0 because the electrons and holes can be captured independently in a single QD [83]. Denoting PL intensity as $L = np/\tau_0$ and defining $(\tau_n \tau_p)^{1/2} = \tau$ as the normalized nonradiative lifetime, Eq. (9.9) can be reduced to:

$$\alpha I_{\text{ex}} = L + \sqrt{L\tau_0}/\tau. \tag{9.10}$$

It is easy to see that $L \approx \alpha^2\tau^2 I_{\text{ex}}^2/\tau_0$ when τ_0/τ^2 is sufficiently large, and $L \approx \alpha I_{\text{ex}}$ while τ_0/τ^2 is sufficiently small. Providing the nonradiative lifetime τ varies from dot to dot, the different excitation density dependence of PL intensity can be understood. In Fig. 9.18(a), it is clear that the results at 10 K are well fitted by Eq. (9.10) with different τ for different peaks. The dot dependent nonradiative lifetime is thought to be the result of different number of nonradiative centers in different dots. Considering the chemical processing before QD regrowth, the nonradiative centers might be some impurities or defects in the vicinity of QDs, which are survived from the incomplete surface cleaning. The existence of good QDs, *e.g.* dot a of little nonradiative center, implies that the impurities/defects are extrinsic in the present site-controlled QDs and thus can be cleaned away by some method.

The effects of impurities/defects can be further studied by the temperature dependence of the single dot emission. The symbols other than solid ones in Fig. 9.18(a) show the excitation density dependence at elevated temperatures for dot a. It is seen that increasing temperature leads to more and more quadratic excitation density dependence of PL intensity. These can also be well fitted by Eq. (9.10) but with τ more and more shortened with increasing temperature. In detail, Fig. 9.18(b) demonstrates that, at a fixed excitation density, the PL intensity of each dot is nearly constant at low temperatures but thermally quenched at higher temperatures. These data are suggestive of an expression of the nonradiative lifetime τ as:

$$\tau^{-1} = \tau_1^{-1} + \tau_2^{-1} e^{-E/kT}, \tag{9.11}$$

where τ_1 and τ_2 are time constants, E is an activation energy, k is the Boltzmann constant and T is the temperature. The solid lines in Fig. 9.18(b) indicate that the experimental results are well fitted to Eq. (9.10) together with (9.11). What is more important, all the site-controlled QDs show an activation energy $E \approx 22.5$ meV, suggestive of phonon scattering as in conventionally self-assembled QDs [84] at higher temperatures. In addition, τ_1 is dot dependent and close to the value of τ at 10K, while α, τ_0 and E are almost dot independent. The parameter τ_2 is also dot dependent with the same trend as but more weakly than τ_1. This may be ascribed to impurities/defects-enhanced phonon scattering [85]. It looks that the effect of

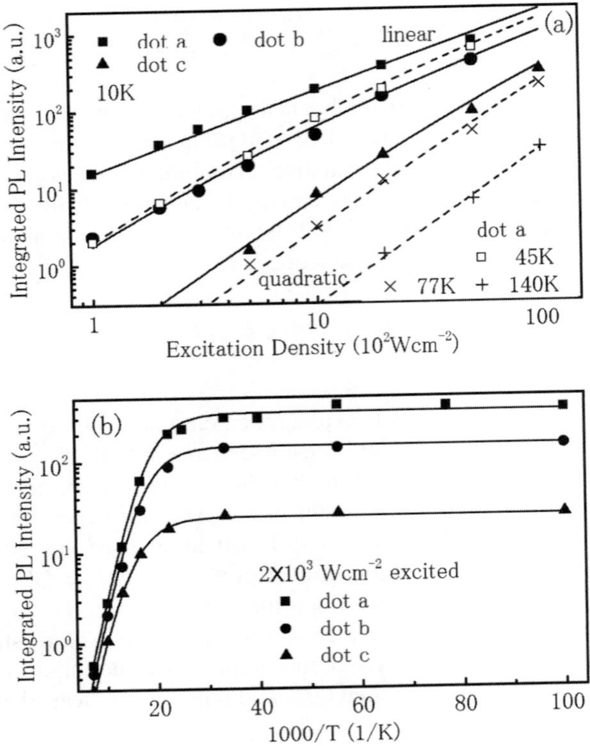

Fig. 9.18 (a) Excitation density and (b) temperature dependences of the single dot PL intensities of site-controlled InAs/InP QDs. Lines show the fitted results using Eq. (9.10) and (9.11)

nanradiative centers does not dominate over the intrinsic phonon scattering at higher temperatures.

On the other hand, the PL from the wetting layer is as normal, *i.e.* it does not show a complicated excitation density dependence but simply a linear behavior in a lower excitation range (not shown). It means that the nonradiative process due to interface impurities/defects does not have an obvious effect in the wetting layer. Therefore, the impurities/defects exist mainly at QD sites. This is probably because there are higher density of steps with dangling bonds at the hole sites. It again indicates that these impurities/defects are not intrinsic in the present site-controlled InAs/InP QDs. They will be well suppressed by improving the fabrication technique, *e.g.* using *in situ* atomic hydrogen irradiation to remove oxide dots, optimizing the annealing condition and finely controlling the AFM oxidation. However, the present QD quality is not below the limit of application in quantum communication because conventionally self-assembled InAlAs/AlGaAs QDs of similar quality have exhibited single photon emission [86]. We are currently struggling to perform single photon transmission using such site-controlled QDs as the source working at telecommunication wavelengths.

9.4 Summary

In summary, this chapter described our recent studies on self-assembled semiconductor QDs as candidate building blocks for prospective quantum information processing. A simultaneous observation of both the microscopic and electronic structures of thinly capped QDs is proved possible by using STM, as a result of a nearly straight relation between the STM signal and the wavefunction of the embedded QD. It provides a way to individually access qubits in a QD-based quantum computer. For fundamental studies of single self-assembled QDs, dot density dilution is realized controllably by post-growth annealing of a subcritical wetting layer. Its controllability relies on the precursors growing well before the 2D–3D growth mode transition and the nucleation from precursors into QDs in growth interruption. In order to exactly define the position and size of QDs in a quantum computer, an AFM-assisted technique is developed with good uniformity in sites and better homogeneity in QD size than in conventional S-K growth. It enables constructing sophisticated QD-qubit structures to be applied in quantum computing. For quantum communication, double-capped InAs/InP QDs demonstrate single-photon emission in telecommunication bands, indicating for the first time the feasibility of using semiconductor QDs as SPE devices for silica-fiber based quantum key distribution. Single dot EL from InAs QDs on GaAs is presented with telecommunication emission wavelength longer than reported before. Together with Stark shift giving significant dipole moment, this result is considered an important step approaching practical QD-based telecommunication SPEs. QD-based SPEs also need to be precisely positioned, for which we realized site-controlled single dot emission over the whole telecommunication bands from InAs/InP QDs. Their optical quality is comparable to conventionally self-assembled InAlAs/AlGaAs QDs, suggesting the future realization of exactly designed and precisely accessed QD-based SPEs working at telecommunication wavelengths.

Acknowledgments We thank Dr. K. Takemoto, T. Miyazawa, Y. Nakata, T. Ohshima, M. Takatsu, and N. Yokoyama of Fujitsu Lab. Ltd., Prof. M. Kawabe and Y. Sakuma of National Institute for Materials Science (NIMS), Prof. Y. Okada of Tsukuba University and Prof. Y. Arakawa of Tokyo University for their contributions to this chapter.

References

1. M. A. Nielsen and I. L. Chuang, *Quantum computation and quantum information*, (Cambridge University Press, Cambridge, England, 2000).
2. D. P. DiVincenzo, chapter 7 in *Semiconductor spintronics and quantum computation*, ed. D. D. Awschalom, D. Loss and N. Samarth, Springer-Verlag Berlin Heidelberg (2002).
3. A. Barenco, D. Deutsch, A. Eckert, and R. Jozsa, Phys. Rev. Lett. **74**, 4083 (1995).
4. D. Loss and D. P. DiVincenzo, Phys. Rev. A **57**, 120 (1998).
5. A. Imamoglu, D. D. Awschalom, G. Burkard, D. P. DiVincenzo, D. Loss, M. Sherwin, and A. Small, Phys. Rev. Lett. **83**, 4204 (1999).
6. T. Ohshima, Phy. Rev. A **62**, 062316 (2000).

7. F. Troiani, U. Hohenester, and E. Molinari, Phys. Rev. B **62**, R2263 (2000).
8. E. Biolatti, I. D'Amico, P. Zanardi, and F. Rossi, Phys. Rev. B **65**, 075306 (2002).
9. B. W. Lovett, J. H. Reina, A. Nazir, and G. A. D. Briggs, Phys. Rev. B **68**, 205319 (2003).
10. X. Li, Y. Wu, D. Steel, D. Gammon, T. H. Stievater, D. S. Katzer, D. Park, C. Piermarocchi, and L. J. Sham, Science **301**, 809 (2003).
11. J. R. Petta, A. C. Johnson, J. M. Taylor, E. A. Laird, A. Yacoby, M. D. Lukin, C. M. Marcus, M. P. Hanson, and A. C. Gossard, Science **309**, 2180 (2005).
12. C. Santori, D. Fattal, J. Vuèkovæ, G. S. Solomon, and Y. Yamamoto, Nature **419**, 594 (2002).
13. Z. Yuan, B. E. Kardynal, R. M. Stevenson, A. J. Shields, C. J. Lobo, K. Cooper, N. S. Beattie, D. A. Ritchie, and M. Pepper, Science **295**, 102 (2002).
14. B. Lounis and W. E. Moerner, Nature **407**, 491 (2000).
15. C. Kurtsiefer, S. Mayer, P. Zarda, and H. Weinfurter, Phys. Rev. Lett. **85**, 290 (2000).
16. S. Strauf, P. Michler, M. Klude, D. Hommel, G. Bacher, and A. Forchel, Phys. Rev. Lett. **89**, 177403 (2002).
17. M. H. Baier, E. Pelucchi, E. Kapon, S. Varoutsis, M. Gallart, I. Robert-Philip, and I. Abram, Appl. Phys. Lett. **84**, 648 (2004).
18. J. Hours, S. Varoutsis, M. Gallart, J. Bloch, I. Robert-Philip, A. Cavanna, I. Abram, F. Laruelle, and J. M. Gérard, Appl. Phys. Lett. **82**, 2206 (2003).
19. I. Marcikic, H. de Riedmatten, W. Tittel, H. Zbinden, and N. Gisin, Nature **421**, 509 (2003).
20. T. Yamaguchi, Y. Matsuba, L. Bolotov, M. Tabuchi, and A. Nakamura, Appl. Phys. Lett. **77**, 4368 (2000).
21. T. K. Johal, R. Rinaldi, A. Passaseo, R. Cingolani, A. Vasanelli, R. Ferreira, and G. Bastard, Phys. Rev. B **66**, 075336 (2002).
22. O. Millo, D. Katz, Y. W. Cao, and U. Banin, Phys. Rev. B **61**, 16773 (2000).
23. Y. Okada, M. Miyagi, K. Akahane, Y. Iuchi, and M. Kawabe, J. Appl. Phys. **90** , 192 (2000).
24. H. Z. Song, M. Kawabe, Y. Okada, R. Yoshizaki, T. Usuki, Y. Nakata, T. Ohshima, and N. Yokoyama, Appl. Phys. Lett. **85**, 2355 (2004).
25. H. Yamaguchi and Y. Hirayama, Jpn. J. Appl. Phys. **37**, L899 (1998).
26. M. Schmid, S. Crampin and P. Varga, J. Electr. Spec. Rel. Phen. **71–84**, 109 (2000).
27. R. M. Feenstra, Phys. Rev. B **50**, 4561 (1994).
28. G. S. Solomon, J. A.Trezza, A. F. Marshall, and J. S. Harris, Jr., Phys. Rev. Lett. **76**, 952 (1996).
29. H. Z. Song, K. Akahane, S. Lan, H. Z. Xu, Y. Okada, and M. Kawabe, Phys. Rev. B **64**, 085303 (2001).
30. G. Hörmandinger and J. B. Pendry, Surf. Sci. **295**, 34 (1993).
31. H. Z. Song, T. Usuki, Y. Nakata, N. Yokoyama, H. Sasakura, and S. Muto, Phys. Rev. B **73**, 115327 (2006).
32. CH. Heyn, Phys. Rev. B **66**, 075307 (2002).
33. P. Finnie, B. J. Riel, and Z. R. Wasilewski, J. Vac. Sci. Technol. B **20**, 2210 (2002).
34. N. Saucedo-Zeni, A. Yu. Gorbatchev, and V. H. Méndez-García, J. Vac. Sci. Technol. B **22**, 1503 (2004).
35. F. Patella, S. Nufris, F. Arciprete, M. Fanfoni, E. Placidi, A. Sgarlata, and A. Balzarotti, Phys. Rev. B **67**, 205308 (2003).
36. A. Polimeni, A. Patanè, M. Capizzi, F. Martelli, L. Nasi, and G. Salviati, Phys. Rev. B **53**, R4213 (1996).
37. B. J. Spencer, P. W. Voorhees, and S. H. Davis, Phys. Rev. Lett. **67**, 3696 (1991).
38. F. Long, S. P. A. Gill, and A. C. F. Cocks, Phys. Rev. B **64**, R121307 (2001).
39. H. R. Eisenberg and D. Kandel, Phys. Rev. Lett. **85**, 1286 (2000).
40. C. Priester and M. Lannoo, Phys. Rev. Lett. **75**, 93 (1995).
41. Y. Chen and J. Washburn, Phys. Rev. Lett. **77**, 4046 (1996).
42. T. J. Krzyzewski, P. B. Joyce, G. R. Bell, and T. S. Jones, Phys. Rev. B **66**, R121307 (2002).
43. H. T. Dobbs, D. D. Vvedensky, A. Zangwill, J. Johansson, N. Carlsson, and W. Seifert, Phys. Rev. Lett. **79**, 897 (1997).

44. S. Suo and Z. Zhang, Phys. Rev. B **58**, 5116 (1998).
45. V. G. Dubrovskii, G.. E. Cirlin, and V. M. Ustinov, Phys. Rev. B **68**, 075409 (2003).
46. T. Oda, K. One, M. Stopa, T. Hatano, S. Tarucha, H. Z. Song, Y. Nakata, T. Miyazawa, T. Ohshima, and N. Yokoyama, Phys. Rev. Lett. **93**, 066801 (2004).
47. T. Oda, M. Rontani, S. Tarucha, Y. Nakata, H. Z. Song, T. Miyazawa, T. Usuki, M. Takatsu, and N. Yokoyama, Phys. Rev. Lett. **95**, 236801 (2005).
48. T. Ishikawa, T. Nishimura, S. Kohmoto, and K. Asakawa, Appl. Phys. Lett. **76**, 167 (2000).
49. H. Lee, J. A. Johnson, M. Y. He, J. S. Speck, and P. M. Petroff, Appl. Phys. Lett. **78**, 105 (2001).
50. D. Chithrani, R. L. Williams, J. Lefebvre, P. J. Poole, and G. C. Aers, Appl. Phys. Lett. **84**, 978 (2004).
51. H. Z. Song, T. Ohshima, Y. Okada, K. Akahane, T. Miyazawa, M. Kawabe, and N. Yokoyama, *Proceedings of the 26th ICPS*, Edinburgh, 29 July- 2 August 2002, p. 32.
52. T. Ohshima, H. Z. Song, Y. Okada, K. Akahane, T. Miyazawa, M. Kawabe, and N. Yokoyama, Phys. Stat. Sol. (c) **4**, 1364 (2003).
53. U. F. Keyser, H. W. Schumacher, U. Zeitler, R. J. Haug, and K. Zberl, Appl. Phys. Lett. **76**, 457 (2000).
54. H. Z. Song, T. Usuki, S. Hirose, K. Takemoto, Y. Nakata, N. Yokoyama, and Y. Sakuma, Appl. Phys. Lett. **86**, 113118 (2005).
55. A. Hirai and K. M. Itoh, Physica E **23**, 248 (2004).
56. Y. Okada, Y. Iuchi, M. Kawabe, and J. S. Harris, Jr., J. Appl. Phys. **88**, 1136 (2000).
57. H. Z. Song, Y. Nakata, Y. Okada, T. Miyazawa, T. Ohshima, M. Takatsu, M. Kawabe, and N. Yokoyama, Physica E **21**, 625 (2004).
58. Zh. M. Wang, Y. I. Mazur, Sh. Seydmohamadi, G. J. Salamo, and H. Kissel, Appl. Phys. Lett. **87**, 213105 (2005)
59. H. Z. Song, S. Lan, K. Akahane, K. Y. Jang, Y. Okada, and M. Kawabe, Solid State Communications **115**, 195 (2000).
60. D. L. Huffaker and D. G. Deppe, Appl. Phys. Lett. **73**, 366 (1998).
61. M. B. Ward, D. C. Unitt, Z. Yuan, P. See, R. M. Stevenson, K. Cooper, P. Atkinson, I. Farrer, D. A. Ritchie, and A. J. Shields, Physica E **21**, 390 (2004).
62. T. Mensing, L. Worschech, R. Schwertberger, J. P. Reithmaier, and A. Forchel, Appl. Phys. Lett. **82**, 2799 (2003).
63. C. Paranthoen, N. Bertru, O. Dehaese, A. Le Corre, S. Loualiche, B. Lambert, and G. Patriarche, Appl. Phys. Lett. **78**, 1751 (2001).
64. S. Raymond, S. Studenikin, S. J. Cheng, M. Pioro-Ladriére, M. Ciorga, P. J. Poole, and M. D. Robertson, Semicond. Sci. Technol. **18**, 385 (2003).
65. Y. Sakuma, K. Takemoto, S. Hirose, T. Usuki, and N. Yokoyama, Physica E **26**, 81 (2005).
66. Y. Sakuma, M. Takeguchi, K. Takemoto, S. Hirose, T. Usuki, and N. Yokoyama, J. Vac. Sci. Technol. B **23**, 1741 (2005).
67. K. Takemoto, Y Sakuma, S. Hirose, T. Usuki, N. Yokoyama, T. Miyazawa, M. Takatsu, and Y. Arakawa, Jpn. J. Appl. Phys. **43**, L993 (2004).
68. K. Takemoto, Y. Sakuma, S. Hirose, T. Usuki, and N. Yokoyama, Jpn. J. Appl. Phys. **43**, L349 (2004).
69. R. Hanbury Brown and R. Q. Twiss, Nature **178**, 1447 (1956).
70. T. Miyazawa, K. Takemoto, Y. Sakuma, S. Hirose, T. Usuki, N. Yokoyama, M. Takatsu, and Y. Arakawa, Jpn. J. Appl. Phys. **44**, L620 (2005).
71. A. Imamoglu and Y. Yamamoto, Phys. Rev. Lett. **72**, 210 (1994).
72. X. Xu, D. A. Williams, and J. R. A. Cleaver, Appl. Phys. Lett. **85**, 3238 (2004).
73. A. J. Bennett, D. C. Unitt, P. See, A. J. Shields, P. Atkinson, K. Cooper, and D. A. Ritchie, Appl. Phys. Lett. **86**, 181102 (2005).
74. T. Miyazawa, J. Tatebayashi, S. Hirose, T. Nakaoka, S. Ishida, S. Iwamoto, K. Takemoto, T. Usuki, N. Yokoyama, M. Takatsu, and Y. Arakawa, Jpn. J. Appl. Phys. **45**, 3621 (2006).
75. J. Tatebayashi, M. Nishioka, and Y. Arakawa, Appl. Phys. Lett. **78**, 3469 (2001).
76. I. E. Itskevich, S. I. Rybchenko, I. I. Tartakovskii, S. T. Stoddart, A.Levin, P. C. Main, L. Eaves, M. Henini, and S. Parnell, Appl. Phys. Lett. **76**, 3932 (2000).

77. B. Alloing, C. Zinoni, V. Zwiller, L. H. Li, C. Monat, M. Gobet, G. Buchs, and A. Fiore, Appl. Phys. Lett. **86**, 101908 (2005).
78. P. W. Fry, I. E. Itskevuch, D. J. Mowbray, M. S. Skolnick, J. J. Fineley, J. A. Barker, E. P. O'Reilly, L. R. Wilson, I. A. Larkin, P. A. Maksym, M. Hopkinson, M. Al-Khafaji, J. P. R. David, A. G. Cullis, G. Hill, and J. C. Clark, Phys. Rev. Lett. **84**, 733 (2000).
79. S. Raymond, J. P. Reynolds, J. L. Merz, S. Fafard, Y. Feng, and S. Charbonneau, Phys. Rev. B **58**, R13415 (1998).
80. F. Findeis, M. Baier, E. Beham, A. Zrenner, and G. Abstreiter, Appl. Rev. Lett. **78**, 2958 (2001).
81. M. Baier, F. Findeis, A. Zrenner, M. Bichler, and G. Abstreiter, Phys. Rev. B **64**, 195326 (2001).
82. W. Feng, Y. Wang, J. Wang, W. K. Ge, Q. Huang, and J. M. Zhou, Appl. Phys. Lett. **72**, 1463 (1998).
83. K. F. Karlsson, E. S. Moskalenko, P. O. Holtz, B. Monemar, W. V. Schoenfled, J. M. Garcia, and P. M. Petroff, Appl. Phys. Lett. **78**, 2952 (2001).
84. I. V. Ignatiev, I. E. Kozin, S. V. Nair, H. -W. Ren, S. Sugou, and Y. Masumoto, Phys. Rev. B **61**, 15633 (2000).
85. M. Pepper, J. Phys. C **13,** L709 (1980).
86. S. Kimura, H. Kumano, M. Endo, I. Suemune, T. Yokoi, H. Sasakura, S. Adachi, S. Muto, H. Z. Song, S. Hirose, and T. Usuki, Jpn. J. Appl. Phys. **44**, L793 (2005).

Chapter 10
Stress Relaxation Phenomena in Buried Quantum Dots

N.A. Bert, V.V. Chaldyshev, A.L. Kolesnikova, and A.E. Romanov

Abstract We report on the results of experimental and theoretical investigation of mechanical stress relaxation in heterostructures with buried quantum dots. Quantum dot is viewed as a dilatational inclusion with eigenstrain (transformation strain) caused by crystal lattice mismatch between the dot and matrix materials. Stresses and energies for spheroid inclusions in an infinite medium, in a half-space, and in a plate of finite thickness, are presented. We propose and develop three models of the stress relaxation in buried quantum dots, which correspond to different types of the dot/matrix interfaces and local environment of the dot. The first mechanism corresponds to the case when there are no material-conservation-related restrictions on formation of a misfit dislocation loop at the dot/matrix interface. The second mechanism additionally takes into account the local material conservation law, which leads to formation of a pair of prismatic dislocation loops. One of them is a misfit dislocation loop that lies at the dot/matrix interface, whereas the other is a satellite dislocation loop that locates in the adjacent matrix. The third mechanism also accounts for the local materials conservation by formation of a satellite dislocation loop, however, a reduction of the initial dot eigenstrain is considered instead of the formation of a misfit dislocation loop. For each mechanism the energy criteria are derived, which govern the formation of misfit and satellite dislocation loops. We determine the critical radii of stressed dots, which are required to initiate the relaxation processes, as well as the dependence of the satellite loop size on the quantum dot diameter. The model calculations are compared to the relevant experimental data. The developed mechanisms seem to be appropriate to describe relaxation in a large variety of the quantum dots self-organized in the bulk of different matrices. We demonstrate that the third model gives qualitatively and quantitatively correct results for buried AsSb quantum dots formed in the bulk of GaAs films. The first and second models are discussed to be applicable to sub-critical quantum dots, which are self-organized on the surface and then buried by epitaxial overgrowth.

10.1 Introduction

It is well known that due to the crystal lattice mismatch between the materials of a quantum dot (QD) and surrounding matrix, considerable elastic strains can be generated inside and in a vicinity of a QD *(Devis 1998, Andreev et al. 1999, Shchukin*

Z. M. Wang, *Self-Assembled Quantum Dots.*

et al. 2002a). This strain has a strong impact on the electronic properties of the semiconductor QD by changing the energy levels of localized electrons and holes and their wave functions (*Romanov et al. 2005*). In polar semiconductors it also induces piezoelectric polarization charges that lead to strong electrostatic fields (*Devis 1998, Andreev and O'Reilly 2000*). The strain induced by a QD in the surrounding matrix modifies the physical properties of the medium. In the ensemble of QDs their long-range elastic interaction (transmitted through the matrix) leads to self-ordering phenomena, which are important for the electronic properties and performance of the device structures (*Shchukin and Bimberg 1999, Shchukin et al. 2002a, Tichert 2002, Stangl et al. 2004*).

Stress relaxation phenomena in QDs may be considered by analogy with the processes in thin films grown on mismatched substrates, which are related to formation of misfit dislocations (MDs) at the film/substrate interface, e.g. *(Freund and Suresh 2003)*. One may expect the emerging and development of similar relaxation phenomena in strained QDs. As a critical thickness exists for strained thin films, for QDs there should be a critical size for the onset of relaxation processes and the appearance of additional structural defects. These additional defects formed at the dot/matrix interface or in the QD vicinity will modify physical properties of the material. In many cases the presence of these defects is undesirable due to crucial impact on the material crystallinity and corresponding device performance. Eventually, relaxation processes change the above mentioned strain-dependent properties of semiconductors and lead to degradation and failure of semiconductor devices.

Stress relaxation phenomena in buried quantum dots possess their own peculiarities that make them different from the relaxation processes in surface islands (unburied QDs) and thin films. In fact, the free surface of thin films and islands always participates in the material transport accompanied stress relaxation. In the case of buried QDs the elastic field is localized inside the dot and in the adjacent matrix. As a result the material transport in the course of relaxation can be locked in a close vicinity of the inclusion. The key issue of this chapter is to account for these peculiarities of stress relaxation in buried QDs. We propose and develop new physical mechanisms of stress relaxation in buried quantum dots, which have a general meaning for a wide variety of semiconductor materials.

The chapter is organized as follows. In *Section 10.2* we briefly describe some experimental observations of the relaxation phenomena and structure transformation for QDs of different origin. We consider the InAs/GaAs and Ge/Si heteroepitaxial systems where the QDs are self-organized on the surface, as well as the As, MnAs and AsSb QDs that are self-organized in the bulk of GaAs films. *Section 10.3* provides the basis of stress and elastic energy calculations for buried QDs, where the results for spheroidal dilatational inclusions are delivered. *Section 10.4* considers three mechanisms of stress relaxation in buried quantum dots, which correspond to different types of the dot/matrix interfaces and local environment of the dots. *Section 10.5* presents experimental verification of the developed stress relaxation models. It is based on a comprehensive experimental study of AsSb QDs in GaAs, which provides sufficient information for calculations with a single verifiable parameter. In Conclusions we discuss relevancy and limitations of the developed approach.

10.2 Experimental Observations of Relaxation Phenomena in Quantum Dots

Among the variety of QD systems we briefly consider just several examples in this section, which, being important for device applications, show main characteristic features of the atomic structure, mechanical properties and relaxation phenomena. The selected examples are subdivided in two groups based on difference in self-organization processes resulting in QD formation. We distinguish self-organization phenomena on the surface and in the bulk. In the former case, the atomic transport is much faster, and self-organization is possible during the film growth with or without short-term growth interruptions. When the self-organization is realized in the bulk, a two-step procedure is required. At the first step a metastable medium should be produced, where self-organization of QDs occurs under special treatment at the second step.

10.2.1 Misfit Defects in Surface-nucleated Quantum Dots

One of the most attractive paths to form QDs is the use of epitaxial techniques, mainly molecular-beam epitaxy (MBE) or metal-organic chemical vapor deposition (MOCVD), that enable to control deposit amount at the level of monolayer (ML) and can produce nano-scale islands from atomic constituents and/or molecular precursors (*Shchukin et al. 2002b*). Self-organization of QDs is a consequence of the lattice mismatch between the matrix (or substrate) material and the dot or epilayer material (*Bimberg et al. 1995, Shchukin et al. 2002a*). The minimization of misfit strain energy results in the formation of three-dimensional islands, typically with a pyramidal or domelike shape. For most epitaxial QDs, these three-dimensional islands occur during the Stranski-Krastanow growth mode, which is characterized by initial growth of a thin two-dimensional, pseudo-morphic (or wetting) layer followed by its transformation into the three-dimensional QDs (free-standing or surface QDs), which continue to grow during subsequent deposition of the dot material. After overgrowth with the substrate material the surface QDs occur buried into the matrix.

Most efforts have been dedicated to the growth and the properties of coherent QDs that are attractive for various device applications. In contrast, much less work has been done to understand the strain evolution and relaxation in growing QDs. Investigations of such a kind should involve techniques with a high spatial resolution, which allow one to directly monitor small, nanometer-scale features. Defects in QDs can be observed directly by scanning tunnel microscopy (STM), atomic force microscopy (AFM), and transmission electron microscopy (TEM). TEM including high-resolution mode (HREM) is mainly used to visualize and analyze defects in detail.

Among different material combinations investigated so far, QDs in the InAs/GaAs material system are most comprehensively studied and are often considered to be a model for fundamental studies. An ensemble of perfect InAs

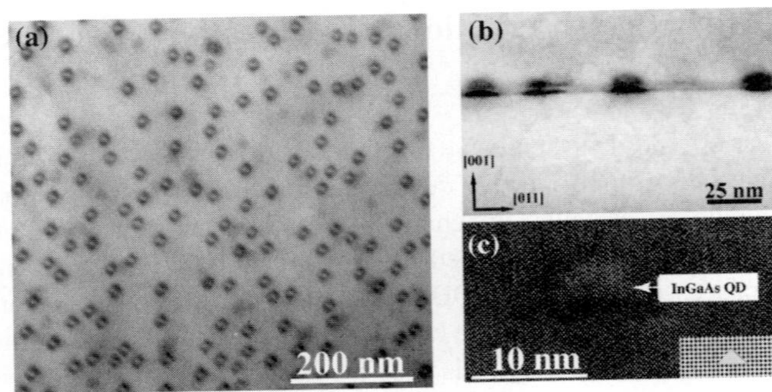

Fig. 10.1 Coherent InAs QDs embedded in GaAs: (a) plan-view TEM image in [001] zone; (b) (110) cross-section 002 bright-field TEM image (note strain contrast around QDs), (c) (110) cross-section high-resolution image. Part (c) is taken from *Ledentsov et al. 1996*

QDs embedded in the GaAs matrix is exemplified in Fig. 10.1(a). The dots seem to have rectangular base with their sides being arranged along <110> directions. The averaged lateral size of the QDs in this ensemble is about 12 nm. Close look at the cross-section of QD (Fig. 10.1(b) and (c)) enables to evaluate QD height as around 4 nm. Remarkable feature clearly seen through TEM under two-beam diffraction conditions (Fig. 10.1(b)) is a strong dark contrast surrounding QD image, which is indicative of a high strain inside the dot and in the adjacent GaAs matrix region.

There are no defects associated with the QD in Fig. 10.1 as evidenced by the TEM images taken under different diffraction conditions. However, other growth procedures that result in either coarsening or coalescence of QDs can lead to generation of various defects.

For surface-nucleated InGaAs/GaAs QDs, misfit dislocations are often injected from the sides of the InGaAs islands. Various extended defects are observed in the islands near or at the interface with the substrate. These defects are associated with the intrinsic stacking faults (SF) which lie along {111} planes (*Guha et al. 1990, Jin-Phillipp and Phillipp 1999*). Figure 10.2 demonstrates a large free-standing InGaAs QD formed on GaAs (001) by MBE in Stranski-Krastanow growth mode. Careful inspection of HREM images revealed the Burgers vectors, ***b***, of the dislocations involved. The determined values are $a/3<111>$, which may result from the sum of the Burgers vectors $\boldsymbol{b} = a/2<110>$ of a perfect 60° dislocation and that of a 30° partial dislocation. Additionally, perfect dislocations of Lomer type are often observed at the middle of the QD at InGaAs/GaAs interface. The Lomer dislocations are typical misfit dislocations (MD) in strongly mismatched thin films (*Dregia and Hirth 1991*, *Schwartzman and Sinclair 1991*).

InAs QDs embedded in GaAs can undergo strain relaxation through generation of MD loops clearly seen in week-beam TEM images. Figure 10.3(a) shows a plan-view TEM micrograph of MOCVD grown InAs/GaAs QDs taken under weak-beam conditions and exhibiting a specific white contrast of the images of some QDs (*Zakharov et al. 2001*). The MD loops often engird QD, however, sometimes dislocation loops appear to be growing outwards the QD layer (*Stewart et al. 2003*) that

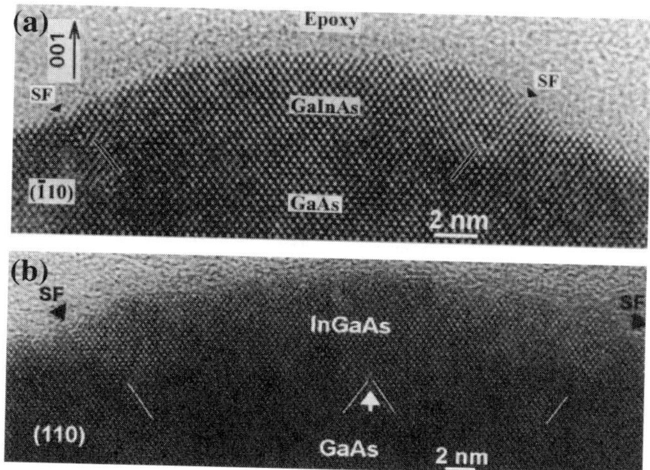

Fig. 10.2 Cross-section HREM images of InGaAs QD formed by deposition of nominal 8 ML of $In_{0.6}Ga_{0.4}As$ onto GaAs (001) (*Jin-Phillipp and Phillipp 1999*): (a) (110) cross-section view showing defect complexes formed on both sides of the QD; extra half-planes of atoms are indicated with dark lines, stacking faults are marked with arrow heads; (b) (110) cross-section view showing a Lomer dislocation (arrowed) in addition to defect complexes on both sides of the QD; extra half-planes of atoms are indicated with white lines

can be seen in Fig. 10.3(b). Formation of the engirdling MD loops and dislocation loop outwards the QDs seem to be due to different strain relief mechanisms.

Besides dislocation loops, V-shaped defects and stacking faults are observed extending from the QD layer into the cap GaAs layers (*Stewart et al. 2003, Sears et al. 2006*). Figure 10.4 represents such defects appeared in TEM plan-view and cross-section images of the sample in which the InAs QDs were deposited onto a thin $In_{0.15}Ga_{0.85}$ As buffer layer, leading to indium enriched islands, and covered with GaAs. The V-shaped defects typically have a 55° incline to the (100)

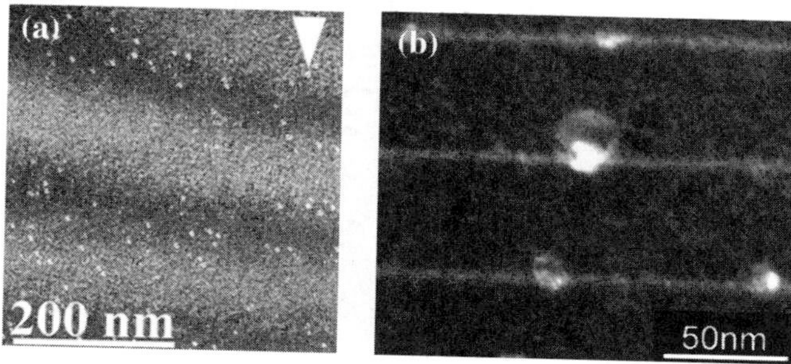

Fig. 10.3 TEM images of dislocated InAs/GaAs QDs: (a) weak-beam plan-view image showing dislocation loops assosiated with some QDs; arrow marks one of such QDs (*Zakharov et al. 2001*); (b) cross-section image of a GaAs sample with few layers of InAs QDs, a QD at the center of the image is associated with a dislocation loop (*Stewart et al. 2003*)

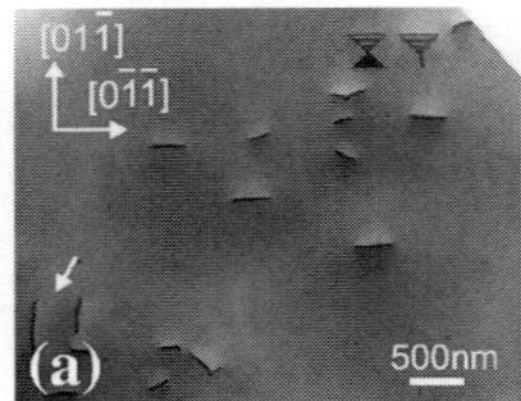

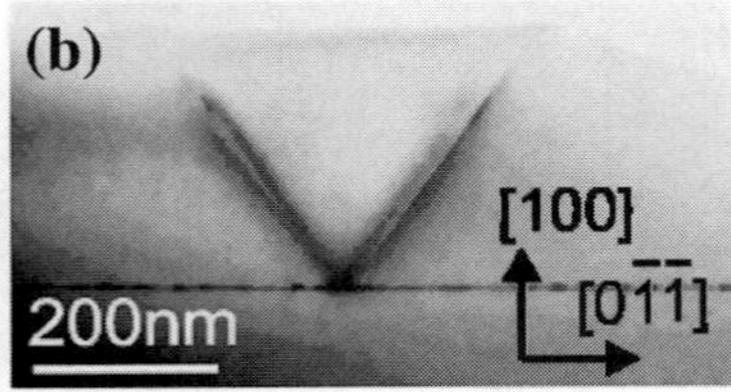

Fig. 10.4 TEM images of V-shaped defects associated with InAs QDs deposited onto a nominally 7 ML $In_{0.15}Ga_{0.85}As$ buffer layer (*Sears et al. 2006*): (a) plan-view image taken under on-axis, [100], bright field conditions; (b) (011) cross-section view

plane, indicating that they are gliding in opposite directions from the buried QD on the adjacent (111) slip planes, viewed edge-on in cross-section TEM images, through the GaAs capping layer to the surface. The V-shaped defects are found (*Sears et al. 2006*) to typically consist of pairs of 60 ° Shockley partials originating from an edge-type dislocation. It is important to note that the dislocations that form V-shaped defect are symmetrically injected near either ends of the growing island (*Guha et al. 1990, Jin-Phillipp and Phillipp 1999*) and traveling in opposite directions from the buried QD to the growth surface.

In InAs QDs embedded in the GaAs matrix dislocations have been revealed inside QD *(Zakharov et al. 2001, Wang et al. 2003)* additionally to those settled close to the dot/matrix interface. An example can be seen in Fig. 10.5 that is a cross-section EM image of a large InAs island in GaAs taken at high resolution. Analysis of the image in Fig. 10.5 using Fourier filtering yielded the dislocations most probably to be 60 °-type with their Burgers vectors $\boldsymbol{b} = a/2\langle 110\rangle$ that is common for GaAs structure. The dislocations often form a dipole configuration (AB and CD in Fig. 10.5). The total Burgers vector of all dislocations in this particular case is equal to zero but it is not always so. In some cases one can observe threading dislocations stretched from large QDs.

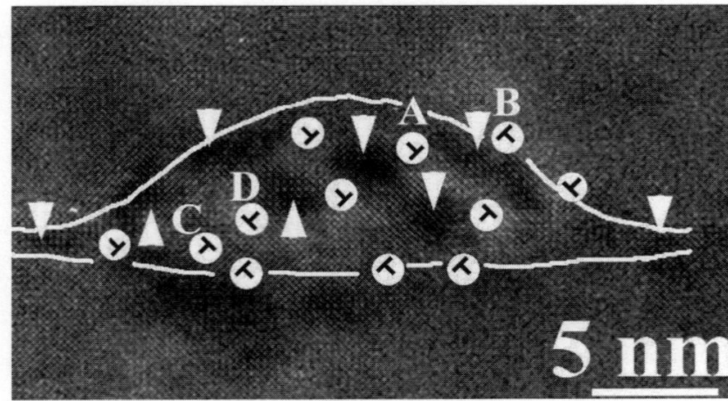

Fig. 10.5 (110) cross-section HREM image of InAs QD buried in GaAs *(Zakharov et al. 2001)*. Dislocations are marked by ⊥, A-B and C-D depict dislocation dipoles, arrows mark inhomogeneities in In distribution across the QD

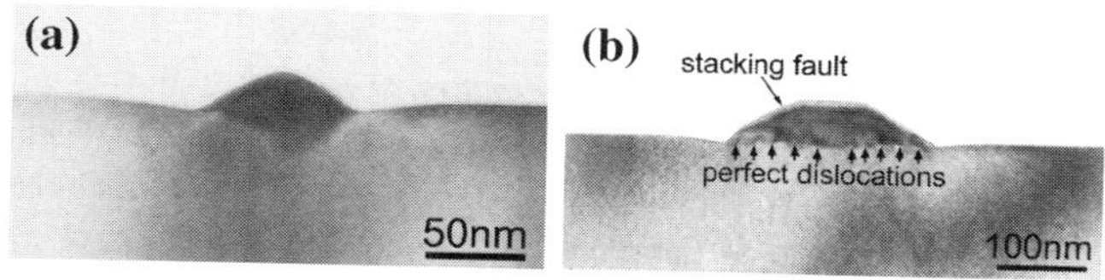

Fig. 10.6 Bright-field cross-sectional TEM images *(Zou et al. 2002)* showing Ge island on Si: (a) coherent; (b) incoherent

Another material system comprehensively studied from the standpoint of QD formation is Ge/Si. Here the stress relaxation is also observed by several characterization techniques. A set of coaxial rings (a tree-ring structure) was discovered (*Merdzhanova et al. 2006*) at the bottom of large Ge islands (also called "superdomes") by combination of selective wet chemical etching and AFM. The tree-ring structure is interpreted as a footprint of dislocation generation, the number of rings being the number of dislocations introduced in the island during its growth. Large surface islands contain perfect dislocations near the dot/matrix interface, as well as stacking faults (*Zou et al. 2002*). An example of such QD is presented in Fig. 10.6. It is important to mention that sessile 90° dislocations have been observed (*LeGoues et al. 1995*) to form directly and not by the reaction of two 60° dislocations. As sessile 90° dislocations are immobile and cannot glide, their generation should be associated with material transfer.

Similar transformation from coherent to dislocated state with increasing QD size have been found in different material systems: three-dimensional islands of $In_xGa_{1-x}N$ self-assembled on GaN(0001) (*Liu et al. 2005*), InP QDs in GaP (*Kurtenbach et al. 1995*), GaAs QDs in GaAsSb (*Toropov et al. 2004*) and others.

10.2.2 Relaxation in Quantum Dots Self-organized in the Bulk

The other self-organization process, we consider, is used to obtain QDs directly in the bulk of an insulating or semiconducting medium. First, a supersaturated medium is obtained by growing the material far from equilibrium conditions, by exposure to a gas containing the impurity atoms, or by direct ion implantation into the matrix. During subsequent annealing, the supersaturation gives rise to nucleation and growth of a second phase, which in some cases is self-organized in compact QDs.

An extensively studied example of such technology is GaAs grown by MBE at low (around 200 °C) temperature (LT-GaAs). The main feature of LT-GaAs is a high arsenic excess that provides unique properties of this material, in particular, a very short life-time of non-equilibrium charge carriers and exceptionally high specific resistance, see review by *Krotkus and Coutaz 2005*. The arsenic excess can be varied in a wide range by the growth temperature and As/Ga flux ratio, see review by *Lavrent'eva et al. 2002*. Under annealing excess As conglomerates to form As QDs distributed over the GaAs film bulk.

Owing to close interplanar distances in GaAs and As, the precipitates occur to be semicoherently built in the matrix (*Claverie and Liliental-Weber 1993*) despite different lattice types (cubic in GaAs and rhombohedral, being commonly described in hexagonal axes, in As). Figure 10.7 presents an example of coherent As QDs embedded in the GaAs matrix. Arsenic inclusions show no relaxation even for considerably large size, i.e. 10÷20 nm in diameter (that is the case of annealing at high temperatures, 700–800 °C).

During the low-temperature MBE the GaAs films can be doped with different impurities, which are incorporated in a non-equilibrium way. The most interesting impurities in respect of the buried As QD system are Mn (*DeBoeck et al. 1996*) and Sb (*Bert et al. 1999*). Both impurities can be obtained by LT MBE in the GaAs matrix in high concentrations.

The LT-GaAs films remain misfit defect-free when few percent of Mn is incorporated into it during low-temperature MBE (*Moreno et al. 2004*). Under subsequent high-temperature annealing, nanoscale MnAs inclusions are self-organized in the GaAs matrix (Fig. 10.8). Being ferromagnetic the MnAs QDs are greatly attractive in the field of magneto-optics and spin-optoelectronics. The LT-GaAs films stuffed with MnAs QDs are shown (*Akinaga et al. 2000*) to be good candidates to achieve giant magnetoresistance effects and to serve as a tool for spin manipulation. The MnAs inclusions have well-defined orientation with respect to the matrix, which is quite similar to the orientation relationship between pure As QDs and GaAs matrix. The MnAs QDs hold a large anisotropic strain experimentally determined by *Moreno et al. 2005*. Nevertheless, both MnAs QDs and adjacent GaAs matrix are observed (*Moreno et al. 2004*) to be dislocation-free even for big QD size, well above 20 nm. An example of a big MnAs magnetic dot is represented in Fig. 10.8.

The LT-GaAs films also remain misfit defect-free when few percent of isovalent Sb impurity is incorporated into it during low-temperature MBE, whether randomly or in the form of δ-layers. Antimony is found (*Bert et al. 1999*) to segregate into the

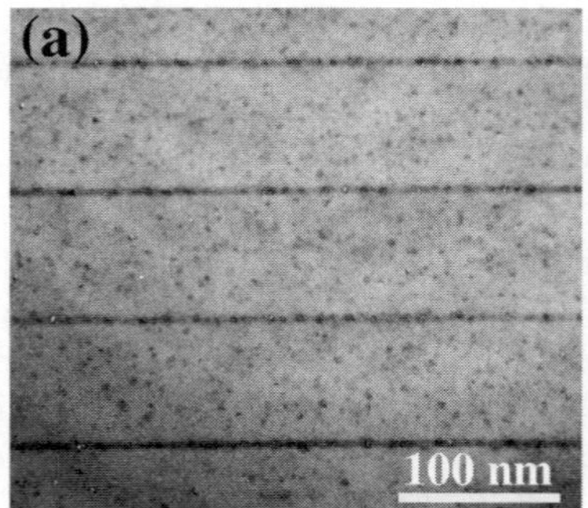

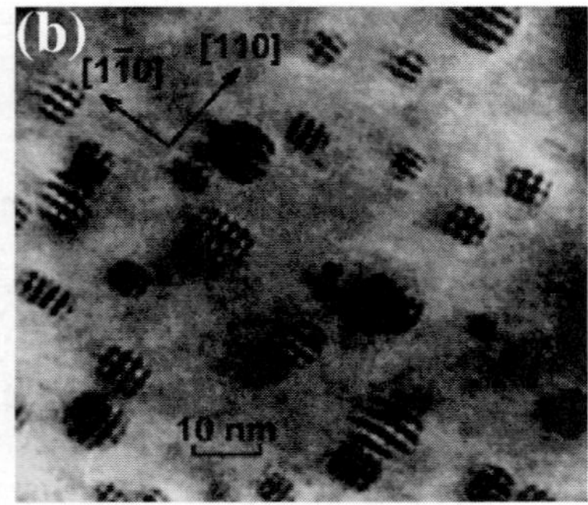

Fig. 10.7 TEM micrographs of As clusters coherently built in GaAs grown at 200 °C and annealed at 600 °C for 15 min: (a) (110) cross-section 002 bright-field image of LT-GaAs δ-doped with In to accumulate As clusters (*Bert et al. 1997*); (b) [001] on-zone image showing Moire fringes indicative of crystalline state of the clusters (*Bert and Chaldyshev 1996*)

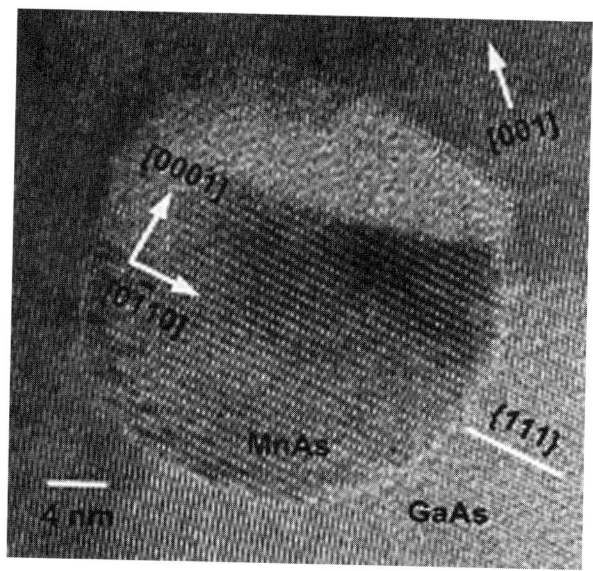

Fig. 10.8 HREM image of a MnAs cluster embedded in GaAs *(Moreno et al. 2004)*. The incident electron beam is along the [110] direction of GaAs

self-organized QDs under post-growth annealing, which appeared to be AsSb alloy. The alloying results in a large QD/matrix lattice mismatch. The AsSb precipitates were found to be similar to the pure As QDs observed in Sb-free LT-GaAs: they were of rounded shape and had rhombohedral crystal structure with c-axis of a cluster being along one of the $\langle 111 \rangle$ directions in the GaAs matrix.

Figure 10.9 shows micrographs of a LT-GaAs sample δ-doped with Sb and annealed at 600 °C. The TEM images of AsSb precipitates demonstrate expanded period of Moire fringes as compared with that for pure As ones (Fig. 10.7). Calcu-

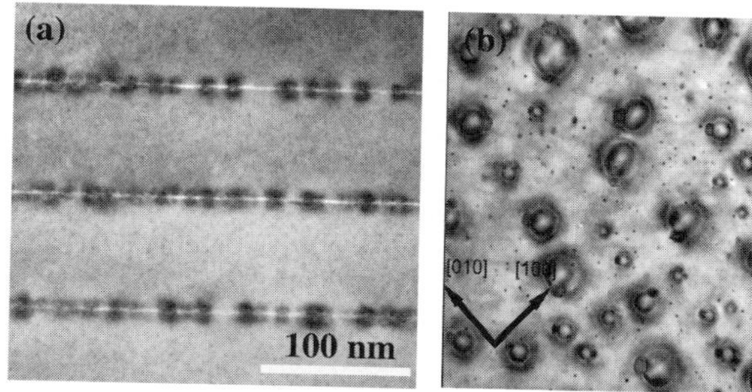

Fig. 10.9 TEM micrographs representing AsSb inclusions in LT-GaAs (a) (110) cross-section view at small (less than 8 nm) clusters located on Sb δ-layers and exhibiting a strong strain contrast *(Bert et al. 1999)*; and (b) (001) plan-view on-zone image of big clusters accosiated with dislocation loops *(Chaldyshev et al. 2002)*

lation of interplanar distances in the inclusions from Moire fringe period as well as direct energy-dispersive X-ray microanalysis both yield Sb concentration in the QDs as approximately 20% to 30%. TEM images of small AsSb QDs taken in bright-field mode are surrounded by a heavy dark contrast exposing strong local strains in the adjacent matrix region. Each precipitate exceeding 7-8 nm in size is seen to be joined with a dislocation loop unlikely to the pure As precipitates in LT-GaAs where no dislocation loops were revealed even for much bigger sizes.

The small QDs seen on Sb δ-layers in cross-section view (Fig. 10.9(a)) exhibit a pronounced strain contrast unlikely to those of pure As QDs (compare with Fig. 10.7(a)) that indicates an increased dot/matrix misfit. When the QD size become as big as more than 8 nm dislocation loops associated with AsSb inclusions appear in the TEM image (Fig. 10.9(b)). The dislocation loops are seen to have a ring shape and lying on (001) plane inconveniently for diamond or zincblende structures. They extend along one of two orthogonal <110> directions from the associated clusters. It should be noted that the dislocation loops are always associated with the AsSb inclusions and have never been observed alone. More details on the AsSb QDs in LT-GaAs are given in Section 10.5.

In general, the above observations demonstrate that the mechanisms of the stress relaxation in QDs are different from that in mismatched thin films in many respects. Even for the surface QDs, misfit defect generation often involves atomic diffusion rather then gliding of dislocations from the free surface. In buried QDs the TEM investigation revealed engirdling misfit dislocation loops and satellite dislocation loops in a close vicinity of individual QDs. Some buried QDs show considerably strong apparent internal elastic strains without clear signature of the misfit dislocations at the dot/matrix interface. These key features are considered in the specific mechanisms developed below for relaxation in buried QDs.

10.3 Mechanics of Quantum Dots

For understanding the stress relaxation processes taking place in heterostructures with quantum dots, it is important to determine the full elastic fields in the dots and surrounding matrix. With relation to elasticity problem the quantum dot is an inclusion with transformation strain or eigenstrain (*Eshelby 1957, 1959, Mura 1987*) due to lattice mismatch between the dot and matrix material. The elastic fields of inclusion depend on its eigenstrain, the elastic properties of both the inclusion and the matrix, the inclusion shape, and the position of inclusion with respect to free surfaces and interfaces. The subject of our investigation is the inclusion embedded into the matrix.

One of the first works dealing with elastic behavior of the ellipsoidal inclusion in an isotropic infinite body was published by *Goodier 1937*. In his case the eigenstrain was a thermal strain. Thermoelastic spherical inclusion near free surface was studied by *Mindlin and Cheng 1950*. The general inclusion problem was extensively developed in the pioneering investigations by *Eshelby 1957, 1959*. Later micromechanics of defects including elasticity of inclusions was advanced by *Mura 1987*. At present, three main methods have been applied to determine the elastic strains and stresses of inclusions, namely: (i) theory of inclusions based on the analytical or numerical solution of elasticity equations ("Eshelby-like" or related approaches), (ii) finite element method (FEM), and (iii) atomistic modeling. Methods (i) to (iii) have their own advantages and disadvantages. FEM is very effective for specific cases but does not provide general solutions. Atomistic models, *e.g.* molecular dynamics simulation, require accurate interatomic potentials and are further restricted to small systems of atoms in comparison with real QD sizes and the surrounding matrix. The advantage of Eshelby-like approach lays in the possibility to provide integral expressions for inclusion elastic fields which then can be effectively analyzed.

In the application of QDs it is relevant to study the elasticity of simple shaped inclusions such as spheroid *(Yoffe 1974, Teodosiu 1982, Bert et al. 2002, Fischer and Böhm 2005, Fischer et al. 2006, Duan et al. 2005, 2006)*, ellipsoid *(Eshelby 1959, Tanaka and Mori 1972, Mura 1987)*, cuboid *(Chiu 1977)*, or pyramid (*Glas 2001, Jogai 2001)*. Note that more complex or even exotic inclusion shapes were already considered, for example finite cylindrical *(Wu and Du 1995a,b)*, superspherical *(Onaka 2001)*, hemispherical *(Wu and Du 1999)*, and doughnut-like inclusions *(Onaka et al. 2002)*. Among simple inclusion shapes spheroid inclusion has the advantage of explicit analytical solutions for elastic fields and energy that allows pertaining in the inclusion related problems on qualitative and in many cases quantitative physical levels. Therefore bellow we are dealing with the case of spheroid inclusions.

10.3.1 Quantum Dot as a Dilatational Nanoinclusion. Technique to Calculate the Elastic Fields of Inclusion

We model the quantum dot as a dilatational nanoinclusion, *i.e.* as selected part of a material undergoing the volume change, *e.g.* expansion or contraction. Figure 10.10 describes Eshelby procedure for dilatational inclusion formation (*Eshelby 1957, 1959*). The procedure involves the following steps: (i) create a cut along the surface surrounding an imaginary inclusion, as shown in Fig. 10.10(a); (ii) take out the internal part of the material and subject it to stress-free strain, as shown in the upper section of Fig. 10.10(b); (iii) elastically deform inclusion and/or the hole and insert the inclusion inside the hole shown in the low section in Fig. 10.10(b); (iv) release all loads applied to the inclusion and/or the hole; as a result strained DI embedded in an elastic matrix is formed as shown in Fig. 10.10(c).

In general the transformation strain or eigenstrain ε^* of the inclusion is described with the help of strain transformation matrix:

$$\boldsymbol{\varepsilon}^* = \begin{pmatrix} \varepsilon_{\mathrm{m}xx} & \varepsilon_{\mathrm{m}xy} & \varepsilon_{\mathrm{m}xz} \\ \varepsilon_{\mathrm{m}yx} & \varepsilon_{\mathrm{m}yy} & \varepsilon_{\mathrm{m}yz} \\ \varepsilon_{\mathrm{m}zx} & \varepsilon_{\mathrm{m}zy} & \varepsilon_{\mathrm{m}zz} \end{pmatrix} \delta(\Omega), \qquad (10.1)$$

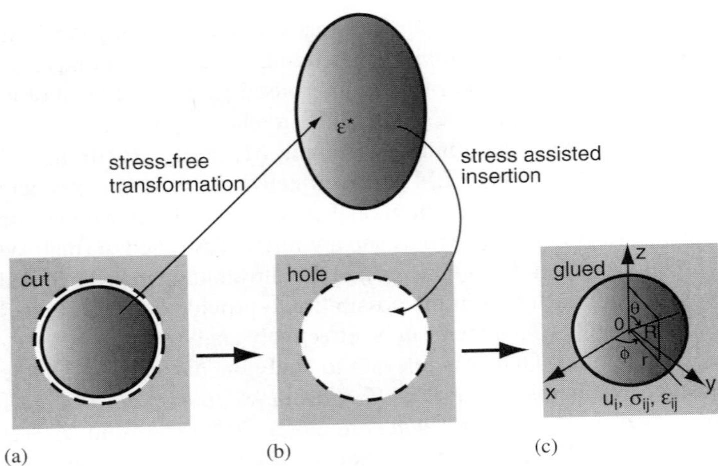

Fig. 10.10 Eshelby procedure of a spherical inclusion formation (see comments in the text)

where the diagonal terms correspond to dilatational mismatch of crystal lattices of DI and matrix and another terms mean the shear mismatch of the crystal lattices, $\delta(\Omega) = \begin{cases} 1, \mathbf{r} \in \Omega \\ 0, \mathbf{r} \notin \Omega \end{cases}$, Ω – is the area occupied by inclusion, (x, y, z) is Cartezian coordinate system.

For crystal lattice mismatch ε_{m} between the materials of inclusion and matrix the diagonal terms in Eq (10.1) acquire the mining the misfit parameters:

$$\varepsilon_{\mathrm{m}} = \frac{a_{\mathrm{inc}} - a_{\mathrm{mat}}}{a_{\mathrm{inc}}}, \tag{10.2}$$

where a_{inc} is crystal lattice parameter of inclusion, a_{mat} is the crystal lattice parameter of the surrounding matrix. Note that the sign convention used in Eq. (10.2) is opposite to those applied when defining the misfit parameter for heteroepitaxial thin films on thick substrates (see, for example, *Beanland et al. 1996*).

In order to determine the elastic fields of inclusion, one can use the relation connecting the total displacements and eigenstrains in the form given by *Mura 1987*:

$$u_m(r') = -i \int \int_{-\infty}^{\infty} \int \xi_l C_{jikl} L_{mk} \hat{\varepsilon}^*_{ij} \exp(i\, \boldsymbol{\xi} \cdot \mathbf{r}) d\xi_x d\xi_y d\xi_z, \tag{10.3}$$

where C_{jikl} are the elastic constants;L_{mk} and $\hat{\varepsilon}^*_{ij}$ are Fourier transforms of Green's function G_{km} and eigenstrain ε^*_{ij}, correspondently, $\boldsymbol{\xi} \cdot \mathbf{r} = \xi_x x + \xi_y y + \xi_z z$. However for analytical calculations formula (10.3) shows restricted abilities. This is because of the facts that Green function and its Fourier transform have analytical representation for a restricted number of elasticity problems. In addition the possible complete shape of the inclusion, which is reflected in eigenstrain and its Fourier transform, males it difficult to proceed the analytical calculations.

For an isotropic infinite medium the integrants in Eq. (10.3) are:

$$C_{jikl} = \frac{2G\nu}{1-2\nu}\delta_{ji}\delta_{kl} + G(\delta_{ik}\delta_{jl} + \delta_{il}\delta_{jk}), \tag{10.4a}$$

$$L_{mk} = \frac{1}{(2\pi)^{3/2}}\frac{2(1-\nu)\xi^2\delta_{mk} - \xi_m\xi_k}{2(1-\nu)G\xi^4}, \tag{10.4b}$$

where G is shear modulus, ν is Poisson's ratio, δ_{km} is Kroneker symbol, $\xi^2 = \xi_x^2 + \xi_y^2 + \xi_z^2$. Eqs. (10.3) and (10.4) can be used to obtain the displacements and the related elastic fields for inclusions and various defects such as dislocations and disclinations in infinite isotropic media.

10.3.2 Mechanical Stresses of a Dilatational Spheroid in Infinite Isotropic Media

For a spheroid inclusion in isotropic infinite media, elastic fields can be determined in explicit form for arbitrary dilatational eigenstrains $\varepsilon_{\mathrm{m}xx} \neq \varepsilon_{\mathrm{m}yy} \neq \varepsilon_{\mathrm{m}zz}$, $\varepsilon_{\mathrm{m}ij} = 0$, $i \neq j$. Stresses are written as (*Mura 1987, Kolesnikova et al. 2007*):

$$\sigma_{xx}^{(in)} = -\frac{2G}{15(1-\nu)}\left[8\,\varepsilon_{\mathrm{m}xx} + \varepsilon_{\mathrm{m}yy}(5\nu+1) + \varepsilon_{\mathrm{m}zz}(5\nu+1)\right], \tag{10.5a}$$

$$\sigma_{ij}^{(in)} = 0, \quad (i \neq j), \tag{10.5b}$$

$$\begin{aligned}
\sigma_{xx}^{(out)} =& \frac{G}{15(1-\nu)\tilde{R}^9}\cdot \\
&\left[\varepsilon_{mxx}\left(24\tilde{x}^4 - 40\tilde{x}^6 - 72\tilde{x}^2\tilde{y}^2 + 9\tilde{y}^4 + 45\tilde{x}^2\tilde{y}^4 + 5\tilde{y}^6 - 72\tilde{x}^2\tilde{z}^2 + \right.\right.\\
&+ 18\tilde{y}^2\tilde{z}^2 + 90\tilde{x}^2\tilde{y}^2\tilde{z}^2 + 15\tilde{y}^4\tilde{z}^2 + 9\tilde{z}^4 + 45\tilde{x}^2\tilde{z}^4 + 15\tilde{y}^2\tilde{z}^4 + \left.5\tilde{z}^6\right) + \\
&+ \varepsilon_{myy}\left(-12\tilde{x}^4 + 10\tilde{x}^6 + 81\tilde{x}^2\tilde{y}^2 - 45\tilde{x}^4\tilde{y}^2 - 12\tilde{y}^4 - 45\tilde{x}^2\tilde{y}^4 + 10\tilde{y}^6 - \right.\\
&- 9\tilde{x}^2\tilde{z}^2 + 15\tilde{x}^4\tilde{z}^2 - 9\tilde{y}^2\tilde{z}^2 - 45\tilde{x}^2\tilde{y}^2\tilde{z}^2 + 15\tilde{y}^4\tilde{z}^2 + 3\tilde{z}^4 - 5\tilde{z}^6 - 10\nu\,\tilde{x}^6 - \\
&- 30\nu\,\tilde{x}^4\tilde{y}^2 - 30\nu\,\tilde{x}^2\tilde{y}^4 - 10\nu\,\tilde{y}^6 + 30\nu\,\tilde{x}^2\tilde{z}^4 + 30\nu\,\tilde{y}^2\tilde{z}^4 + \left.20\nu\,\tilde{z}^6\right) + \\
&+ \varepsilon_{mzz}\left(-12\tilde{x}^4 + 10\tilde{x}^6 + 81\tilde{x}^2\tilde{z}^2 - 45\tilde{x}^4\tilde{z}^2 - 12\tilde{z}^4 - 45\tilde{x}^2\tilde{z}^4 + 10\tilde{z}^6 - \right.\\
&- 9\tilde{x}^2\tilde{y}^2 + 15\tilde{x}^4\tilde{y}^2 - 9\tilde{y}^2\tilde{z}^2 - 45\tilde{x}^2\tilde{y}^2\tilde{z}^2 + 15\tilde{z}^4\tilde{y}^2 + 3\tilde{y}^4 - 5\tilde{y}^6 - 10\nu\,\tilde{x}^6 \\
&\left.\left. -30\nu\,\tilde{x}^4\tilde{z}^2 - 30\nu\,\tilde{x}^2\tilde{z}^4 - 10\nu\,\tilde{z}^6 + 30\nu\,\tilde{x}^2\tilde{y}^4 + 30\nu\,\tilde{z}^2\tilde{y}^4 + 20\nu\,\tilde{y}^6\right)\right],
\end{aligned} \tag{10.5c}$$

$$\sigma_{xy}^{(out)} = \frac{G\tilde{x}\tilde{y}}{(1-\nu)\tilde{R}^9}\Big[\varepsilon_{mxx}\left(4\tilde{x}^2 - 4\tilde{x}^4 - 3\tilde{y}^2 - 3\tilde{x}^2\tilde{y}^2 + \tilde{y}^4 - 3\tilde{z}^2 - 3\tilde{x}^2\tilde{z}^2 + 2\tilde{y}^2\tilde{z}^2 + \tilde{z}^4\right)$$
$$+\varepsilon_{myy}\left(4\tilde{y}^2 - 4\tilde{y}^4 - 3\tilde{x}^2 - 3\tilde{x}^2\tilde{y}^2 + \tilde{x}^4 - 3\tilde{z}^2 - 3\tilde{y}^2\tilde{z}^2 + 2\tilde{x}^2\tilde{z}^2 + \tilde{z}^4\right)$$
$$+\varepsilon_{mzz}\left(-\tilde{x}^2 + \tilde{x}^4 - \tilde{y}^2 + 2\tilde{x}^2\tilde{y}^2 + \tilde{y}^4 - 6\tilde{z}^2 - 3\tilde{x}^2\tilde{z}^2 - 3\tilde{y}^2\tilde{z}^2 - 4\tilde{z}^4\right.$$
$$\left.-2\nu\tilde{x}^4 - 4\nu\tilde{x}^2\tilde{y}^2 - 2\nu\tilde{y}^4 - 4\nu\tilde{x}^2\tilde{z}^2 - 4\nu\tilde{y}^2\tilde{z}^2 - 2\nu\tilde{z}^4\right)\Big]. \quad (10.5d)$$

where Cartesian coordinates (x, y, z) are connected with center of spheroid, $\tilde{R} = \frac{R}{R_{sph}}$, $\tilde{x} = \frac{x}{R_{sph}}$, $\tilde{y} = \frac{y}{R_{sph}}$, $\tilde{z} = \frac{z}{R_{sph}}$, $R^2 = r^2 + 2^2$, $r^2 = x^2 + y^2$, R_{sph} is a radius of the spheroid. Upper index (in), (out) notes internal or external area of spheroid, correspondently. Stress components $\sigma_{yy}^{(in),(out)}$, $\sigma_{zz}^{(in),(out)}$, $\sigma_{zx}^{(out)}$ and $\sigma_{yz}^{(out)}$ are determined by cyclic permutation of indexes and coordinates (x, y, z) in Eqs. (10.5a, 10.5c, 10.5d).

Here we illustrate the use of the general stress relations for two practical cases: (i) uniaxial dilatation $\varepsilon_{mzz} = \varepsilon_m$ and (ii) equiaxial dilatation $\varepsilon_{mxx} = \varepsilon_{myy} = \varepsilon_{mzz} = \varepsilon_m$. For case (i) stresses inside and outside dilatational spheroid were first derived by *Bert et al. 2002*. Stress maps for different components are shown in Fig. 10.11.

For case (ii) stresses have the simplest representation in spherical coordinates (R, θ, φ) (Fig. 10.10(c)):

$$\sigma_{RR}^{(in)} = \sigma_{\theta\theta}^{(in)} = \sigma_{\varphi\varphi}^{(in)} = -\frac{4}{3}\frac{(1+\nu)G\varepsilon_m}{(1-\nu)}, \quad (10.6a)$$

$$\sigma_{RR}^{(out)} = -\frac{4}{3}\frac{(1+\nu)G\varepsilon_m}{(1-\nu)}\cdot\frac{R_{sph}^3}{R^3}, \quad (10.6b)$$

$$\sigma_{\theta\theta}^{(out)} = \sigma_{\varphi\varphi}^{(out)} = \frac{2}{3}\frac{(1+\nu)G\varepsilon_m}{(1-\nu)}\cdot\frac{R_{sph}^3}{R^3}, \quad (10.6c)$$

where the designations are the same as in Eqs. (10.5).These expressions are well-known in physics and mechanics of dilatational inclusions (see, for example, *Teodosiu* 1982, Mura 1987).

In a far field ($R >> R_{sph}$) the internal structure of the dilatational inclusion becomes irrelevant for its elastic properties. The inclusion can be viewed as a *point stressor* with the strength $S = \varepsilon_m V$, where V is the inclusion volume, *i.e.* $V = \frac{4}{3}\pi R_{sph}^3$ for spheroid. The stresses of a point dilatational stressor $\sigma_{ij}^{PS}(\mathrm{r})$ in infinite isotropic elastic media are given by the same expressions of Eq. (10.6), where the multiplier ε_m should be replaced by $\frac{3}{4\pi R_{sph}^3}S$. As a result, in Cartesian coordinate system (x_1, x_2, x_3), stresses of the point dilatational source can be expressed in a simple form:

$$\sigma_{\mathrm{ij}}^{\mathrm{PS}} = S\frac{G(1+\nu)}{2\pi(1-\nu)}\frac{\partial^2}{\partial x_{\mathrm{i}}\partial x_{\mathrm{j}}}\left(-\frac{1}{R}\right), \; i, j = 1, 2, 3, \quad (10.7)$$

where $R = \sqrt{x_1^2 + x_2^2 + x_3^2}$.

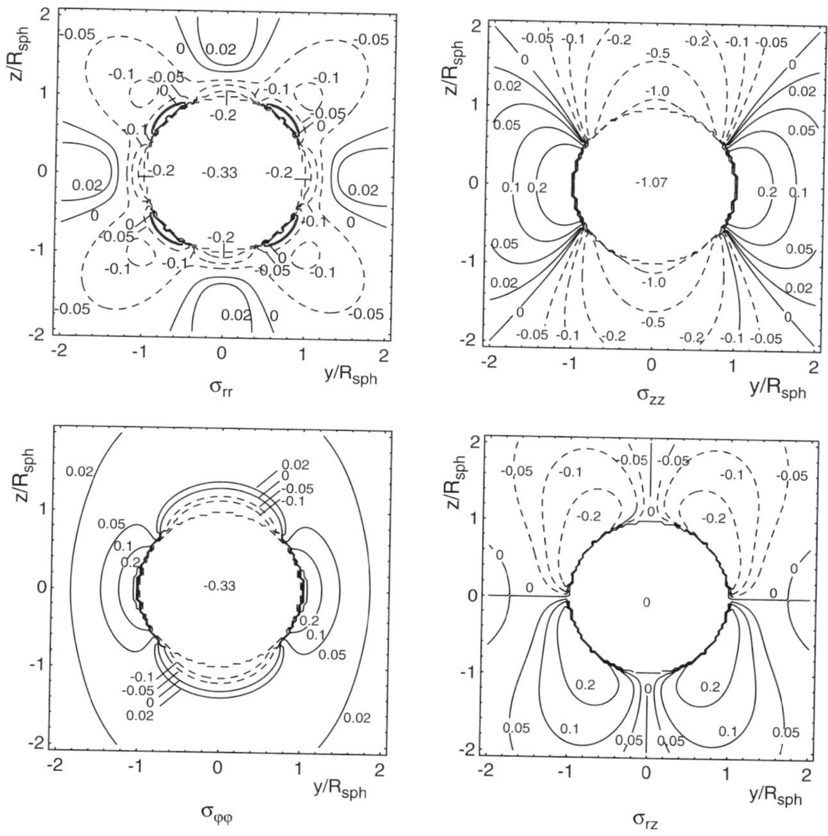

Fig. 10.11 Distribution of the elastic stress components at a spherical inclusion with uniaxial dilatation $\varepsilon^*_{zz} = \varepsilon_m$. Stresses are in the $\frac{G\varepsilon_m}{(1-\nu)}$ units, where G is the shear modulus, ν is Poisson ratio, R_{sph} is the radius of spheroid, and $\nu = 0.3$

10.3.3 Inclusions Near Free Surfaces and Interfaces

Presence of the free surface in the finite size elastic bodies or interfaces in multi component elastic bodies modifies inclusion elastic fields. There exists a number of papers dealing with the elastic behavior of the different inclusions in media with internal and external boundaries *(Mindlin 1950, Seo and Mura 1979, Hu 1989, Yu and Sandy 1992, Tsuchida et al. 2000, Glas 2001, Jogai 2001, Romanov et al. 2001, Kolesnikova and Romanov 2004a)*. The elastic field modifications (or distortions) are known to be responsible for a number of physical effects related to inclusions. The configurational forces of the inclusion interaction with free surface or interfaces appear and the interaction between crystal lattice defects and inclusions changes. For example, in infinite media dilatational inclusions produce no pressure (the trace of stress tensor σ_{ij} given by Eqs. (10.6) or (10.7) vanishes) and therefore do not interact with dilatational point defects. Being placed in subsurface layer dilatational inclusions become the sources of non-zero pressure and therefore will interact with point defects and among themselves. The other physical properties, which depend on pressure, *e.g.* the position of the conduction band edge and the magnitude of the band gap in semiconductors *(Romanov et al. 2005)*, will be also affected by dilatational inclusions in subsurface layers.

Various techniques can be explored for the solution of elasticity boundary value problems for dilatational inclusions. In case the solution $\sigma_{\mathrm{ij}}^{\mathrm{PSboundary}}$ for point dilatational stressor satisfying the boundary conditions is known, it can be integrated over finite size inclusion interior to provide the resulting stress $\sigma_{\mathrm{ij}}^{\mathrm{DI}}$:

$$\sigma_{\mathrm{ij}}^{\mathrm{DI}}(\mathrm{r}) = \int\limits_{V^{\mathrm{DI}}} \frac{1}{V^{\mathrm{PS}}} \sigma_{\mathrm{ij}}^{\mathrm{PSboundary}}(\mathbf{r} - \mathbf{r}')dV', \tag{10.8}$$

where V^{PS} is the effective volume of the point dilatational stressor and the integration is performed over the volume V^{DI} of the finite size dilatational inclusion. However, the boundary value problem solutions for point stressor are known for limited number cases, *e.g.* near planar free surface of isotropic half-space (see below).

Other technique, which was successfully used for the determination of the inclusion stresses under the influence of boundary conditions, is know as virtual defect method *(Kolesnikova and Romanov 2004a)*.

In the framework of this method the elastic field p_{ij} (displacements, distortions, strains, and stresses) of the real defect placed in the medium with free surfaces and interfaces is presented as:

$$p_{ij} = {}^{\infty}p_{ij} + {}^{i}p_{ij}, \tag{10.9}$$

where ${}^{\infty}p_{ij}$ is the field of the real defect in infinite medium without interfaces and ${}^{i}p_{ij}$ is the additional field, due to ensembles of virtual defects. Virtual defects are distributed continuously outside the medium, in which their elastic fields act. The boundary conditions at free surfaces and interfaces $p_{ij}\big|_S = \psi$ are then re-written with respect to Eqs. (10.9) and become the integral equations for unknown distribution functions. The problem exists how to choose the appropriate defects as virtual ones. For cylindrical or spherical symmetry problems, which are related to spheroid dilatational inclusions, the external and internal boundary surfaces can be plane, cylindrical, spherical and the dislocation-disclination loops can be taken as virtual defects *(Kolesnikova and Romanov 1986, 1987, 2004a)*.

To illustrate the use of the virtual defect technique we consider here the solution of the boundary value problem for the spheroid dilatational inclusion placed in the plate with thickness t as it shown in Fig. 10.12(a).

In the chosen geometry the elastic stresses ${}^{\infty}\sigma_{ij}$ of the inclusion in the infinite media (see Eq. (10.6)) can be rewritten in the cylindrical coordinate system (r, φ, z) (Fig. 10.12(a)):

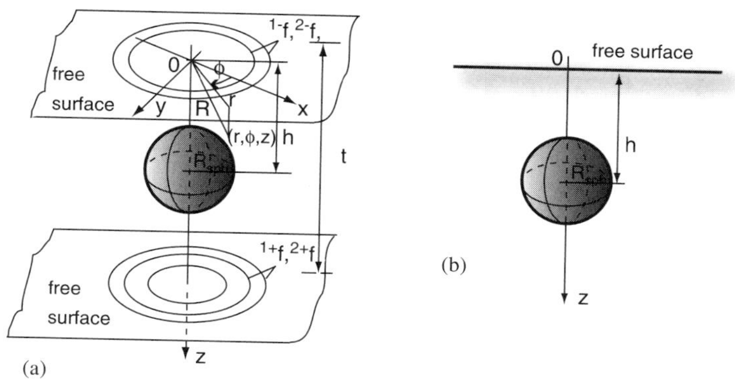

Fig. 10.12 Spherical inclusion in a plate (a) and in a half-space (b). Distributions of virtual loop defects are shown on the plate surfaces

$$^{\infty}\sigma_{rr}^{(in)} = {}^{\infty}\sigma_{\varphi\varphi}^{(in)} = {}^{\infty}\sigma_{zz}^{(in)} = -\frac{4G\varepsilon_m(1+\nu)}{3(1-\nu)}, \tag{10.10a}$$

$$^{\infty}\sigma_{rz}^{(in)} = {}^{\infty}\sigma_{r\varphi}^{(in)} = {}^{\infty}\sigma_{z\varphi}^{(in)} = 0, \tag{10.10b}$$

$$^{\infty}\sigma_{rr}^{(out)} = \frac{2G\varepsilon_m(1+\nu)}{3(1-\nu)}\left(-\frac{2}{(\tilde{r}^2+(\tilde{z}-\tilde{h})^2)^{3/2}} + \frac{3(\tilde{z}-\tilde{h})^2}{(\tilde{r}^2+(\tilde{z}-\tilde{h})^2)^{5/2}}\right), \tag{10.10c}$$

$$^{\infty}\sigma_{\varphi\varphi}^{(out)} = \frac{2G\varepsilon_m(1+\nu)}{3(1-\nu)} \cdot \frac{1}{(\tilde{r}^2+(\tilde{z}-\tilde{h})^2)^{3/2}}, \tag{10.10d}$$

$$^{\infty}\sigma_{zz}^{(out)} = \frac{2G\varepsilon_m(1+\nu)}{3(1-\nu)}\left(\frac{1}{(\tilde{r}^2+(\tilde{z}-\tilde{h})^2)^{3/2}} - \frac{3(\tilde{z}-\tilde{h})^2}{(\tilde{r}^2+(\tilde{z}-\tilde{h})^2)^{5/2}}\right), \tag{10.10e}$$

$$^{\infty}\sigma_{rz}^{(out)} = \frac{2G\varepsilon_m(1+\nu)}{3(1-\nu)}\left(\frac{-3\tilde{r}(\tilde{z}-\tilde{h})}{(\tilde{r}^2+(\tilde{z}-\tilde{h})^2)^{5/2}}\right), \tag{10.10f}$$

$$^{\infty}\sigma_{r\varphi}^{(out)} = {}^{\infty}\sigma_{z\varphi}^{(out)} = 0. \tag{10.10g}$$

On the free surface of the plate the following boundary conditions for resulting stress field $\sigma_{ij} = {}^{\infty}\sigma_{ij} + {}^{i}\sigma_{ij}$ must held:

$$\sigma_{jz}\Big|_{\left\{\begin{smallmatrix} z=0 \\ z=t \end{smallmatrix}\right\}} = 0, \quad j = r, \varphi, z \tag{10.11}$$

To satisfy the boundary conditions we assume that the additional field ${}^{i}\sigma_{ij}$ is generated by the distributions of circular prismatic dislocation loops (ensemble 1) and radial disclination loops (ensemble 2), placed on the free surfaces of the plate, see Fig. 10.12. These virtual defects possess σ_{zz} and σ_{rz} stress components and vanishing $\sigma_{\varphi z}$ component. Rewriting Eqs. (10.11) with the help of the fields of the virtual loops we obtain four integral equations with respect to four unknown functions of loop distributions with radial coordinate.

One can find these distributions and therefore derive the elastic fields of virtual dislocation and disclination loops. Finally, the stresses of dilatational inclusion have the following form *(Kolesnikova and Romanov 2004a)*:

$$\begin{aligned}
\sigma_{rr} = {}^{\infty}\sigma_{rr} + \frac{2G\,\varepsilon_m(1+\nu)R_{sph}^3}{3(1-\nu)}\Bigg[&\frac{1-2\nu}{r}\int_0^\infty J_1(r\beta)\left({}^{1-}\tilde{H}\,E_2 + {}^{1+}\tilde{H}\,\frac{E_0}{E_2}\right)d\beta + \\
&+\int_0^\infty J_0(r\beta)\left({}^{1-}\tilde{H}\,z\beta E_2 + {}^{1+}\tilde{H}\,(t-z)\beta\frac{E_0}{E_2}\right)\beta\, d\beta \\
&-\int_0^\infty J_0(r\beta)\left({}^{1-}\tilde{H}\,E_2 + {}^{1+}\tilde{H}\,\frac{E_0}{E_2}\right)\beta\, d\beta - \\
&-\frac{1}{r}\int_0^\infty J_1(r\beta)\left({}^{1-}\tilde{H}\,z\beta E_2 + {}^{1+}\tilde{H}(t-z)\beta\,\frac{E_0}{E_2}\right)d\beta \\
&+\frac{2(\nu-1)}{r}\int_0^\infty J_1(r\beta)\left({}^{2-}\tilde{H}\,E_2 - {}^{2+}\tilde{H}\,\frac{E_0}{E_2}\right)d\beta - \\
&-\int_0^\infty J_0(r\beta)\left({}^{2-}\tilde{H}\,z\beta E_2 - {}^{2+}\tilde{H}(t-z)\beta\,\frac{E_0}{E_2}\right)\beta\, d\beta \\
&+2\int_0^\infty J_0(r\beta)\left({}^{2-}\tilde{H}\,E_2 - {}^{2+}\tilde{H}\,\frac{E_0}{E_2}\right)\beta\, d\beta + \\
&+\frac{1}{r}\int_0^\infty J_1(r\beta)\left({}^{2-}\tilde{H}\,z\beta E_2 - {}^{2+}\tilde{H}(t-z)\beta\,\frac{E_0}{E_2}\right)d\beta\Bigg],
\end{aligned} \tag{10.12a}$$

$$\begin{aligned}
\sigma_{\varphi\varphi} = {}^{\infty}\sigma_{\varphi\varphi} + \frac{2G\,\varepsilon_m(1+\nu)R_{sph}^3}{3(1-\nu)}\Bigg[&\frac{2\nu-1}{r}\int_0^\infty J_1(r\beta)\left({}^{1-}\tilde{H}\,E_2 + {}^{1+}\tilde{H}\,\frac{E_0}{E_2}\right)d\beta - \\
&-2\nu\int_0^\infty J_0(r\beta)\left({}^{1-}\tilde{H}\,E_2 + {}^{1+}\tilde{H}\,\frac{E_0}{E_2}\right)\beta\, d\beta \\
&+\frac{1}{r}\int_0^\infty J_1(r\beta)\left({}^{1-}\tilde{H}\,z\beta E_2 + {}^{1+}\tilde{H}(t-z)\beta\,\frac{E_0}{E_2}\right)d\beta \\
&+\frac{2(1-\nu)}{r}\int_0^\infty J_1(r\beta)\left({}^{2-}\tilde{H}\,E_2 - {}^{2+}\tilde{H}\,\frac{E_0}{E_2}\right)d\beta
\end{aligned}$$

$$+ 2\nu \int_0^\infty J_0(r\beta)({}^{2-}\tilde{H}\, E_2 - {}^{2+}\tilde{H}\, \frac{E_0}{E_2})\beta\, d\beta -$$

$$\left. - \frac{1}{r} \int_0^\infty J_1(r\beta)({}^{2-}\tilde{H}\, z\beta\, E_2 - {}^{2+}\tilde{H}(t-z)\beta\, \frac{E_0}{E_2})\, d\beta \right], \tag{10.12b}$$

$$\sigma_{zz} = {}^{\infty}\sigma_{zz} - \frac{2G\,\varepsilon_m(1+\nu)R_{sph}^3}{3(1-\nu)} \left[\int_0^\infty J_0(r\beta)({}^{1-}\tilde{H}\, E_2 + {}^{1+}\tilde{H}\, \frac{E_0}{E_2})\beta\, d\beta + \right.$$

$$+ \int_0^\infty J_0(r\beta)({}^{1-}\tilde{H}\, z\beta\, E_2 + {}^{1+}\tilde{H}\, (t-z)\beta \frac{E_0}{E_2})\beta\, d\beta$$

$$\left. - \int_0^\infty J_0(r\beta)({}^{2-}\tilde{H}\, z\beta\, E_2 - {}^{2+}\tilde{H}\, (t-z)\beta \frac{E_0}{E_2})\beta\, d\beta \right], \tag{10.12c}$$

$$\sigma_{rz} = {}^{\infty}\sigma_{rz} - \frac{2G\,\varepsilon_m(1+\nu)R_{sph}^3}{3(1-\nu)} \left[\int_0^\infty J_1(r\beta)({}^{1-}\tilde{H}\, z\beta\, E_2 - {}^{1+}\tilde{H}(t-z)\beta\, \frac{E_0}{E_2})\beta\, d\beta + \right.$$

$$- \int_0^\infty J_1(r\beta)({}^{2-}\tilde{H}\, z\beta E_2 + {}^{2+}\tilde{H}\, (t-z)\beta \frac{E_0}{E_2})\beta\, d\beta$$

$$\left. + \int_0^\infty J_1(r\beta)({}^{2-}\tilde{H}\, E_2 + {}^{2+}\tilde{H}\, \frac{E_0}{E_2})\beta\, d\beta \right], \tag{10.12d}$$

$$\sigma_{z\varphi} = \sigma_{r\varphi} = 0, \qquad 0 \le z \le t, \quad R_{sph} \le h \le t - R_{sph}. \tag{10.12e}$$

where $E_0 = \exp[-t\beta]$, $E_1 = \exp[-h\beta]$, and $E_2 = \exp[-z\beta]$, $J_0(r\beta)$ and $J_1(r\beta)$ are Bessel functions and ${}^{1-}\tilde{H}$, ${}^{1+}\tilde{H}$, ${}^{2-}\tilde{H}$, ${}^{2+}\tilde{H}$ are the normalized Hankel-Bessel transforms of virtual defect distributions:

$${}^{1-}\tilde{H} = \frac{E_1\beta[-1 + E_0^2\, E_1^{-2}(1+2t\beta) - E_0^4\, E_1^{-2} + E_0^2(1-2t\beta)]}{(1-E_0^2)^2 - 4E_0^2\, t^2\beta^2}, \tag{10.13a}$$

$${}^{1+}\tilde{H} = \frac{E_0^{-1}\, E_1\beta[-E_0^4 - E_0^2\, E_1^{-2} + E_0^4\, E_1^{-2}(1-2t\beta) + E_0^2(1+2t\beta)]}{(1-E_0^2)^2 - 4E_0^2\, t^2\beta^2}, \tag{10.13b}$$

$${}^{2-}\tilde{H} = \frac{E_1\beta[1 + E_0^2\, E_1^{-2}(1-2t\beta) - E_0^4\, E_1^{-2} - E_0^2\,(1+2t\beta)]}{(1-E_0^2\,)^2 - 4E_0^2\, t^2\beta^2}, \tag{10.13c}$$

$${}^{2+}\tilde{H} = \frac{E_0^{-1}\, E_1\beta[E_0^4 - E_0^2\, E_1^{-2} + E_0^4\, E_1^{-2}(1+2t\beta) - E_0^2\,(1-2t\beta)]}{(1-E_0^2\,)^2 - 4E_0^2\, t^2\beta^2}. \tag{10.13d}$$

The stresses Eqs. (12) satisfy boundary conditions at the plate surfaces Eqs. (10.11) and elasticity equilibrium equations.

In the limit case when the plate thickness $t \to \infty$ we find the stresses of the inclusion in a half-space (Fig. 10.12(b)):

$$\sigma_{rr} = {}^{\infty}\sigma_{rr} + \frac{2G\varepsilon_m(1+\nu)}{3(1-\nu)} \left[\frac{2(2\nu-3)}{[\tilde{r}^2+(\tilde{z}+\tilde{h})^2]^{3/2}} - \frac{3(\tilde{z}+\tilde{h})^2 - 12(\tilde{z}+\tilde{h})(3\tilde{z}+\tilde{h})}{[\tilde{r}^2+(\tilde{z}+\tilde{h})^2]^{5/2}} - \frac{30\tilde{z}(\tilde{z}+\tilde{h})^3}{[\tilde{r}^2+(\tilde{z}+\tilde{h})^2]^{7/2}} \right], \tag{10.14a}$$

$$\sigma_{\varphi\varphi} = {}^{\infty}\sigma_{\varphi\varphi} + \frac{2G\varepsilon_m(1+\nu)}{3(1-\nu)} \left[\frac{(3-8\nu)}{[\tilde{r}^2+(\tilde{z}+\tilde{h})^2]^{3/2}} - \frac{6(\tilde{z}+\tilde{h})[(1-2\nu)(\tilde{z}+\tilde{h})-\tilde{h}]}{[\tilde{r}^2+(\tilde{z}+\tilde{h})^2]^{5/2}} \right] \tag{10.14b}$$

$$\sigma_{zz} = {}^{\infty}\sigma_{zz} - \frac{2G\varepsilon_m(1+\nu)}{3(1-\nu)} \left[\frac{1}{[\tilde{r}^2+(\tilde{z}+\tilde{h})^2]^{3/2}} - \frac{3(\tilde{z}+\tilde{h})^2 - 18\tilde{z}(\tilde{z}+\tilde{h})}{[\tilde{r}^2+(\tilde{z}+\tilde{h})^2]^{5/2}} - \frac{30\tilde{z}(\tilde{z}+\tilde{h})^3}{[\tilde{r}^2+(\tilde{z}+\tilde{h})^2]^{7/2}} \right], \tag{10.14c}$$

$$\sigma_{rz} = {}^{\infty}\sigma_{rz} - \frac{2G\varepsilon_m(1+\nu)}{3(1-\nu)} \left[\frac{3\tilde{r}(\tilde{z}+\tilde{h})}{[\tilde{r}^2+(\tilde{z}+\tilde{h})^2]^{5/2}} - \frac{6\tilde{r}\tilde{z}[4(\tilde{z}+\tilde{h})^2 - r^2]}{[\tilde{r}^2+(\tilde{z}+\tilde{h})^2]^{7/2}} \right], \tag{10.14d}$$

$$\sigma_{z\varphi} = \sigma_{r\varphi} = 0, \tag{10.14e}$$

$$Tr\boldsymbol{\sigma} = \frac{8G\varepsilon_m(1+\nu)^2}{3(1-\nu)} \cdot \frac{2(\tilde{z}+\tilde{h})^2 - \tilde{r}^2}{[\tilde{r}^2+(\tilde{z}+\tilde{h})^2]^{5/2}}, \qquad 0 \le z, \quad R_{sph} \le h \tag{10.14f}$$

Here we use the same designations as those in Eqs. (10.5).

The result shown Eqs. (10.14) can be also derived from the general results obtained by *Mindlin 1950* for spherical source of thermal stresses.

Finally, the stresses of a point dilatational stressor placed at the distance h from a free surface can be deduced from Eq. (10.14) and written in Cartesian coordinate system in compact form *(Seo and Mura 1979):*

$$\begin{aligned}\sigma_{\mathrm{ij}}^{\mathrm{PSboundary}} = S\frac{G(1+\nu)}{2\pi(1-\nu)} \bigg[& \frac{\partial^2}{\partial x_{\mathrm{i}}\partial x_{\mathrm{j}}}\left(-\frac{1}{R_1}\right) + \nu\delta_{\mathrm{ij}}\frac{\partial^2}{\partial x_3^2}\left(\frac{4}{R_2}\right) \\ & - x_3\frac{\partial^3}{\partial x_3\partial x_{\mathrm{i}}\partial x_{\mathrm{j}}}\left(\frac{2}{R_2}\right) + (-4\nu+3)(-1+\delta_{3\mathrm{i}}+\delta_{3\mathrm{j}})\frac{\partial^2}{\partial x_{\mathrm{i}}\partial x_{\mathrm{j}}}\left(\frac{1}{R_2}\right) \\ & - \delta_{3\mathrm{j}}\frac{\partial^2}{\partial x_3\partial x_{\mathrm{i}}}\left(\frac{1}{R_2}\right) - \delta_{3\mathrm{i}}\frac{\partial^2}{\partial x_3\partial x_{\mathrm{j}}}\left(\frac{1}{R_2}\right)\bigg],\end{aligned} \tag{10.15}$$

where $R_{1,2}^2 = x_1^2 + x_2^2 + (x_3 \mp h)^2$. It can be easily checked, that the stresses given by Eq. (10.14) or (10.15) produces non-zero pressure in the material:

$$p = -\frac{1}{3}\sum_{k=1}^{3}\sigma_{kk} = -\frac{2S}{3}\frac{G(1+\nu)^2}{\pi(1-\nu)}\frac{\partial^2}{\partial x_3^2}\left(\frac{1}{R_2}\right)$$
$$= -\frac{8G\varepsilon_m(1+\nu)^2}{9(1-\nu)}\cdot\frac{2(\tilde{z}+\tilde{h})^2-\tilde{r}^2}{[\tilde{r}^2+(\tilde{z}+\tilde{h})^2]^{5/2}}. \quad (10.16)$$

10.3.4 Energies of Inclusions in Infinite Medium and a Half-space. Interaction Between Inclusions and Other Defects

The elastic energy of an arbitrary inclusion is defined as *(Mura 1987)*:

$$E = -\frac{1}{2}\int_{\Omega}\sigma_{ij}\varepsilon^*_{ij}dV', \quad (10.17)$$

where Ω is area of the inclusion.

Elastic energy of spheroid with uniaxial dilatation $\varepsilon_{mzz} = \varepsilon_m$ is:

$$E = \frac{32\pi G\varepsilon_m^2}{45(1-\nu)}R_{sph}^3 \quad (10.18)$$

Elastic energy of spheroid with equiaxial dilatation $\varepsilon_{mxx} = \varepsilon_{myy} = \varepsilon_{mzz} = \varepsilon_m$ is:

$$E = \frac{8\pi(1+\nu)G\varepsilon_m^2}{3(1-\nu)}R_{sph}^3 \quad (10.19)$$

Utilizing Eqs. (12) and (10.14) we find the energy of the dilatational inclusion in a plate or in a half-space, correspondently.

The energy of spherical inclusion in a half-space is:

$$E = \frac{8\pi G\varepsilon^{*2}(1+\nu)R_{sph}^3}{3(1-\nu)} - \frac{4\pi G\varepsilon^{*2}(1+\nu)^2R_{sph}^3}{9(1-\nu)}\cdot\frac{1}{\tilde{h}^3},\ R_{sph} \le h. \quad (10.20)$$

Here the first term is the energy of inclusion in infinite medium and the second term is the interaction energy between inclusion and a free surface.

The elastic interaction energy between two defects, i.e. inclusions, dislocations, or others can be written as *(Mura 1987)*:

$$E_{\text{int}} = -\int_{\Omega_{II}}\sigma^{I}_{ij}\varepsilon^{*II}_{ij}dV' = -\int_{\Omega_I}\sigma^{II}_{ij}\varepsilon^{*I}_{ij}dV' \quad (10.21)$$

where Ω_{I}, Ω_{II} are the areas occupied by the defects I and II correspondently, $\varepsilon^{*\text{I}}_{ij}$, $\varepsilon^{*\text{II}}_{ij}$ are the eigenstrans of defects, σ^{I}_{ij}, σ^{II}_{ij} are the stresses due to the defects.

10.4 Mechanisms of Stress Relaxation in Buried Quantum Dots

For a buried QD in the coherent state, the elastic energy is proportional to its volume (see Eqs. (10.17), (10.18), (10.19)). Hypothetically this stored energy can reach enormously high values. As a result at a critical magnitude of QD radius R_c some physical mechanisms providing the energy release will come into play. Similarly to well known stress relaxation phenomena in thin films, *e.g. Beanland et al. 1996, Freund and Suresh 2003*, energy release in strained QDs is related to the material transport and defect formation inside and/or in the vicinity of the QD.

In thin strained films and surface islands (unburied QDs) the material transport (i.e. plastic deformation) always involves a free surface of the film or the island. Dislocations, which are the carries of plastic deformation, nucleate at the free surface, propagate through the film and form misfit dislocation at the film/substrate interface. Threading dislocation motion and diffusion-assisted plasticity should be also considered accounting for the proximity of the free surface. For example, for diffusion mass transport, free surface will serve as an effective source/sink. In the case of buried QDs the situation can differ drastically. Elastic field of the QD is localized inside the strained inclusion and in adjacent matrix material. Therefore, dislocation or diffusion assisted material transport can be locked in a closed vicinity of the inclusion. However, under the influence of additional factors such as temperature, the diffusion length or dislocation motion can be influenced. In the following we account for these peculiarities of stress relaxation at buried QDs.

Necessary energy criterion for the operation of a stress relaxation mechanism is:

$$E^{initial} \geq E^{\mathrm{final}}, \tag{10.22}$$

where E^{initial}, E^{final} is the energy of the QD/matrix system before and after relaxation. The initial energy E^{initial} is simply the elastic energy of strained QD as given, for example, by Eqs. (10.18), (10.19). The final system energy includes the energy of all newly formed defects. Following the results obtained by *Kolesnikova et al. 2007* we will consider a number of possibilities for release of the elastic energy of a strained QD by nucleation of dislocation loops.

10.4.1 Single Misfit Dislocation Loop Nucleation at Strained Quantum Dot

For a QD with equiaxial eigenstrain $\varepsilon^*_{ii} = \varepsilon_{\mathrm{m}}$, $\varepsilon^*_{ij} = 0$ $(i \neq j; \quad i, j = x, y, z)$ misfit dislocation (MD) loop can be formed in the equatorial plane of the QD (see Fig. 10.13). The formation of misfit dislocations at the dot/matrix interface is similar to the formation of misfit dislocations at the film/substrate interface for planar heterostructures. For compressed QD, *i.e.* if the crystal lattice parameter of the dot material is larger than those in the matrix, the misfit dislocation loop of vacancy type is formed. The formation of this loop should be accompanied by material transport

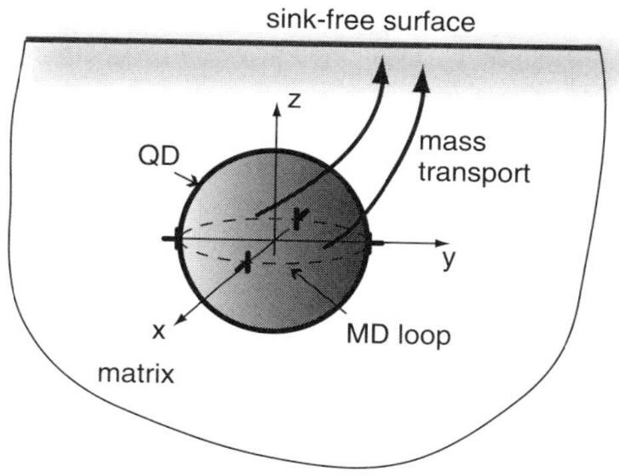

Fig. 10.13 Nucleation of misfit dislocation loop MD at a quantum dot QD. MD-nucleation is accompanied by atom transport to external sink

in the form of atom flow from the QD. Here we assume that these atoms go far away from the dot and reach sinks, e.g. free surface. They will play no role in the energy balance given by Eq. (10.22).

In the final state (after misfit dislocation formation) the energy of the system consists of the final elastic energy of the dot $E_{\mathrm{QD}}^{\mathrm{final}}$, self energy of the dislocation loop E_{MD} and the interaction energy between the QD and loop $E_{\mathrm{QD-MD}}$. Supposing the elastic energy of the QD after relaxation does not change $E_{\mathrm{QD}}^{\mathrm{initial}} = E_{\mathrm{QD}}^{\mathrm{final}}$, Eq. (10.22) gives the condition for energy release for such mechanism:

$$0 \geq E_{\mathrm{MD}} + E_{\mathrm{QD-MD}}. \tag{10.23}$$

The analysis of this condition was performed by *Kolesnikova and Romanov 2004b*, where the explicit analytical expressions for the self and interaction terms was used. This gives the following threshold QD radius R_{c} at which the formation of the MD-loop becomes favorable:

$$R_{\mathrm{c}} = \frac{3b}{8\pi(1+\nu)\varepsilon_{\mathrm{m}}}\left(\ln\frac{\alpha\, R_{\mathrm{c}}}{b}\right), \tag{10.24}$$

where b is the magnitude of the misfit dislocation Burgers vector, parameter α takes into account the energy of the dislocation core and has the value from 1 to 4 (*Hirth and Lothe 1982*).

The dependence of R_{c} on misfit strain is presented in Fig. 10.14 together with the results for quantum wires (cylindrical inclusions) *Kolesnikova and Romanov 2004b* and strained planar layers *Beanland et al. 1996*. Obviously, nucleation of misfit dislocations at strained inclusions is more difficult than for wires and planar films.

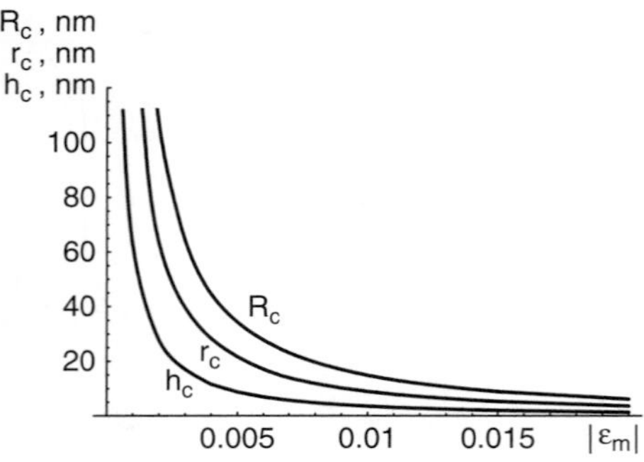

Fig. 10.14 Critical radius of quantum dot R_c for misfit dislocation formation versus misfit parameter ε_m. For comparison, critical radius r_c and critical thickness h_c of misfit dislocation formation in quantum wire and planar heterolayer, are shown. For this plots Poisson ratio $\nu = 0.3$, $b = 0.3\,\text{nm}$, $\alpha = 4$

10.4.2 Nucleation of Dislocation Loop Pair: Misfit Dislocation Loop and Satellite Dislocation Loop

Assume now that stress relaxation process includes formation of prismatic misfit dislocation (MD) loop at QD/matrix interface and a prismatic satellite dislocation (SD) loop in the QD vicinity (as shown in Fig. 10.15). Such mechanism proceeds via local transport of the material. In contrast to the previous model, now the atoms do not go far away from the inclusion to an external sink, instead they remain in a vicinity of the QD and form the satellite dislocation loop *(Kolesnikova et al. 2007)*.

When the QD energy E_{QD} reaches some threshold value, the coherent system "QD/matrix" transforms to a semi-coherent system "QD/ misfit dislocation loop MD/ satellite dislocation loop SD/ matrix". To analyze this case quantitatively, the condition of Eq. (10.22) can be rewritten as:

$$E_{\text{QD}} \geq E_{\text{QD}} + E_{\text{MD}} + E_{\text{SD}} + E_{\text{QD-MD}} + E_{\text{QD-SD}} + E_{\text{MD-SD}}. \qquad (10.25)$$

Here E_{MD}, E_{SD} are the energies of a misfit dislocation loop and a satellite dislocation loop correspondently; $E_{\text{QD-MD}}$, $E_{\text{QD-SD}}$ are the interaction energies of QD with the loops; $E_{\text{MD-SD}}$ is the energy of the interaction between the loops.

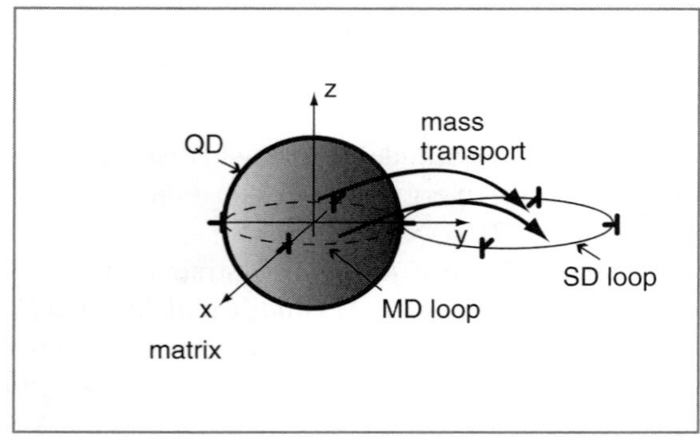

Fig. 10.15 Formation of misfit dislocation MD and satellite dislocation SD loops at strained quantum dot QD

In addition to the energy balance of equation (10.25), the material conservation law should be taken into account:

$$b_{\mathrm{MD}}\, S_{\mathrm{MD}} = b_{\mathrm{SD}}\, S_{\mathrm{SD}}. \tag{10.26}$$

where b_{MD}, b_{SD} are the Burgers vectors of the corresponding loops; $S_{\mathrm{MD}} = \pi\, r_{\mathrm{MD}}^2$, $S_{\mathrm{SD}} = \pi\, r_{\mathrm{SD}}^2$ are the areas occupied by MD- and SD-loops, respectively. From Eq. (10.26) we can obtain the SD radius as function of MD-radius (or QD-radius) $r_{\mathrm{SD}} = \beta\, r_{\mathrm{MD}}$, $\beta = \sqrt{b_{\mathrm{MD}}/b_{\mathrm{SD}}}$.

The elastic interaction of two defects (in our case the defects are MD and SD loops or QD and loop) is obtained from expression (10.21). To estimate the critical radius R_{c} at which the relaxation of the type shown in Fig. 10.15 can be triggered we accept $r_{\mathrm{MD}} = R_{\mathrm{QD}}$ and $b_{\mathrm{MD}} = b_{\mathrm{SD}} = b$, and neglect the energies of the interactions SD-QD and SD-MD. In this case the critical radius R_{c} can be derived analytically:

$$R_{\mathrm{c}} \approx \frac{3b}{4\pi\,(1+\nu)\varepsilon_{\mathrm{m}}}\left(\ln \frac{\alpha\, R_{\mathrm{c}}}{b}\right), \tag{10.27}$$

Note that Eq. (10.27) is an approximate formula, which is correct with accuracy of about 20% in practical important ε_{m} range.

In Fig. 10.16 (curve 1) the critical radius R_{c} as the function of misfit parameter ε_{m} is shown. In the same figure the results of a single misfit dislocation loop formation at strained QD are also given (curve 2).

10.4.3 Formation of Incoherent Inclusion and Satellite Dislocation Loop

For QDs with uniaxial eigenstrain $\varepsilon^*_{zz} = \varepsilon_m$, $\varepsilon^*_{xx} = \varepsilon^*_{yy} = 0$ another relaxation mechanism can be considered *(Chaldyshev et al. 2005)*. In this case a SD loop attached to the nanoinclusion (see Fig. 10.17) is nucleated but no localized MD is present at AQ/matrix interface. The atoms used for the SD formation arrive from various region of the inclusion or surrounding matrix providing the formation

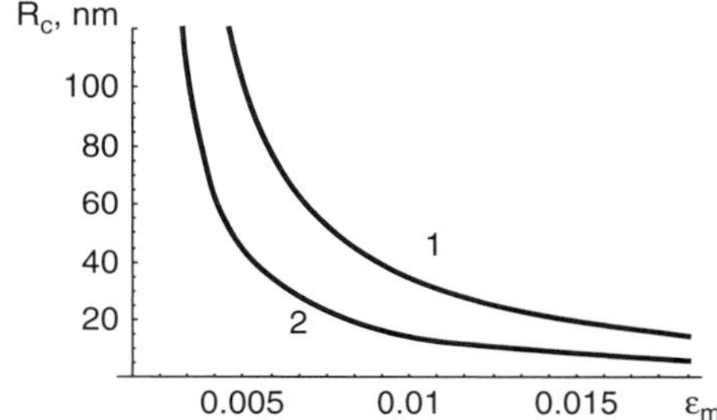

Fig. 10.16 Critical radius of quantum dot R_{c} versus misfit parameter ε_m. 1-for the model with pair dislocation loop formation; 2-for the model with single misfit dislocation loop nucleation. Poisson ratio $\nu = 0.3$, $b_{\mathrm{MD}} = b_{\mathrm{SD}} = b = 0.3\,\mathrm{nm}$

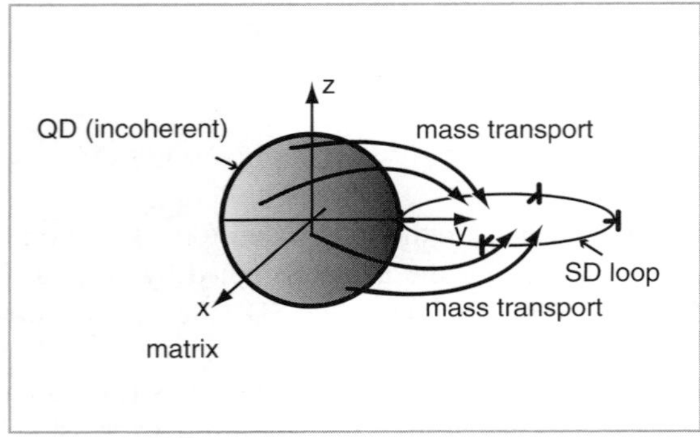

Fig. 10.17 Single satellite dislocation SD loop formation at strained quantum dot QD

incoherent interface between QD and matrix. Such defect configurations were identified in experiments with doped GaAs *(Chaldyshev et al. 2002, 2005)*.

In the framework of the developed approach the threshold condition from the Eq. (10.22) operates with the initial elastic energy of system E^{initial}, which is the elastic energy of QD, $E_{\text{QD}}^{\text{initial}}$, and final energy of system E^{final}, which is a sum of the final elastic energy of the QD, $E_{\text{QD}}^{\text{final}}$, elastic energy of the satellite prismatic dislocation loop, E_{SD}, and their interaction energy, $E_{\text{QD-SD}}$. In this case the energy condition of Eq. (10.22) may be written as:

$$E_{\text{QD}}^{\text{initial}} \geq E_{\text{QD}}^{\text{final}} + E_{\text{SD}} + E_{\text{QD-SD}}. \tag{10.28}$$

Transport of atoms from the matrix nearby QD or from the QD interior causes a change in the misfit parameter ε_m for the nanoinclusion from $\varepsilon_{\text{m}}^{\text{initial}}$ to $\varepsilon_{\text{m}}^{\text{final}}$. Taking into account the law of material conservation we can write:

$$(\varepsilon_{\text{m}}^{\text{initial}} - \varepsilon_{\text{m}}^{\text{final}})V_{\text{QD}} = bS_{\text{SD}}, \tag{10.29}$$

where $V_{\text{QD}} = \frac{4\pi}{3} R_{\text{QD}}^3$, $S_{\text{SD}} = \pi\, r_{\text{SD}}^2$. From equation (10.29) the difference between the initial and final misfit parameters is determined as $\Delta\varepsilon_{\text{m}} = 3b\, r_{\text{SD}}^2/(4R_{\text{QD}}^3)$.

From the condition of the system energy minimum one can find the SD-loop radius as a function of QD radius (*Chaldyshev et al. 2002, 2005*), which is shown in Fig. 10.18 for the set of model parameters relevant to the experiment (see next section).

10.5 Experimental Verification of the Models

Although there are many experimental observations of stress relaxation in different QDs (Section 10.1), comparison of the available experimental data with the models developed in Section 10.4 appears to be a very difficult task. This difficulty is due to the fact that the majority of the experimental investigations were neither comprehensive, nor systematic. Since formation of extra dislocations is far from the

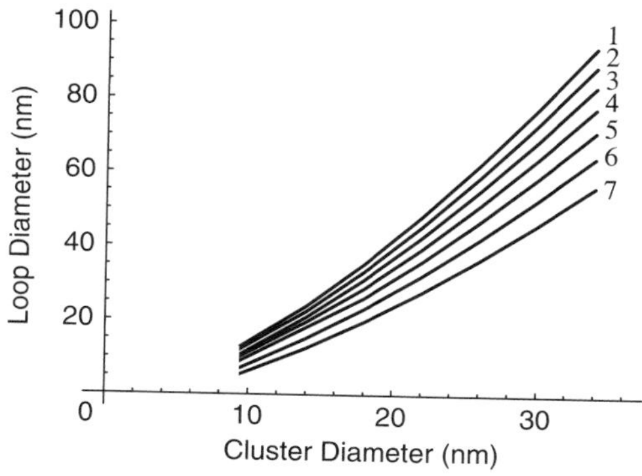

Fig. 10.18 Calculated dependences of dislocation loop diameter on the nanoinclusion diameter for different values of the initial relative misfit parameter $\varepsilon_m = 0.1$ (1), 0.09 (2), 0.08(3), 0.07 (4), 0.06 (5), 0.05 (6), 0.04 (7). Burgers vector magnitude $b = 0.28$ nm, and the Poisson ratio $\nu = 0.3$

initial purpose of a typical QD-related research, this phenomenon usually initiates intuitive empirical changes in the technology rather then in-depth analysis. Actually, we know only one experimental system, namely AsSb QDs in GaAs, in which the experimental data are comprehensive enough for quantitative comparison with the theoretical calculations. In this section we do an examination of the experimental studies of this system. It is followed by justification of the theoretical model and parameters and fit of the model calculation to the experimental data.

Self-organization of the AsSb QDs is realized in the bulk of the crystalline GaAs matrix through a two-stage procedure (*Chaldyshev 2002*). At the first stage a Sb-doped GaAs film is grown by molecular-beam epitaxy (MBE) at low temperature (150–250° C). Then the material is subjected to an annealing at 400 °C or higher temperatures for several minutes. Thus obtained material is usually denoted as LT-GaAs. Let us consider the both stages in more detail.

The purpose of the first stage is formation of a metastable crystalline medium with supersaturation on group V componets. It should be noted that a rather large As/Ga flux ratio of 3-30 is typical for the MBE, however, at the optimal growth temperature, which is around 600 °C, the redundant arsenic atoms desorb from the growth surface, and the common GaAs films are stoichiometric and contain a very low concentration of the native point defects. Reduction of the growth temperature results in an increase in the concentration of native point defects. When the growth temperature is reduced to 200 °C and the As/Ga flux ratio is remained high, the material becomes remarkably off-stoichiometric (see Fig. 10.19); the concentration of the point defects, mainly arsenic antisites, As_{Ga}, (*Liu et al. 1995*), reaches 1 10^{20} cm^{-3}. It is important that in spite of such a huge density of the point defects, the crystal lattice of the LT-GaAs grown at 200 °C remains perfectly regular, as evidenced by numerous x-ray diffraction studies and TEM examinations (see, for instance, *Faleev et al. 1998, Bert et al. 1993*). The 200 °C grown LT-GaAs films remain free of dislocations and other extended defects being as thick as 1.5 μm and over. This makes it possible to grow the large variety of heterostructures with composition and doping profiles required for physical research and device applications.

The concentration of the excess arsenic in LT-GaAs as a function of the growth temperature is plotted in Fig. 10.19. The highest arsenic excess of about 1.5 at.% can

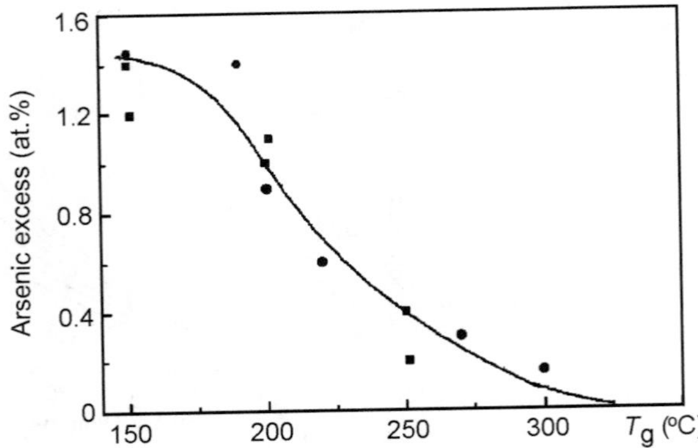

Fig. 10.19 Excess arsenic concentration in LT-GaAs films vs. the growth temperature T_g. The data points are collected from publications of different research groups: (*Bert et al. 1993*), (*Yu Kin Man et al. 1992*) and (*Luysberg et al. 1998*). Line is a guide for eyes

be reached in the LT-GaAs films grown at 150 °C. Farther reduction of the growth temperature leads to lost of crystallinity, except very thin layers and low growth rates (*Shimogishi et al. 2002*), due to critical shortage in the surface diffusivity of the absorbed atoms. Elevation of the growth temperature from 200 to 300 °C and up leads to a rapid decrease in the concentration of excess arsenic in the LT-GaAs films.

It is important that the arsenic excess in LT-GaAs can be easily evaluated. The lattice parameter of the LT-GaAs epitaxial film, even undoped, is larger than that of normal GaAs. This enlargement originates from the fact that As-As tetragonal bond length (2.65 Á) is larger than the As-Ga bond length in GaAs (2.45 Á) (*Liu et al. 1995*). The lattice mismatch between normal and LT-GaAs was calibrated, in order to evaluate the concentration of antisites [As_{Ga}] and corresponding arsenic excess [As_{ex}] from routine x-ray diffraction measurements (*Liu et al. 1995*):

$$\frac{a_L - a_S}{a_S} = 1.24 \cdot 10^{-23}[As_{Ga}(cm^{-3})] = 0.27[As_{ex}], \tag{10.30}$$

where a_L and a_S stand for the usually measured lattice periods of the layer and substrate in the growth direction. It should be noted that the interface between the LT-GaAs film and stoichiometric GaAs substrate is coherent and, therefore, the LT-GaAs film is tetragonally strained.

Another common technique for evaluation of the As_{Ga} concentrations bases on measurements of near infrared optical absorption. The absorption coefficient for these defects was calibrated at 1.0 and 1.06 μm by *Martin 1981*:

$$\alpha_{1\mu m}(cm^{-1}) = 1.3 \cdot 10^{-16}[As_{Ga}(cm^{-3})]. \tag{10.31}$$

Of particular interest in this paper are the LT-GaAs films doped with isovalent Sb impurity. Two types of such films have been studied (*Bert et al. 1999, Chaldyshev et al. 2001, Chaldyshev et al. 2005*). In one of them the Sb doping was uniform within the film bulk. The Sb concentration in these films was varied in the range from 0.1 to 1 mol.%. The films of the other type were delta-doped with Sb. The nominal thickness of the delta-layers was varied from 0.001 to 1 monolayer. In order to produce the delta-layers in the most reliable manner, the As molecular source was cooled down during deposition of Sb and warm up to grow undoped LT-GaAs layers in between.

Precise evaluation of the excess arsenic concentration from x-ray diffraction curves for LT-GaAsSb is a complicated task, since both As-Sb alloying and off-stoichiometry modify the lattice parameter of the film. Nevertheless, careful x-ray diffraction study and optical measurements led to the conclusion that the Sb-doped LT-GaAs films contained approximately the same amount of excess arsenic as the Sb-free ones grown under the same conditions (*Vasyukov et al. 2001*).

Deposition of 1 monolayer thick GaSb delta-layers on GaAs at any temperature did not result in formation of misfit dislocations as documented by TEM study (*Bert et al. 1999*). Due to roughness of the growth surface, the GaSb delta-layers sandwiched between LT-GaAs buffers were imaged as thick as 4 monolayers (*Chaldyshev et al. 2001*). The effective thickness of the Sb-containing region did not depend on the actual Sb concentration in the delta-layers. In the case of uniform Sb doping during the low-temperature MBE, the Sb concentration was uniform over the whole film bulk. TEM study has never revealed dislocations in the as-grown LT-GaAsSb with Sb concentration of 1 mol. % or less.

The purpose of the second stage – post growth anneal – is formation of compact nanoinclusions from the off-stoichiometric As and Sb atoms distributed over the GaAs matrix as individual antisite defects. The amount of the excess arsenic incorporated into the LT-GaAs films during the low-temperature MBE is far beyond the equilibrium solubility limit. However, this supersaturation is frozen until certain temperature due to low diffusivity of all the native point defects.

A post-growth anneal makes the excess arsenic-related point defects movable and gives rise to phase transformations and self-organization of nanoscale As QDs. The process can be monitored by relaxations of As_{Ga}-related optical absorption and lattice expansion. Plotted in Fig. 10.20 is the difference between the lattice parameters of undoped LT-GaAs films and stoichiometric GaAs substrates for several samples grown and annealed at various temperatures. Before annealing the difference is always positive due to arsenic excess and its value depends on the growth temperature. The post-growth annealing at 400 °C and higher temperatures causes a decrease in the lattice mismatch. After anneal at 400 °C the a_L-a_S is still noticeably positive, after anneal at 500 °C it is close to zero and remains close to zero or slightly negative after anneal at 600 °C and higher temperatures. The vanished or even inverted film/substrate lattice mismatch and almost total disappearance of the As_{Ga}-related optical absorption show that the majority of excess-arsenic-related point defects transform into nanoinclusions.

The typical array of As nanoinclusions formed in GaAs matrix by annealing at 600 °C for 15 minutes is shown in Fig. 10.7(b), Section 10.2. The bright-field TEM imaging in the (001) zone axis reveals two sets of Moire fringes, originating

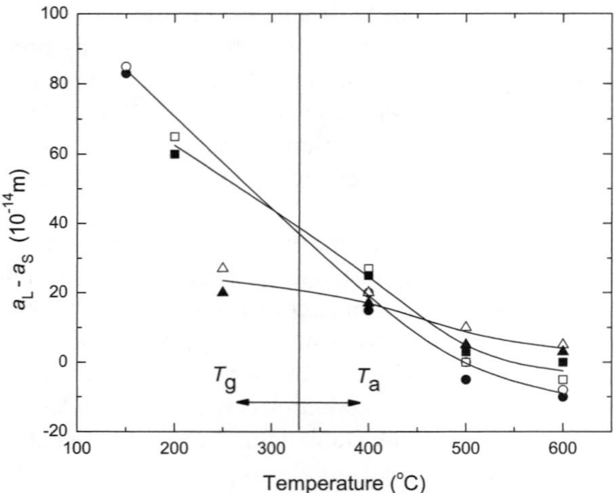

Fig. 10.20 Difference between the lattice parameters of the epitaxial LT-GaAs layers, a_L, and stoichiometric GaAs substrate, a_S, for the samples grown at different temperatures, T_g, before (left window) and after (right window) annealing at various temperatures, T_a. The annealing duration was 15 min. Open and filled symbols correspond to two different values of the JAs/JGa flux ratio. Lines are guides for eyes. The data are taken from *Bert et al. 1993*

from the double diffraction of the electron beam at the crystal lattices of the As nanoinclusion and GaAs matrix. The detailed analysis of the Moire fringes in the As/GaAs system has been done by *Claverie and Liliental-Weber 1993* and *Bert and Chaldyshev 1996*. The As nanoinclusion were found to have rhombohedral A7 structure often referred to as hexagonal. The atomic planes of the As nanoinclusions are oriented in the GaAs matrix as $\{0003\}_{As}||\{111\}_{GaAs}$ and $\{-12\text{-}10\}_{As}||\{011\}_{GaAs}$. The atomic microstructure is directly viewed in Fig. 10.21, which is a HREM micrograph taken in the cross-sectional (110) projection of the LT-GaAs film. The As/GaAs interface is semicoherent with continuous major parallel planes mentioned above and discontinuity or tilts of most others.

The analysis of the Moire fringes shows that strain within the As nanoinclusions has different sign along their c-axes and in the basal plane. The strain depends on their size and reaches 3% along the $[0001]_{As}$ axes. The hydrostatic stress is low. Due to the described nature of the local internal stress within the As inclusions and much higher stiffness of GaAs compared to that of As, the local strain in the surrounding GaAs matrix is normally not detectable in TEM. A small average shrinking in the lattice parameter of the annealed LT-GaAs film with built-in As nanoinclusions is, however, observable in x-ray diffraction. It can be deduced from Fig. 10.20 that the lattice parameter of LT-GaAs annealed at 600 °C is reduced by about 0.01 % for the volume fraction of As inclusions of about 1 %. This residual negative strain results from averaging of the local fields from the inclusions with four equivalent orientations of their c-axes relatively to the host matrix and originates from the fact that the molecular volume in hexagonal arsenic (43 Á^3) is slightly less than corresponding volume in GaAs (45 Á^3).

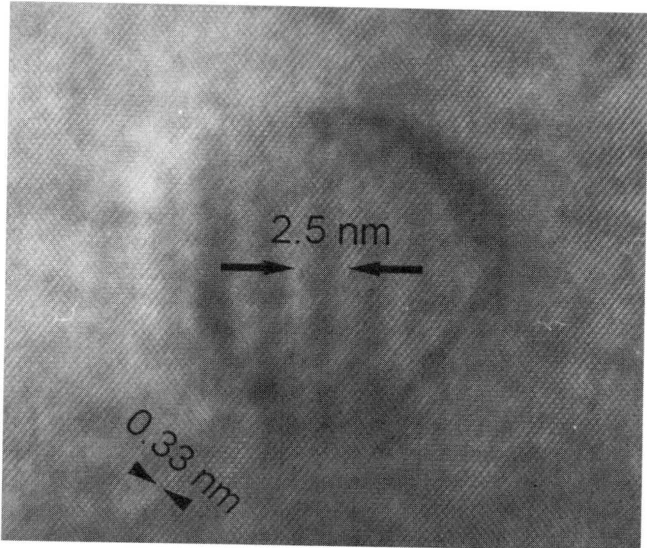

Fig. 10.21 High-resolution cross-sectional (110) transmission electron micrograph of a large As QD built in GaAs matrix. The distance of 3.3 nm between the resolved (333) planes in the matrix is marked by arrows. The Moire fringes with the period of 2.5 nm on the QD contrast originate from double diffraction at the $(2\text{-}20)_{GaAs}||(2\text{-}1\text{-}10)_{As}$ atomic planes

Precipitation in LT-GaAsSb shows several features when compared to the conventional Sb-free material. Bert et al. 1999 revealed that the precipitation rate is strongly enhanced at GaSb delta-layers built in the LT-GaAs matrix or in the whole film when it is uniformly doped with Sb at the level of about 1.0 mol.%. The precipitates in LT-GaAsSb appeared to be AsSb alloy with Sb concentration of 20–40% evidenced by the energy dispersive X-ray microanalysis.

Since the atomic covalent radius of Sb is considerably larger then that of As, the mismatch of the AsSb QDs to GaAs matrix is much stronger compared to the As QDs. This results in strong deformation fields within and around the AsSb inclusions. Figure 10.22 is a cross-sectional TEM micrograph of an LT-GaAs sample with built-in 1-monolayer-thick GaSb delta-layers. The Sb-containing regions of the GaAs matrix are seen as bright lines. The dark spots near these lines are due to local stresses surrounded the AsSb precipitates that are located at GaSb delta-layers. The pure As inclusions in Sb-free LT-GaAs spacers in between GaSb delta-layers are also seen in Fig. 10.22. The pure As inclusions are much smaller than AsSb ones and do not induce noticeable deformation in the surrounding matrix.

Figure 10.23 presents HREM micrographs of the atomic structure of the AsSb inclusions in LT-GaAs at different stages of their development. It is interesting that at the initial stage of self-organization of the AsSb inclusions at GaSb delta-layers, they have a lens shape (Fig. 10.23(a)), which may originate from relatively soft Ga-Sb bonds along (001) plane, when compared to Ga-As bonds in other directions. Ripening of the AsSb inclusions transform their shape to spherical (Fig. 10.23(a)),

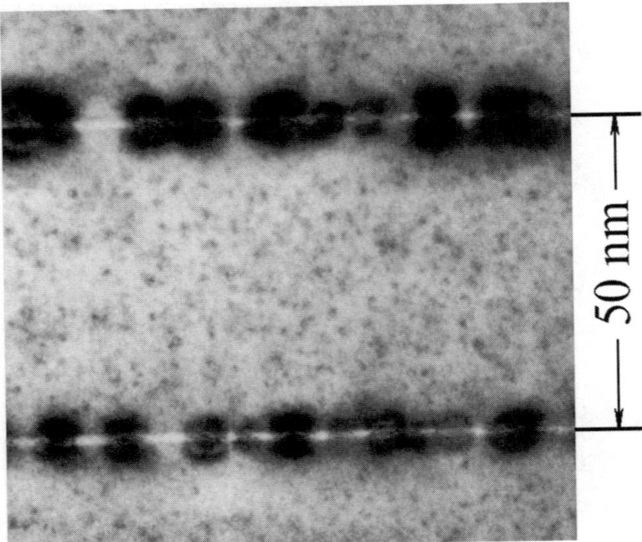

Fig. 10.22 Cross-sectional 002 dark-field micrograph of the LT-GaAs sample with two Sb delta-layers. The sample was grown at 200 °C and annealed at 500 °C for 10 min. The Sb delta-layers exhibit a bright chemical contrast and are separated by 50 nm thick spacer. The pronounced large spots of dark contrast are induced by strain around the AsSb QDs located at the Sb delta layers. The As QDs in the GaAs spacer are seen as small dark specks

which is common to pure As inclusions. The atomic structure of the developed AsSb QDs was found to be hexagonal, quite similar to that of pure As QDs. The orientation relationships between the AsSb inclusionand the GaAs matrix also remain $\{0003\}_{AsSb}||\{111\}_{GaAs}$ and $\{\text{-}12\text{-}10\}_{AsSb}||\{022\}_{GaAs}$.

While the incorporation of the substantial amount of Sb into the As inclusions does not change their crystalline symmetry, it increases the inter-planar distances, when compared to the pure As inclusions. This improves the inclusion-matrix match for the atomic planes passing the inclusion *c*-axes such as the $\{\text{-}12\text{-}10\}_{AsSb}$ and

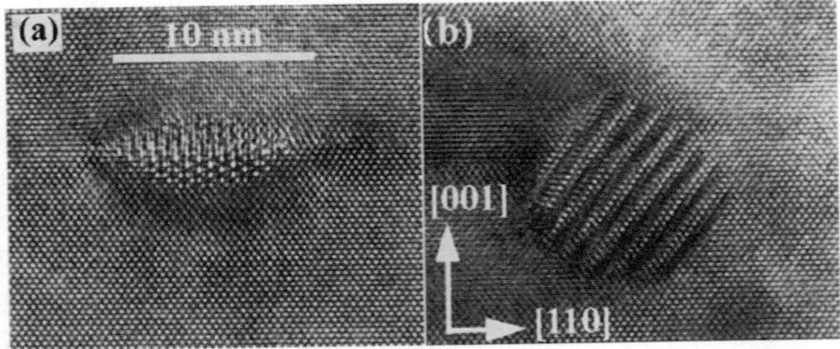

Fig. 10.23 High-resolution transmission electron images of the QDs attached to the Sb delta-layers in LT-GaAs (*Bert et al. 1999*): (a) the lens-shaped AsSb QD at the early formation stage and (b) the AsSb QD with a developed rhombohedral structure exhibiting Moire fringes

corresponding $\{220\}_{GaAs}$ planes. At the same time, it increases the mismatch along the inclusion c-axes, since the (0003) inter-plane distance even in pure As is larger than the (111) inter- plane distance in GaAs (0.352 nm against 0.332, correspondingly). These changes in matching conditions have been revealed by analysis of the Moire patterns in the TEM images (*Chaldyshev et al. 2002, 2005*). The quantitative analysis showed that the eigenstrain associated with the As-Sb QDs in GaAs is mainly uniaxial, along the c-axes of the inclusion. The absolute value of the eigenstrain can be estimated as $\varepsilon_m = 0.05 \pm 0.2$.

Ripening of the stressed AsSb inclusions in GaAs causes relaxation phenomena, which can be detected by x-ray diffraction (*Vasyukov et al. 2001*) and observed by TEM (*Bert et al. 1999, Chaldyshev et al. 2002, 2005*). At the late stage of Ostwald ripening, when the As-Sb inclusions become larger than 7 nm, satellite dislocation (SD) loops have been observed nearby the large clusters. An ensemble of the SD loops and inclusions is presented in Fig. 10.9(b), Section 10.2, which is a plan-view electron micrograph from an LT-GaAs film δ-doped with Sb and annealed at 600○C. One can see that the SD loops are attached to the QDs, and the loop diameter correlates with the accompanied QD diameter. Note that the SD loops nucleate when the inclusion system is already well developed. As well, such loops have never been observed in Sb-free LT-GaAs. In the LT-GaAs films δ-doped with Sb all the SD loops lie on the (001) plane, *i.e.* are aligned along δ layers. In the LT-GaAs films doped with Sb uniformly, the SD loops were observed to lie on all three orthogonal {100} planes (*Chaldyshev et al. 2002*).

The SD loops were always stretched out the associated clusters in one of the two orthogonal $\langle 110 \rangle$ directions, *i.e.* along or perpendicular to the projection of the c axis of the AsSb inclusion onto the (001) plane. Comparison of the Moire fringe arrangement in the inclusion images with the location of the associated SD loops showed no correlation between the two possible loop-inclusion mutual orientations.

To determine the orientation of the SD loop Burgers vector we exploited strong two-beam diffraction conditions. Figure 10.24 shows TEM images of the satellite dislocation loops taken at two different orientations of the operating diffraction vector. As can be seen for $\mathbf{g} = 220$ and 040, the contrast elimination line is perpendicular to the diffraction vector in both cases. Such behavior of the contrast elimination

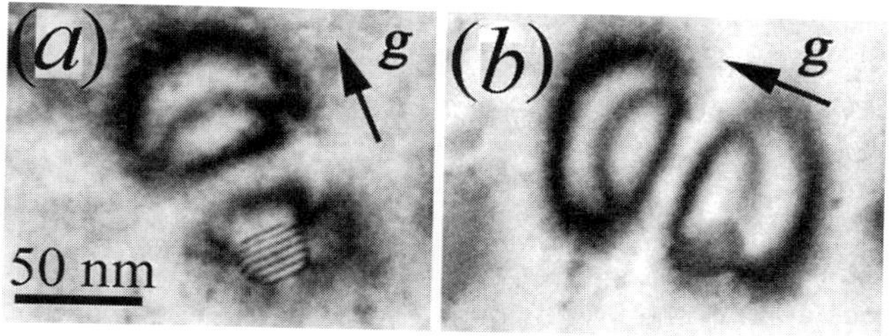

Fig. 10.24 Plan-view bright-field TEM micrographs of an AsSb QD with a satellite dislocation loop in LT-GaAs under different diffraction conditions (*Chaldyshev et al. 2002*). The diffraction vector **g** is along [220] (α) and [040] (β)

line is known to be characteristic of the prismatic dislocation loops where the Burgers vector orientation is perpendicular to the plane of the loop, *i.e.* directed along the [001] axis in our case. Analysis of the contrast behavior in the bright-dark-field images of the loops located at various distances from the foil surface and observed edge-on suggested the SD loops are of interstitial type.

The conventional visibility-invisibility analysis appeared to be not applicable to determine the Burgers vector magnitude directly. Due to a small size of the loops it is hardly possible to reliably distinguish a specific contrast from the fault defect inside the SD loop, which would manifest itself in large loops under certain diffraction conditions if the Burgers vector is less than the [001] GaAs lattice period. However, a specific two-lobe contrast of the loops was revealed under strong two-beam diffraction imaging at $\boldsymbol{g} = 400$. Taking into account the $\boldsymbol{g}\cdot\boldsymbol{b} = 2$ criteria, this observation allows us to suggest that the Burgers vector is equal to $a/2$ [001].

It should be mentioned that the faulted dislocation loops located on the {001} lattice planes are unusual for the zinc blende structure, the {111} set of planes is much more typical. Similar dislocation loops aligned on {001} were once reported for InGaAsP degraded laser heterostructures (*Chu and Nakahara 1990*). Those loops were as large as several microns and were not accompanied with clusters or inclusions. The TEM investigation by *Chu and Nakahara 1990* gave a reliable evidence for the $a/2$ [001] Burgers vector of such dislocation loops.

The orientation relationship between a AsSb QD, attached satellite dislocation loop and its Burgers vector is schematically demonstrated in Fig. 10.25.

As can be seen in Fig. 10.9, Section 10.2, all the large precipitates in Sb-doped LT-GaAs are associated with the SD loops, whereas small ones are not. The threshold inclusion diameter required for the formation of the SD loop was found to be 7-8 nm. The correlation between the AsSb inclusion diameter and the SD loop diameter was experimentally investigated by *Bert et al. 2002* and *Chaldyshev et al. 2002* and shown in Fig. 10.26. The experimental points in Fig. 10.26 are collected from the samples annealed at different temperatures. It is evident that variation of the annealing temperature does not affect the threshold of the SD loop formation and nonlinear correlation between SD loop diameter and AsSb QD diameter. The obvious effect of the elevation of the annealing temperature is an

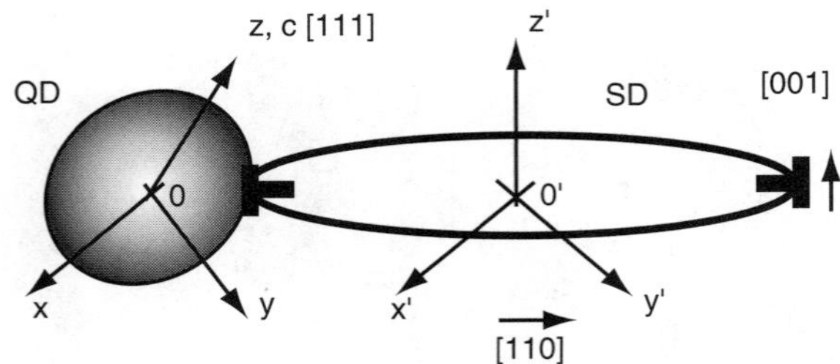

Fig. 10.25 Sketch of a AsSb QD buried in the GaAs matrix and associated with a prismatic SD loop. Direction of the distortion (z) in the AsSb QD coincides with its hexagonal c-axis and lies along [111] of zinc-blende GaAs structure. The Burgers vector $\boldsymbol{b}$ of the loop is perpendicular to its (001) plane

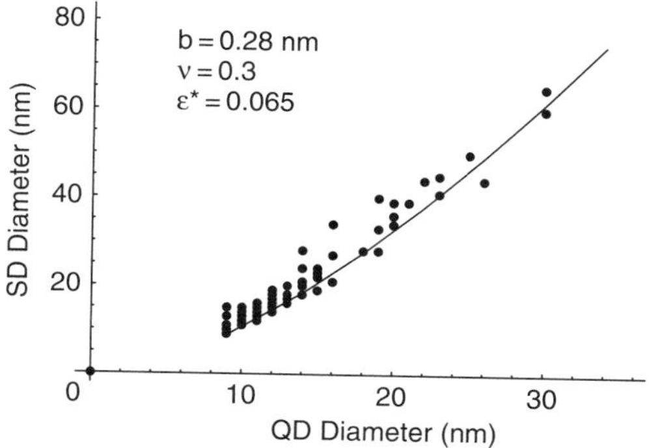

Fig. 10.26 Experimental points and theoretical curve representing the SD loop diameter as a function of the AsSb QD diameter in annealed Sb delta-doped LT GaAs. The experimental points were obtained by TEM analysis of the two parts of the Sb-delta-doped LT-GaAs sample annealed at 500 and 600 °C. The plotted solid curve is the best fit obtained within theoretical model 3 by variation of the relative plastic elongation ε_{m}. All other parameters and orientation relationships were taken from experiment, as shown in Fig. 10.9. The magnitude of the SD Burgers vector b = 0.28 nm (a) and the Poisson ratio $\nu = 0.3$. (*Chaldyshev et al. 2005*)

increase in the diameter and a reduction in the density of both AsSb QD and their SD loops. Since the correlation between the diameters of a AsSb QD and its SD loop is statistically well defined and reproducible for various annealing conditions, we can conclude that a AsSb QD and its SD loop are in equilibrium at each stage of growth. The formation of a SD loop in a vicinity of the cluster is apparently a factor that reduces the elastic stresses localized near the AsSb QDs and the mechanical energy of the system as a whole. Note that loops not associated with AsSb QDs have never been observed in such materials.

The comprehensive experimental observations described above make system of AsSb QDs in GaAs suitable for quantitative theoretical consideration. While the AsSb QDs are buried deeply in the GaAs bulk, the theoretical models with material conservation conditions are adequate to the experimental situation. The As/GaAs and AsSb/GaAs interfaces were observed to be not perfectly coherent. Therefore, the model with a single satellite dislocation loop and non-local distribution of the mismatch at the inclusion interface (Section 10.4.3) seems to be most applicable.

The theoretical calculation has been done within the model (Section 10.4.3) with experimentally determined SD and QD parameters and orientation relationships (Fig. 10.25). The only fitting parameter of the model was the initial plastic distortion in AsSb QDs, which was varied in the theoretical calculations to obtain the best fit to the experimentally observed nonlinear correlation between the SD loop diameter and AsSb QD diameter (Fig. 10.26). For the SD loop Burgers vectors $\boldsymbol{b} = \frac{1}{2}\,a[001]$ a good agreement between the theoretical calculations and experimental data was obtained with the values of the uniaxial eigenstrain: $\varepsilon_m = 0.065$, which is reasonably close to the experimental estimate $\varepsilon_m = 0.05 \pm 0.02$. The calculation also accounts for the non-linear behavior of the correlation observed. So, a quantitative agreement has been achieved between the theory and experiment.

10.6 Conclusions

Summarizing the above theoretical and experimental analysis we can draw a very general conclusion that the main mechanisms of strain relaxation in buried quantum dots differ from that known for thin films. The reason for this difference is localization of the mechanical fields within and in a close vicinity of the buried QD, whereas a mismatched thin film on a thick substrate is stressed as a whole with nearly constant stress and strain.

We developed three models of the stress relaxation in buried QDs. One of these models is proven above to be qualitatively and quantitatively correct, when applied to AsSb QDs embedded in the GaAs matrix. It is important that these QDs are self-organized in the film bulk, i.e. are truly buried. Another example of buried QDs are MnAs magnetic nanoinclusions briefly described in Section 10.2. These objects seem to be relevant to the theoretical model developed here. There are many other examples of QDs self-organized in the bulk due to supersaturation achieved by ion implantation, non-equilibrium low-temperature epitaxy, high-temperature crystal growth, etc. Unfortunately, the experimental data available on these objects are insufficient to make quantitatively-exact theoretical calculations.

The developed models may also be relevant to the QDs self-organized on the surface, such as InGaAs QDs on GaAs or Ge QDs on Si described in Section 10.2. Relaxation of such QDs can occur either in the free-standing state or when they are buried in the bulk by epitaxial overgrowth. In the latter case, the driving force of the relaxation is the change in the mechanical energy of the QD (see Eq.10.20, Section 10.3) depending on its distance to the free surface. I.e., a subcritical QD can become overcritical when buried. The fingerprint of this process is formation of dislocation loops, single MD or MD+SD pair. Two corresponding mechanisms are developed in Section 10.4.

The case of relaxation in non-buried surface QDs is beyond the scope of this chapter. Some examples given in Section 10.2 show that dislocation loops can also form in this case by direct nucleation of the misfit dislocations at the interface during the ripening of the free-standing QD. However, in most situations the misfit dislocation ends do not meet, giving rise to threading dislocation in the cap layers during overgrowth.

All the presented theoretical models consider relaxation of a single stressed QD. As shown in Section 10.3, the stress induced by a QD in the matrix diminishes as $(R/R_{\mathrm{QD}})^{-3}$, where R is the distance and R_{QD} is the radius of QD. Therefore, the approach is valid for a QD array if the distance between QDs exceeds at least $2 \cdot R_{\mathrm{QD}}$. This requirement is met for the majority of systems of QDs nucleated either on the surface or in the bulk, as described in Section 10.2.

Another important assumption of the models is a local quasi-equilibrium between QD, MD and SD in the final relaxed state that corresponds to a minimum of the free energy. While many QD systems are metastable, the local mass transfer between a QD and associated defects is normally much faster than the corresponding process between different dots. Therefore, the utilized quasi-equilibrium thermodynamic approach is valid not only for the systems near global equilibrium, but also for the metastable arrays of QDs at each moment of the Ostwald ripening. An example, where the assumption was validated, is the metastable system of AsSb QDs in GaAs described in Section 10.5.

In some QD systems, however, the kinetics of mass transfer and collective effects may play an important role in stress relaxation phenomena. This is the scope of our future work.

Acknowledgments The authors are grateful to the Russian Academy of Sciences and the Russian Foundation for Basic Research for long term financial support.

References

Akinaga H, Mizuguchi M, Ono K, Oshima M (2000) Room-temperature thousandfold magnetoresistance change in MnSb granular films: Magnetoresistive switch effect. Applied Physics Letters 76: 357–359

Andreev AD, Downes JR, Faux DA, O'Reilly EP (1999) Strain distributions in quantum dots of arbitrary shapes. Journal of Applied Physics 86: 297–305

Andreev AD, O'Reilly EP (2000) Theoretical study of the electronic structure of self-organized GaN/AlN QDs. Nanotechnology 11: 256–262

Beanland R, Dunstan DJ, Goodhew PJ (1996) Plastic relaxation and relaxed buffer layers for semiconductor energy. Advances in Physics, 45: 87–146

Bert NA, Chaldyshev VV (1996) Changes in the Moire patterns in electron-microscope images of As clusters in LT-GaAs as their size decreases. Semiconductors 30: 988–989

Bert NA, Chaldyshev VV, Faleev NN, Kunitsyn AE, Lubyshev DI, Preobrazhenskii VV, Semyagin BR, Tret'yakov VV (1997)Two-dimensional precipitation of As clusters due to indium delta-doping of GaAs films grown by molecular beam epitaxy at low temperature. Semiconduct. Science and Technology 12: 51–54

Bert NA, Chaldyshev VV, Suvorova AA, Preobrazhenskii VV, Putyato MA, Semyagin BR, Werner P (1999) Enhanced precipitation of excess As on antimony delta layers in low-temperature-grown GaAs. Applied Physics Letters 74: 1588–1590

Bert NA, Kolesnikova AL, Romanov AE, Chaldyshev VV (2002) Elastic behaviour of a spherical inclusion with given uniaxial dilatation. Physics of the Solid State 44: 2240–2250

Bert NA, Veinger AI, Vilisova MD, Goloshchapov SI, Ivonin IV, Kozyrev SV, Kunitsyn AE, Lavrent'eva LG, Lubyshev DI, Preobrazhenskii VV, Semyagin BR, Tretyakov VV, Chaldyshev VV, Yakubenya MP (1993) Gallium-arsenide grown by molecular-beam epitaxy at low temperature: crystalline structure, properties, superconductivity. Physics of the Solid State 35: 1289–1297

Bimberg D, Grundmann M, Ledentsov NN, Ruvimov SS, Werner P, Richter U, Gösele U, Heydenreich J, Ustinov VM, Kop'ev PS, Alferov ZI (1995) Self-organization processes in MBE-grown quantum dot structures. Thin Solid Films 267: 32–36

Chaldyshev VV (2002) Two-dimensional organization of As clusters in GaAs. Materials Science & Engineering B 88: 85–94

Chaldyshev VV, Bert NA, Musikhin YG, Suvorova AA, Preobrazhenskii VV, Putyato MA, Semyagin BR, Werner P, Gosele U (2001) Enhanced As-Sb intermixing of GaSb monolayer superlattice in low-temperature-grown GaAs. Applied Physics Letters 79: 1294–1296

Chaldyshev VV, Bert NA, Romanov AE, Suvorova AA, Kolesnikova AL, Preobrazhenskii VV, Putyato MA, Semyagin BR, Werner P, Zakharov ND, Claverie A (2002) Local stresses induced by nanoscale As-Sb clusters in GaAs matrix. Applied Physics Letters 80: 377–379

Chaldyshev VV, Kolesnikova AL, Bert NA, Romanov AE (2005) Investigation of dislocation loops with As-Sb nanoclusters in GaAs. Journal of Applied Physics 97: 024309(1-10)

Chiu YP (1977) On the stress field due to initial strain in a cuboid surrounded by an infinite elastic space. Journal of Applied Mechanics 44: 587–590

Chu SNG and Nakahara S (1990) ½ <100> {100} dislocation loops in zinc blend structure. Applied Physics Letters 56: 434–436

Claverie A, Liliental-Weber Z (1993) Extended defects and precipitates in LT-GaAs, LT-InAlAs and LT-InP. Material Science and Engineering B 22: 45–54

DeBoeck J, Oesterholt R, VanEsch A, Bender H, Bruynseraede C, VanHoof C, Borghs G (1996) Nanometer-scale magnetic MnAs particles in GaAs grown by molecular beam epitaxy. Applied Physics Letters 68: 2744–2746

Devis JH (1998) Elastic and piezoelectric fields around a buried quantum dot: a simple picture. Journal of Applied Physics 84: 1358–1365

Dregia SA, Hirth, JP (1991) A rebound mechanism for Lomer dislocation formation in strained layer structures. Journal of Applied Physics 69: 2169–2175

Duan HL, Karihaloo BL, Wang J, Yi X (2006) Compatible composition and critical sizes of alloyed quantum dots. Physical Review B 74: 195328(4)

Duan HL, Wang J, Huang ZP (2005) Eshelby formalism for nano-inhomogeneities. Proceedings of the Royal Society of London A 461: 3335–3353

Eshelby JD (1957) The determination of the elastic field of an ellipsoidal inclusion. Proceedings of the Royal Society of London A 241: 376–396

Eshelby JD (1959) The elastic field outside in ellipsoidal inclusion. Proceedings of the Royal Society of London A 252: 561–569

Faleev NN, Chaldyshev VV, Kunitsyn AE, Tret'yakov VV, Preobrazhenskii VV, Putyato MA, Semyagin BR (1998) High-resolution x-ray diffraction study of InAs-GaAs superlattices grown by molecular-beam epitaxy at low temperature. Semiconductors 32: 19–25

Fischer FD, Böhm HJ (2005) On the role of the transformation eigenstrain in the growth or shrinkage of spheroidal isotropic precipitations. Acta Materialia 53: 367–374

Fischer FD, Böhm HJ, Oberaigner ER, Waitz T (2006) The role of elastic contrast on strain energy and the stresses state of a spheroidal inclusion with a general eigenstrain state. Acta Materialia 54: 151–156

Freund LB, Suresh S (2003) Thin film materials: Stress, defect formation and surface evolution. Cambrige University Press, Cambridge

Glas F (2001) Elastic relaxation of truncated pyramidal quantum dots and quantum wires in a half space: An analytical calculation. Journal of Applied Physics 90: 3232–3241

Goodier JN (1937) On the integration of the thermo-elastic equations. Philosophical Magazine 23: 1017–1032

Guha S, Madhukar A, Rajkumar KC (1990) Onset of incoherency and defect introduction in the initial stages of molecular beam epitaxical growth of highly strained $In_xGa_{1-x}As$ on GaAs(100) Applied Phys.ics Letters 57: 2110–2112

Hirth JP, Lothe J (1982) Theory of dislocations. Wiley, New-York

Hu SM (1989) Stress from a parallelepipedic thermal inclusion in a semispace. Journal of Applied Physics 66: 2741–2743

Jin-Phillipp NY, Phillipp F (1999) Defect formation in self-assembling quantum dots of InGaAs on GaAs: a case study of direct measurements of local strain from HREM. Journal of Microscopy 194: 161–170

Jogai B (2001) Tree-dimensional strain field calculations in multiple InN/AlN wurzite quantum dots. Journal of Applied Physics 90: 699–704

Kolesnikova AL, Romanov AE (1986) Circular dislocation-disclination loops and their application to boundary problem solution in the theory of defects. Physico-Technical Institute Preprint No 1019, Leningrad (in Russian)

Kolesnikova AL, Romanov AE (1987) Edge dislocation perpendicular to the surface of a plate. Soviet Technical Physics Letters 3: 272–274

Kolesnikova AL, Romanov AE (2004a) Virtual circular dislocation-disclination loop technique in boundary value problems in the theory of defects. Journal of Applied Mechanics 71: 409–417

Kolesnikova AL, Romanov AE (2004b) Misfit dislocation loops and critical parameters of quantum dots and wires. Philosophical Magazine Letters 84: 043510(1–4)

Kolesnikova AL, Romanov AE, Chaldyshev VV (2007) Elastic-energy relaxation in heterostructures with strained nanoinclusions. Physics of the Solid State 49: 667–674

Krotkus A, Coutaz JL (2005) Non-stoichiometric semiconductor materials for terahertz optoelectronics applications. Semiconductor Science and Technology 20: S142–S150

Kurtenbach A, Eberl K, and Shitara T (1995) Nanoscale InP islands embedded in InGaP. Applied Physics Letters 66: 361–363

Lavrent'eva LG, Vilisova MD, Preobrazhenskii VV, Chaldyshev VV (2002) Low-temperature molecular-beam epitaxy of GaAs: effect of excess arsenic on the structure and properties of the GaAs layers. Russian Physics Journal 45: 735–752

Ledentsov NN, Bohrer J, Bimberg D, Kochnev IV, Maximov MV, Kop'ev PS, Alferov ZhI, Kosogov AO, Ruvimov SS, Werner P, Go/sele U (1996) Formation of coherent superdots using metal-organic chemical vapor deposition. Applied Physics Letters 69: 1095–1097

LeGoues F K, Tersoff J, Reuter MC, Hammar M, Tromp R (1995) Relaxation mechanism of Ge islands/Si(001) at low temperature. Applied Physics Letters 67: 2317–2319

Liu X, Prasad A, Nishio J, Weber ER, Liliental-Weber Z, Walukiewicz W (1995) Native point defects in low-temperature-grown GaAs. Applied Physics Letters 67: 279–281

Liu Y, Cao YG, Wu HS, Xie MH (2005) Coherent and dislocated three-dimensional islands of $In_xGa_{1-x}N$ self-assembled on GaN(0001) during molecular-beam Epitaxy. Physical Review B 71: 153406 (1–4)

Luysberg M, Sohn H, Prasad A, Specht P, Liliental-Weber Z, Weber ER, Gebauer J, Krause-Rehberg R (1998) Effects of the growth temperature and As/Ga flux ratio on the incorporation of excess As into low temperature grown GaAs. Jornal of Applied Physics 83: 561–566

Martin GM (1981) Optical assessment of the main electron trap in bulk semi-insulating GaAs. Applied Physics Letters 39: 747–749

Merdzhanova T, Kiravittaya S, Rastelli A, Stoffel M, Denker U, and Schmidt OG (2006) Dendrochronology of strain-relaxed islands. Physical Review Letters 96: 226103

Mindlin RD, Cheng DH (1950) Thermoelastic stress in the semi-infinite solid. Journal of Applied Physics 21: 931–933

Moreno M, Kaganer VM, Jenichen B, Trampert A, Daweritz L, Ploog KH (2005) Micromechanics of MnAs nanocrystals embedded in GaAs. Physical Review B 72: 115206

Moreno M, Trampert A, Daweritz L, Ploog KH (2004) MnAs nanoclusters embedded in GaAs: synthesis and properties. Applied Surface Science 234: 16–21

Mura T (1987) Micromechanics of defects in solids. Martinus Nijhoff, Dordrecht

Onaka S (2001) Averaged Eshelby tensor and elastic strain energy of a super spherical inclusion with uniform eigenstrains. Philosophical Magazine Letters 81: 265–272

Onaka S, Sato H, Kato M (2002) Elastic states of doughnut-like inclusions with uniform eigenstrains treated by averaged Eshelby tensors. Philosophical Mahazine Letters 82: 1–7

Romanov AE, Beltz GE, Fischer WT, Petroff PM, Speck JS (2001) Elastic fields of quantum dots in subsurface layers. Journal of Applied Physics 89: 4523–4531

Romanov AE, Waltereit P, Speck JS (2005) Buried stressors in nitride semiconductors: influence in electronic properties. Journal of Applied Physics 97: 043708(1–13)

Shimogishi F, Mukai K, Fukushima S, Otsuka N (2002) Hopping conduction in GaAs layers grown by molecular-beam epitaxy at low temperatures. Physical Review B 65: 165311(1–5)

Schwartzman AF, Sinclair R (1991) Metastable and equilibrium defect structure of II-VI/GaAs interfaces. Journal of Electronic Materials 20: 805–814

Shchukin VA, Bimberg D (1999) Spontaneous ordering of nanostructures on crystal surfaces. Reviews of Modern Physics 71: 1125–1171

Shchukin VA, Ledentsov NN, Bimberg D (2002a) Epitaxy of Nanostructures. Springer, Berlin

Shchukin VA, Ledentsov NN, Bimberg D (2002b) Entropy effects in the self-organized formation of nanostructures. In: Kotrla M, Papanicolaou NI, Vvedensky DD (eds) Atomistic Aspects of Crystal Growth", Kluwer, New-York, pp 397–409

Sears K, Wong-Leung J, Tan HH, Jagadish C (2006) A transmission electron microscopy study of defects formed through the capping layer of self-assembled InAs/GaAs quantum dot samples. Journal of Applied Physics 99: 113503 (1–8)

Seo K, Mura T (1979) The elastic field in a half space due to ellipsoidal inclusions with uniform dilatational eigenstrains. Journal of Applied Mechanics 46: 568–572

Stangl J, Holy V, Bauer G (2004) Structural properties of self-organized semiconductors nanostructures. Reviews of Modern Physics 76: 725–783

Stewart K, Buda M, Wong-Leung J, Fu L, Jagadish C, Stiff-Roberts A, Bhattacharya P (2003) Influence of rapid thermal annealing on a 30 stack InAsÕGaAs quantum dot infrared photodetector. Journal of Applied Physics 94: 5283–5289

Tanaka K, Mori T (1972) Note on volume integrals of the elastic field around an ellipsoidal inclusion. Journal of Elasticity 2: 199–200

Teodosiu C (1982) Elastic models of crystal defects. Springer, Berlin

Tichert C (2002) Self-organisation of nanostructures in semiconductor heteroepitaxy. Physics Reports 365: 335–432

Toropov AA, Lyublinskaya OG, Meltser BYa, Solov'ev VA, Sitnikova AA, Nestoklon MO, Rykhova OV, Ivanov SV, Thonke K, Sauer R (2004) Tensile-strained GaAs quantum wells and quantum dots in a $GaAs_xSb_{1-x}$ matrix. Physical Review B 70: 205314(1-8)

Tsuchida E, Arai Y, Nakazawa K, Jasiuk I. (2000) The elastic stress field in half-space containing a prolate spheroidal inhomogeneity to pure shear eigenstrain. Materials Science and Engineering A 285: 338–344

Vasyukov DA, Baidakova M.V, Chaldyshev VV, Suvorova AA, Preobrazhenskii VV, Putyato MA, Semyagin BR (2001) Structural transformations in low-temperature grown GaAs:Sb. Journal of Phys.ics D: Applied Physics 34: A15–A18

Wang YQ, Wang ZL, Shen JJ, Brown A (2003) Effect of dissimilar anion annealing on structures of InAs/GaAs quantum dots. Journal of Crystal Growth 252: 58–67

Wu LZ, Du SY (1995a) The elastic field caused by a circular cylindrical inclusion. 1. Inside the region $x_1^2+x_2^2<a^2$, $-\infty<x_3<\infty$ where the circular cylindrical inclusion is expressed by $x_1^2+x_2^2\leq a^2$, $-h<x_3<h$. Journal of Applied Mechanics 62: 579–584

Wu LZ, Du SY (1995b). The elastic field caused by a circular cylindrical inclusion. 2. Inside the region $x_1^2+x_2^2>a^2$, $-\infty<x_3<\infty$ where the circular cylindrical inclusion is expressed by $x_1^2+x_2^2\leq a^2$, $-h<x_3<h$. Journal of Applied Mechanics 62: 585–589

Wu LZ, Du SY (1999) The elastic field with a hemispherical inclusion. Proceedings of the Royal Society of London A 455: 879–891

Yoffe EH (1974) Calculation of elastic strain-spherical particle in a cubic material, Philosophical magazin 30: 923–933

Yu HY, Sandy SC (1992) Center of dilatation and thermal stresses in an elastic plate. Proceedings of the Royal Society of London A 438: 103–112

Yu KM, Kaminska M, Liliental-Weber Z (1992) Characterization of GaAs layers grown by low temperature molecular beam epitaxy using ion beam techniques. Journal of Applied Physics 72: 2850–2856

Zakharov ND, Werner P, Gösele U, Ledentsov NN, Bimberg D, Cherkashin NA, Bert NA, Volovik BV, Ustinov VM, Maleev NA, Zhukov AE, Tsatsul'nikov AF (2001) Reduction of defect density in structures with InAs-GaAs quantum dots grown at low temperature for 1.55 μm range. Material Research Society Symposium Proceedings 672: O8.5.1–O8.5.6

Zou J, Liao XZ, Cockayne DJH, Jiang ZM (2002) Alternative mechanism for misfit dislocation generation during high temperature Ge(Si)/Si (001) island growth. Applied Physics Letters 81: 1996–1998

Chapter 11
Capacitance-Voltage Spectroscopy of InAs Quantum Dots

D. Reuter

11.1 Introduction

Semiconductor quantum dots (QDs) have been studied intensively in the past [1, 2] as model systems for three-dimensional carrier confinement as well as building blocks of novel devices. Because they can be fabricated in high quality and provide strong confinement for electrons and holes, self-assembled InAs QDs – mostly grown by molecular beam epitaxy – are arguably the most intensively studied QD system. One of the most important consequences of the complete carrier confinement are the atomically sharp energy levels due to three-dimensional quantization. The detailed knowledge of the energy level structure, which determines the electrical and optical properties of the QDs, is very interesting from a fundamental point of view as well as for possible applications. Compared to the single particle picture, the energetic situation becomes more complicated, if the QDs are charged with more than one carrier because then interaction energies have to be taken into account [3, 4]. If the confinement length is small, as for example in InAs QDs, the interaction energies can be comparable or even larger than the quantization energies.

To study the energy level structure in QDs, several methods have been employed. On the one hand, there are optical methods [2] with photoluminescence (PL) as the most prominent one and on the other hand electrical methods as tunneling spectroscopy [5–8] and capacitance-voltage (C-V) spectroscopy [9–38]. The optical methods give a wealth of information but measure in general the conduction and the valence band system together. Also, the investigated QDs are in general electrically neutral although optical investigations of charge-tuneable QDs are possible in special heterostructures [39]. In contrast, electrical methods give information of the conduction or the valance band level structure separately. Also, the interaction of carriers in few-carrier QDs can be studied.

One of the most widely used electrical characterisation methods is C-V spectroscopy, for which several variations exist [9–38]. In general, the QDs have to be embedded into a doped heterostructure, so the charge in the QDs can be tuned by an external gate bias. Either p-n-junctions [27–29], n- [3, 9, 11–20, 28–37] as well as p-type [21–26] Schottky diodes or n-i-n structures [38] have been employed.

Z. M. Wang, *Self-Assembled Quantum Dots.*

Depending on the heterostructure used, one can distinguish roughly three types of capacitance techniques, which give different information:

a) Structures used for deep level transient spectroscopy (DLTS), were the InAs-QDs are embedded in a homogenously doped layer [27–34]. For DLTS, the QDs are charged by a short voltage pulse and then the capacitance is recorded as a function of time. The capacitance is related to the charging state of the QDs so this measurement allows to follow the de-charging process. By performing these measurements temperature dependent, one can determine the emission energies from the QDs. DLTS experiments have been performed by several groups for the electron as well as the hole system of InAs QDs. Very recently, high resolution measurements became feasible by an optimized layer sequence [30].
b) C-V depth profiling, where the QDs are embedded in a homogenously doped part of Schottky diode [35–38]. Here, the carrier depth profile reflects the position of the QD plane. E. g., one finds that small QDs result in an increased carrier concentration whereas bigger dots that contain dislocations cause strong carrier depletion. In such depth profiling measurements, the energy level structure due to the confinement is not resolved.
c) C-V spectroscopy employing a Schottky diode with the QDs embedded in the intrinsic region between back contact and gate, allows the charging of the QDs with individual carriers, e. g. measure the addition spectrum [3, 9–26]. This charging spectroscopy with and without applying a magnetic field has been employed by several groups in the past and from the addition spectra for electrons and holes valuable information on the conduction and valence band level structure as well as the carrier-carrier interaction has been obtained.

All three techniques probe, with few exceptions, dot ensembles with $\sim 10^5$ to 10^7 QDs. As a consequence, a narrow distribution in size and composition is required because otherwise interesting features might be obscured by inhomogeneous broadening. By sophisticated measurement techniques [11] smaller ensembles can be measured but it is so far not feasible to do capacitance spectroscopy of InAs QDs on a single dot level. The information gained by the three varieties of capacitance experiments are complementary, especially for a) and c). Unfortunately, different sample structures are necessary to obtain optimal results for DLTS and charging spectroscopy, so these measurements cannot be performed on the same QD ensemble. One has to grow different layer sequences and keep the growth conditions for the QDs as similar as possible to measure similar QD ensembles. Photoluminescence spectroscopy can be used to check for differences between the individual samples.

To review the work done with DLTS and C-V depth profiling is beyond the scope of this contribution and I will limit myself to charging spectroscopy of InAs QDs as described in c). Before discussing some aspects in more detail, I want to give a brief summary of milestones: C-V spectroscopy was first used by Drexler and co-workers [9] to investigate the conduction band states of self-assembled InAs QDs.

They were the first, who could demonstrate the controlled charging of individual electrons into the QDs and identify the role of Coulomb interaction for the charging spectra. By performing C-V spectroscopy in a perpendicular magnetic field (magnetic field perpendicular to the base plane of the QDs), it was possible to identify the nature of the individual charging peaks [3, 10, 11] and a basic model for the conduction band energy level structure could be derived [3]. Medeiros-Ribeiro and co-workers investigated the spin-splitting in the ground state and could determine the electron g-factor [18]. Luyken and co-workers [17], Medeiros-Ribeiro et al. [14], as well as Horiguchi and co-workers [15] investigated the dynamics of the charging, which occurs via tunnelling. Based on this work, Wibbelhoff et al. were able to map the momentum space wave functions corresponding to the individual electron charging peaks [20] by applying an in-plane magnetic field. High quality samples allowing C-V spectroscopy of the valence band states with good resolution became available only recently [22, 23] but already the first data showed the significant difference between the electron and the hole system. Magneto-C-V spectroscopy gave further insight in the nature of the individual charging peaks [24, 25] and together with wave function mapping experiments [26] it became clear that the situation is much more complex than for the conduction band system and sophisticated theories are needed to account for the experimental results [4, 39, 40]. In recent experiments, a two-dimensional electron gas was used as back contact and the charging of the QDs could be detected via careful Hall measurements [41].

11.2 Basics

11.2.1 Sample Design

Figure 11.1 shows a typical sample design for charging studies by C-V spectroscopy. The samples are generally grown by molecular beam epitaxy (MBE). The relevant part of the layer sequence starts with a doped back contact (doping concentration usually larger than $10^{18}\,\mathrm{cm}^{-3}$). To study conduction band states, an n-doped back contact has to be employed whereas for valence band state investigations, a p-type layer is required. This implies that electron and hole charging cannot be studied in the same samples with this type of layer sequence. One can only try to grow n- and p-type samples with QDs as similar as possible, which can be checked by photoluminescence measurements.

After the back contact, an undoped GaAs layer of thickness t_1 is deposited, which serves as tunnelling barrier. For n-type samples, t_1 is usually in the range of 25 nm but can be as large as 45 nm for the QD charging to occur on a reasonable timescale. Due to their higher effective mass, the tunnelling barrier for hole charging has to be significantly thinner. Typical values are in the range from 17 nm to 20 nm.

The tunnelling barrier is followed by a single layer of self-assembled InAs QDs. As already mentioned in chapter 1, the chosen growth conditions should result in an ensemble as homogeneous as possible. Over the years significant progress has

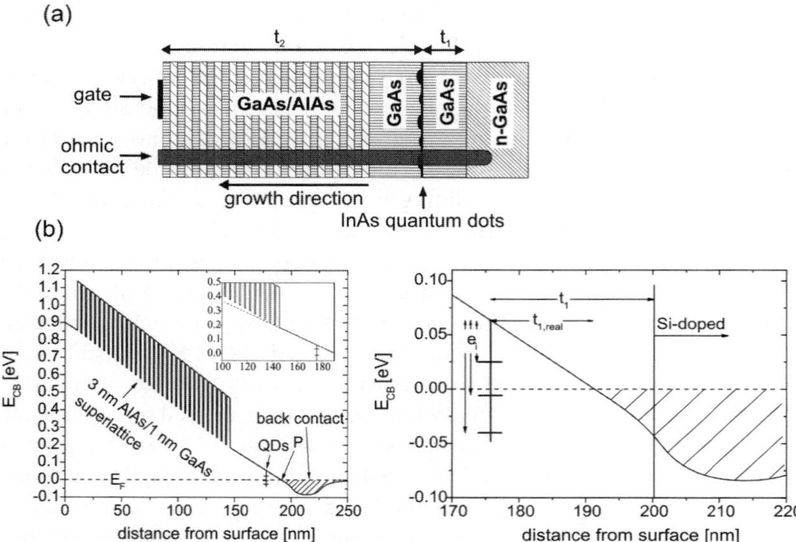

Fig. 11.1 (a) shows schematically a typical n-type heterostructure employed for charging spectroscopy of InAs QD conduction band states. In (b) the corresponding conduction band for zero gate bias as obtained by one-dimensional calculations [44] is depicted. The insert in the left panel shows a blow-up which illustrates the kink in the band due to the different permittivities of GaAs and AlAs. In the right panel, the transition region between doped back contact and the tunnel barrier can be seen in more detail. It is clearly visible that free carrier spill over into the undoped region reducing the effective thickness of the tunnelling barrier to $t_{1,\mathrm{real}}$. Please note that only the band structure of the Schottky diode without the QDs was simulated. The QD energy levels are only sketches as guide to the eyes. The dimensions of the individual layers correspond to the n-type sample for which the charging spectrum is shown in Fig. 11.2a

been made. Whereas in the first n-type samples, the first two charging peaks where not very well resolved it is now possible to distinguish up to six individual charging peaks in high quality samples (see Fig. 11.2 for an example).

The QDs are overgrown by GaAs with a typical thickness of 8 to 30 nm. This layer is followed by an AlAs/GaAs superlattice and a GaAs cap layer. Often uncapped InAs QDs are grown on the sample surface for atomic force microscopy imaging. The thickness of the superlattice is chosen such that the overall distance $t_1 + t_2$ is roughly 6 to 12 times larger than the thickness t_1 of the tunnelling barrier. The superlattice has in the final device the function to block the current flow to the top gate. This could in principle also be reached by employing pure AlAs or a random $\mathrm{Al_xGa_{1-x}As}$-alloy but several groups found that a superlattice with a high average Al content (50 % or larger) minimizes the leakage currents best. Charging spectroscopy employing structures without the AlAs/GaAs superlattice is possible but the gate voltage range is limited in forward direction.

From the heterostructures described above, Schottky diodes are fabricated by the processing of ohmic contacts and metallic top gates employing standard techniques as photolithography and lift-off. As gate metallization, Au or Cr/Au is usually

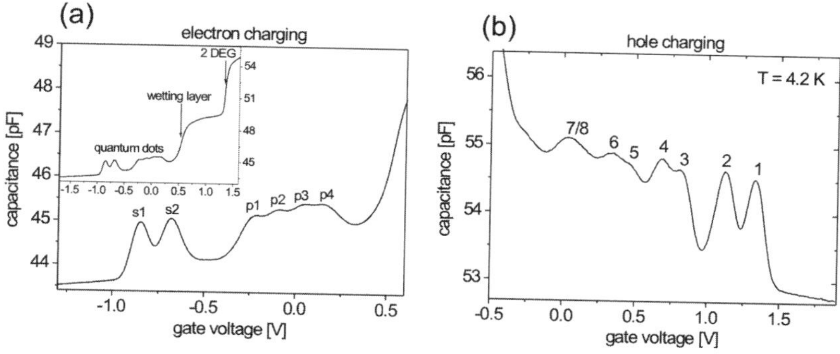

Fig. 11.2 Charging spectra for conduction (a) and valence (b) band states of InAs QDs measured at 4.2 K. The insert in (a) shows a larger gate bias range so that the charging of the wetting layer and the two-dimensional electron gas (2DEG) can be seen. Please note that the sign of the gate voltage corresponds to a grounded back contact. The labels for the individual charging peaks are kept throughout this article

employed and the lateral gate dimensions are typically in the range of several 10 μm to a few hundred μm.

11.2.2 Measurement Principle

Experimentally, one performs charging spectroscopy by measuring the device capacitance as function of an applied DC gate bias employing a small AC bias superimposed on the DC component. For certain DC bias values charging peaks, which are related to the QD energy levels, occur. The experiments are performed at low temperature so that the relevant energies are much smaller than kT. For InAs QDs, 4.2 K is sufficient and no extremely low temperatures are required. In the following, I will discuss in more detail, for which bias values charging is observed, how the voltage values are related to relevant energies of the QDs, and which other information can be extracted from a C-V spectrum.

In Fig. 11.1b, the conduction band diagram for a typical n-type sample is shown. By applying a DC bias to the top gate, the band in the intrinsic region is, in first approximation, tilted with the point P as a pivot. Due to this band tilting, the energy levels in the QDs shift with respect to the Fermi level in the back contact. Under resonance conditions, a carrier can tunnel into the QD. Due to the small AC component of the bias, the QD is periodically charged and de-charged, which results in a charging current corresponding to an increase in the device capacitance (see Fig. 11.2 for sample spectra). Because all levels below the Fermi energy are occupied, tunnelling into an excited QD state does not probe the single particle energies but determines the addition spectrum. The features related to the QDs are superimposed on a large background due to the voltage dependent capacitance of the Schottky diode. The following aspects of charging spectra as shown in Fig. 11.2 give information about the QDs:

a. One central point is the position of the charging peaks on the gate voltage scale. By converting the gate voltage scale to an energy scale, one obtains information on the quantization and the interaction energies. By measuring in magnetic field, one can observe the magnetic field dispersion, which might give information on the effective mass and can help to identify the nature of the charging peak.
b. The area under the charging peaks is related to the number of QDs under the gate. With known gate area, the QD density can quite easily be calculated by integrating over the first charging peak (or the first two, if the resolution is not sufficient). For the integration, one has to take into account that not the full gate voltage V_G is effective in the plane of the QDs but only $t_1/(t_1 + t_2)V_G$.
c. At high frequencies, the height of the charging peaks as function of the measurement frequency gives also information of the tunnelling rate into the QDs. By measuring at frequencies comparable to the tunnelling rate between back contact and QD layer, the charging dynamics can be determined. Furthermore, one becomes sensitive to the tunnelling rate, which can be used to perform wave function mapping. This is a rather new development and it will be discussed in more detail in chapter 4.

To extract energy values from the C-V spectra, one has to discuss the resonance condition in more detail. I will do this exemplary for the conduction band states. The valence band states can be treated analogously. The QD will be charged with the Nth electron when the gate voltage dependent overall energy of the N-electron state $E_N(V_G)$ is equal to the energy $E_{N-1}(V_G)$ of the (N-1)-electron state [2, 3], i.e. at the value for V_G for which this condition is fulfilled the Nth charging peak will occur.

Under the assumption that the quantization and the interaction contributions are separable ("constant interaction model"), $E_N(V_G)$ can be written in the following form:

$$E_N(V_G) = \sum_{i=1}^{N} e_i + \sum E_{C,QD} + \sum E_{ES}(V_G) + \sum E_{mirror} + \sum E_{C,QD-QD} \tag{11.1}$$

Here the first term is the sum over the single particle energies e_i (measured from the edge of the *GaAs* conduction band, see Fig. 11.1b) for the N electrons. Each of the levels is two-fold spin-degenerated but several e_i, e_j might be degenerated due to the QD symmetry. The second term contains the interactions (Coulomb and exchange) between electrons in the same QD and the third term describes the gate voltage dependent electrostatic energy due to the field in the Schottky diode. The forth term takes the mirror charge effects due to the interaction between the electrons in the QD and the gate/back contact into account. The Coulomb repulsion between electrons in different QDs is described by the last term of Eq. (11.1). The first three terms are most important whereas terms four and five bring corrections that can amount to up to 30 % but are usually smaller [13]. Especially the last term can only be treated by numerical simulation [13], so I will omit it for the following quantitative discussion and discuss it qualitatively afterwards. Using this approach, I will exemplarily discuss the first three charging peaks for the electron system, where the single particle levels are filled sequentially as known from measurements in magnetic field [3, 11]. The energies up to N = 3 can be written as:

$$\begin{aligned} E_0 &= 0 \\ E_1 &= -e_{1,e} + E_{ES}(V_G) - E_{mirror,1} \\ E_2 &= -2e_{1,e} + E_C^{11} + 2E_{ES}(V_G) - E_{mirror,2} \\ E_3 &= -2e_{1,e} - e_{2,e} + E_C^{11} + 2E_C^{12} - E_{ex}^{12} + 3E_{ES}(V_G) - E_{mirror,3} \end{aligned} \tag{11.2}$$

As discussed above, the first electron will tunnel into the dot for a gate voltage $V_{G,1}$ for which E_0 is equal E_1. With (11.2), this results in

$$e_{1,e} = E_{ES}(V_{G,1}) - E_{mirror,1} \tag{11.3}$$

To first order the mirror charge effect can be neglected and one has to determine $E_{ES}(V_{G,1})$ to obtain e_1. In the literature, different approaches to calculate E_{ES} as function of the gate voltage are used. The most common approach is the lever arm method [3, 9, 13]. Assuming the interface between back contact and tunnelling barrier as centre of rotation and neglecting the charge in the QD layer, one obtains from simple geometrical considerations for E_{ES}:

$$\begin{aligned} E_{ES} &= e\frac{t_1}{(t_1+t_2)}(V_{bi} - V_G) \quad \text{for electrons} \\ E_{ES} &= e\frac{t_1}{(t_1+t_2)}(V_{bi} + V_G) \quad \text{for holes} \end{aligned} \tag{11.4}$$

Here e is the elementary charge and V_{bi} is the built-in voltage. Please note that in (11.4) the built-in voltage is measured with respect to the conduction band for electrons and with respect to the valence band for holes. Although, the lever arm law is the most widely used approach, there are some pitfalls: The exact value for V_{bi} is not known and differs in different publications [19, 22, 42]. As seen a little later, this uncertainty does not matter if only energy differences between the charging peaks are of interest because then V_{bi} cancels out, so normally literature values for V_{bi} are used. When absolute energy values with respect to the GaAs conduction band edge are to be determined, an error in V_{bi} gives inaccurate values for the energies. Therefore, Bock and co-workers as well as Granados and Garcia have taken special care to determine the built-in voltage in their samples carefully. Bock and co-workers used selective etching to produce sample with different t_2 and could determine V_{bi} employing (11.4). Granados and Gracia have grown a reference sample without QDs and from the flat-band voltage, they could determine V_{bi}. Both approaches are a clear progress in enhancing the accuracy of the V_{bi} value and it seems that the uncertainty is $\sim 5\%$. Unfortunately, both methods require significant additional experimental work.

The lever arm law as expressed in (4) and generally used in literature neglects the difference in dielectric permittivity ε_r between GaAs and the AlAs/GaAs superlattice. This difference causes a kink in the band diagram as can be seen clearly in the insert of Fig. 11.1b. The different ε_r can be taken into account by approximating the Schottky diode by a plate capacitor filled with different dielectrics. This approach gives the following relation:

$$E_{ES} = e\frac{t_1}{d_1 + \frac{\varepsilon_{GaAs}}{\varepsilon_{SPS}}d_2 + d_3}(V_{bi} - V_G) \text{ with } \varepsilon_{SPS} = \frac{(d_{GaAs} + d_{AlAs})\varepsilon_{GaAs}\varepsilon_{AlAs}}{\varepsilon_{GaAs}d_{AlAs} + \varepsilon_{AlAs}d_{GaAs}} \tag{11.5}$$

Here d_1 is the thickness of the GaAs-cap layer, d_2 the thickness of the superlattice and d_3 the distance between back contact and superlattice. d_{GaAs} and d_{AlAs} denoted the thickness of the individual layers in the superlattice. ε_{GaAs} and ε_{SPS} are the permittivity of GaAs and the effective permittivity of the superlattice, respectively. Assuming a constant permittivity ε_r as usually done, gives larger values for E_{ES} compared to the results of (5), because AlAs has a smaller dielectric permittivity than GaAs. The differences are, as can be seen in Table 11.1, for typical samples in the 20–30 meV range for the absolute values of E_{ES} and 10–15% when energy differences between the charging peaks are considered. To incorporate the differences in ε_r is hampered by the fact the permittivites at low temperatures are not very well known. One has usually to resort to extrapolation formulas [43].

Another problem with the lever arm approach is that it does not take the voltage dependent depletion of the back contact into account. This problem is difficult to

Table 11.1 Electrostatic energies corresponding to the individual charging peaks shown in Fig. 11.2a and 11.2b. The values in brackets give the energy difference between the Nth and the (N–1)th charging peak. Results for three different approaches as discussed in the text are given. A built-in voltage of $V_{bi} = 0.9$ V (0.765 V) was employed for electrons (holes). An ε_r of 12.4 for GaAs and 9.61 for AlAs, respectively, was used [43]

charging peak electrons	V_G/[V]	E_{ES}/[meV] lever arm	E_{ES}/[meV] lever arm + different ε	E_{ES}/[meV] full simulation
s1	−0.847	216.7	187.3	152.8
s2	−0.666	195.4 (21.3)	168.9(18.4)	133.4(19.4)
p1	−0.217	139.4 (56.0)	120.5(48.4)	84.4(49.0)
p2	−0.086	123.1 (16.3)	106.4(14.1)	70.7(13.6)
p3	0.037	107.6 (15.4)	93.0(13.4)	58.2(12.5)
p4	0.178	90.1 (17.5)	77.9(15.1)	44.6(13.6)
charging peak holes				
1	1.242	207.7	180.3	190.5
2	1.036	186.4 (21.3)	161.9(18.4)	168.9(21.6)
3	0.743	156.0 (30.4)	135.5(26.4)	138.6(30.3)
4	0.590	140.3 (15.7)	121.8(13.7)	123.2(15.4)
5	0.376	118.1 (22.2)	102.5(19.3)	101.8(21.4)
6	0.231	103.1 (15.0)	89.5(13.0)	87.5(14.3)

tackle, because of the complicated situation at the interface between back contact and the tunnelling barrier (see Fig. 11.1b). Due to the rather high doping concentrations generally employed in the back contact and the corresponding large energy difference between the conduction (valence) band and the Fermi level (degenerate semiconductor), usually free charge spills over into the undoped region. This means a region with positive as well with negative space charge exists and any approach, which neglects this fact, is questionable. The charge spill-over results in most cases in a distance $t_{1,real}$ between free carriers in the back contact and the QDs that is smaller than t_1 used in the lever arm law. Thus, it is evident that Eq. (11.4) as well as (11.5) in general gives too large values for E_{ES}. There is no analytical expression to take the charge spill-over into account but one can use one-dimensional simulations solving the Poisson equation numerically for a given value of V_G and determine E_{ES} from the calculated band diagram. We did this exemplarily for the two samples from Fig. 11.2 employing a freeware Poisson solver by G. Snider [44]. The results are listed in Table 11.1 together with the values calculated by the lever arm law (11.4) and the results of Eq. (11.5). Comparing the results of Eq. (11.5) to the simulation results, it can be seen that the spill over has different effects on the n- and p-type samples. Whereas the full simulation gives even lower energies for n-type samples the values for E_{ES} are larger for p-type samples. This is due to the fact that the spill over is much smaller for the p-type sample because for a given dopant concentration, the difference between the Fermi level and the band edge is much smaller for holes because of the significantly larger density of states in the valence band. Interesting is the behaviour of the energy differences between charging peaks: For the full simulation, the differences become larger compared to the results of (11.5) approaching the values obtained by the simple lever arm law (n-typ) or even exceeding them slightly (p-type). The discrepancies between full simulation and simple lever arm law are for typical samples up to 25 % for the absolute values of E_{ES} and 10 % when considering energy differences. The deviations might be larger for disadvantageous layer sequences (high doping level, small t_1). For the extraction of $e_{1,e}$ from the C-V data according to (3) the variations in E_{ES} directly translate to variations in $e_{1,e}$ (see Table 11.2). The values for e_1 obtained for our samples presented here exemplarily are in good agreement with values determined by other groups [11, 19, 22] for similar QDs. The results also agree well with theoretical calculations [4].

Let us now see, how large the corrections due to the mirror charge effect are. The usual approximation to account for $E_{mirror,N}$ is to assume a point charge of Ne at

Table 11.2 Ground state energies e_1 and ground state Coulomb energies E_c^{11} for the n- and p-type samples from Fig. 11.2. The energies are extracted from the E_{ES} values given in Table 11.1 determined by a full simulation of the conduction band diagram.

energy	electron /[meV] no mirror charge	electron /[meV] with mirror charge	holes /[meV] no mirror charge	holes /[meV] with mirror charge
e_1	152.8	151.7	190.5	188.8
E_c^{11}	19.4	22.9	21.6	26.7

a distance t_l from a metal plate (highly conductive back contact) [13]. It is a good approximation, to consider only the mirror charge in the back contact, because the top gate is much further away. In this case $E_{mirror,N}$ can be written as

$$E_{mirror,N} = \frac{(Ne)^2}{16\pi\,\varepsilon_0\varepsilon_r t_1} \tag{11.6}$$

From (6) it becomes clear that the mirror charge effect becomes more important for thinner tunnelling barriers. In Table 11.2, the e_1-values for the two samples from Fig. 11.2 are listed. By taking the mirror charge effect into account, the values for e_1 become smaller but only around one percent. The effect is more pronounced for the p-type sample because here t_1 is smaller (17 nm compared to 25 nm).

As already mentioned above, the effect of Coulomb interaction between electrons in different QDs is difficult to treat quantitatively. Medeiros-Ribeiro and co-workers did sophisticated calculations [13] and the reader is referred to their work for an in-depth discussion. In the following, I will only summarize their main findings and give qualitative arguments: The Coulomb repulsion between carriers in different QDs will result in a smaller value for $E_{ES}(V_{G,l})$ for which charging occurs, i. e. e_l will be underestimated. This effect is with a few meV only in the 1 % range. Because the inter-dot Coulomb interaction increases with increasing charge on the dot, it is stronger for the second charging peak. This difference in shifts leads to an overestimation of the Coulomb blockade energies $E_{C,QD}$ within the QD. Also, the fluctuations in the inter-dot distances will cause fluctuations of the inter-dot Coulomb interaction, which will result in a peak broadening in addition the broadening due to the different e_i. All the effects discussed are quite small (~1 % for absolute energies, < 10 % for peak width and energy differences) if the tunnelling barriers are 25 nm or thinner because in this case the Coulomb interaction is screened quite effectively by the highly conductive back contact. For thicker tunnelling barriers the effect of inter-dot Coulomb interaction might be as large as 30 % of the intra-dot interactions, e. g., the ground state Coulomb blockade discussed in the following paragraphs. In general, also a lower QD density helps to reduce the effect of inter-dot interactions. For the reasons discussed above, we will neglect the inter-dot interactions in all following discussions.

The charging of the first carrier into the dot is not affected by any Coulomb interaction $E_{C,QD}$ but this changes drastically, when the second carrier is added. From the resonance condition and (2) one obtains

$$E_{ES}(V_{G,2}) = e_{1,e} - E_C^{11} + E_{mirror,2} - E_{mirror,1} \tag{11.7}$$

Combining (3) and (7), one sees that the energy difference between charging peak one and two is given by the Coulomb interaction (Coulomb blockade) between the two carriers in the QD ground state corrected by the mirror charge. In Table 11.2, the ground state Coulomb blockade energies are given for our two standard samples with and without mirror charge effects. The values for the holes are slightly larger than for the electron system which points to a spatially more confined wave function for the hole ground state. This is consistent with the higher hole mass. Without considering E_{mirror} the Coulomb energy is underestimated. Whereas for n-type sample the effect is ~15 %, it is around 22 % for the p-type sample. The larger effect for the p-type sample is due to the smaller t_1 in these layer structures, which is required by the slower tunnelling dynamics (larger mass). The values for E_C^{11} given in Table 11.2 are similar to those reported by other groups [11, 19, 22].

The experimentally determined Coulomb energies are important quantities that can be compared to theory. In a simple picture [45], it can be described by the energy required to add another carrier to a small island with the capacitance C_{s-QD}. One obtains

$$E_C^{11} = {e^2}/{C_{s-QD}} \tag{11.8}$$

This simple model explains well, why a Coulomb blockade exists and allows to estimate the order of magnitude but it cannot account correctly for all experimental observations. E. g., ${E_C}^{11}$ should be the same for electrons and holes, but the values for holes are approximately 15 % larger (see Table 11.2). To account for this, more complicated models, which take the wave function shape corresponding to the individual levels into account have to be used [3, 4].

In analogy to the calculations presented above, one can determine when the third carrier is charged. One obtains

$$E_{ES}(V_{G,3}) = e_{2,e} - 2E_C^{12} + E_{ex}^{12} + E_{mirror,3} - E_{mirror,2} \tag{11.9}$$

From (7) and (9), one can extract the meaning of the energy difference between the charging peaks 2 and 3:

$$\begin{aligned} E_{ES}(V_{G,3}) - E_{ES}(V_{G,2}) = e_{2,e} - e_{1,e} - 2E_C^{12} + E_{ex}^{12} + E_C^{11} + E_{mirror,3} \\ - 2E_{mirror,2} + E_{mirror,1} \end{aligned} \tag{11.10}$$

Obviously, this energy difference is composed from the quantization energy between the ground state and the first excited state as well as various Coulomb contributions. This means that C-V spectroscopy gives in principle information about the quantization energies but they can only be extracted when the Coulomb contributions are known. Here, comparisons with theory or comparison to other experiments measuring directly the quantization energies, e. g. intraband absorption spectroscopy [10] are necessary. Qualitatively, it can be concluded that the quantization energy $e_1 - e_2$ is significantly larger for electrons than for holes because – as can be seen in Table 11.1 – the energy difference between peaks 2 and 3 is almost two times larger for the electron system although the Coulomb contributions should be quite similar (${E_C}^{11}$ differs only 15 %). The larger quantization energy for electrons is expected because of their smaller effective mass.

To discuss the charging peaks for N > 3 in a general way is not possible. One has to assume a concrete filling sequence, to establish which exchange and direct

Coulomb terms have to be taken into account. Additional experimental evidence has to be used to deduce a reasonable filling sequence. This will be discussed in detail for the electron and the hole system in chapter 3. Very nice discussions of the energies for the different N-carrier configurations are given in [3] and [4].

11.3 Magneto-C-V-Spectroscopy for Conduction and Valence Band States

11.3.1 Charging Spectra for Electrons Revisited

The charging spectra for conduction band states have been investigated first by Drexler and co-workers [9] in 1994. They have already observed charging spectra similar to those shown in Fig. 11.2 but the individual peaks were not so well resolved. Already in this work, measurements in a perpendicular magnetic field, i.e., field direction perpendicular to the base plane of the QDs/sample surface, were performed to identify the nature of the individual charging peaks. This work was extended Fricke and co-workers [10] as well as by Miller et al. [11] with better samples and higher magnetic fields. They concluded that the lateral confinement potential can be approximated by a two-dimensional oscillator and that due to the rotational symmetry, the single particle levels can be classified by their orbital angular momentum in analogy to atomic physics [3, 10, 11]. This is by and large also confirmed by more sophisticated calculations [4]. The dispersion in magnetic field is then given mainly by the dispersion of the single particle levels, because the Coloumb energies depend only weakly on the magnetic field [3]. The dispersion with perpendicular magnetic field of the single particle levels for a two-dimensional harmonic oscillator is long known [46] and allows to describe the magnetic field behaviour depicted in Fig. 11.3: The first two charging peaks show almost no dispersion and can be assigned to the charging of the twofold degenerated s-like ground state. The next two peaks shift downwards in energy whereas peaks 5 and 6 shift upwards in energy. This is consistent with the charging of the fourfold degenerated p-shell (orbital quantum number $l = \pm 1$). The downwards-shifting branches belong to $l = -1$. An open question remains the filling sequence at B = 0: For a perfect rotation symmetry in the confinement potential, one expects a filling according to Hund's rule ($l = -1$ followed by $l = 1$) whereas for an asymmetry in the potential, the p-level would split and favour a filling $l = -1$ followed by $l = -1$. From first sight, the dispersion measurements seem to support the second case (first two p-peaks shift downwards) but Warburton and co-workers [3] have calculated that already at very small fields of $\sim$ 1 T, the magnetic field forces a filling sequence $l = -1$ followed by $l = -1$. This small kink could probably be not resolved in the measurements. Also the larger distance between p2 and p3 compared to the distances p1–p2/p3–p4 should be observed for both filling sequences and cannot serve as a proper criterion. Because the filling sequences are energetically quite close and no large anisotropies are required,

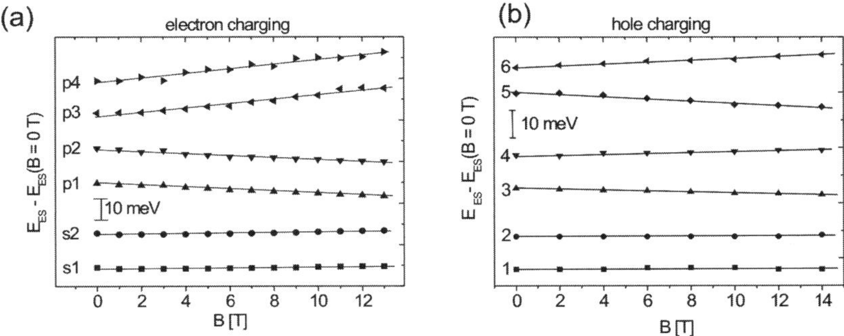

Fig. 11.3 $E_{ES} - E_{ES}(B = 0\,T)$ corresponding to the individual charging peaks as function of a perpendicular magnetic field. (a) shows the data for electrons and (b) for holes (for charging spectra of the samples see Fig. 11.2). E_{ES} was calculated by the simple lever arm law as described in the text. The lines are guide to the eyes

probably both cases exist depending on the exact sample details. Wave function maps as discussed in section 11.4 could help to decide this question for a particular QD ensemble but it is difficult to measure this with sufficient resolution for the p-states.

The Coulomb energies are slightly magnetic field depended because the wave functions depend on the magnetic field (magnetic confinement) [3]. In general, the Coulomb interactions increase with increasing magnetic field due to the additional confinement. For a detailed discussion of the magnetic field measurements, their description within a harmonic oscillator model and the dispersion of the Coulomb interactions, the reader is referred to [3].

11.3.2 Addition Spectra for Holes in InAs QDs

As one can see in Fig. 11.2b, for hole charging no shell-like structure with a fourfold degenerated p-shell as for electrons seems to be present. Rather, the charging peaks appear in pairs. The magneto-C-V data shown in Fig. 11.3b are needed to deduce the nature of the individual charging peaks. The first two peaks do not shift with magnetic field, which is consistent with the charging of an s-like ground state. Peaks 3–6 shift upward and downward in energy in an alternating fashion. Arguing along the same line as used for the conduction band states (effective mass approximation, rotational symmetry of the confinement potential, no spin-orbit coupling), the data seem at first sight to point to the filling of a fourfold degenerated p-shell according to Hund's rule but this in contrast to the large energy difference between peak 4 and 5 which seems to be a hint to a strongly split p-shell. In addition, a quantitative analysis shows that peaks 5 and 6 shift twice as strong as peaks 3 and 4, which is unexpected, if all four charging peaks belong to the same orbital angular momentum. We proposed an incomplete shell filling were peaks 5 and 6 belong to the filling of the d-shell and the filling within the shells would occur according to Hund's rule [25]. Wave function maps showed that this explanation is not completely satisfying [26]: One observes that the wave functions corresponding to peaks 3 and 4 show the same symmetry (see chapter 4 for a detailed discussion) but under the assumptions made above, one expects orthogonal wave functions for these two peaks. Climente and co-worker [40] as well Bester and co-workers [47] showed that one has to consider spin-orbit coupling as well light-heavy hole mixing, to account for the dispersion in perpendicular field as well as for the symmetry of the wave functions. Climente and co-workers predict sequential shell filling, i. e., a completely filled p-shell for six holes, whereas Bester and co-workers assign peaks 5 and 6 to charging of a d-like state, i. e., calculate a non-sequential shell filling. Because both groups can reproduce the magnetic field dispersion, we cannot discriminate between these two scenarios from our data. Therefore, the question if non-sequential shell filling occurs for hole charging is open at the moment. Maybe calculations of the wave functions might give more insight because it might be possible to discriminate between p- and d-like wave functions.

11.4 Wave Function Mapping by C-V Spectroscopy

11.4.1 Measurements Principle

As discussed in section 11.2, the charging and de-charging of the QDs occurs via tunnelling. Therefore, one expects that the height of the charging peaks should decrease if the AC frequency becomes comparable to the tunnelling rate. This was experimentally observed first by Medeiros-Rebeiro and co-workers [14], Horiguchi et al. [15] as well as Lyken and co-workers [17] for electron charging and by us for holes [23]. In Fig. 11.4, frequency dependent measurements for a p-type sample showing charging spectra very similar to those depicted in Fig. 11.2b are shown. It can be seen that the ground state peaks are suppressed at lower frequencies than the excited states. This is due to the fact that because of the band tilting, the height of the tunnelling barrier becomes smaller for the excited states. Also the effective length $t_{1,real}$ of the tunnelling barrier becomes smaller for tunnelling into excited states because of the voltage dependent spill over effect discussed in chapter 2. By a quantitative evaluation of the frequency dependence, the tunnelling rate τ^{-1} can be determined as outlined in [17]. In a good approximation, τ for the individual charging peaks can be determined from the frequency $f_{0.5}$, for which the height of the capacitive signal C_{QD} of the corresponding charging peaks has decreased to half the value observed at very low frequencies (f $\rightarrow$ 0). The following relation is valid:

$$\tau = {}^{1}\!/_{2\pi f_{0.5}} \tag{11.1}$$

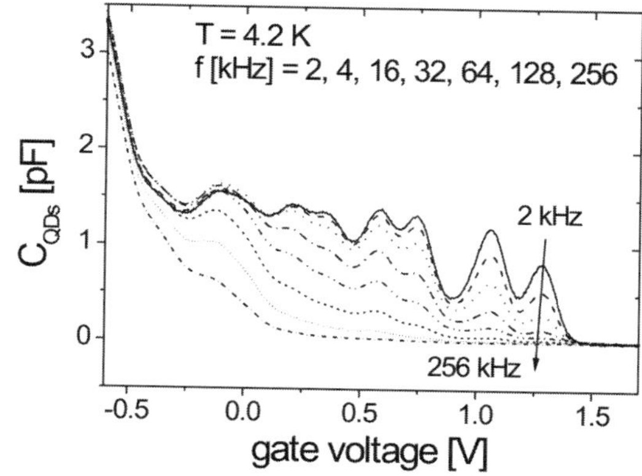

Fig. 11.4 C-V spectra for a p-type sample for different frequencies. The tunnelling barrier for this sample is $t_1 = 19\,\text{nm}$ and a linear background was subtracted from the data

Comparing the tunnelling rates for n- and p-type samples, one finds much lower tunnelling rates for holes. A quantitative evaluation shows that the holes tunnel with predominantly heavy hole character [23].

The correlation between tunnelling rate and height of the capacitance signal can also be exploited in a different way: A measurement frequency of approximately $f_{0.5}$ is chosen and the height of the capacitance signal is measured as function of an in-plane magnetic field $\mathbf{B}_{||}$. Because the signal height is directly correlated to τ^{-1}, one maps by this kind of magneto-C-V spectroscopy τ^{-1} as a function of $\mathbf{B}_{||}$. For small changes in C_{QD}, the signal height is directly proportional to τ^{-1} but for large variations a more complicated relationship is valid [17].

In analogy to magneto-tunnelling spectroscopy, one can show that the tunnelling rate τ^{-1} is proportional to the square of the momentum space wave function of the corresponding charging peak [5, 6]. The in-plane magnetic field adds an additional in-plane momentum (see Fig. 11.5 for a sketch of the geometry) and determines at which $\mathbf{k}_{||}$-vector the QD wave function is probed. The following relation holds approximately:

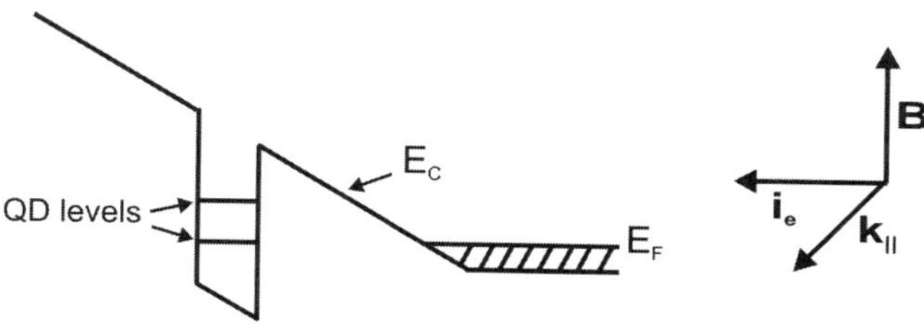

Fig. 11.5 Sketch of the directions relevant for the in-plane momentum generation by an in-plane magnetic field in case of electron tunnelling. For tunnelling holes, $k_{||}$ points in the opposite direction due to the different sign of the charge. Please note that i_e means here the particle current and not the electrical current

$$C_{QD}(B_x, B_y) \sim {}^1\!/_{\tau(k_x,k_y)} \sim \left|\phi_{QD}(k_x,k_y)\right|^2 \quad \text{with } k_{||} = \frac{et_1}{\hbar}B \tag{11.2}$$

In (2), C_{QD} is the height of an individual charging peak and ϕ_{QD} is the wave function that corresponds to this charging peak. Whereas in magneto-tunnelling spectroscopy, it is quite clear that the single particle momentum space wave function is probed this is not the case for charging spectroscopy where in general the difference of multi-carrier states are probed. This is discussed by Rontani and Molinari [48]. Their calculations show that for strongly confined systems, as InAs QDs, the probability distribution probed by charging spectroscopy is very close to the single particle value. In addition, it is possible to calculate directly the quasi-particle probability distribution measured by C-V spectroscopy and compare it to experimental results.

In Fig. 11.6, the wave function maps for the conduction band states are shown. For the first two charging peaks, they are at maximum for $k_{||} = 0$ and decrease monotonically towards larger $k_{||}$ values. This is expected for an s-like ground state with no orbital momentum and is consistent with the dispersion behaviour in perpendicular magnetic field. In addition, one sees that the wave function in momentum space representation is elongated along the [11] (real space) crystal axis. This means that the wave function in real space representation is elongated along the [0–11] direction. Such an anisotropy was deduced by other authors indirectly from absorption measurements [10] but only wave function maps visualise it directly. For the p1p2 and p3p4 charging peaks, a node like structure around $k_{||} = 0$ and maximum signal at finite k-values is observed as expected for an excited state. The bone like pattern points also to an anisotropic confinement potential because otherwise a distribution with rotational symmetry should be observed. p1p2 are orthogonal to p3p4 which is a hint that p1p2 belong to the same orbital state so the filling sequence is not according to Hund's rule as already speculated in section 11.3. The wave function symmetries as well as the magnetic field dispersion can be well understood by assuming an asymmetric two-dimensional harmonic confinement potential, a single effective mass and neglecting any spin-orbit interaction [3, 20]. This is not the case for valence band states as discussed in the following.

For holes, the first two charging peaks (see Fig. 11.7) look qualitatively very similar to those for electrons. They belong to charging of an s-like ground state as already concluded from magneto-C-V spectroscopy in a perpendicular magnetic field. The hole ground state wave function shows the same anisotropy as the corresponding electron wave function. This points to a structural reason for this anisotropy, either due to an elliptical shape of the dots or due to the ZnS crystal structure. In case of a strong piezoelectrical contribution to the confinement potential as discussed in literature [49], the electron and hole wave functions should show orthogonal anisotropy, which is not observed for our QDs [26].

For the charging peaks 3 to 6 a node like structure around $k = 0$ and maximum signal for finite **k**-vectors is observed. This agrees with the assignment as excited states with finite orbital angular momentum based on the magnetic field dispersion discussed in section 11.3. However, when trying to interpret our data in analogy

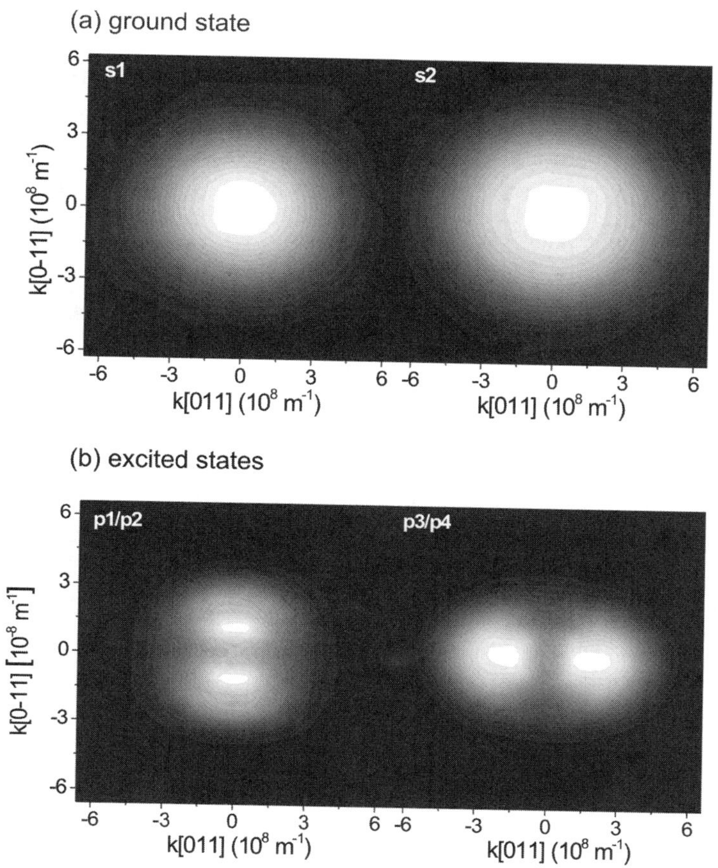

Fig. 11.6 The height of the capacitance signal corresponding to the individual electron charging peaks is shown as function of the in-plane momentum $k_{||}$ generated by the in-plane magnetic field up to 11 T. Bright corresponds to a high and dark to a low signal height. (a) shows the results for the electron ground state (s1, s2) and (b) depicts the data for the excited states (p1/p2 and p3/p4). For this sample the resolution was not sufficient to measure all four p-peaks separately

to the electron system in a two-dimensional effective mass approximation without including spin-orbit coupling, we discover a severe contradiction: As for the electron system, one would expect an orthogonal symmetry for the wave functions corresponding to peaks 3 and 4 because of the shift in opposite direction in perpendicular magnetic field. However, in the experiments, we observe the same symmetry in the wave functions for peaks 3 and 4. Similar arguments are also valid for peaks 5 and 6. The apparent discrepancy between the results obtained by the wave function mapping and the measurements in perpendicular field can only be resolved by taking spin orbit coupling into account [39, 40]. Qualitatively the argumentation is the following: The charging peaks 3 and 4 belong to a two-fold degenerated state with the overall angular momentum $F_z = \pm 1/2$. This explains directly why one observes very similar wave functions because the same single particle state is charged. The

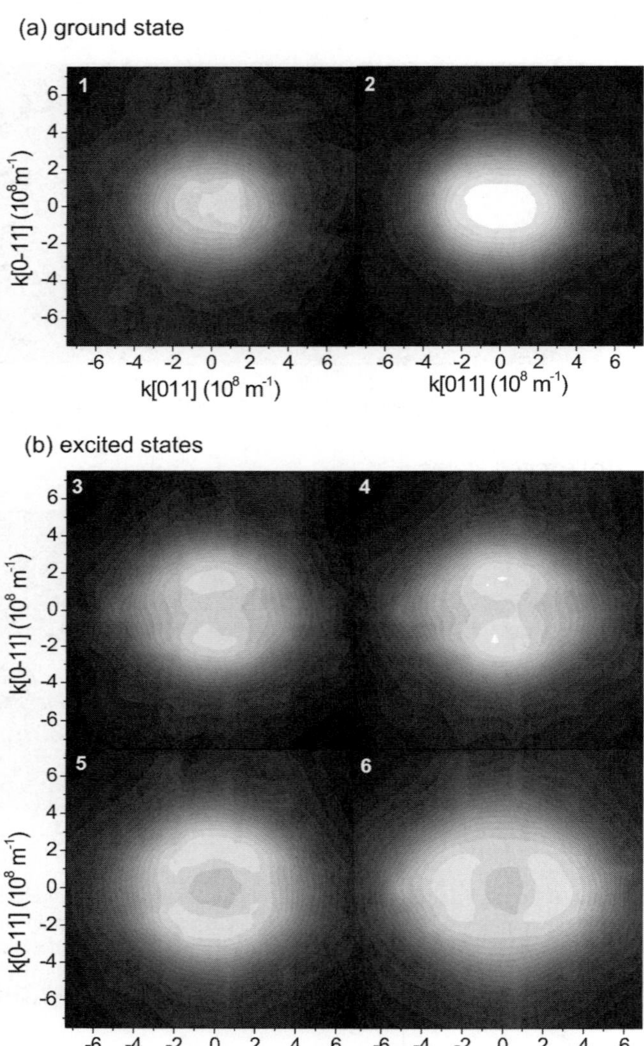

Fig. 11.7 Wave function maps for valence band states in InAs QDs for in-plane magnetic field up to 25 T. The height of the capacitance signal corresponding to the individual charging peaks is shown as function of the in-plane momentum $k_{||}$ generated by the in-plane magnetic field. Bright corresponds to a large and dark to a small signal. (a) shows the results for the hole ground state (1,2) and (b) depicts the data for the excited states (3–6). For charging spectra of this sample see Fig. 11.4. Please note that, because of the thinner tunnelling barrier t_1, for holes significantly higher magnetic fields were required to transfer the same in-plane momentum $k_{||}$ as for the n-type samples

$F_z = 1/2$ (–1/2) state corresponds to the orbital angular momentum $l = -1$ (1) and the hole Bloch character $J_z = 3/2(-3/2)$. In a perpendicular magnetic field, the dispersion is mainly determined by l, which means the $F_z = 1/2$ component will shift downwards whereas the $F_z = -1/2$ will shift upwards in energy.

From the above discussion, it can be seen that wave function mapping by C-V spectroscopy gives valuable information. Especially, it allows direct determination of anisotropies in the lateral confinement potential.

11.5 Summary

I have presented C-V spectroscopy as a powerful method to gain information on the energy level structure and the Coulomb interaction in InAs QDs by measuring the charging peaks. A certain disadvantage of the method is that the measurements have to be done on an ensemble and cannot be done on a single dot level. On the other hand, this ensures that the properties of typical QDs are measured. To obtain the energetic position of the charging peaks, one has to convert the gate voltage scale to an energy scale. I have discussed different methods and while the absolute energy values can differ quite significantly, the energy differences between the charging peaks are rather independent of the evaluation method. Therefore, the differences are better suited for comparison with theoretical calculations. Another important task when interpreting charging spectra is the assignment of the individual charging peaks to the corresponding single particle level. Here, experiments in perpendicular magnetic field are of inestimable help.

By performing C-V spectroscopy at AC frequencies comparable to the tunnelling rate, one becomes sensitive to the tunnelling dynamics. In combination with an in-plane magnetic field, this allows to map the in-plane momentum space wave functions corresponding to the individual charging peaks. This rather new measurement approach gives additional information for assignment of the individual charging peaks. Moreover, it visualises directly anisotropies in the confinement potential which are hard to access with other methods. This is one of the strength of the wave function mapping.

Acknowledgments I gratefully acknowledge the fruitful collaboration with P. Schafmeister, P. Kailuweit, R. Roescu, A. Ludwig, M. Richter, U. Zeitler, J. C. Maan, O. Wibbelhoff, and A. Lorke on the C-V spectroscopy of InAs QDs. I want to thank A. D. Wieck for giving me the opportunity to perform these experiments in his group. Financial support by the DFG and the BMBF is gratefully acknowledged.

References

1. for a review see: D. Bimberg, M. Grundmann, and N. N. Ledentsov, Quantum Dot Heterostructures (Wiley, New York, 1999).
2. for a review see: Semiconductor Quantum Dots, Eds. Y. Masamoto and T. Takagahara, (Springer, Berlin, 2002).

3. R. J. Warburton B. T. Miller, C. S. Dürr, C. Bödefeld, K. Karrai, J. P. Kotthaus, G. Medeiros-Ribeiro, and P. M. Petroff, Phys. Rev. B **58**, 16221 (1998).
4. L. He and A. Zunger, Phys. Rev. B **73**, 115324 (2006).
5. E. E. Vdovin, A. Levin, A. Patanè, L. Eaves, P. C. Main, Yu. N. Khanin, Yu. V. Dubrovskii, M. Henini, and G. Hill, Science 290, 122 (2000).
6. A. Patanè, R. J. A. Hill, L. Eaves, P. C. Main, M. Henini, M. L. Zambrano, A. Levin, N. Mori, C. Hamaguchi, Yu. V. Dubrovskii, E. E. Vdovin, D. G. Austing, S. Tarucha, and G. Hill, Phys. Rev. B **65**, 165308 (2002).
7. K. H. Schmidt, M. Versen, C. Bock, U. Kunze, D. Reuter, and A. D. Wieck, Physica E **7**, 425 (2000).
8. I. Hapke-Wurst, U. Zeitler, H. Frahm, A. G. M. Jansen, R. J. Haug, and K. Pierz, Phys. Rev. B **62**, 12621 (2000).
9. H. Drexler, D. Leonard, W. Hansen, J. P. Kotthaus, and P. M. Petroff, Phys. Rev. Lett. **73**, 2252 (1994).
10. M. Fricke, A. Lorke, J. P. Kotthaus, G. Medeiros-Ribeiro, and P. M. Petroff, Europhys. Lett. **36**, 197 (1996).
11. B. T. Miller, W. Hansen, S. Manus, R. J. Luyken, A. Lorke, and J. P. Kotthaus, Phys. Rev. B. **56**, 6764 (1997).
12. K. H. Schmidt, G. Medeiros-Ribeiro, M. Oestreich, P. M. Petroff, and G. H. Döhler, Phys. Rev. B **54**, 11346 (1996).
13. G. Medeiros-Ribeiro, F. G. Pikus, P. M. Petroff, and A. L. Efros, Phys. Rev. B **55**, 1568 (1997).
14. G. Medeiros-Ribeiro, J. M. Garcia, and P. M. Petroff, Phys. Rev. B **56**, 3609 (1997).
15. N. Horiguchi, T. Futatsugi, Y. Nakata, and N. Yokoyama, Jpn. J. Appl. Phys. **36**, L1247 (1997).
16. K. H. Schmidt, G. Medeiros-Ribeiro, and P. M. Petroff, Phys. Rev. B. **58**, 3597, 1998.
17. R. J. Luyken, A. Lorke, A. O. Govorov, J. P. Kotthaus, G. Medeiros-Ribeiro, and P. M. Petroff, Appl. Phys. Lett. **74**, 2486 (1999).
18. G. Medeiros-Ribeiro, M. V. B. Pinheiro, V. L. Pimentel, and E. Marega, Appl. Phys. Lett. **80**, 4229 (2002).
19. D. Granados and J. M. Garcia, Nanotechnology **16**, S282 (2005).
20. O. S. Wibbelhoff, A. Lorke, D. Reuter, and A. D. Wieck, Appl. Phys. Lett. **86**, 92104 (2005), Appl. Phys. Lett. **88**, 129901 (2006).
21. G. Medeiros-Ribeiro, D. Leonard, and P. M. Petroff, Appl. Phys. Lett. **66**, 1767 (1995).
22. C. Bock, K. H. Schmidt, U. Kunze, S. Malzer, and G. Döhler, Appl. Phys. Lett. **82**, 2071 (2003).
23. D. Reuter, P.Schafmeister, P. Kailuweit, and A. D. Wieck, Physica E **21**, 445 (2004).
24. D. Reuter, P. Kailuweit, A. D. Wieck, U. Zeitler, and J. C. Maan, Physica E **26**, 446 (2005).
25. D. Reuter, P. Kailuweit, A. D. Wieck, U. Zeitler, O. Wibbelhoff, C. Meier, A. Lorke, and J. C. Maan, Phys. Rev. Lett. **94**, 26808 (2005).
26. P. Kailuweit, D. Reuter, A. D. Wieck, O. Wibbelhoff, A. Lorke, U. Zeitler, and J. C. Maan, Physica E **32**, 159 (2006), D. Reuter, P. Kailuweit, R. Roescu, A. D. Wieck, O. S. Wibbelhoff, A. Lorke, U. Zeitler, and J. C. Maan, phys. stat. sol.(b), **243**, 3942 (2006). The directions for the hole wave function anisotropy are wrong in these publications. After careful experimental checks, we are sure that the directions as given in this publications are the correct ones.
27. C. M. A. Kapteyn, F. Heinrichsdorf, O. Stier, R. Heitz, M. Grundmann, N. D. Zakharov, D. Bimberg, and P. Werner, Phys. Rev. B 60, 14265 (1999).
28. C. M. A. Kapteyn, M. Lion, R. Heitz, D. Bimberg, P. N. Brunkov, B. V. Volovik, S. G. Konnikov, A. R. Kovsh, and V. M. Ustinov, Appl. Phys. Lett. **76**, 1573 (2000).
29. H. L. Wang, F. H. Yang, S. L. Feng, H. J. Zhu, D. Ning, H. Wang, and X. D. Wang, Phys. Rev. B **61**, 5530 (2000).
30. S. Schulz, S. Schmüll, Ch. Heyn, and W. Hansen, Phys. Rev. B **69**, 195317 (2004), S. Schulz, A. Schramm, C. Heyn, and W. Hansen, Phys. Rev. B **74**, 33311 (2006).
31. J. Ibáñez, R. Leon, D. T. Vu, S. Chaparro, S. R. Johnson, C. Navarro, and Y. H. Zhang, Appl. Phys. Lett. **79**, 2013 (2001).
32. V. V. Ilchenko, S. D. Lin, C. P. Lee, and O. V. Tretyak, J. Appl. Phys. **89**, 1172 (2001).

33. J. S. Wang, J. F. Chen, J. L. Huang, P. Y. Wang, and X. J. Guo, Appl. Phys. Lett. 77, 3027 (2000).
34. M. M. Sobolev, A. R. Kovsh, V. M. Ustinov, A. Yu. Egorov, A. E. Zhukov, and Yu. G. Musikhin, Semiconductors **33**, 157 (1999).
35. P. N. Brunkov, A. Patanè, A. Levin, L. Eaves, P. C. Main, Yu. G. Musikhin, B. V. Volovik, A. E. Zhukov, V. M. Ustinov, and S. G. Konnikov, Phys. Rev. B **65**, 85326 (2002).
36. E. Gombia, R. Mosca, P. Frigeri, S. Franchi, S. Amighetti, and C. Ghezzi, Mat. Sci. Engineering B **91–92**, 393 (2003), E. Gombia, R. Mosca, S. Franchi, P. Frigeri, and C. Ghezzi, Mat. Sci. Engineering C **26**, 867 (2006).
37. A. J. Chiquito, Yu. A. Pusep, S. Mergulhão, J. C. Galzerani, N. T. Moshegov, and D. L. Miller, J. Appl. Phys. **88**, 1987 (2000).
38. P. M. Martin, A. E. Belyaev, L. Eaves, P. C. Main, F. W. Sheard, T. Ihn, and M. Henini, Semiconductor Physics **1**, 7 (1998).
39. R. J. Warburton, C. Schäflein, D. Haft, F. Bickel, A. Lorke, K. Karrai, J. M. Garcia, W. Schoenfeld, and P. M. Petroff, Nature **405**, 926 (2000).
40. J. I. Climente, J. Planelles, M. Pi, and F. Malet, Phys. Rev. Lett. **72**, 233305 (2005).
41. M. Ruÿ, C. Meier, A. Lorke, D. Reuter, and A. D. Wieck, Phys. Rev. B **73**, 115334 (2006).
42. H. Künzel, K. Graf, M. Hafendörfer, A. Fischer, and K. Ploog, Technisches Messen **48**, 295 (1981).
43. L. Pavesi and M. Guzzi, J. Appl. Phys. **75**, 4779 (1994).
44. G. Snider, University of Notre Dame, USA (2002).
45. L. P. Kouwenhoven, D. G. Austing, and S. Tarucha, Rep. Prog. Phys. **64**, 701 (2001).
46. V. Fock, Z. Phys. **47**, 446 (1928),
C. G. Darwin, Proc. Cambridge Philos. Soc. **27**, 86 (1930).
47. L. He, G. Bester, and A. Zunger, Phys. Rev. Lett. **95**, 246804 (2005).
48. M. Rontani and E. Molinari, Jpn. J. Appl. Phys. **45**, 1966 (2006).
49. O. Stier, M. Grundmann, and D. Bimberg, Phys. Rev. B 59, 5688 (1999).

Chapter 12
In(Ga)As/GaAs Quantum Dots Grown by MOCVD for Opto-electronic Device Applications

K. Sears, S. Mokkapati, H. H. Tan, and C. Jagadish

12.1 Introduction

This chapter focuses on the self-assembled growth of In(Ga)As/GaAs quantum dots using Metal Organic Chemical Vapor Deposition (MOCVD) and their application to quantum dot lasers and photonic integrated circuits.

Quantum dots (QDs) have attracted intense interest over the past decade due to their 3D confinement of carriers and unique properties. One application that has gained much attention is the semiconductor QD laser. Quantum dots offer a number of potential benefits for laser performance such as improved efficiencies, improved thermal stability and reduced threshold currents (Arakawa and Sakaki 1982, Dingle and Henry 1976). Since the first demonstration of a quantum dot laser in 1994 (Kirstaedter et al. 1994), many of these predicted advantages have been realized. For example, low threshold currents on the order of 20 A/cm^2 per dot layer (Liu et al. 2000, Sellers et al. 2004, Kovsh et al. 2002) and high thermal stability with characteristic temperatures as high as 300 K at room temperature (Kovsh et al. 2002, Shchekin et al. 2002, Maximov et al. 1997) have been reported. Lasing at 1.3 and 1.5 μm has also been achieved by capping InAs quantum dots with thin InGaAs capping layers (Sellers et al. 2004, Karachinsky et al. 2005, Fehse et al. 2003, Ledentsov et al. 2003, Zhukov et al. 1999b). However it is still difficult to achieve all of the above characteristics in the one device, with only a few such reports (Kovsh et al. 2002, Shchekin et al. 2002). Furthermore, the majority of QD laser studies have focused on growth using molecular beam epitaxy (MBE). In comparison, there are very few reports of In(Ga)As/GaAs quantum dot lasers grown using metal-organic chemical vapor deposition (MOCVD) (Maximov et al. 1997, Sellin et al. 2001, Kaiander et al. 2004, Sellin et al. 2003, Heinrichsdorff et al. 1997b, Tatebayashi et al. 2005, Heinrichsdorff et al. 2000, Tatebayashi et al. 2003, Nuntawong et al. 2005, Walter et al. 2002, Sears et al. 2007, Lever et al. 2004a) and the production of high performance, long wavelength quantum dot lasers using MOCVD remains an elusive goal. This is despite the fact that MOCVD is the growth technique of choice for light emitting devices because of its scalability to high volume production and its ability to grow a diverse range of semiconductor compounds.

Z. M. Wang, *Self-Assembled Quantum Dots.*

This chapter starts with a brief description of the MOCVD growth technique. The Stranski-Krastanow growth mode used for the self-assembled growth of In(Ga)As QDs is then described followed by a detailed discussion of the key MOCVD growth parameters and their influence on the nucleation process. Stacked QD growth is necessary for many device applications and this is reviewed in Section 12.4 before discussing the important lasing characteristics of QD lasers.

The integration of QD lasers with other components such as waveguides and modulators is also of interest. Such photonic integrated circuits (PICs) offer a number of advantages such as:

- reduced volume,
- reduced packaging costs,
- improved reliability, and
- reduced losses associated with fiber coupling.

A promising technique for the production of QD photonic integrated circuits is selective area epitaxy (SAE). Unlike other post growth processing techniques, SAE does not alter the quality of the QD active region which is a distinct advantage for active devices such as QD lasers. The final sections of this chapter discuss the basic principles of SAE and demonstrate how it can be used to achieve controlled nucleation and band gap tuning of QDs.

12.2 Basic Principles of Metal Organic Chemical Vapor Deposition (MOCVD)

There are two main epitaxial growth techniques used to grow self-assembled III-V semiconductor quantum dots. These are Metal Organic Chemical Vapor Deposition (MOCVD), also known as metal organic vapor phase epitaxy (MOVPE), and Molecular Beam Epitaxy (MBE). These techniques differ considerably in the sources used and surface growth kinetics. MOCVD is particularly complex with many factors such as precursor chemistry, growth kinetics, hydrodynamics, mass transport and surface chemistry influencing growth.

All of the samples and devices discussed in this chapter were grown on semi-insulating (001) GaAs substrates using a low pressure (100mbar) AIXTRON 200/4 horizontal flow MOCVD system. The MOCVD growth technique uses the group III metal alkyls (e.g. trimethyl-gallium, $Ga(CH_3)_3$) and group V hydrides (e.g. arsine, AsH_3) to grow thin layers of III-V semiconductor materials. These precursors flow over a heated substrate where they react to form a solid. As an example, the overall reaction for GaAs can be written as:

$$Ga(CH_3)_3(g) + AsH_3(g) \rightarrow GaAs(s) + 3CH_4(g)$$

By mixing different metal alkyls and hydrides a wide range of binary, ternary and even quaternary III-V semiconductors can be formed. MOCVD is therefore highly versatile and capable of growing a wide range of III-V semiconductor compounds including both P and Al containing ones. The growth rate is generally dictated by the group III flow rate.

Figure 12.1 shows a schematic of the system used and the available sources. The sources used to grow the samples discussed in this chapter are trimethyl-indium (TMIn), trimethyl-gallium (TMGa), trimethyl-aluminium (TMAl) and arsine (AsH_3). Silane (SiH_4) and carbon tetrachloride (CCl_4) were used for n and p type doping, respectively. Diethyl-zinc (DEZn) is also often used as a p-type dopant. However Zn diffuses rapidly in GaAs and therefore CCl_4 was used instead.

The metal-organics (with the exception of trimethyl-indium) are liquids and are kept in bubblers. Palladium purified hydrogen is used as a carrier gas and is bubbled through the sources. The amount of metal-organics 'picked up' by the carrier gas is determined by the bath temperature and source pressure. The hydrides are stored in high pressure cylinders and supplied directly to the reactor. Mass flow controllers are used to deliver accurate gas flows to the reaction chamber. In order to avoid precursor pre-reactions, the hydrides and metal-organics are delivered separately in the Hydride and MO run lines, respectively, before entering the reaction chamber. In the reaction chamber, the substrate sits on a gas foil rotation plate within a graphite susceptor which is heated by infrared lamps. The sources are thermally decomposed within the reaction chamber by the elevated susceptor temperatures. The unreacted

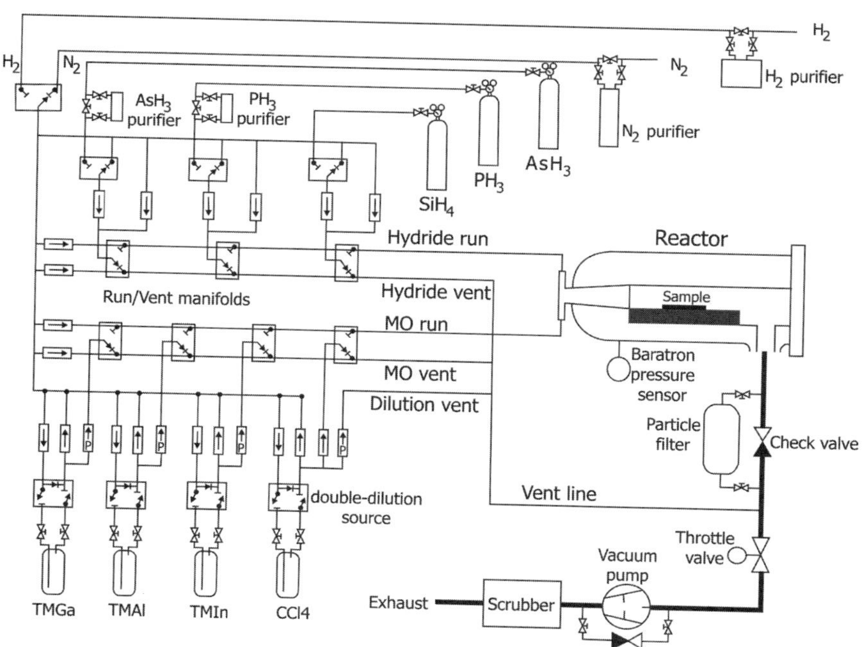

Fig. 12.1 Schematic of the Aixtron 200/4 MOCVD system used to grow the QD samples and devices

sources and byproducts are passed through a dry charcoal scrubber before being exhausted to atmosphere.

For nano-scale and quantum dot growth, abrupt interfaces between layers of different composition and precise control on the monolayer (ML) scale are needed. This is achieved through a combination of techniques. Firstly, the schematic in Fig. 12.1 shows two additional lines called the Hydride and MO vent lines. This enables the sources to be vented and the gas flows to stabilise, prior to switching them into the reaction chamber. The differential pressure between the vent and run lines is kept as low as possible, again aiding faster switching. Secondly, the electronically controlled pneumatic valves allow precise and automated control of the gas switching, reproducible to within 0.1 s. Finally, the total reactor gas flow is kept constant (8 slm in our case) by switching on/off so called "Dummy lines" to compensate for the changing number of source flows entering the reaction chamber.

In this chapter the deposition times are specified to an accuracy of 0.1 s, coverage to 0.025 ML, the temperature to 10 ℃. We would like to stress that these accuracies are nominal and dependent upon external instrumentation as well as our specific reactor setup. Therefore the values given should be taken as a guide only and more value placed upon the overall trends observed.

12.3 Self-assembled Growth of In(Ga)As/GaAs Quantum Dots Using MOCVD

12.3.1 The Stranski-Krastanow Growth Mode

Initial efforts at quantum dot growth involved patterning, etching and regrowth of quantum well structures, but suffered from poor interface quality and low surface coverage. The major break-through came in 1985 with the discovery that defect free semiconductor quantum dots with strong luminescence could be formed in a self-assembly process via the Stranski-Krastanow growth mode (Goldstein et al. 1985).

There are three different modes in which epitaxial growth can proceed as illustrated in Fig. 12.2. The Frank-Van der Merve mode (Fig. 12.2A) occurs for lattice matched systems and consists of 2 dimensional (2D), layer by layer growth. The second mode is called the Volmer-Weber growth mode (Fig. 12.2B). This takes place for systems with a large lattice mismatch or high interfacial energy, which drives the direct formation of islands. These islands tend to be dislocated and therefore not of use for device applications.

To form device quality In(Ga)As/GaAs quantum dots, the Stranski-Krastanow mode is typically used (Fig. 12.2C). This mode relies on a slight lattice mismatch between the epilayer and substrate, $\sim$7.2% for InAs on GaAs and 3.6% for $In_{0.5}Ga_{0.5}As$ on GaAs, which drives the formation of islands after an initial period of layer by layer growth. Initially the deposited In(Ga)As compresses to fit the smaller GaAs lattice and 2D growth proceeds. However, with each additional layer, the strain energy accumulates until it becomes energetically favorable for islands to

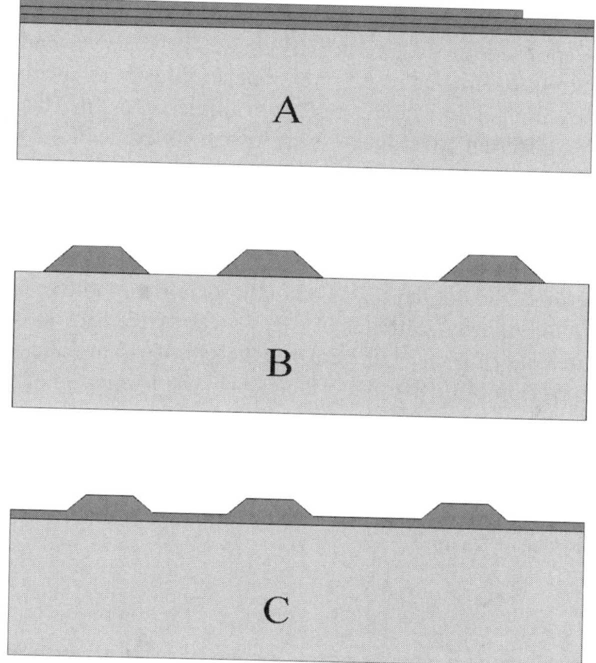

Fig. 12.2 Schematics of the three epitaxial growth modes: (A) Frank-Van der Merve, (B) Volmer-Weber, and (C) Stranski-Krastanow. Reproduced from (Sears 2006)

form. The quantum dots or islands therefore sit on top of a thin 2D layer, called the wetting layer (which is essentially a very thin quantum well). The layer thickness at which the 2D-3D growth transition occurs is called the critical thickness and depends on the lattice mismatch. For the $In_{0.5}Ga_{0.5}As$ /GaAs system the critical thickness is 4 monolayers (ML), where as for the InAs/GaAs system, the critical thickness is much smaller, only 1.7 ML, due to the larger lattice mismatch. For PL measurements and device applications these quantum dots are typically capped with a GaAs layer.

Whilst the driving force for island nucleation is clearly a thermodynamic one, the nucleation process and final QD characteristics are highly dependent on the growth parameters. This is especially the case for metal organic chemical vapor deposition where growth kinetics play a much greater role. In the following sections the key growth parameters, such as coverage, the V/III ratio, growth temperature and growth interruptions will be discussed.

12.3.2 Coverage

The amount of material deposited, or coverage, is one of the most important parameters for QD formation. If too much material is deposited, large coalesced islands form, which are detrimental for device performance. On the other hand, if too little material is deposited only a low density of islands (or no islands) form. These trends and the Stranski-Krastanow growth mode are illustrated nicely by the atomic force microscopy images in Fig. 12.3. This example is for $In_{0.5}Ga_{0.5}As$ quantum dots grown using different coverages. At the lowest coverage of 3.3 ML, the critical thickness has not yet been exceeded so that only a thin wetting layer has formed. The monolayer steps can be clearly observed. At 4.7 ML, the critical thickness has only just been exceeded so that a low density ($8 \times 10^9\ cm^{-2}$) of shallow islands form. These islands have a height of only ~2 nm and a base diameter of ~50 nm. As more material is deposited, the dot density and height increase up until a saturation point. In this example the density saturates at a coverage of 5.7 ML, at which stage the islands have a density of $\sim 5 \times 10^{10}\ cm^{-2}$, a height of ~5 nm and a base diameter of ~20 nm. Deposition of further material leads to the formation of large coalesced islands and a reduced island density and height. One large coalesced island can be seen in the lower left hand corner of Fig. 12.3(d). These large coalesced islands are typically dislocated and act as material sinks, growing at the expense of the surrounding islands. Fig. 12.4 shows an extreme example where many of these large coalesced islands have formed.

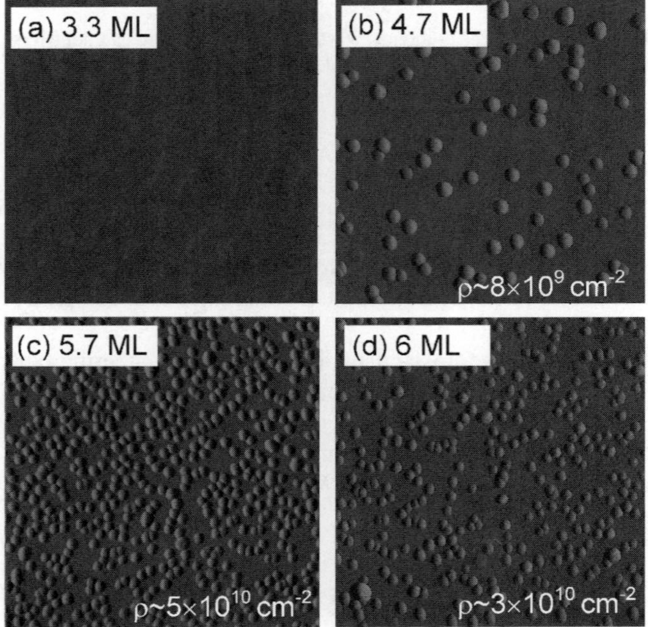

Fig. 12.3 $1 \times 1\ \mu m^2$ deflection AFM images of $In_{0.5}Ga_{0.5}As/GaAs$ QDs grown using coverages of (a) 3.3, (b) 4.7, (c) 5.7, and (d) 6 ML. The other growth parameters were kept constant (1.7 ML/s, V/III ratio = 15, 550 °C). Adapted from (Tan et al. 2007)

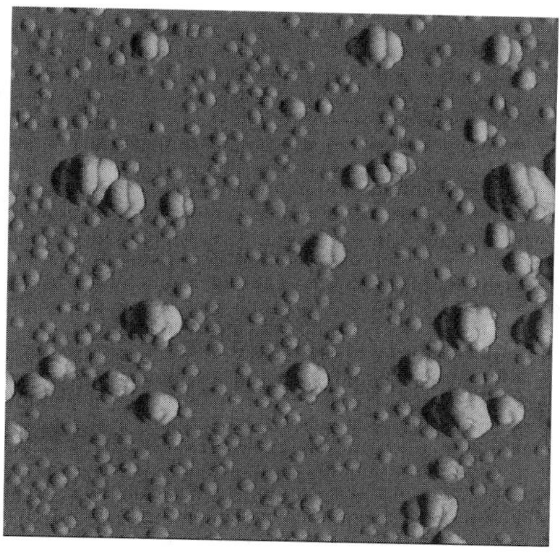

Fig. 12.4 1 × 1 μm^2 deflection AFM image of an $In_{0.5}Ga_{0.5}As$/GaAs QD sample formed using too much material. A high density of coalesced islands is observed

Figure 12.5 shows PL spectra for the same $In_{0.5}Ga_{0.5}As$ QD samples discussed above. At 3.3 ML the PL spectrum is centered at 890 nm with a relatively narrow line width, characteristic of a quantum well. As more material is deposited, the photoluminescence shifts to longer wavelengths (red-shifted) as islands form and grow in size. The PL line width also increases substantially which is due to the random nucleation process and resulting large island size distribution. At 8.3 ML the PL intensity drops as more and more large coalesced islands are created.

The effect of coverage on InAs/GaAs QD formation is very similar except that the higher lattice mismatch (7.2%) leads to a much faster nucleation process. This

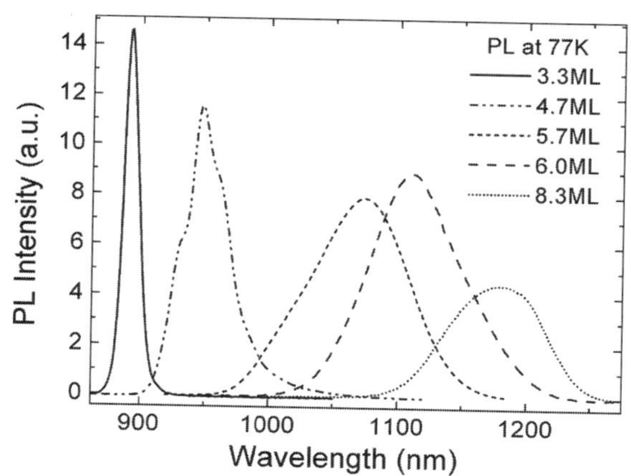

Fig. 12.5 77 K photoluminescence spectra of $In_{0.5}Ga_{0.5}As$/GaAs QDs grown using coverages of 3.3, 4.7, 5.7, 6.0 and 8.3 ML. The other growth parameters were kept constant (1.7 ML/s, V/III ratio = 15, 550 ℃). Adapted from (Sears et al. 2006b)

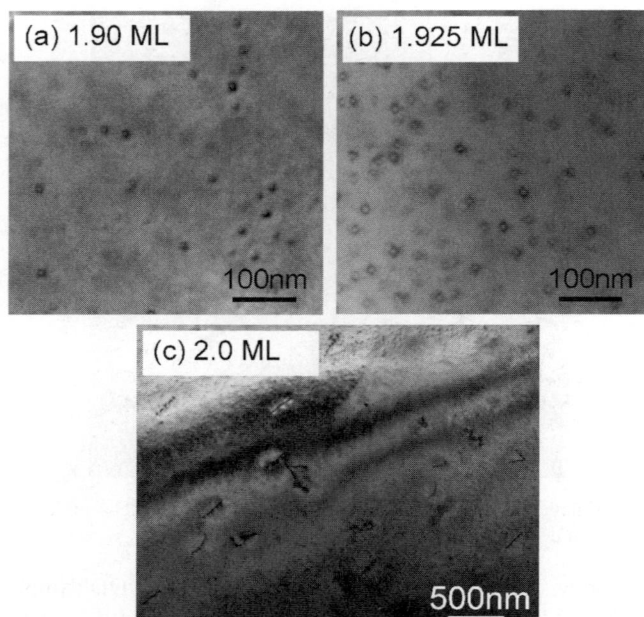

Fig. 12.6 Plan view TEM images of InAs/GaAs quantum dots grown using coverages of 1.90, 1.925 and 2.0 ML. The other growth parameters were kept constant (0.25 ML/s, V/III ratio = 40, 520 °C). The QDs are also capped by a 300 nm GaAs layer. Adapted from (Sears et al. 2004)

is illustrated by Figs. 12.6(a), (b) and (c) which show plan-view TEM images of InAs QD samples formed using 1.90, 1.925 and 2.0 ML. In these plan-view TEM images each dot appears as a 'donut' like contrast. At a coverage of 1.90 ML only a low density of islands has formed. However, when the coverage is increased by only 0.025 ML, a high density of islands with minimal defects and strong PL (not shown) results, while a further 0.075 ML leads to dislocations. Clearly the nucleation process is extremely fast in the case of InAs QDs making it a more challenging system to grow. Due to the more challenging nature of the InAs/GaAs QD system, the selective area epitaxy results discussed later in Section 12.6 concentrate on the use of $In_{0.5}Ga_{0.5}As$/GaAs QDs. This example also stresses how important it is to avoid the formation of large coalesced islands which are very susceptible to dislocations.

12.3.3 Growth Temperature

Temperature affects the adatom mobility and therefore the density and size of the QDs. Each nucleation site acts as a sink for the surrounding material, essentially creating a depleted zone within which the adatoms are collected and further dot nucleation is inhibited. Therefore, at higher temperatures, a lower density of larger QDs tends to form due to an increased diffusion length and vice versa. These

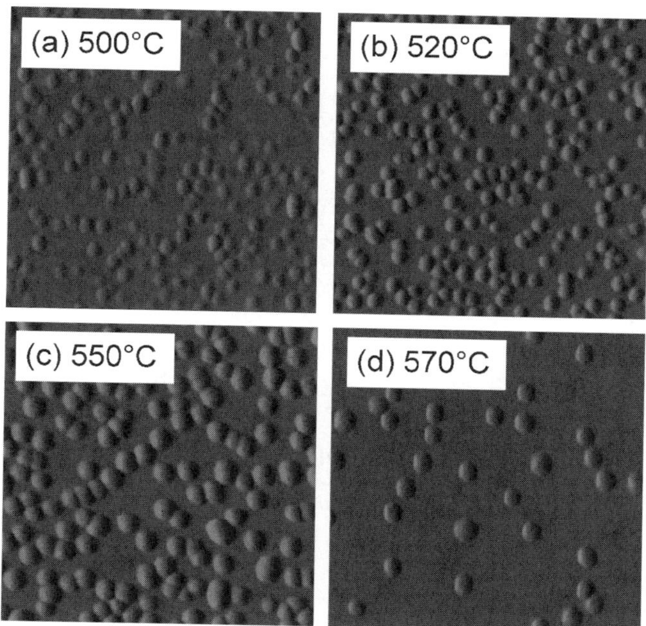

Fig. 12.7 $500 \times 500\ nm^2$ deflection AFM images of $In_{0.5}Ga_{0.5}As$/GaAs QDs grown at temperatures of 500, 520, 550 and 570 °C. Adapted from (Tan et al. 2007)

trends are illustrated in Fig. 12.7 which shows AFM images of $In_{0.5}Ga_{0.5}As$ QDs deposited at different temperatures. With increasing growth temperature, the QD height increases from ~1 nm to 5.5 nm and the QD dot density decreases, consistent with an increased adatom mobility. At the highest temperature of 570 °C, the dot density decreases dramatically as large coalesced islands form.

The effects of growth temperature are also nicely illustrated by the PL spectra in Fig. 12.8. With increasing growth temperature, the PL emission shifts to longer wavelengths, characteristic of larger QDs. The PL intensity initially increases, which is likely due to improved material quality. However it then begins to drop at the highest growth temperature which we suspect is due to defects/dislocations forming in the larger QDs. Due to the temperature dependent adatom mobility and cracking efficiency of the precursors (discussed below), the optimum coverage varies with growth temperature. Therefore a study of coverage was made for each QD growth temperature and an optimum temperature of 520 and 550 °C identified for the InAs and $In_{0.5}Ga_{0.5}As$ QDs respectively.

The above studies have concentrated on QD growth temperatures between ~500 and 550 °C. These growth temperatures are relatively low, especially for MOCVD growth and are needed in order to reduce the adatom mobility and hence the tendency for islands to coalesce. However these low growth temperatures also introduce added complexities for QD growth by MOCVD. Temperatures between 600–750 °C are typically used in MOCVD growth to ensure proper cracking of

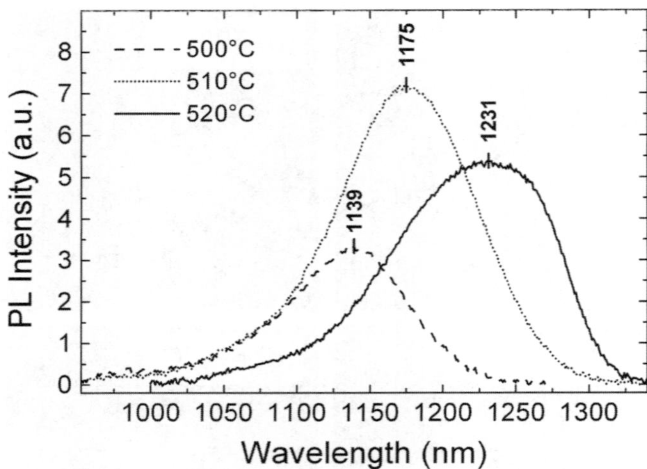

Fig. 12.8 Room temperature PL spectra of InAs/GaAs QDs grown at different temperatures. Adapted from (Sears et al. 2004)

the standard precursor gases (TMIn, AsH_3 and TMGa). For example, In(Ga)As is typically deposited at 600–650 ℃. At lower growth temperatures, improper cracking of the precursors can lead to (i) reduced deposition rates and (ii) degraded material quality due to the presence of larger incompletely dissociated metal-organic molecules. Despite this we have achieved excellent quality InAs and $In_{0.5}Ga_{0.5}As$ QDs at optimum temperatures of 520 and 550 ℃ respectively.

Some groups have also used alternative precursors such as tertiarybutyl-arsine (TBA) and triethyl-gallium (TEGa) which decompose at much lower temperatures. For example, TBA fully decomposes at typical quantum dot growth temperatures whereas AsH_3 is only 50–80% decomposed (in the presence of TMGa) (Stringfellow 1999). Good quality quantum dots and quantum dot lasers have been reported using both TBA and TEGa (Sellin et al. 2003, Yang et al. 2004).

12.3.4 Growth Interrupts

Another method of controlling QD nucleation is to use a growth interrupt (GRI). A growth interrupt involves pausing the growth for a short period of time, prior to capping the QDs. Typically all of the gases are switched off during the growth interrupt. Later, in Section 12.3.5, the effect of maintaining an AsH_3 flow is discussed in detail. Typically, the growth is paused just before island formation or just as a low density of islands begin to form (i.e. shortly after the 2D-3D transition). Due to the high indium mobility and the high strain of these systems, especially in the case of InAs QDs, material continues to redistribute even after the gases have been switched off (Kamins et al. 1999, Pötschke et al. 2004, Lee et al. 1997,

Heinrichsdorff et al. 1998). This enables a high density of islands to develop during the growth interrupt.

The effects of a growth interrupt are demonstrated by Fig. 12.9 which shows PL spectra of InAs QD samples formed with and without a 15 s GRI. For both samples 1.9 ML of InAs deposited. When no GRI is used, a low density of small islands form with low intensity PL centered at 1158 nm. However, by introducing a 15 s growth interrupt, a high density of islands develop with strong PL centered at 1209 nm. The WL intensity at 944 nm is also reduced and blue-shifted confirming that material is redistributed from the WL to the QDs during the growth interrupt.

It has also been shown by several groups that a growth interrupt after capping the QDs with a thin layer can selectively dissolve larger, dislocated islands (Steimetz et al. 2000, Ledentsov et al. 2000). This technique is depicted in the inset to Fig. 12.10. First a thin capping layer is deposited which fully covers the smaller, defect free islands while the larger islands are only partially capped. A growth interrupt (with AsH_3) is then introduced and the sample typically heated. During this growth interrupt, the material from the larger exposed islands migrates to the surface, forming a second wetting layer.

Figure 12.10 shows PL spectra for two InAs QD samples and demonstrates how the PL intensity can be improved using the above GRI technique. In this example two QD samples were formed using conditions that led to a high density of coalesced islands. The QDs were then capped with either a 7.5 or 30 nm GaAs layer before interrupting the growth for 2.5 minutes and heating the sample from 520 to 600 ℃ with a constant AsH_3 flow of 4×10^{-5} mol/min. The 30 nm GaAs layer fully covers both the smaller coherent islands and the large coalesced islands, which leads to poor PL intensity. However, in the case of the 7.5 nm layer, the

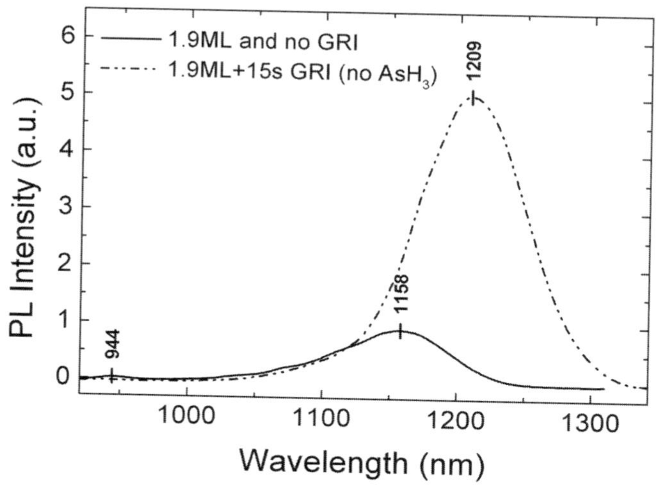

Fig. 12.9 Room temperature PL spectra for InAs/GaAs QDs formed with and without a 15 s GRI (No AsH_3). Adapted from (Tan et al. 2007)

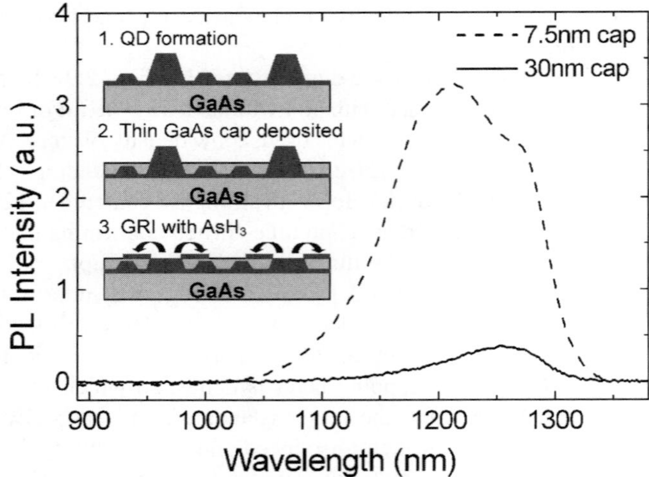

Fig. 12.10 Room temperature PL for two InAs/GaAs QD samples formed using identical QD growth conditions and then capped with either 7.5 nm or 30 nm of GaAs before interrupting the growth for 2.5 minutes. During this growth interrupt the sample was heated from 520 to 600 °C. The rest of the 300 nm cap was then grown at 600 °C. The inset illustrates how larger partially capped islands are dissolved during a GRI with AsH_3. Adapted from (Sears et al. 2006c)

large coalesced islands are only partially capped and therefore dissolved during the subsequent growth interrupt. This clearly leads to much stronger PL intensity. It is also important to note that the small coherent InAs QDs typically have heights between 3–4 nm and are therefore fully covered by the thinner 7.5 nm GaAs layer and not dissolved.

12.3.5 The V/III Ratio

The V/III ratio is the ratio of the group V (AsH_3) flow rate to the group III (TMIn+TMGa) flow rate. It is standard procedure in MOCVD growth to supply the group V in excess because As is the more volatile species. In the present study the group III flow is kept constant (i.e. constant growth rate) while the group V flow rate is varied so that an increased V/III ratio is equivalent to an increased AsH_3 flow.

The AsH_3 flow has a pronounced effect on island nucleation both during growth and afterwards during a growth interrupt. We have found that higher AsH_3 flows result in a faster QD nucleation process and greater island ripening. Therefore, in order to achieve high densities of defect free QDs, it is critical to use low AsH_3 flows (in our case V/III ratios between 10 and 40). We have also found that this ripening effect is particularly pronounced for the InAs/GaAs system compared to the $In_{0.5}Ga_{0.5}As$ /GaAs system. Therefore in the following examples we focus on the InAs/GaAs material system.

The exact mechanism by which AsH_3 encourages ripening is still unclear. Two theories have been proposed in the literature. One involves atomic hydrogen. In MOCVD growth using hydrides as the group V source, an increased V/III ratio is also always associated with an increased concentration of atomic hydrogen. This has led some groups to propose that atomic hydrogen destabilizes the wetting layer, leading to these ripening effects (Heinrichsdorff et al. 1997a, Steimetz et al. 1997). On the other hand, Chung et al. (Chung et al. 2005) have recently suggested that higher concentrations of As free radicals in the gas phase may aid faster incorporation of indium into existing islands instead of new island nucleation.

12.3.5.1 AsH_3 Flow During Growth

Figure 12.11 shows room temperature PL spectra for three InAs QD samples grown using identical growth conditions except for the V/III ratio (AsH_3 flow) which was set to 10 (6.7×10^{-5} mol/min), 40 (2.6×10^{-4} mol/min) and 70 (4.5×10^{-4} mol/min). Increasing the V/III ratio clearly results in a strong red-shift of the PL. For example, the PL peak red-shifts by 60 nm when the V/III ratio is increased from 40 to 70. This strong red-shift indicates that larger islands form with higher V/III ratios.

The island density also varies with V/III ratio. At the lowest V/III ratio of 10, the QD PL is very weak and a peak from the wetting layer is observed, suggesting a low dot density. In contrast, plan-view TEM revealed high dot densities ($\sim 3 \times 10^{10}\,cm^{-2}$) for V/III ratios of 40 and 70. Both the increased island density and size suggest that higher AsH_3 flows generally lead to a faster nucleation process.

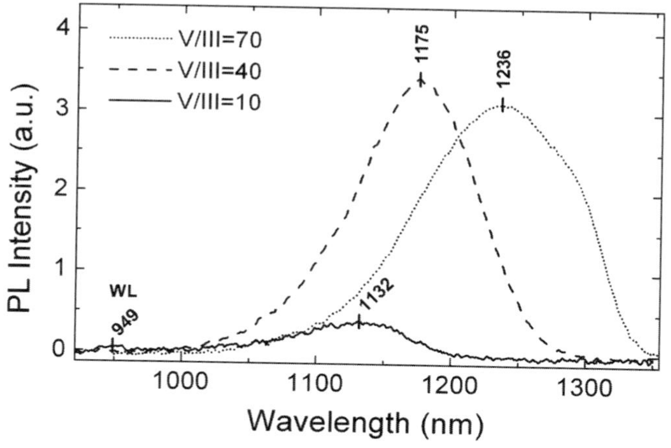

Fig. 12.11 Room temperature PL spectra for InAs/GaAs QDs grown using V/III ratios of 10, 40, and 70. The other growth parameters were kept constant (0.25 ML/s, 1.950 ML, 520 °C). Adapted from (Sears et al. 2006c, Sears et al. 2004, Tan et al. 2006)

12.3.5.2 AsH_3 Flow During Growth Interrupts

The presence of AsH_3 during a growth interrupt also results in strong island ripening. This is clearly revealed by the AFM image in Fig. 12.12(a) which shows uncapped InAs QDs which were cooled down after growth under an AsH_3 flow of 1.3×10^{-3} mol/min. Essentially the cool-down period is one long growth interrupt with AsH_3. The conventional approach in MOCVD growth is to maintain an AsH_3 flow during cooling to avoid As out diffusion. In the case of our MOCVD system, it takes ~4 minutes to cool from 520 °C (the growth temperature) to 400 °C (the temperature at which the AsH_3 is switched off). During this cool-down period the island density and morphology change drastically. As shown by the 3 by 3 μm^2 AFM image in Fig. 12.12(a), a low dot density of $\sim 1 \times 10^9\,cm^{-2}$ results and the islands have poor size uniformity and are extremely large. In contrast, plan-view TEM images of dots formed under identical conditions but capped immediately with GaAs, show a high density ($3 \times 10^{10}\,cm^{-2}$) of small islands with base diameters of 10–12 nm and an absence of large ripened islands (see Fig. 12.12 (b)). It is also important to note that the TEM image in Fig. 12.12 (b) is 1/36th the size of the AFM image in Fig. 12.12 (a). Clearly, the uncapped islands continue to ripen during cooling under AsH_3. We suspect that some islands plastically relax during this ripening process. These islands can then behave as material sinks, quickly growing at the expense of the smaller islands so that the island morphology is drastically changed.

The PL intensity is also degraded by a GRI with AsH_3, most likely because larger defective islands are formed. This is clearly illustrated by Fig. 12.13 which shows PL spectra for two InAs QDs samples. In one case a 5 s GRI with no AsH_3 was used, which resulted in strong PL centered at 1196 nm. In the second sample a 5+10 s GRI was used in which the AsH_3 flow was switched off for the first 5 s and then reintroduced during the final 10 s (flow rate = 4×10^{-4} mol/min). Clearly the PL intensity is degraded by the GRI with AsH_3. The PL is also red-shifted, consistent with the idea that AsH_3 flows encourage island ripening.

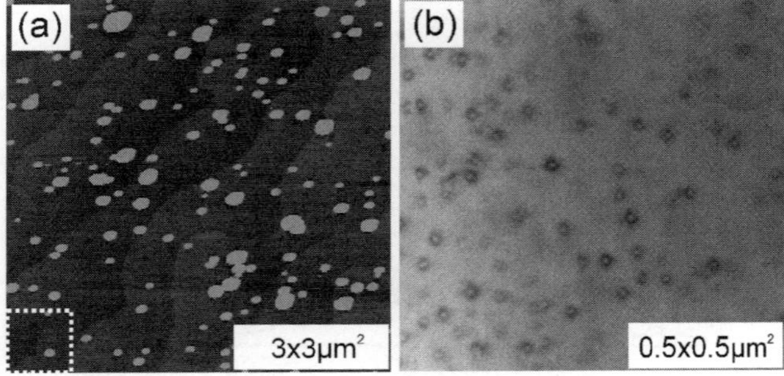

Fig. 12.12 (a) 3 by 3 μm^2 AFM height image of uncapped InAs/GaAs QDs cooled down under AsH_3 (1.3×10^{-3} mol/min) and (b) 0.5 by 0.5 μm^2 planview TEM image of a buried QD layer formed without a GRI. Identical QD growth conditions (V/III = 40, 0.25 ML/s, 520 °C, 1.925 ML) were used for both samples. The dotted square in (a) illustrates the relative size of the plan-view TEM image in (b). Adapted from (Sears et al. 2006c)

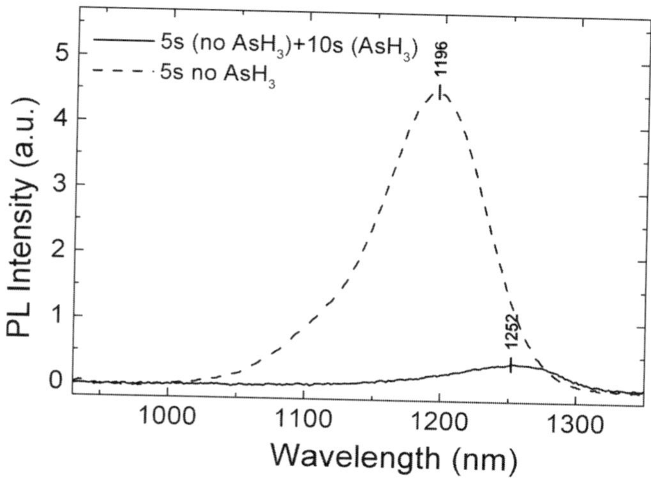

Fig. 12.13 Room temperature PL for InAs/GaAs QDs formed using (i) a 5s growth interrupt with no AsH_3, and (ii) a 5+10s growth interrupt in which the AsH_3 flow was switched off for the first 5 s and then switched on for the last 10 s. In both cases 1.7 ML of InAs was deposited at a V/III ratio of 10. Adapted from (Sears et al. 2006c)

12.3.6 InAs QDs with Thin InGaAs Capping Layers

One potential application of quantum dot lasers is long haul optical communications which requires sources at 1.31 and 1.55 μm. However, it is difficult to achieve PL from In(Ga)As QDs at wavelengths beyond 1.2 μm because of the tendency for larger islands to plastically relax. One promising technique for achieving longer wavelengths, involves embedding or covering InAs quantum dots in an $In_xGa_{1-x}As$ quantum well. Most QD lasers emitting at 1.3 μm or longer consist of such a structure (Sellers et al. 2004, Shchekin et al. 2002, Zhukov et al. 1999b, Heinrichsdorff et al. 1997b, Liu et al. 1999, Park et al. 2000b). By varying the indium composition of the thin $In_xGa_{1-x}As$ layer, the PL can also be tuned over a wide range of wavelengths (Tatebayashi et al. 2001, Ustinov et al. 1999, Maximov et al. 2000, Stintz et al. 2000, Liu et al. 2003). However the thickness and composition of the $In_xGa_{1-x}As$ layer need to be carefully controlled in order to avoid degradation of the quantum dots. Indium compositions between 5 and 20% typically give good results while higher compositions lead to PL degradation (Tatebayashi et al. 2001, Liu et al. 2003).

This shift to longer wavelength is illustrated nicely by Fig. 12.14(a) which shows PL spectra from InAs/GaAs QDs capped with a thin, nominally 5 nm thick, $In_xGa_{1-x}As$ layer ($x = 0, 0.1, 0.13$ and 0.15). As shown by Fig. 12.14(b), a linear trend is observed with increasing indium composition. Fig. 12.14(c) shows a schematic of the structure.

The shift to longer wavelength with $In_xGa_{1-x}As$ capping is attributed to several factors. It is generally accepted that the InGaAs capping layer relieves some of the

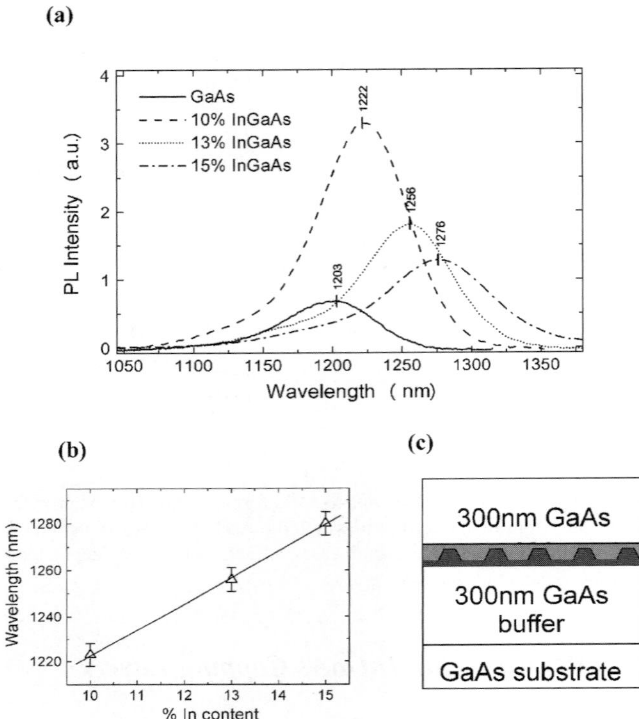

Fig. 12.14 (a) Room temperature PL spectra for InAs QDs capped with a nominally 5 nm $In_xGa_{1-x}As$ layer (x=0, 0.1, 0.13 and 0.15). (b) shows the PL peak wavelength versus the indium composition. A schematic of the structure is shown in (c). Adapted from (Sears 2006)

lattice mismatch and compressive strain in the quantum dot. This reduces the InAs bandgap and hence leads to longer wavelength PL. This seems to be the primary cause of the wavelength shift. For example, PL at much longer wavelengths has been reported for uncapped quantum dot structures (El-Emawy et al. 2003, Miao et al. 2005, Yeh et al. 2000). Some reports also attribute the red-shift to a larger island size as a result of indium migration from the $In_xGa_{1-x}As$ capping layer to the QDs (Maximov et al. 2000).

In Fig. 12.14(a) it is also worth noting that the PL intensity increases strongly for the $In_{0.10}Ga_{0.90}As$ capping layer. This is likely the result of increased carrier capture into the QDs. Quantum dot growth in the Stranski-Krastanow mode leads to low surface coverages (~10–20%) in comparison to quantum wells. This means that the quantum dots have a relatively small cross-section for carrier capture. By embedding them or capping them in a thin $In_xGa_{1-x}As$ layer, the carriers can first be trapped and confined by the two dimensional (2D) $In_xGa_{1-x}As$ layer, enhancing carrier capture into the dots. Reports have shown that this can lead to significant improvements in laser performance such as lower threshold currents, ground state lasing at shorter cavity lengths and improved internal quantum efficiency (Qui et al. 2001, Park et al. 2000b, Lester et al. 1999).

In Fig. 12.14(a) the PL intensity decreases slightly as the indium composition is further increased up to 15% (i.e. $In_{0.15}Ga_{0.85}As$), although it is still more intense than that observed for GaAs capped dots. The reason for this reduced intensity is still unclear.

12.3.7 Capping Temperature

In(Ga)As/GaAs QDs are very sensitive to the temperature used to grow the overlayers. During growth of these overlayers at elevated temperatures the QDs are essentially annealed. Figure 12.15 shows PL spectra for InAs QD samples grown using identical conditions and capped with GaAs at different temperatures. The first 10 nm of the capping layer was grown at 520 ℃, the same temperature used to deposit the QDs. This is sufficient material to fully cover the QDs. The growth was then interrupted and the sample heated to the desired temperature before depositing the rest of the GaAs cap. The GaAs capping layer thickness was 300 nm for all of the samples except the 520 ℃ one. This sample only has a 10 nm capping layer. Fig. 12.15 clearly shows that at temperatures above 600 ℃ the PL emission blue-shifts significantly due to intermixing of Ga-In at the QD/GaAs interface during the heating step and $\sim$8 minute capping layer growth. The PL intensity was also reduced for the higher capping growth temperatures which is likely due to a reduced quantum dot confinement potential, although defect generation may also be possible. In the case of $In_{0.5}Ga_{0.5}As$ QDs, temperatures up to 650 ℃ were possible before a significant blue-shift was observed.

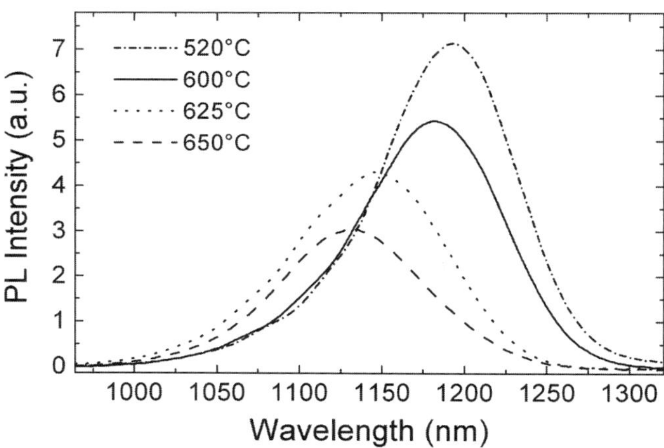

Fig. 12.15 Room temperature PL spectra for InAs QDs capped with a 300 nm GaAs layer at different temperatures. The first 10 nm of the capping layer was always deposited at 520 ℃, the growth temperature used for the QDs. The growth was then interrupted and the sample heated before depositing the rest of the 300 nm capping layer. Adapted from (Sears et al. 2004)

This also has consequences for laser devices, where the AlGaAs cladding layers above the QDs must be deposited at temperatures ~100 ℃ below optimum in order to avoid a significant blue-shift of the QDs.

12.4 Growth of Stacked Quantum Dot Structures

Stacked QD layers are needed to ensure sufficient ground state gain for lasing. QDs formed in the Stranski-Krastanow growth mode typically have densities between 10^9 and $10^{11}\,cm^{-2}$. These densities correspond to a low surface coverage, ~10–20% that of a quantum well. Furthermore, because of the large QD size distribution, only a subset of the islands contributes to the gain at any particular wavelength. Consequently, the gain provided by a single layer of QDs is often insufficient to achieve ground state lasing and stacks of three to five QD layers are needed (Zhukov et al. 1999b, Schmidt et al. 1996, Heinrichsdorff et al. 1997b, Tatebayashi et al. 2005).

Stacked quantum dot structures, especially those grown by MBE, have been extensively studied in the literature. These studies have identified three important factors which affect island nucleation in the upper layers and ultimately determine the properties of the final stacked QD structure. These three factors are:

(i) the thickness of the GaAs spacer layer separating each QD layer,
(ii) the QD growth conditions, and
(iii) the spacer layer surface morphology,

Each of these factors is discussed in the following sections.

12.4.1 Spacer Layer Thickness

The spacer layer thickness has a strong impact on the stacking process and the electronic properties of the final structure. For spacer layer thicknesses less than ~20 nm, strong vertical alignment of the quantum dots is observed as illustrated schematically in Fig. 12.16(b) (Sugiyama et al. 1997, Xie et al. 1995, Solomon et al. 1996, Bruls et al. 2003, Le Ru et al. 2002). This has been attributed to the strain field generated by the underlying dot layer which modifies the adatom mobility and/or nucleation process so that islands preferentially nucleate above islands in the previous layer (Xie et al. 1995, Howe et al. 2004, Tersoff et al. 1996). At very small spacer layer thicknesses of 1–5 nm, the quantum dots are so close that they become electronically coupled and each column of dots forms a single superdot (Sugiyama et al. 1997). This strong electronic coupling also results in a red-shift of the photoluminescence which may be beneficial for reaching emission at 1.31 and 1.55 μm for telecommunications (Sugiyama et al. 1997, Solomon et al. 1996, Le Ru et al. 2002). However a draw back of the strong coupling regime is that the

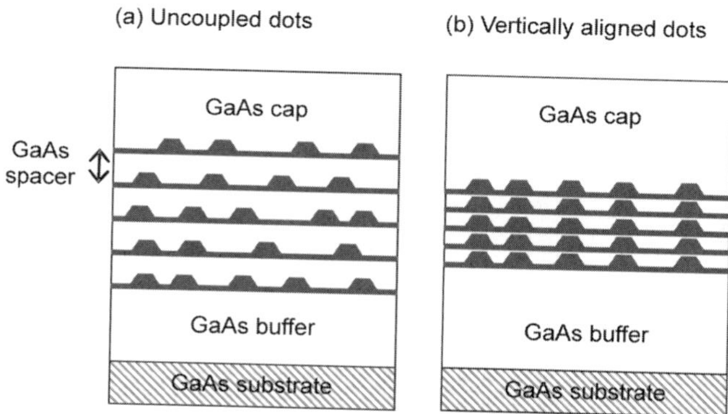

Fig. 12.16 Schematics of stacked QD structures with (a) wide spacer layers so that each QD layer nucleates randomly and (b) narrow spacer layers so that the QDs align vertically. Reproduced from (Sears 2006)

overall dot density is not necessarily increased (Le Ru et al. 2002). At large spacer thicknesses, greater than ~20 nm, the strain field is diminished, resulting in a return to random nucleation and no vertical correlation as illustrated in Fig. 12.16(a). In our work we have concentrated on stacked QD structures separated by 30 nm GaAs spacers. This spacer layer thickness results in no vertical correlation or coupling between the dot layers which greatly simplifies the growth. At the same time, a thickness of 30 nm is still small enough to achieve good overlap between the optical field of the laser structure and each of the QD layers.

12.4.2 Surface Smoothing

The surface morphology of the spacer layer is extremely important because it affects both island nucleation and the optical scattering losses of a device. The islands are most relaxed at their top, where the lattice can bend and expand more readily and this affects how the GaAs overgrowth proceeds. Typically a rough surface develops (Songmuang et al. 2003, Hasegawa et al. 2003, Joyce et al. 2001), especially in the case of InAs QDs where the strain field is much greater. Fig. 12.17(a) shows an example of the rough surface that develops above a single layer of InAs QDs when no surface smoothing is applied. The hillocks that form are strongly elongated along one of the [011] directions. Similar anisotropic hillock morphology has been observed by other groups (Hasegawa et al. 2003, Joyce et al. 2001, Sellin et al. 2000) and is attributed to the anisotropic diffusion of Ga.

A rough surface such as shown in Fig. 12.17(a) will clearly affect nucleation of the next QD layer. Typically the upper QD layers have a reduced density and a greater size. The formation of larger islands in the upper layers can also increase the probability of defect generation. Fig. 12.18 shows AFM images of the final $In_{0.5}Ga_{0.5}As$ quantum dot layer of a 3 stacked structure grown (a) without and (b) with surface smoothing techniques. With no surface smoothing, the top dot layer has

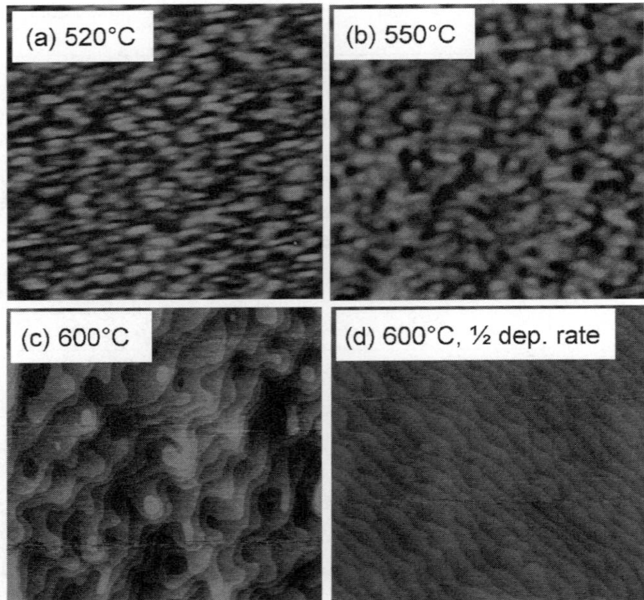

Fig. 12.17 AFM height images of InAs QDs capped by a 30 nm GaAs layer grown at (a) 520, (b) 550, and (c) (d) 600 °C. The first 10 nm of GaAs was deposited at 520 °C, the same temperature used to deposit the QDs. The growth was then interrupted and the sample heated to the desired temperature. In (d) the deposition rate is also reduced from 6 to 3 Å/s for the final 20 nm of the GaAs cap. All of the images are $4 \times 4\,\mu m^2$. Adapted from (Stewart et al. 2005a)

a density half that of a single layer. However with surface smoothing techniques, the dot density in the top layer is nearly identical to that of a single layer.

In general, surface smoothing techniques rely on increasing the adatom mobility. For example, growth interrupts and higher growth temperatures are commonly used

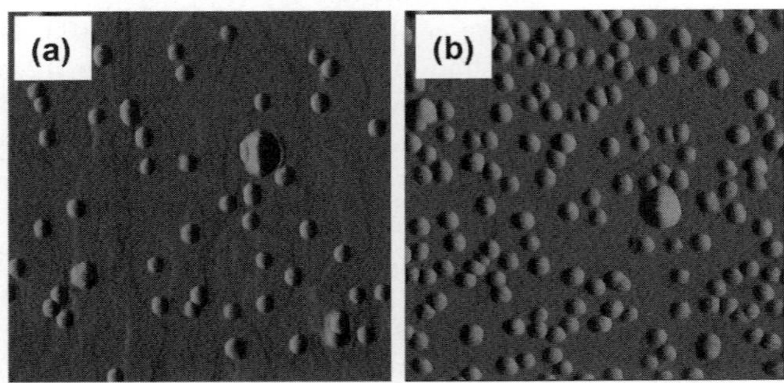

Fig. 12.18 $500 \times 500\,nm^2$ AFM deflection images of the top $In_{0.5}Ga_{0.5}As$ QD layer in a 3 stacked structure (a) with no surface smoothing and (b) with surface smoothing. Adapted from (Sears et al. 2006b)

(Sellin et al. 2000, Liu et al. 2004, Le Ru et al. 2002). Fig. 12.17 shows AFM images of a single layer of InAs quantum dots capped with a 30 nm GaAs layer at different temperatures and growth rates. In each case the first 10 nm of the capping layer was grown at the same temperature as the InAs QDs (520 ℃), in order to avoid indium desorption/intermixing effects. The growth was then paused for ~2.5 minutes and the sample heated to the desired temperature before growing the final 20 nm of GaAs. Clearly the higher growth temperature and increased adatom mobility enables a planarized surface to be reconstructed. For InAs QDs the best smoothing procedure relied on (i) increasing the growth temperature from 520 ℃ (that of the quantum dots) to 600 ℃ and (ii) reducing the GaAs depositing rate from 6 to 3Å/s, for the final 20nm of the GaAs spacer layer. In the case of $In_{0.5}Ga_{0.5}As$ quantum dots, the introduction of a 1.5 minute growth interrupt after capping the dots with 7 nm of GaAs was sufficient to replanarize the surface. It is also worth noting that at smaller spacer layer thicknesses, in the strong coupling regime, Le Ru et al. (2002) found that annealing did not replanarize the surface because the strain field from the underlying dot layers was too strong. However the use of thin GaP and InGaP strain compensating layers appears to give helpful smoothing effects (Lever et al. 2004c, Tatebayashi et al. 2006).

12.4.3 The QD Growth Parameters

It is vital to choose QD growth parameters which avoid the creation of defects/dislocations. Once formed, a dislocation tends to propagate throughout the entire stacked structure, often multiplying to create large V-shaped or avalanche type defects (Shiramine et al. 1999, Stewart et al. 2003, Ledentsov et al. 2000, Roh et al. 2001, Liu et al. 2004). Fig. 12.19 shows an example of a large V-shaped defect which was observed in a 30-stack, MBE grown InAs QD structure. This defect originates at one of the QD/GaAs interfaces and then travels upwards through the GaAs spacers from one QD layer to the next. We have observed that islands preferentially nucleate at the surface termination sites of dislocations and this may explain the propagation of defects from one layer to the next. As discussed later in Section 12.5.3, such V-shaped defects are detrimental for diode laser performance. It has also been reported that defect generation and propagation can be reduced through the use of growth interrupts to dissolve larger, dislocated islands (see section 2.4) (Ledentsov et al. 2000, Sellin et al. 2001).

Not only are the defects themselves detrimental for device performance but they also result in a rough surface which will increase device scattering losses and affect island nucleation in the upper layers. Fig. 12.20 shows an AFM image of a 4 stacked InAs QD structure. The final dot layer has been capped with a 30 nm GaAs layer. An extremely rough surface with "pit-like" features is observed. We and others have associated these pits with the formation of large defects/dislocations (Stewart et al. 2005b, Ng and Missous 2006). Clearly defect generation in stacked structures has a number of undesirable effects.

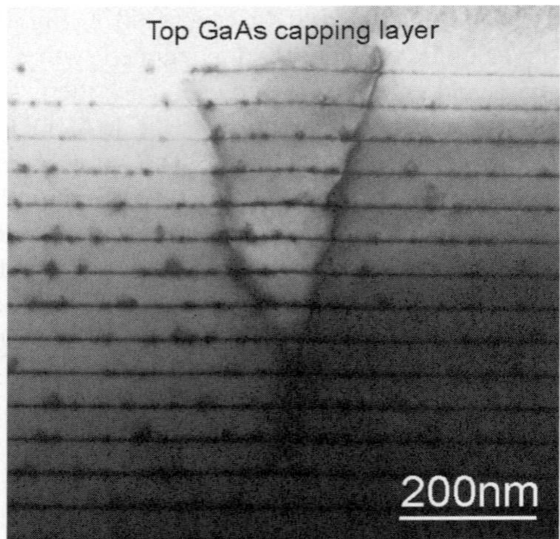

Fig. 12.19 (011) cross sectional transmission electron micrograph of a V-shaped defect through a stacked InAs QD structure. Adapted from (Stewart et al. 2003)

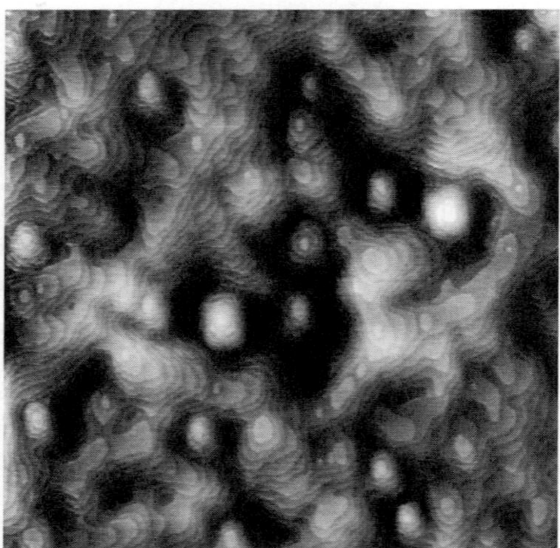

Fig. 12.20 $8 \times 8\,\mu m^2$ AFM height image of a 4-stack QD structure. The last QD layer was capped with a 30 nm GaAs layer. Adapted from (Sears 2006)

12.5 In(Ga)As QD Lasers

This section discusses the lasing characteristics of edge emitting diode lasers incorporating stacked layers of quantum dots in the active region. Both the InAs/GaAs and $In_{0.5}Ga_{0.5}As$ /GaAs quantum dot lasers showed similar lasing characteristics and here we concentrate primarily on the InAs/GaAs QD system. A brief overview of the laser design is first given. The lasing characteristics of devices with three and five stacked layers of QDs are then compared and finally the influence of the QD growth parameters on laser performance is discussed.

12.5.1 Laser Design and Growth Details

Thin p-clad laser structures were grown and fabricated into 4 μm ridge waveguides using standard lithography and wet chemical etching. Fig. 12.21 shows a schematic of the final structure. The mirror facets were left as cleaved (no reflectivity coatings applied) and the devices tested at 7 ℃ in pulsed mode with a duty cycle of 5% (25 kHz, 2 μs pulse). Either 3 or 5 stacked QD layers were incorporated into the active region. Each QD layer was separated by a 30 nm GaAs spacer layer so there is no vertical correlation of the QDs.

Figure 12.22 gives further details on the thin p-clad laser structure. It is based on an asymmetric graded-index separate-confinement heterostructure (GRINSCH) design. A detailed description of GRINSCH lasers is beyond the scope of this discussion but can be found in a number of excellent textbooks (Coldren and Corzine 1995, Sze 1985). The quantum dot active region is sandwiched between two 0.16 μm undoped AlGaAs graded index (GRIN) layers. In order to minimize the time spent by the QDs at elevated temperatures, the top p-cladding layer is kept thin and consists of a 0.45 μm $Al_{0.45}Ga_{0.55}As$ confinement layer and a 0.1 μm highly doped GaAs contact layer. As discussed earlier in Section 12.3.7 , the QDs are highly sensitive to the overlayer growth temperature. The bottom, n-cladding layer is much thicker and consists of a 1.8 μm $Al_{0.45}Ga_{0.55}As$ and 0.25 μm $Al_{0.3}Ga_{0.7}As$

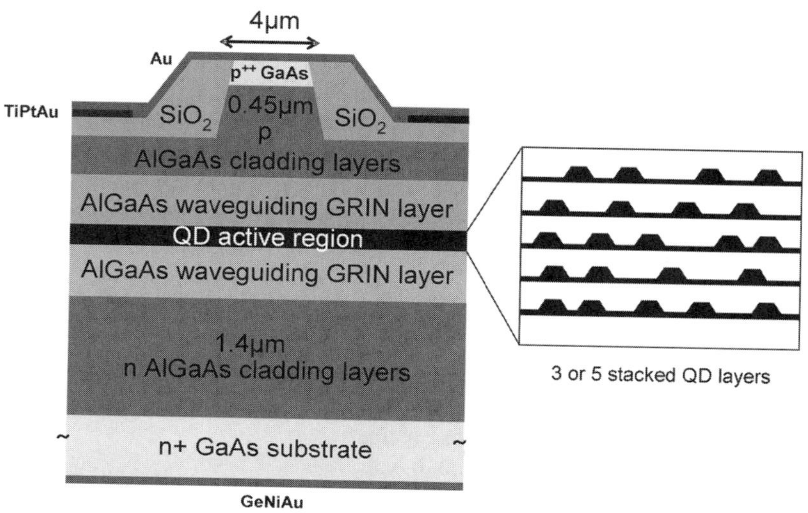

Fig. 12.21 Schematic of the thin p-clad laser structure after processing. Reproduced from (Sears et al. 2007)

	Composition	Thickness (μm)	Doping (cm^{-3})	Temperature (°C)
p-contact layer	GaAs	0.10	1e19	600
p-cladding	$Al_{0.45}Ga_{0.55}As$	0.45	5e17	600
GRIN layer	$Al_{0.3}Ga_{0.7}As$ GaAs	0.16	Intrinsic	600
Active region	3-5 QD layers	0.12-0.18	Intrinsic	520
GRIN layer	GaAs $Al_{0.3}Ga_{0.7}As$	0.16	Intrinsic	750-650
n-cladding	$Al_{0.3}Ga_{0.7}As$	0.25	3e17	750
n-cladding	$Al_{0.45}Ga_{0.55}As$	0.3	3e17	750
n-cladding	$Al_{0.45}Ga_{0.55}As$	1.5	2e18	750
n-buffer layer	GaAs	0.25	2e18	750
	n+ GaAs substrate			

Fig. 12.22 Schematic of the laser structure. Each layer's composition, doping level and growth temperature are specified. Adapted from (Sears 2006)

layer. This design gives a small spot size of 0.375 μm which increases the overlap between the optical field and QD gain region and helps to minimize the effects of gain saturation (discussed later). Further details on the laser structure can be found elsewhere (Buda et al. 2003).

In order to avoid degradation of the QD active region, the AlGaAs cladding layers above the InAs ($In_{0.5}Ga_{0.5}As$) QDs were grown at 600 (650) ℃. This is considerably lower than the optimum growth temperature of ~700 ℃. Studies on reference QW lasers show that this low growth temperature increases the device threshold current density by at least a factor of two (Sears et al. 2007). This is likely due to an increased concentration of nonradiative recombination centers in the upper AlGaAs cladding layers grown at lower temperature. All the layers below the QD active region were grown at their optimum temperature of 700 ℃.

Where current densities are quoted they were calculated using the area of the ridge waveguide. They therefore overestimate the current density in the 4 μm wide ridge waveguides by a factor of 2–2.5 as a result of current spreading.

12.5.2 Lasing Characteristics of 3 and 5-stack QD Lasers

12.5.2.1 Ground and Excited State Lasing

One commonly reported characteristic of QD lasers is their susceptibility to excited state lasing. Each quantum dot state can only accommodate a finite number of carriers. This in combination with the low surface densities (10^{10}–10^{11} cm^{-2}) and the large size distribution typical of the Stranski-Krastanow growth mode, limits the available ground state gain. Quantum dot devices therefore show a tendency to lase from the excited states, which typically have higher degeneracies and can therefore provide greater gain (Zhukov et al. 1999a, Shoji et al. 1997, Ustinov et al. 2003, Bimberg et al. 2000, Tatebayashi et al. 2005, Smowton et al. 2001). Several groups have also suggested that (i) incomplete filling of the QD states (Park et al. 2000a, Matthews et al. 2002), and (ii) a low wavefunction overlap within the dots (Asryan et al. 2001), may further enhance the tendency for QD devices to lase from excited states.

This behavior is illustrated by Fig. 12.23(a), which shows lasing spectra for 1.4 and 5 mm long devices with 3-stacked QD layers in the active region. The PL spectrum for the same sample prior to device fabrication is also shown for comparison. Both devices lase at much shorter wavelength than the PL peak emission, which indicates that they lase from excited states. Saturation of the ground state gain becomes particularly apparent at shorter device lengths, such as the 1.4 mm device shown in Fig. 12.23(a). At these short cavity lengths the cavity losses are much higher so that greater gain is needed for lasing. In order to avoid excited state lasing it is important to provide a sufficient number of stacked QD layers. In our case, 5-stacked QD layers were needed to achieve lasing from the QD ground state for laser cavity lengths down to 1.5 mm. Fig. 12.23(b) shows a lasing spectrum from a 5 stack QD laser (4.5 mm) as well as a PL spectrum from the same sample prior to device fabrication. Clearly the lasing spectrum is closely aligned with the peak PL wavelength, indicating ground state lasing.

Excited state lasing is also generally associated with increased threshold currents and reduced efficiencies (Zhukov et al. 1999b, Shoji et al. 1997, Ustinov et al. 2003, Park et al. 2000a) and is therefore undesirable. For example, the 3-stack QD devices which lased from the excited state had threshold currents twice that of the 5-stack QD devices. Fig. 12.24 also illustrates nicely the correlation between excited state lasing and increased threshold currents. This figure plots the threshold current density and lasing wavelength for a 5-stack QD laser as a function of device length. These 5-stack QD devices lase predominantly from the QD ground state for cavity lengths down to 1.5 mm. At shorter cavity lengths, the ground state gain can no longer sustain lasing and the excited states become populated. This leads to a sharp drop in the lasing wavelength. At the same time a rapid increase in threshold current density is observed.

One reason for the increased threshold current can be intuitively understood with reference to Fig. 12.25. This figure shows modeled gain curves for the 3- and 5-stack QD lasers, as a function of injection current. The model is based on the same principles given in reference (Sears 2007, Dikshit and Pikal 2004) and uses Fermi-Dirac population statistics. It assumes a Gaussian distribution with a linewidth of 100 meV for the QDs, with a ground and excited state transition at 1.092 and 1.19 eV, respectively. Degeneracies of 1 and 1.5 (not including spin) are assumed for the ground and excited state transitions, respectively. For both the 3 and 5 stack lasers, the modeled gain increases strongly at first and then starts to saturate as the QD ground states are filled. Eventually at higher injection currents, the gain increases rapidly for a second time as the excited state is populated. A horizontal line is drawn in

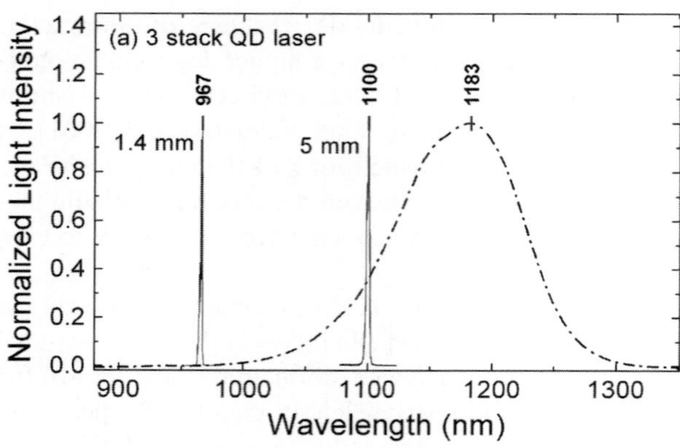

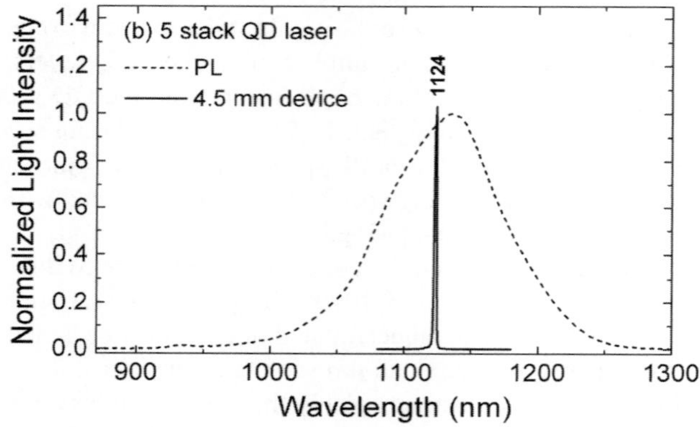

Fig. 12.23 Lasing spectra at 1.1 I_{th} for (a) a 1.4 and 5 mm long device with 3 stacked layers of InAs QDs, and (b) a 4.5 mm long device with 5 stacked layers of InAs QDs. PL spectra of the same samples prior to device fabrication are also shown for comparison. Adapted from (Sears et al. 2007)

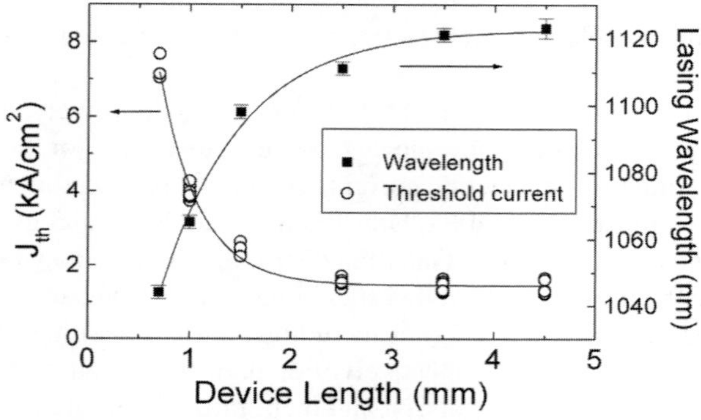

Fig. 12.24 Threshold current density and lasing wavelength versus device length for a 5 stack InAs QD laser. The line fits are shown as a guide only. Adapted from (Sears et al. 2006b)

Fig. 12.25 at a typical cavity loss of $10\,cm^{-1}$. Due to the greater gain volume provided by the 5-stack structure, this cavity loss can be overcome before the ground state gain saturates. This considerably reduces the threshold current in comparison to the 3-stack QD laser where population of the excited states is needed. Several groups have also suggested that population of the excited states may increase the number of non-radiative recombination pathways, further increasing the threshold current (Sears et al. 2007, Deppe et al. 1999, Sandall et al. 2006).

12.5.2.2 Electroluminescence Spectra and Modeling of the Observed Blueshift

Electroluminescence measurements can give further insight into how lasing develops and the gain saturation/state filling effects. Fig. 12.26(a) shows electroluminescence spectra for a 3 stacked InAs QD laser with increasing injection current up to threshold. As the injection current increases, a gradual blue-shift and saturation of the QD ground state electroluminescence is observed. This is followed by the onset of at least one excited state from which lasing eventually develops. For comparison, Fig. 12.26(b) shows electroluminescence spectra for a 5-stack QD laser with increasing injection current. Clearly, the extra QD layers enable lasing to develop directly from the QD ground state.

Even in the case of the 5-stack QD lasers, a slight blue-shift of the electroluminescence spectra is observed with increasing injection current. The inset to Fig. 12.26(b) shows the shift in peak wavelength with injection current. Modeling has shown that this shift is primarily a consequence of the large QD size distribution and Fermi-Dirac like carrier population of the dots. Essentially, larger islands with lower lying energy states are populated first. With increasing injection current (and quasi-Fermi levels), smaller dots with higher lying energy levels are then progressively filled, resulting in the observed blue-shift. The good qualitative agreement

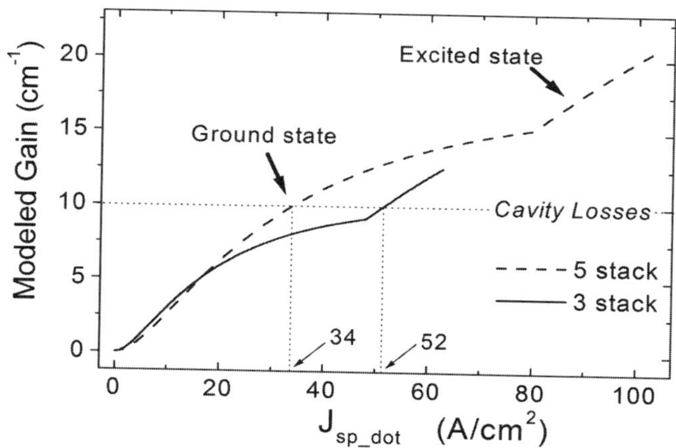

Fig. 12.25 Modeled gain for a 3- and 5-stack QD laser. The model assumes a QD ground and excited state transition. Adapted from (Sears 2006)

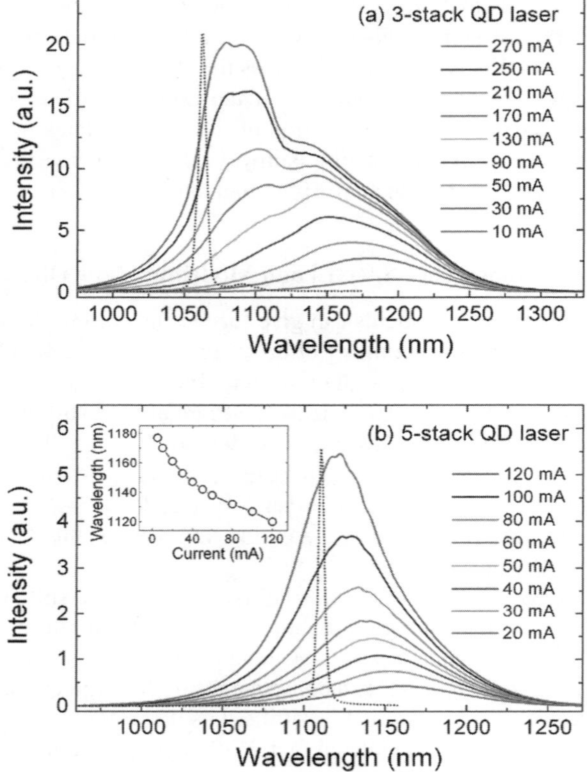

Fig. 12.26 Electroluminescence spectra (broad curves) and a lasing spectrum (dotted curve) at 1.1 I_{th} for (a) a 3-stack (2.7 mm) and (b) a 5-stack (2.5 mm) QD laser. The intensity of the lasing spectra has been normalized. The inset to figure (b) plots the peak electroluminescence wavelength as a function of injection current. Adapted from (Sears et al. 2006a)

between the modelling and experimental data can be seen in Fig. 12.27, which shows modeled electroluminescence spectra assuming a Gaussian distribution for the QDs, centred at an energy of 1.092 eV with a linewidth of 100 meV. The inset to Fig. 12.27 also shows the shift in peak wavelength with injection current and clearly demonstrates that a shift on the order of 40 nm is possible, although this will vary with the amount of inhomogeneous broadening.

A number of groups have reported a built in electric dipole in self-assembled quantum dots (Passaseo et al. 2001, Passaseo et al. 2003, Fry et al. 2000, Hsu et al. 2001). This built in electric dipole can also reduce the ground state gain due to a lower wavefuction overlap between the electrons and holes, and cause a blue-shift of the electroluminescence spectra relative to the QD photoluminescence (Passaseo et al. 2001). The magnitude and direction of this built in dipole depends strongly on the QD shape (i.e. pyramid, lens shape, truncated pyramid), aspect ratio, and indium composition and profile (Passaseo et al. 2003, Sheng and Leburton 2003,

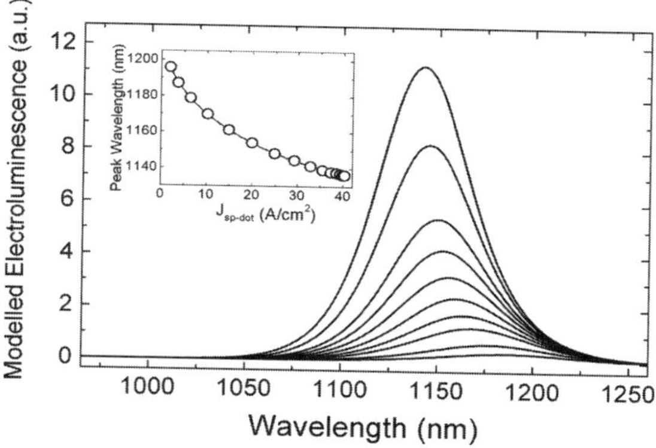

Fig. 12.27 Modeled electroluminescence spectra assuming a Gaussian distribution for the QDs with a linewidth of 100 meV, a single transition at 1.092 eV and Fermi-Dirac carrier population of the dots. The inset shows the shift in the peak wavelength with injection current. Adapted from (Sears et al. 2006a)

Fry et al. 2000) and as such will depend on the growth parameters used. We have observed a negligible built-in dipole for our InAs QDs and a small built in electric dipole of ~20 kV/cm for our $In_{0.5}Ga_{0.5}As$ /GaAs QDs (Lever et al. 2004b). In our case the blue-shift of the electroluminescence is therefore attributed primarily to the large QD size distribution and Fermi-Dirac like carrier population statistics.

12.5.3 *Effect of Quantum Dot Growth Conditions on Laser Performance*

QD growth conditions that avoid defect formation are extremely important because even a low defect density severely degrades laser performance. This is demonstrated by Fig. 12.28, which shows plots of inverse differential efficiency versus device length for two 5-stack QD lasers. Laser #1 was formed using the optimized QD growth conditions which led to PL centered at 1135 nm and excellent device performance. For example a linear fit to the data shown in Fig. 12.28 gives an optical loss of 8 cm^{-1} and a high quantum internal efficiency of close to 100%. In contrast, laser #2 was fabricated using conditions that led to defect density of $\sim 1(\pm 0.5) \times 10^7$ cm^{-2} (based on pit density observed by AFM – see Section 12.4.3). A linear fit to the data for laser #2 in Fig. 12.28 gives a very high slope, which is an indication of poor performance and high optical losses. Furthermore, the threshold current density for laser #2 was consistently higher than that of laser #1 by a factor of 2 or more.

As shown previously in Fig. 12.19, the defects that form tend to nucleate at a dot layer and then thread from one dot layer to the next, terminating at the sample surface. As such, these threading defects may act as short circuits for the injected current, reducing the amount of current reaching the QDs in the active region,

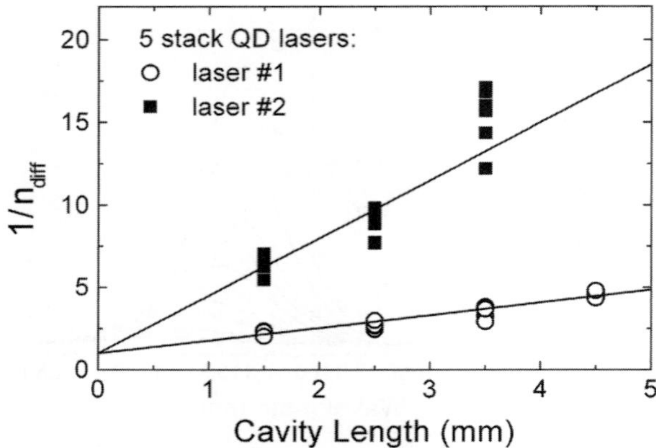

Fig. 12.28 Inverse differential efficiency versus device length for two 5-stack QD lasers. Laser #1 was grown using our optimized growth conditions, while Laser #2 was grown using growth conditions that led to a defect density of $\sim 1(\pm 0.5) \times 10^7$ cm^{-2}. Adapted from (Sears 2006)

even above threshold. This effect would also reduce the differential efficiency, and contribute to the high slope observed in Fig. 12.28. These defects may also lead to increased optical absorption as well as greater surface roughness and scattering losses.

12.6 Selective Area Epitaxy for Optoelectronic Device Integration

Photonic device integration requires the ability to realize epitaxial layers with different band gaps on the same substrate. Post growth processing techniques like impurity free vacancy disordering (IFVD) and ion-implantation induced intermixing have been used to demonstrate band gap tuning of quantum dots (Lever et al. 2003 and 2004d and references therein). Although the simplicity of these techniques makes them very attractive, they rely on introducing defects into the active region, which is not desirable for device performance. In contrast, selective area epitaxy (SAE) is a band gap tuning technique that does not involve post growth processing and thus does not alter the quality of epitaxial layers in the active region of the devices. Another advantage of SAE compared to intermixing techniques, is that it does not rely on high temperature heat treatments. This is a distinct advantage when it comes to quantum dots. As mentioned in Section 12.3.7, the quantum dot emission blue-shifts considerably during high temperature annealing. This can be avoided through the use of SAE while still achieving large bandgap tuning across the wafer. SAE also enables us to simultaneously grow both QDs and quantum wells (QWs) (controlled nucleation of QDs), which is very important for the integration of lasers with low loss passive waveguides. This section briefly outlines the basic principle of SAE and illustrates how it can be used to achieve controlled nucleation and band gap tuning of QDs, essential for optoelectronic device integration.

12.6.1 Basic Principle of SAE

The term SAE is broadly used to describe growth techniques in which material deposition in certain regions of the semiconductor substrate is prevented by use of a mask. During MOCVD growth, if the semiconductor substrate is patterned with a dielectric mask such as SiO_2 or Si_3N_4, in the unmasked regions where semiconductor surface is exposed it catalyses the pyrolysis of the group V source molecules. The adsorbed group III species can then react with the catalyzed group V species to form a stable compound and epitaxial growth proceeds. However in regions masked with the dielectric, much higher temperatures are required for pyrolysis of the group V sources (Coleman 1997 and the references therein). Hence, under suitable conditions, growth only takes place in the unmasked regions.

Figure 12.29 illustrates how this behavior can be used to advantage for photonic integrated devices. In Fig. 12.29 the substrate is patterned with a pair of SiO_2 stripes. As material is deposited, the group III species adsorbed on the dielectric stripes migrate to the openings between the stripes, where they can then react with group V species. Also at temperatures where desorption of the reactant species on the dielectric prevails over their decomposition, additional group III species diffuse through the gas phase into nearby unmasked regions. This gas phase diffusion and surface migration enhances the growth rate in the unmasked openings between the pairs of stripes compared to regions far away from the stripes. Hence by varying the dimensions of the stripes and opening, the growth rate enhancement can be controlled.

12.6.2 Controlled Nucleation of QDs

Figure 12.30 demonstrates how SAE can be used to form areas with and without QDs on the same substrate in a single growth. In this example, only part of the substrate is patterned by pairs of SiO_2 stripes. As discussed in Section 3.2, QDs only form after a certain critical thickness has been exceeded. In this example, the growth conditions have been carefully chosen so that away from the SiO_2 stripes, the critical thickness is not exceeded and only a thin quantum well forms. However

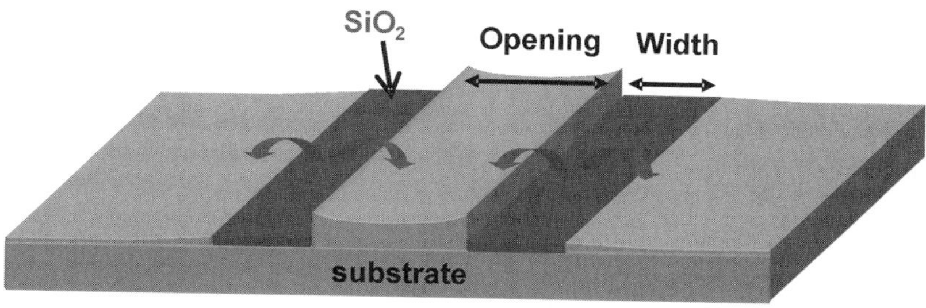

Fig. 12.29 Schematic illustrating the basic principle of selective area epitaxy. Growth does not take place on the SiO_2 stripes and instead the adatoms migrate to the stripe opening enhancing the growth rate there

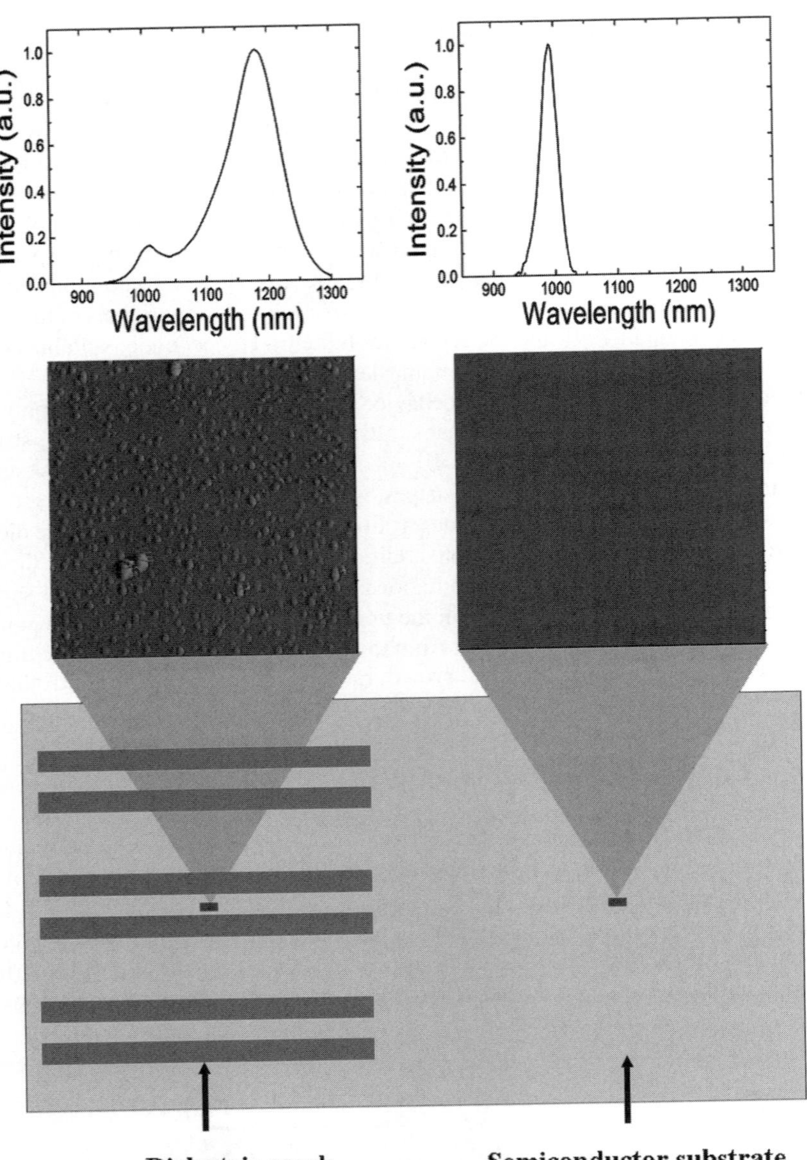

Fig. 12.30 Deflection AFM images showing $In_{0.5}Ga_{0.5}As$ QDs and planar QWs grown simultaneously on the same substrate using SAE. The mask pattern used for patterning the substrate is also shown. The photoluminescence from the selectively grown QDs and QWs is shown above the corresponding AFM images

between the stripes, the growth enhancement enables the critical thickness to be exceeded and a high density of QDs form. The formation of QDs between the stripes, and a planar surface away from the stripes is clearly shown in the AFM images in Fig. 12.30. At the top of Fig. 12.30, PL spectra of the QD and QW regions are also given. The band gap of the QDs is clearly much smaller than that of the QWs. The QWs will therefore be transparent to light emitted by the QDs making these selectively grown QDs and QWs ideal for the integration of a QD laser with a low-loss passive (QW) waveguide.

12.6.3 Band Gap Tuning of QDs

The energy level separation in QDs depends on their size and composition. SAE offers us selective control on these two properties by adjusting the mask dimensions in different regions of a substrate. This is demonstrated by Fig. 12.31 which shows AFM images of $In_{0.5}Ga_{0.5}As$ QDs formed between SiO_2 stripes with a constant opening of 50 μm and different stripe widths. At the smallest stripe width (12 μm) the critical thickness for island formation has only just been exceeded so that shallow QDs of low density have formed. As the stripe width (and hence growth enhancement) is increased, the QD density and height also increases and then eventually saturates. If the stripe width is further increased the island density decreases. This is due to the formation of dislocated clusters (circled in Fig. 12.31) which act as material sinks. This evolution in dot properties (size and density) is similar to that described in Section 12.3.2 for increasing coverage. However in this case, the different dots are all formed on the same substrate in a single growth.

Figure 12.32 shows the peak wavelength from cathodoluminescence (CL) measurements on $In_{0.5}Ga_{0.5}As$ QDs grown between pairs of SiO_2 stripes with different stripe widths. The opening between the stripes was kept constant at 50 μm. Initially, as the stripe width is increased, the peak wavelength redshifts. This redshift is a convolution of two effects: namely a variation in the dot size and composition (Mokkapati et al. 2005). The composition variation occurs due to differential incorporation of InAs and GaAs into the QDs (Mokkapati et al. 2005). As the stripe width is further increased the peak wavelength begins to saturate. This saturation is correlated with the formation of large, dislocated clusters (see Fig. 12.31) which, as mentioned earlier, act as material sinks for any additional deposited material. This example clearly demonstrates that the QD peak wavelength can be tuned between 1150 to 1200 nm simply by increasing the stripe width. This is in spite of a very narrow growth parameter window in which defect free quantum dots are formed (see Section 12.3). In the following section (Section 12.6.4) we discuss how this growth scheme can be used for fabricating multiple wavelength QD lasers on the same substrate.

12.6.4 Device Fabrication

Before discussing our SAE QD device results it is first necessary to discuss the issue of material pile up which occurs at the interface between the dielectric mask and the

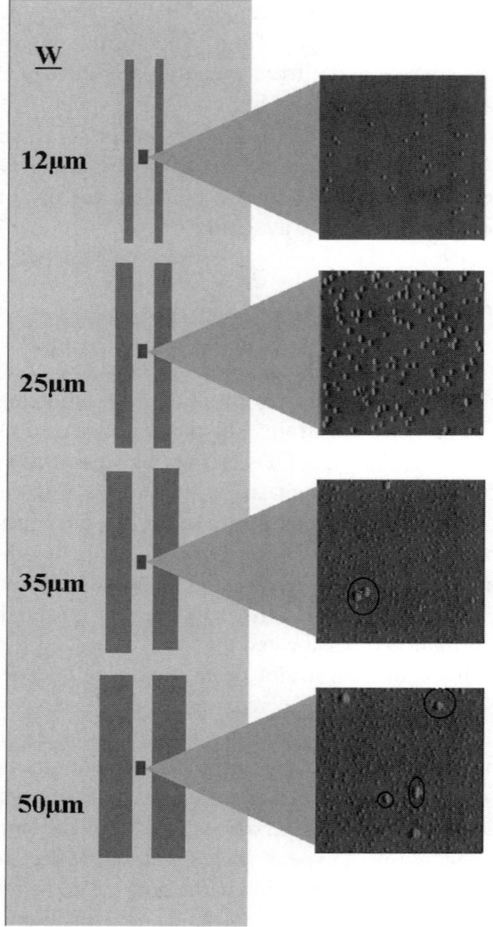

Fig. 12.31 Deflection AFM images showing $In_{0.5}Ga_{0.5}As$ QDs grown in different regions of a patterned GaAs substrate. The mask pattern consists of pairs of SiO_2 stripes. The openings between the stripes is constant (50μm) and the width of the stripes is varied as indicated. A few large coalesced islands have been circled

mask opening. This pile up is shown pictorially in Fig. 12.29. The V/III ratio used during the growth process affects this pile-up and its elimination in MOCVD is not trivial. As a result of this pile-up, we have enhanced growth rates at the edges of the openings. This means that while dislocation free, uniform high density dots may form at the center of the opening, dislocations can still form at the edges due to the material pile up there. Fig. 12.33(a) and (b) demonstrate the non-uniform growth profile that develops across the opening as a result of this pile up and dislocation formation at the edge.

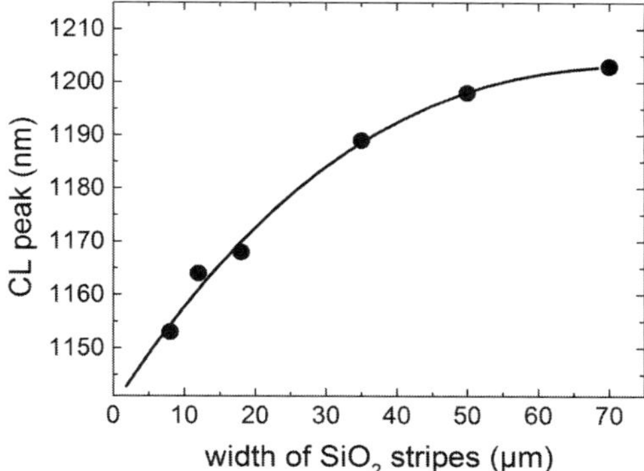

Fig. 12.32 Variation of cathodoluminescence peak wavelength as a function of stripe width for $In_{0.5}Ga_{0.5}As$ QDs grown between pairs of SiO_2 stripes. The opening between the stripes was kept constant at 50 μm and only the stripe width varied. Adapted from (Mokkapati et al. 2005)

In addition, clear facet formation has been observed at the edge of the openings. Due to different growth rates on different facets, no QDs are formed on these facets. Fig. 12.33(c) illustrates the facet formation at the edges. The above mentioned problems can be avoided by growing QDs in larger openings than required for device fabrication. The devices described in this section are fabricated from 4μm wide regions of uniform QDs grown in the centre of 50 or 25μm wide stripe openings.

For the fabrication of multi-wavelength lasers and photonic integrated circuits only the bandgap of the active region needs to be varied across the substrate. The SAE mask is therefore only needed for growth of the active regions. In contrast, it is preferable to grow the surrounding n- and p-type cladding layers without any patterning so that we have the same device structure, apart from the active region, across the entire substrate. The device structure is the same as that described in Section 12.5.1. The general procedure for growing the SAE laser structure is as follows. The bottom n-type cladding layers are first grown and the sample then removed from the growth chamber for patterning. In order to avoid exposure of Al containing layers to atmospheric oxygen during patterning, the growth is terminated with a thin layer of GaAs (Cockerill et al. 1994). The active region is then deposited and the sample again removed from the growth chamber in order to etch away the mask. Finally the upper p-type cladding layers are grown. All of the SAE devices discussed here were based on 5 stacked layers of $In_{0.5}Ga_{0.5}As$ QDs.

Since the growth of SAE QD lasers requires midgrowth processing and epitaxial regrowth, it is important to first compare the performance of these SAE lasers with those grown on unpatterned substrates. Fig. 12.34 shows plots of the inverse differential efficiency vs. cavity length for devices fabricated from unpatterned QDs and selectively grown QDs. Both devices have similar internal losses, indicating that the extra processing steps do not introduce significant additional losses (Mokkapati et al. 2006). Furthermore the SAE QD lasers operate from QD ground states.

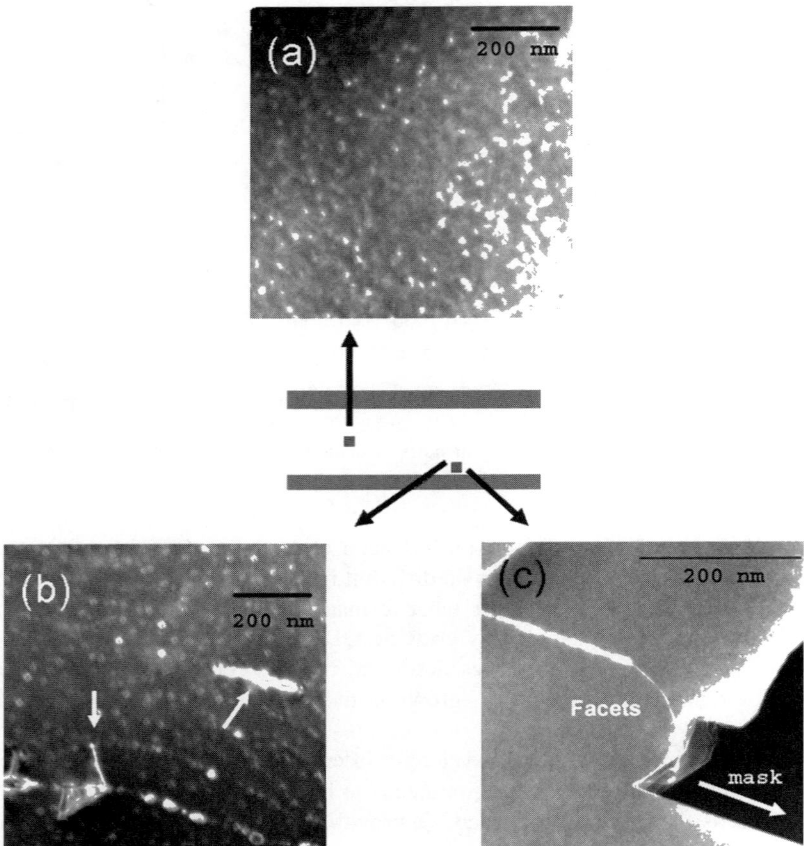

Fig. 12.33 Plan view TEM images of InAs QDs formed at (a) the center and (b) the edge of a 50 μm opening between SiO_2 stripes of width 50 μm. These figures demonstrate the non-uniform growth profile that develops across the opening. White arrows in (b) highlight the formation of dislocations at the edge. (c) is a cross-sectional TEM image showing the facet formation at the edge of the opening

In Section 12.6.2 we discussed the ability to form QDs in one region of the substrate and QWs in another. This has been used to integrate a QD laser with a passive (QW) waveguide (Mokkapati et al. 2006). QDs were formed only in the laser section of the integrated devices and QWs were formed elsewhere on the substrate. The laser optical cavity was defined by the overall length of the laser and waveguide. The waveguide section was electrically isolated from the laser section; and only the laser section of the integrated device was electrically pumped. These devices have very low passive losses of $\sim 3\,cm^{-1}$. The low losses in the waveguide are possible because of the large bandgap energy difference between the QWs in the waveguide section and the QDs in the laser section. This large bandgap difference is evident from Fig. 12.35 which shows lasing spectra of devices fabricated separately

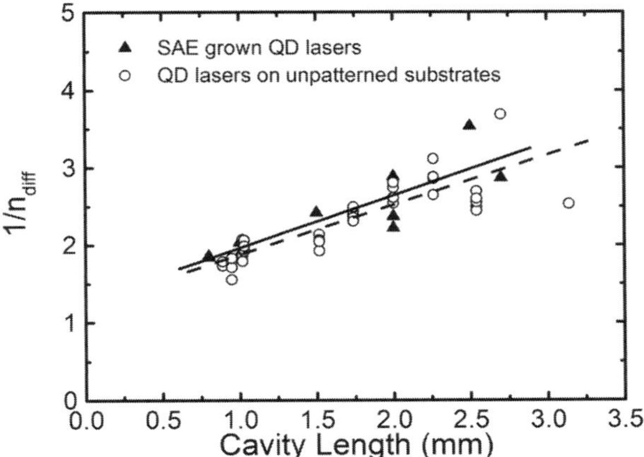

Fig. 12.34 Plots of inverse differential efficiency vs. cavity length for thin p-clad lasers fabricated from selectively grown $In_{0.5}Ga_{0.5}As$ QDs and QDs grown on unpatterned substrates. Both devices have similar losses of ~6cm^{-1}. Adapted from (Mokkapati et al. 2006)

from the selectively grown QD and QW regions. Such a large bandgap difference is difficult to achieve with post growth processing techniques, without leaving any residual defects in the active region of the devices.

In Section 12.6.3 we showed how SAE can be used to tune the QD bandgap in different areas of the same substrate. This ability has been used to fabricate multi-wavelength lasers on the same substrate. The substrate was patterned with pairs of SiO_2 stripes separated from each other by 50 μm and the width of the stipes was

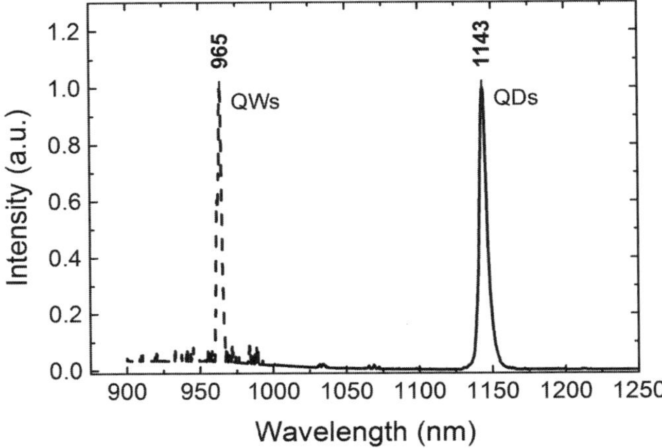

Fig. 12.35 Lasing spectra of devices fabricated from selectively grown QDs and QWs that form the active and passive sections, respectively of a QD laser integrated with a passive waveguide. A large bandgap energy shift (~200 meV) between the epitaxial layers in the two regions enables very low passive losses in waveguide. Adapted from (Mokkapati et al. 2006)

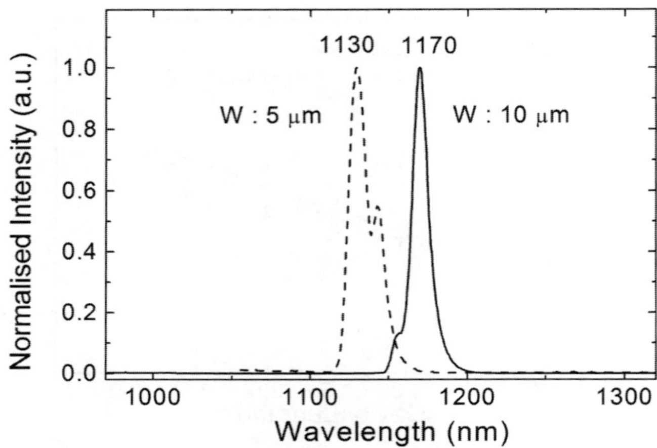

Fig. 12.36 Lasing spectra of 4 mm long lasers fabricated from QDs selectively grown on different regions of the same substrate. The mask pattern used for growing the QDs consists of pairs of SiO_2 stripes. The openings between the stripes are 50 μm and the stripe width is either 5 μm or 10 μm

either 5μm or 10 μm. As discussed in section 12.6.3, this allows us to form QDs with different bandgap energies on different regions across the substrate. Devices with 4μm wide ridges were fabricated from these QDs; and the lasing spectra of the two devices are shifted from each other by ~40 nm as shown in Fig. 12.36. Although similarly large shifts may be obtained by other post growth processing techniques, the heat treatments used by these techniques inevitably blueshift the QD emission. To our knowledge, the ability to tune the lasing emission by 40 nm, while maintaining the long lasing wavelengths depicted in Fig. 12.36, has only been achieved with SAE.

12.7 Summary

The aim of this chapter was to give an overview of (i) the self-assembled growth of In(Ga)As/GaAs QDs using MOCVD, (ii) the basic lasing characteristics of QD lasers and (iii) the use of selective area epitaxy for fabrication of photonic integrated circuits and multi-wavelength lasers.

The various MOCVD growth parameters and their effect on In(Ga)As/GaAs island formation have been discussed in detail. Due to the high strain involved, almost all of the growth parameters have a significant influence on the nucleation process. However three growth parameters, the V/III ratio (or AsH_3 flow), the growth temperature and the coverage have especially strong effects. In order to achieve high densities of defect free QDs it is important to use:

(i) low AsH_3 flows because high flows lead to strong ripening effects,
(ii) low growth temperatures in order to minimize the tendency for islands to coalesce, and

(iii) coverages which give high QD densities but avoid formation of large dislocated clusters/islands.

Most QD lasers use stacked QD structures (3–10 layers) in order to increase the overall QD density and hence ground state gain. As stressed in this chapter, QD lasers show a high susceptibility to excited state lasing because of their limited ground state gain and the higher gain provided by the excited states and even the WL. Such excited state lasing is undesirable due to its associated high threshold currents and reduced efficiencies. However by stacking sufficent QD layers with the use of overlayer smoothing procedures, ground state lasing with good device performance can be achieved.

We have also demonstrated that electroluminescence measurements are a simple way to gain insight into the lasing characteristics and QD state-filling effects. The electroluminescence spectra blue-shifted with increasing injection current. Modelling has indicated that this blue-shift is primarily a consequence of the large QD size distribution and Fermi-Dirac like carrier population of the quantum dots. Although, depending on the QD shape, size, aspect ratio and composition profile, a built in electric dipole may also contribute.

One major challenge for QD formation in the Stranski-Krastanow mode is the broad size distribution that results. QD formation in the S-K growth mode has been a major breakthrough in the realization of QD lasers. It has enabled many of their predicted advantages to be demonstrated and their lasing characteristics to be explored. However their performance is ultimately limited by the large QD size distribution. Ideally, a close-packed array of identical quantum dots is needed to fully benefit from the 3D carrier confinement and discrete energy levels provided by QDs. This remains an ongoing challenge. Luckily research into QD lasers is still an extremely exciting area with a number of groups actively working towards better spatial and size control of islands formed in the Stranski-Krastanow mode, as well as alternative methods of forming QDs such as by patterning with e-beam lithography.

Finally, this chapter also discussed the use of selective area epitaxy for achieving QD-based photonic integrated circuits. We have successfully used SAE to spatially control quantum dot nucleation on a GaAs substrate in a single growth. The ability to form areas with and without QDs on the same substrate has been used to integrate a low loss waveguide with a QD laser, while the ability to tune the QD bandgap energy has been exploited to produce multi-wavelength QD lasers on the same substrate. Clearly SAE is an extremely promising technique for QD photonic device integration.

Acknowledgments Thanks are due to Dr. Manuela Buda and Mr. Michael Aggett for fruitful discussions and expert technical advice, respectively. We also wish to acknowledge Dr Jenny Wong-Leung for her help with TEM imaging, and Dr Vince Craig, Dr. Tim Senden and Dr John FitzGerald for their expert advice on AFM and TEM respectively. Dr Penny Lever is also acknowledged for providing the data for Figs. 12.3, 12.5 and 12.7. We would also like to acknowledge Ms. K. E. McBean and Prof. M. R. Phillips for assistance with the CL measurements presented in this chapter. The Australian Research Council is gratefully acknowledged for its financial support.

References

Arakawa Y, Sakaki H (1982) Multidimensional quantum well laser and temperature dependence of its threshold current. Appl. Phys. Lett. 40(11): 939–941

Asryan LV, Grundmann M, Ledentsov NN, Stier O, Suris RA, Bimberg D (2001) Maximum modal gain of a self-assembled InAs quantum-dot laser. J. Appl. Phys. 90(3): 1666–1668

Bimberg D, Grundmann M, Heinrichsdorff F, Ledentsov NN, Ustinov VM, Zhukov AE, Kovsh AR, Maximov MV, Shernyakov Yu M, Volovik BV, Tsatsul'nikov AF, Kop'ev PS, Alferov Zh I (2000) Quantum dot lasers: breakthrough in optoelectronics. Thin Solid Films 367: 235–249

Bruls DM, Koenraad PM, Salemink HWM, Wolter JH, Hopkinson M, Skolnick MS (2003) Stacked low-growth-rate InAs quantum dots studied at the atomic level by cross-sectional scanning tunneling microscopy. Appl. Phys. Lett. 82(21): 3758–3760

Buda M, Hay J, Tan HH, Wong-Leung J, Jagadish C (2003) Low loss, thin p-clad 980-nm InGaAs semiconductor laser diodes with an asymmetric structure design. IEEE J. Quantum Electron. 39(5): 625–633

Chung T, Walter G, Holonyak N Jr (2005) Growth mechanism of InAs quantum dots on GaAs by metal-organic chemical-vapor deposition. J. Appl. Phys. 97: 53510–53513

Cockerill TM, Forbes DV, Han H Turkot BA, Dantzig JA, Robertson IM, Coleman JJ (1994) Wavelength tuning in strained layer InGaAs-GaAs-AlGaAs quantum well lasers by selective-area MOCVD. J. Electron. Mater. 23(2):115–119

Coldren LA, Corzine SW (1995), Diode lasers and photonic integrated circuits. John Wiley and Sons, USA

Coleman JJ (1997) Metalorganic chemical vapor deposition for optoelectronic devices. Proceedings of the IEEE 85(11):1715–1729

Deppe DG, Huffaker D, Csutak S, Zou Z, Park G, Shchekin OB (1999) Spontaneous emission and threshold characteristics of 1.3 μm InGaAs-GaAs quantum-dot GaAs-based lasers. IEEE J. Quantum Electron. 35(8): 1238–1246

Dikshit AA, Pikal JM (2004) Carrier distribution, gain, and lasing in 1.3-μm InAs-InGaAs quantum-dot lasers. IEEE J. Quantum Electron. 40(2): 105–112

Dingle R, Henry CH (1976) Quantum effects in heterostructure lasers. U.S. Patent (3982207)

El-Emawy AA, Birudavolu S, Huang S, Xu H, Huffaker DL (2003) Selective surface migration for defect-free quantum dot ensembles using metal organic chemical vapor deposition. J. Cryst. Growth 225: 213–219

Fehse R, Marko I, Adams AR (2003) Long wavelength lasers on GaAs substrates. IEE Proc.-Circuits Devices Syst. 150(6): 521–528

Fry PW, Itskevich IE, Mowbray DJ, Skolnick MS, Finley JJ, Barker JA, O'Reilly EP, Wilson LR, Larkin IA, Maksym PA, Hopkinson M, Al-Khafaji M, David JPR, Cullis AG, Hill G, Clark JC (2000) Inverted electron-hole alignment in InAs-GaAs self-assembled quantum dots. Phys. Rev. Lett. 84, (4):733–736

Goldstein L, Glas F, Marzin JY, Charasse MN, Roux G Le (1985) Growth by molecular beam epitaxy ad characterization of InAs/GaAs strained-layer superlattices. Appl. Phys. Lett. 47(10): 1099–1101

Hasegawa S, Suekand O, Takata M, Nakashima H (2003) Scanning tunneling microscopy study of GaAs overgrowth on InAs islands formed on GaAs(001). J. Cryst. Growth 251: 161–165

Heinrichsdorff F, Krost A, Kirstaedter N, Mao M-H, Grundmann M, Bimberg D, Kosogov AO, Werner P (1997a) InAs/GaAs quantum dots grown by metalorganic chemical vapor deposition. Jpn. J. Appl. Phys. 36(6B): 4129–4133

Heinrichsdorff F, Mao M-H, Kirstaedter N, Krost A, Bimberg D, Kosogov AO, Werner P (1997b) Room-temperature continuous-wave lasing from stacked InAs/GaAs quantum dots grown by metalorganic chemical vapor deposition. Appl. Phys. Lett. 71(1): 22–24

Heinrichsdorff F, Krost A, Bimberg D, Kosogov AO, Werner P (1998) Self organized defect free InAs/GaAs and InAs/InGaAs/GaAs quantum dots with high lateral density grown by MOCVD. Appl. Surf. Sci. 123/124: 725–728

Heinrichsdorff F, Ribbat Ch, Grundmann M, Bimberg D (2000) High-power quantum-dot lasers at 1100nm. Appl. Phys. Lett. 76(15): 5556–5558

Howe P, Abbey B, Le Ru EC, Murray R, Jones TS (2004) Strain interaction between InAs/GaAs quantum dot layers. Thin Solid Films 464–465: 225–228

Hsu TM, Chang W-H, Huang CC, Yeh NT, Chyi J-I (2001) Quantum-confined Stark shift in electroreflectance of InAs/$In_xGa_{1-x}As$ self-assembled quantum dots. Appl. Phys. Lett. 78(12) 1760–1762

Joyce PB, Krzyzewski TJ, Steans PH, Bell GR, Neave JH, Jones TS (2001) Shape and surface morphology changes during the inital stages of encapsulation of InAs/GaAs quantum dots. Surf. Sci. 492: 345–353

Kaiander I, Hopfer F, Kettler T, Pohl UW, Bimberg D (2004) Alternative precursor growth of quantum dot-based VCSELs and edge emitters for near infrared wavelengths. J. Cryst. Growth 272: 154–160

Kamins TI, Medeiros-Ribeiro G, Ohlberg DA, Williams RS (1999) Evolution of Ge islands on Si(001) during annealing. J. Appl. Phys. 85(2): 1159–1171

Karachinsky L Ya, Kettler T, Gordeev N Yu, Novikov II, Maximov MV, Shernyakov Yu M, Kryzhanovskaya NV, Zhukov AE, Semenova ES, Vasil'ev AP, Ustinov VM, Ledentsov NN, Kovsh AR, Shchukin VA, Mikhrin SS, Lochmann A, Schulz O, Reissmann L, Bimberg D (2005) Highpower singlemode CW operation of 1.5 μm-range quantum dot GaAs-based laser. Electron. Lett. 41(8): 478–480

Kirstaedter N, Ledentsov NN, Grundmann M, Bimberg D, Ustinov VM, Rumimov SS, Maximov MV, Kop'ev PS, Alferov Zh I, Richter U, Werner P, Gösele U, Heydenreich J (1994) Low threshold, large To injection laser emission from (InGa)As quantum dots. Electron. Lett. 30(17):1416–1417

Kovsh AR, Maleev NA, Zhukov AE, Mikhrin SS, Vasil'ev AP, Shernyakov Yu M, Maximov MV, Livshits DA, Ustinov VM, Alferov Zh I, Ledentsov NN, Bimberg D (2002) InAs/InGaAs/GaAs quantum dot lasers of 1.3 μm range with high (88%) differential efficiency. Electron. Lett. 38(19): 1104–1106

Ledentsov NN, Maximov MV, Bimberg D, Maka T, Sotomayor Torres CM, Kochnev IV, Krestnikov IL, Lantratov VM, Cherkashin NA, Musikhin Yu M, Alferov Zh I (2000) 1.3 μm luminescence and gain from defect-free InGaAs-GaAs quantum dots grown by metal-organic chemical vapor deposition. Semicond. Sci. Technol. 15: 604–607

Ledentsov NN, Kovsh AR, Zhukov AE, Maleev NA, Mikhrin SS,Vasil'ev AP, Semenova ES, Maximov MV, Shernyakov Yu M, Kryzhanovskaya NV, Ustinov VM, Bimberg D (2003) High performance quantum dot lasers on GaAs substrates operating in 1.5 μm range. IEEE Photon. Technol. Lett. 39(15): 1126–1128

Lee H, Lowe-Webb R, Yang W, Sercel P (1997) Formation of InAs/GaAs quantum dots by molecular beam epitaxy: Reversibility of the islanding transition. Appl. Phys. Lett. 71(16): 2325–2327

Le Ru EC, Bennett AJ, Roberts C, Murray R (2002) Strain and electronic interactions in InAs/GaAs quantum dot multilayers for 1300 nm emission. J. Appl. Phys. 91(3): 1365–1370

Lester LF, Stintz A, Li H, Newell C, Pease EA, Fuchs BA, Malloy KJ (1999) Optical characteristics of 1.24 μm InAs quantum-dot laser diodes. IEEE Photon. Technol. Lett. 11(8): 931–933

Lever P, Tan HH, Jagadish C, Reece P, Gal M (2003) Proton-irradiation-induced intermixing of InGaAs quantum dots. Appl. Phys. Lett. 82:2053–2055

Lever P, Buda M, Tan HH, Jagadish C (2004a) Characteristics of MOCVD grown InGaAs quantum dot lasers. IEEE Photon. Technol. Lett. 16(12):2589–2591

Lever P, Buda M, Tan HH, Jagadish C (2004b) Investigation of the blueshift in electroluminescence spectra InGaAs quantum dots. IEEE J. Quantum Electron. 40(10): 1410–6

Lever P, Tan HH, Jagadish C (2004c) InGaAs quantum dots grown with GaP strain compensation layers. J. Appl. Phys. 95(10):5710–5714

Lever P, Tan HH, Jagadish C (2004d) Impurity free vacancy disordering of InGaAs quantum dots. J. Appl. Phys. 96(12):7544–7548

Liu GT, Stintz A, Li H, Malloy KJ, Lester LF (1999) Extremely low room-temperature threshold current density diode lasers using InAs dots in $In_{0.15}Ga_{0.85}As$ quantum well. Electron. Lett. 35(14): 1163–1165

Liu GT, Stintz H Li, Newell TC, Gray AL, Varangis PM, Malloy KJ, Lester LF (2000) The influence of quantum-well composition on the performance of quantum dot lasers using InAs/InGaAs dots-in-a-well (DWELL) structures. IEEE J. Quantum Electron. 36(11): 1272–1279

Liu HY, Hopkinson M, Harrison CN, Steer MJ, Firth R, Sellers IR, Mowbray DJ, Skolnick MS (2003) Optimizing the growth of 1.3 μm InAs/InGaAs dots-in-a-well structure. J. Appl. Phys. 93(5): 2931–2936

Liu HY, Sellers IR, Badcock TJ, Mowbray DJ, Skolnick MS, Groom KM, Gutiérrez M, Hopkinson M, Ng JS, David JPR, Beanland R (2004) Improved performance of 1.3 μm multilayer InAs quantum-dot lasers using a high-growth-temperature GaAs spacer layer. Appl. Phys. Lett. 85(5): 704–706

Matthews DR, Summers HD, Smowton PM, Hopkinson M (2002) Experimental investigation of the effect of wetting-layer states on the gain-current characteristic of quantum-dot lasers. Appl. Phys. Lett. 81(26): 4904–4906

Maximov MV, Kochnev IV, Shernyakov YM, Zaitsev SV, Gordeev N Yu, Tsatsul'nikov AF, Sakharov AV, Krestnikov IL, Kop'ev PS, Alferov Zh I, Ledentsov NN, Bimberg D, Kosogov AO, Werner P, Gösele U (1997) InGaAs/GaAs quantum dot lasers with ultrahigh characteristic temperature (T_o=385K) grown by metal organic chemical vapour deposition. Jpn J. Appl. Phys. 36(6B): 4221–4223

Maximov MV, Tsatsul'nikov AF, Volovik BV, Sizov DS, Shernyakov Yu M, Kaiander IN, Zhukov AE, Kovsh AR, Mikhrin SS, Ustinov VM, Alferov Zh I, Heitz R, Shchukin V, Ledentsov NN, Bimberg D, Musikhin Yu G, Neumann W (2000) Tuning quantum dot properties by activated phase separation of an InGa(Al)As alloy grown on InAs stressors. Phys. Rev. B 62(24): 16671–16680

Miao ZL, Zhang YW, Chua SJ, Chye YH, Chen P, Tripathy S (2005) Optical properties of InAs/GaAs surface quantum dots. Appl. Phys. Lett. 86: 031914–031916

Mokkapati S, Lever P, Tan HH, Jagadish C, McBean KE, Phillips MR (2005) Controlling the properties of InGaAs quantum dots by selective-area epitaxy. Appl. Phys. Lett. 86: 113102–113104

Mokkapati S, Tan HH, Jagadish C (2006) Integration of an InGaAs quantum dot laser with a low-loss passive waveguide using selective-area epitaxy. IEEE Photon. Technol. Lett. 18(15):1648–1650

Ng J, Missous M (2006) Improvements of stacked self-assembled InAs/GaAs quantum dot structures for 1.3μm applications. Microelectronics Journal, 37(12):1446–1450

Nuntawong N, Xin YC, Birudavolu S, Wong PS, Huang S, Hains CP, Huffaker DL (2005) Quantum dot lasers based on a stacked and straincompensated active region grown by metal-organic chemical vapor deposition. Appl. Phys. Lett. 86: 193115–193117

Park G, Shchekin OB, Deppe DG (2000a) Temperature dependence of gain saturation in multilevel quantum dot lasers. IEEE J. Quantum Electron.36(9): 1065–1071

Park G, Shchekin OB, Huffaker DL, Deppe DG (2000b) Low-threshold oxide-confined 1.3 μm quantum-dot laser. IEEE Photon. Technol. Lett. 13(3):230–232

Passaseo A, Maruccio G, De Vittorio M, De Rinaldis S, Todaro T, Rinaldi R, Cingolani R (2001) Dependence of the emission wavelength on the internal electric field in quantum-dot laser structures grown by metal-organic chemical-vapor deposition. Appl. Phys. Lett. 79(10): 1435–1437

Passaseo A, De Vittorio M, Todaro T, Tarantini I, Giorgi MD, Cingolani R, Taurino A, Catalano M, Fiore A, Markus A, Chen JX, Paranthoen C, Oesterle U, Ilegems M (2003) Comparison of radiative and structural properties of 1.3μm $In_xGa_{1-x}As$ quantum dot laser structures grown by metalorganic chemical vapor deposition and molecular beam epitaxy. Appl. Phys. Lett. 82(21): 3632–3643

Pötschke K, Müller-Kirsch L, Heitz R, Sellin RL, Pohl UW, Bimberg D, Zakharov N, Werner P (2004) Ripening of self-organized InAs quantum dots. Physica E 21: 606–610

Qui Y, Gogna P, Forouhar S, Stintz A, Lester LF (2001) High-performance InAs quantum-dot lasers near 1.3 μm. Appl. Phys. Lett. 79(22): 3570–3572

Roh CH, Park YJ, Kim KM, Park YM, Kim EK, Shim KB (2001) Defect generation in multi-stacked InAs quantum dot/GaAs structures. J. Cryst. Growth 226: 1–7

Sandall IC, Smowton PM, Walker CL, Liu HY, Hopkinson M, Mowbray DJ (2006) Recombination mechanisms in 1.3-μm InAs quantum-dot lasers. IEEE Photon. Technol. Lett. 18(8): 965–9677

Schmidt OG, Kirstaedter N, Ledentsov NN, Mao M-H, Bimberg D, Maximov VM, Kop'ev PS, Alferov Zh-I (1996) Prevention of gain saturation by multi-layer quantum dot lasers. Electron. Lett. 32(14): 1302–1303

Sears K, Wong-Leung J, Buda M, Tan HH, Jagadish C (2004) Growth and characterization of InAs/GaAs quantum dots grown by MOCVD. Proceedings of the 2004 Conference on Optoelectronic and Microelectronic Materials and Devices, Brisbane, Australia, pp. 1–4

Sears K (2006) Growth and Characterisation of Self-Assembled InAs/GaAs Quantum Dots and Optoelectronic Devices. Ph.D. thesis, The Australian National University

Sears K, Buda M, Wong-Leung J, Tan HH, Jagadish C (2006a) Growth and characterization of InAs/GaAs quantum dot diode lasers. Proceedings of the 2006 International Conference On Nanoscience and Nanotechnology (ICONN), Brisbane, Australia, in press

Sears K, Mokkapati S, Buda M, Tan HH, Jagadish C (2006b) In(Ga)As/GaAs quantum dots for optoelectronic devices", Proceedings of the SPIE International Symposium on Smart Materials, Nano-, and Micro-Smart Systems, Adelaide, Australia, invited paper, vol. 6415, 641506

Sears K, Tan HH, Wong-Leung J, Jagadish C (2006c) The role of arsine in the self-assembled growth of InAs/GaAs quantum dots by metal organic chemical vapor deposition", J Appl. Phys. 99:044908(1–5)

Sears K, Buda M, Tan HH, Jagadish C (2007) Modeling and characterization of InAs/GaAs quantum dot lasers grown by metal organic chemical vapor deposition. J. Appl. Phys. 101: 013112(1–9)

Sellers IR, Liu HY, Groom KM, Childs DT, Robbins D, Badcock TJ, Hopkinson M, Mowbray DJ, Skolnick MS (2004) 1.3 μm InAs/GaAs multilayer quantum-dot laser with extremely low room-temperature threshold current density. Electron. Lett. 40(22): 1412–1413

Sellin R, Heinrichsdorff F, Ribbat Ch, Grundmann M, Pohl UW, Bimberg D (2000) Surface flattening during MOCVD of thin GaAs layers covering InGaAs quantum dots. J. Cryst. Growth 221: 581–585

Sellin RL, Ribbat Ch, Grundmann M, Ledentsov NN, Bimberg D (2001) Close-to-ideal device characteristics of high-power InGaAs/GaAs quantum dot lasers. Appl. Phys. Lett. 78(9): 1207–1209

Sellin RL, Kaiander I, Ouyang D, Kettler T, Pohl UW, Bimberg D, Zakharov ND, Werner P (2003) Alternative-precursor metalorganic chemical vapor deposition of self-organized InGaAs/GaAs quantum dots and quantum dot lasers. Appl. Phys. Lett. 82(6): 841–843

Sheng W, Leburton J-P (2003) Absence of correlation between built-in electric dipole moment and quantum stark effect in single InAs/GaAs self-assembled quantum dots. Phys. Rev. B, 67:125308 (1–4).

Shiramine K, Horisaki Y, Suzuki D, Itoh S, Ebiko Y, Muto S, Nakata Y, Yokoyama N (1999) TEM observation of threading dislocations in InAs self assembled quantum dot structure. J. Cryst. Growth 205: 461–466

Shoji H, Nakata Y, Mukai K, Sugiyama Y, Sugawara M, Yokoyama NY, Ishikawa H (1997) Lasing characteristics of self-formed quantum-dot lasers with multistacked dot layer. IEEE J. Select. Topics Quantum Electron. 3(2): 188–195

Smowton PM, Herrmann E, Ning Y, Summers HD, Blood P, Hopkinson M (2001) Optical mode loss and gain of multiple-layer quantum-dot lasers. Appl. Phys. Lett. 78(18): 2629–2631

Solomon GS, Trezza JA, Marshall AF, Harris Jr. JS (1996) Vertically aligned and electronically coupled growth induced InAs islands in GaAs. Phys. Rev. Lett. 76(6): 952–955

Songmuang R, Kiravittaya S, Schmidt OG (2003) Shape evolution of InAs quantum dots during overgrowth. J. Cryst. Growth 249: 416–421

Steimetz E, Richter W, Schienle F, Fischer D, Klein M, Zettler J-T (1997) The effect of different group V precursors on the evolution of quantum dots monitored by optical in situ measurements. Jpn. J. Appl. Phys. 37(Part 1(3B)): 1483–1486

Steimetz E, Wehnert T, Kirmse H, Poser F, Zettler J-T, Neumann W, Richter W (2000) Optimizing the growth procedure for InAs quantum dot stacks by optical in situ techniques. J. Cryst. Growth 221: 592–598

Stewart K, Wong-Leung J, Tan HH, Jagadish C (2003) Influence of rapid thermal annealing on a 30 stack InAs/GaAs quantum dot infrared photodetector, J. Appl. Phys. 94(8): 5283–5289

Stewart K, Barik S, Buda M, Tan HH, Jagadish C (2005a) InAs quantum dots for optoelectronic device applications. Proceedings of the 2005 Materials Research Society (MRS) Fall Meeting, Boston, USA, invited paper, vol. 829, B3.4.1

Stewart K, Wong-Leung J, Tan HH, Jagadish C (2005b) InAs quantum dots grown on InGaAs buffer layers by metal-organic chemical vapor deposition. Journal of Crystal Growth, 281: 290–296

Stintz A, Liu GT, Gray L, Spillers R, Delgado SM, Malloy KJ (2000) Characterization of InAs quantum dots in strained $In_xGa_{1-x}As$ quantum wells. J. Vac. Sci. Technol. 18(3): 1496–1501

Stringfellow GB (1999) Organometallic Vapor-Phase Epitaxy. Academic Press, San Diego, USA, second edn

Sugiyama Y, Nakata Y, Futatsugi T, Sugawara M, Awano Y, Yokoyama N (1997) Narrow photoluminescence line width of closely stacked InAs self-assembled quantum dot structures. Jpn. J. Appl. Phys. 36(2A): L158–161

Sze SM (1985) Semiconductor Devices: Physics and Technology. John Wiley and Sons, Singapore

Tan HH, Sears K, Mokkapati S, Fu L, Kim Y, McGowan P, Buda M, Jagadish C (2007) Quantum dots and nanowires for optoelectronic device applications. Invited paper, IEEE J Select. Topics Quantum Electron. 12(6): 1242–1254

Tatebayashi J, Nishioka M, Arakawa Y (2001) Over 1.5 μm emission from InAs quantum dots embedded in InGaAs strain-reducing layer grown by metalorganic chemical vapor deposition. Appl. Phys. Lett. 78(22): 3469–3471

Tatebayashi J, Hatori N, Kakuma H, Ebe H, Sudo H, Kuramata A, Nakata Y, Sugawara M, Arakawa Y (2003) Low threshold current operation of self-assembled InAs/GaAs quantum dot lasers by metal organic chemical vapour deposition. Electron. Lett. 39(15): 1131–1133

Tatebayashi J, Hatori N, Ishida M, Ebe H, Sugawara M, Arakawa Y, Sudo H, Kuramata A (2005) 1.28 μm lasing from stacked InAs/GaAs quantum dots with low-temperature-grown AlGaAs cladding layer by metalorganic chemical vapor deposition. Appl. Phys. Lett. 86: 053107–053109

Tatebayashi J. Nuntawong N. Xin YC. Wong PS. Huang SH. Hains CP. Lester LF. Huffaker DL (2006) Ground-state lasing of stacked InAs/GaAs quantum dots with GaP strain-compensation layers grown by metal organic chemical vapor deposition. Appl. Phys. Lett. 88(22): 221107(1–3)

Tersoff J, Teichert C, Lagally MG (1996) Self-organization in growth of quantum dot superlattices. Phys. Rev. Lett. 76(10): 1675–1678

Ustinov VM, Maleev NA, Zhukov AE, Kovsh AR, Egorov A Yu, Lunev AV, Volovik BV, Krestnikov IL, Musikhin Yu G, Bert NA, Kop'ev PS, Alferov Zh I, Ledentsov NN, Bimberg D (1999) InAs/GaAs quantum dot structures on GaAs substrates emitting at 1.3 μm. Appl. Phys. Lett. 74(19): 2815–2817

Ustinov VM, Zhukov AE, Egorov A YU, Maleev NA (2003) Quantum Dot Lasers. Oxford Science Publications, Oxford, UK, first edn

Walter G, Chung T, Holonyak-Jr N (2002) High-gain coupled InGaAs quantum well InAs quantum dot AlGaAs-GaAs-InGaAs-InAs heterostructure diode laser operation. Appl. Phys. Lett. 80(7): 1126–1128

Xie Q, Madhukar A, Chen P, Kobayashi NP (1995) Vertically self-organized InAs quantum box islands on GaAs(100). Phys. Rev. Lett. 75(13): 2542–2545

Yang T, Tatebayashi J, Tsukamoto S, Nishioka M, Arakawa Y (2004) Narrow photoluminescence linewidth (<17 meV) from highly uniform self-assembled InAs/GaAs quantum dots grown by low-pressure metalorganic chemical vapor deposition. Appl. Phys. Lett. 84(15): 2817–2819

Yeh N-T, Nee T-E, Chyi J-I, Hsu TM, Huang CC (2000) Matrix dependence of strain-induced wavelength shift in self-assembled InAs quantum-dot heterostructures. Appl. Phys. Lett. 76(12): 1567–1569

Zhukov AE, Kovsh AR, Maleev NA, Mikhrin SS, Ustinov VM, Tsatsul'nikov AF, Maximov MV, Volovik BV, Bedarev DA, Shernyakov Yu M, Kop'ev PS, Alferov ZH I, Ledentsov NN, Bimberg D (1999a) Long-wavelength lasing from multiply stacked InAs/InGaAs quantum dots on GaAs substrates. Appl. Phys. Lett. 75(13): 1926–1928

Zhukov AE, Kovsh AR, Ustinov VM, Shernyakov Yu M, Mikhrin SS, Maleev NA, Kondrat'eva E Yu, Livshits DA, Maximov MV, Volovik BV, Bedarev DA, Musikhin Yu G, Ledentsov NN, Kop'ev PS, Alferov Zh I, Bimberg D (1999b) Continuous-wave operation of long-wavelength quantum-dot diode laser on a GaAs substrate. IEEE Photon. Technol. Lett. 11(11): 1345–1347

Chapter 13
Area-selective and Site-controlled InAs Quantum-dot Growth Techniques for Photonic Crystal-based Ultra-small Integrated Circuit

Nobuhiko Ozaki, Shunsuke Ohkouchi, Yoshimasa Sugimoto, Naoki Ikeda, and Kiyoshi Asakawa

13.1 Introduction

So far, semiconductor quantum dots (QDs) as typical nano-structures have provided a variety of attractive electronic/optoelectronic device physics and band-engineering techniques due to their high density-of-states specific to the low-dimensional structures. Some of their research results have been applied to the QD-based laser diodes in the commercial base, while others have still been exploited intensively for advanced electronic/optoelectronic devices.

Recently, a new research scheme has been proposed by combining QDs with another nano-structure, i.e., a photonic crystal (PC), to provide key photonic devices for future advanced telecommunication systems, as shown in Fig. 13.1 (a) [1]. In the left hand wing as a category of an ultra-fast digital photonic network in future, an ultra-small and ultra-fast symmetrical Mach-Zehnder (SMZ)-type [2] all-optical switch (PC-SMZ) has so far been developed in the phase one by using GaAs-based two-dimensional PC slab waveguides embedded with InAs-based QDs (Fig. 13.1(b)) [3]. In phase two, the PC-SMZ is now evolving into a new functional key device, i.e., an ultra-fast all-optical flip-flop (PC-FF) device that is essential for the digital photonic network [1]. As long as a QD technology is concerned all through these two phases, a selective-area-growth (SAG) of QDs has been pointed out to be an important subject for exhibiting an optical nonlinear (ONL) effect in the selective ONL arms only in the PC-SMZ and PC-FF. For this purpose, a metal-mask (MM)/molecular beam epitaxy (MBE) method of InAs QDs has been developed [4].

In the right hand wing in Fig. 13.1 (a), on the other hand, another category of a future quantum information system is described, where a single photon qubit composed of a single QD embedded in a PC-based high-Q cavity is situated as another important product of the PC/QD-combined nano-photonic structure [5–7]. A unique Nano-Jet Probe (NJP) method has been proposed for this purpose because of its potential to position a single QD at the center of the point defect PC high-Q

Z. M. Wang, *Self-Assembled Quantum Dots.*

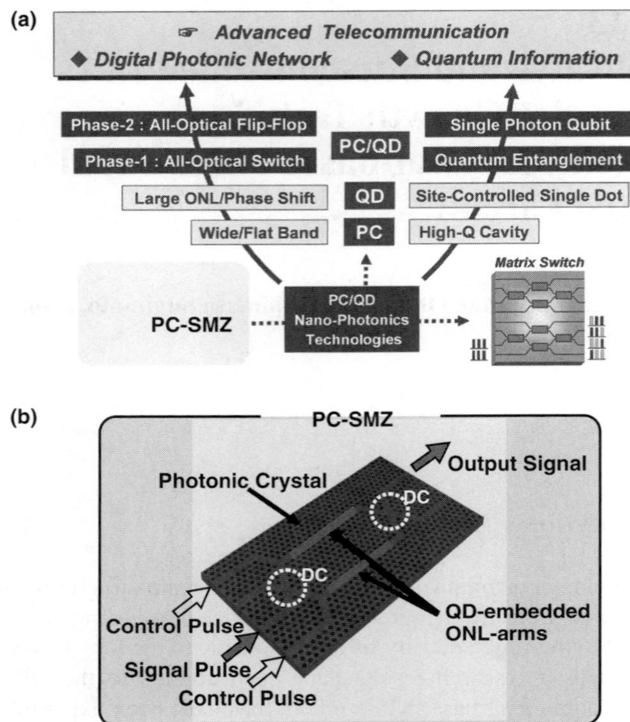

Fig. 13.1 (a) Schematic research scenario of PC/QD combined nano-photonic structures for advanced telecommunication systems. (b) Schematic of the PC-based all-optical switching device, PC-SMZ. QDs embedded partially in the ONL-arms play the role of a phase shifter

cavity [8, 9]. This technique has been initiated from the site-controlled QD growth by using several patterned substrates [10, 11].

In this chapter, two types of unique QD growth techniques have been reviewed. One is the MM method for InAs QDs in the PC-based all-optical switching devices. Another is the advanced techniques for the growth of site-controlled InAs QDs using patterned substrate and the NJP method.

13.2 Selective-area-growth of InAs Quantum Dots Using the Metal-mask (MM)/MBE Method

In this section, we describe a MM/MBE method [4] developed for SAG of InAs-QDs on a GaAs substrate. First, we introduce the developed MM/MBE method and then characterize the SA-grown QDs by using atomic-force-microscopy (AFM) and photoluminescence (PL) measurements, including two-dimensional (2D) PL intensity mapping. As a result, successful SAG of InAs-QDs was clearly demonstrated, and the QD density and homogeneity were almost equal to those of the QDs grown without MM. In addition, we report the controlling of the PL wavelength of InAs-QDs by using a strain-reducing-layer (SRL) [12] inserted on the QDs. These techniques have potential use in developing our proposed photonic-crystal-based all-optical devices, namely, PC-SMZ [3] and PC-FF [1].

13.2.1 MM/MBE Method

We employed a MM to cover the undesired growth area of QDs during conventional MBE growth. A MM is attached on a mask holder that can be mounted on and off in an ultra-high-vacuum (UHV) chamber without exposing the sample to the atmosphere. Figure 13.2 (a) shows an image of the prepared MM attached on the holder.

As illustrated in Fig. 13.2 (b), the MM has a large window (6 mm $\times$ 6 mm) at the center and several small windows of 4 mm $\times$ 0.5 and 1.0 mm around it. The small windows are prepared for SAG of QDs and designed for fitting the PC-SMZ configuration. On the other hand, the large window (6 mm $\times$ 6 mm) is used to optimize growth conditions for RHEED observations and temperature measurements of the sample surface by using a pyrometer during the MBE growth. An electron beam for RHEED observations goes through the tunnels formed on the frame of the MM holder and is diffracted at the surface where molecular beams are irradiated through the large open window during MBE growth (Fig. 13.2(c)).

We have performed the SAG of the QD along a growth sequence shown in Fig. 13.3 (a).

At first, a multi-layer structure of a 50-nm-thick GaAs layer/30-nm-thick $Al_{0.3}Ga_{0.7}As$ layer/GaAs buffer layer was grown on the entire GaAs substrate without the MM at approximately 540°C. Then, the sample was cooled to below 200°C and the MM holder was set on the sample holder in the growth chamber. Following this, the sample was gradually heated up to the InAs-QD growth temperature, typically 470°C–490° C, and 2.6 mono-layer (ML) of InAs was deposited at a rate of 0.2 ML/sec for the QD growth. During this process, we confirmed that the RHEED patterns varied from streaky to spotty patterns, as shown in the photographs in Fig. 13.4 (a) and Fig. 13.4 (b), respectively. This spotty pattern indicates 3D QD growth, and a chevron pattern of each spot originated from the QD facet.

After the sample was slightly cooled by approximately 40°C to avoid the degradation of the QD, a 3-nm-thick InGaAs layer/3-nm-thick GaAs cap layer was sequentially grown. After the SAG of QDs sequence, the MM was removed and a multi-layer structure of 10-nm-thick GaAs cap layer/30-nm-thick $Al_{0.3}Ga_{0.7}As$ layer/50-nm-thick GaAs layer was grown on the entire substrate. The $Al_{0.3}$ $Ga_{0.7}As$ layers below and above the QD layer were inserted as an energy barrier against free-carriers in the photoluminescence (PL) measurement of the QD. The resultant sample structure is shown in Fig. 13.3 (b).

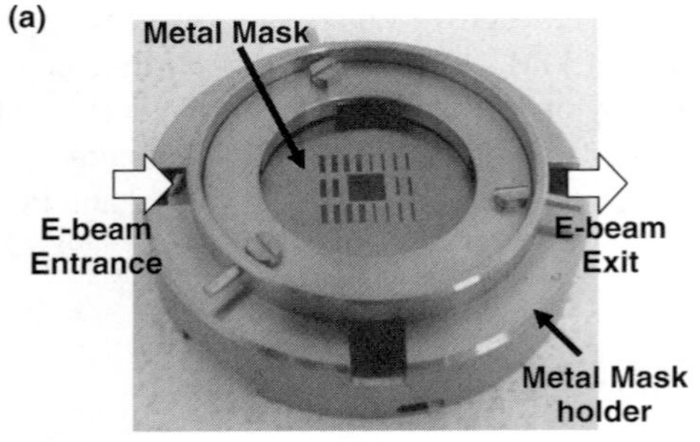

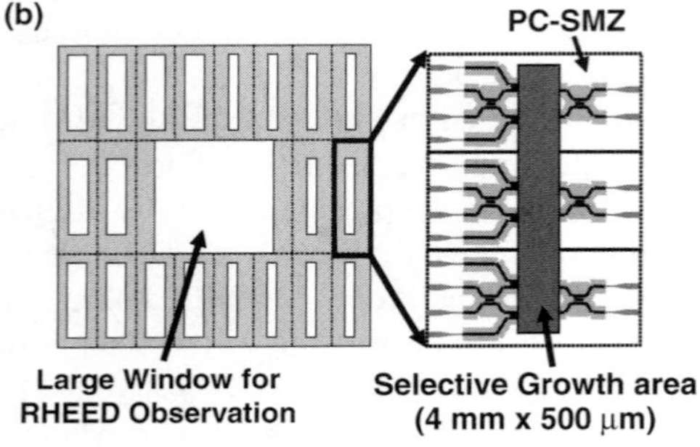

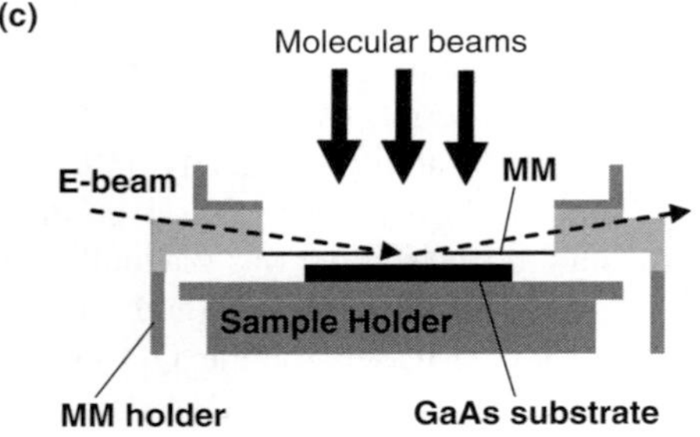

Fig. 13.2 (a) Photographic image of an MM attached on a holder. (b) Schematic of the windows pattern on the MM. A large window of 6 mm × 6 mm at the center and rectangular windows of 4 mm × 1 mm or 0.5 mm around the large window are prepared. (c) Profile illustration of the MM and incident e-beam for RHEED observations

13.2.2 Characterization of SA-Grown QDs

We characterize the SA grown QDs by AFM observations in terms of their structural properties and by PL measurements for their optical quality. For the AFM observations, we grew additional QDs with the MM on a top surface of a sample, since the AFM can only probe the sample surface, and is not useful for embedded QDs. The additional QDs were grown under the same growth conditions as that used for embedded QDs.

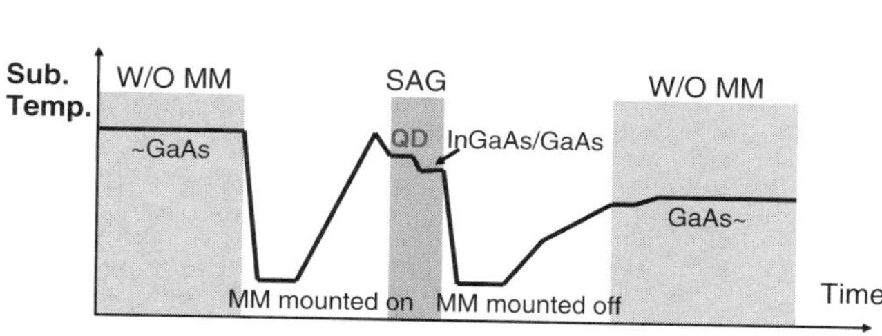

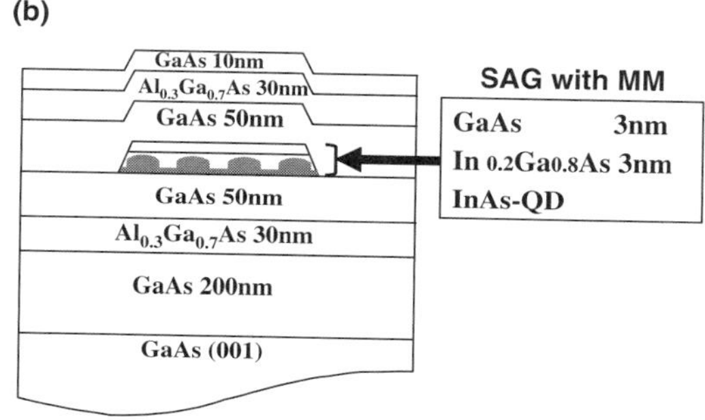

Fig. 13.3 (a) Schematic of the grown sample structure. (b) Growth sequence for SAG of InAs-QDs

Figure 13.5 shows a typical AFM image of the unmasked regions. This image shows a formation of InAs QDs whose density is around $4 \times 10^{10}\text{cm}^{-2}$, whereas no QD was confirmed to be formed on the masked regions of the surface. Such high density is comparable to that of QDs conventionally grown without the MM. The mean lateral size and height of the QD was approximately 40 nm and 5 nm, respectively. These results prove the high structural quality of QDs grown by using the MM.

For investigating the optical quality of SA grown QDs, PL measurements were obtained at room temperature (RT). A He-Ne laser ($\lambda = 633$ nm), focused to a diameter of several microns, was used for QD excitation with a power of 80 μ W. A stepping-motor attached to the sample stage was utilized for positioning the excitation area and 2D mapping of the PL intensity.

PL spectra of the buried QDs in the unmasked and masked regions are shown in Fig. 13.6 (a). The spectra shown with black and gray lines are measured from the

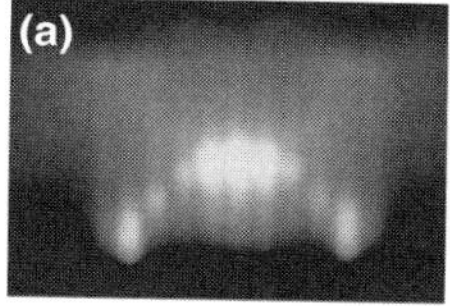

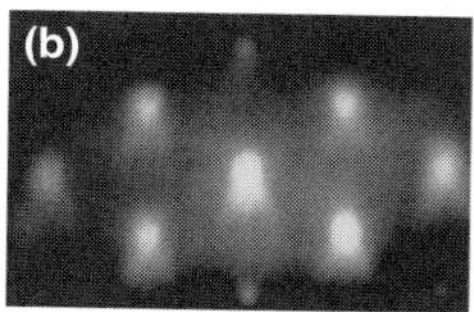

Fig. 13.4 RHEED patterns observed (a) before and (b) after SAG of QDs

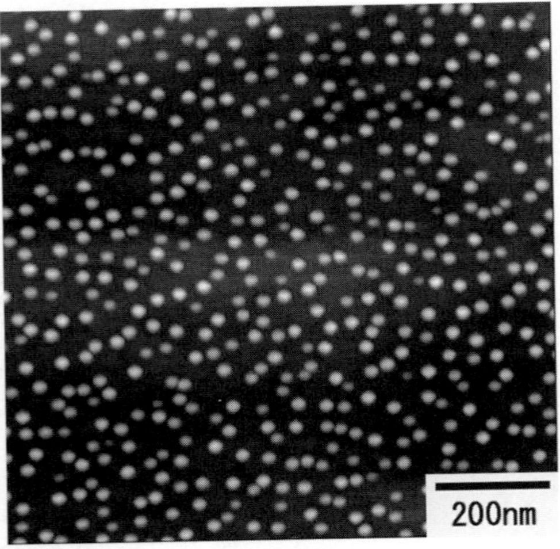

Fig. 13.5 Typical AFM image of QDs grown on unmasked regions by the MM

unmasked regions. The PL peaks of black and gray spectra, attributed to the QDs, result in a center wavelength of 1300 nm (0.95 eV) and a FWHM of approximately 32 meV, which is almost the same optical quality as that of the QD grown without the MM. This small FWHM value obtained at even RT indicates the high homogeneity of the SA grown QD comparable to that of QD grown without the MM. Moreover, these PL spectra measured at different regions show almost identical intensity and linewidth, suggesting that such high density and homogeneous QDs were grown on the selective areas of the substrate without positional dependence. On the other hand, no peak was found in the PL spectrum from the masked region, as shown by a dotted line. Figure 13.6 (b) and (c) show PL intensity mappings of the masked and unmasked areas. High contrast in those images shows high PL intensity from the SA grown QDs. Through these PL intensity mapping images, the successful SAG of the entire region underneath the MM was clearly exhibited. Each QD grown area decreased to approximately 70%–80%, e.g., 750 μm × 3.2 mm SAG area obtained from a window size of 1 mm × 4 mm. This shrinkage of the SAG area is considered to be due to thermal evaporation of QDs near the edge of the windows by the heated MM.

The above characterization results prove the successful SAG of QDs by the MM/MBE method. Here, let us mention the influence of the step-height that results from the SAG for our proposed devices. Considering an application of the SAG to PC-WGs, the step height of the SAG area should be suppressed as much as possible since the step causes the reflection of the propagating light. The SAG method reported here enables us to minimize the step-height at the SAG region to, e.g., around 5 nm. Therefore, this MM method has potential for use in developing the PC-SMZ and PC-FF devices.

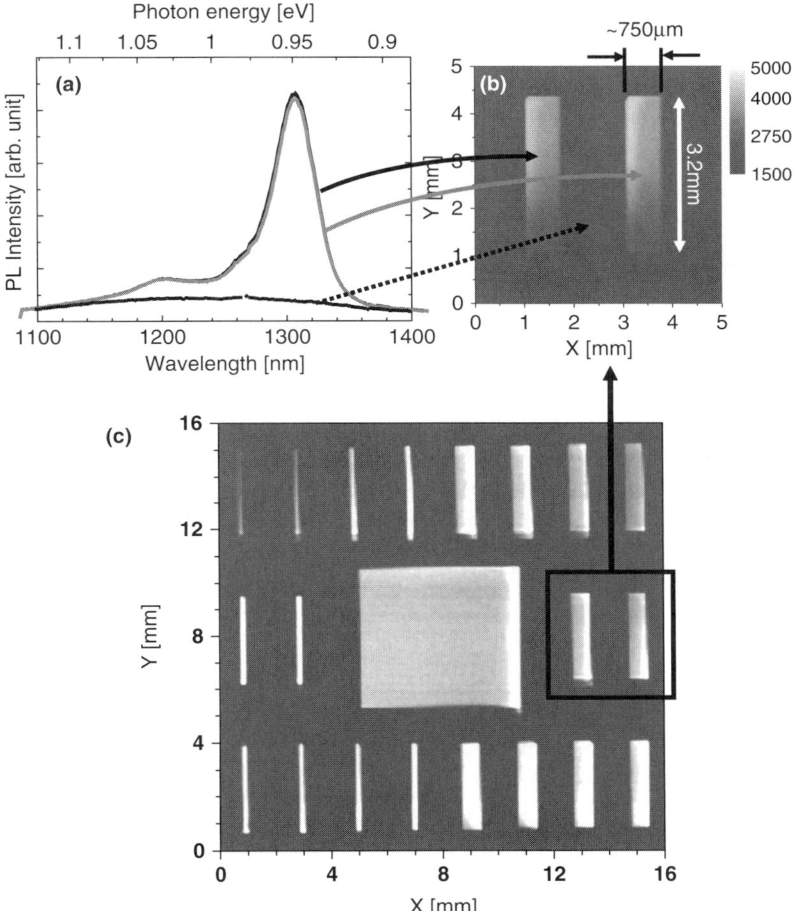

Fig. 13.6 (a) PL spectra at room temperature for the unmasked and masked regions. (b) 2D PL intensity mapping of the region from which PL spectra shown in (a) were obtained. (c) 2D PL intensity mapping of the whole masked and unmasked area

13.2.3 Controlling the Wavelength of QDs

In the previous section, successful SAG of QDs was attained by the MM/MBE method. However, as for applying the MM/MBE method to realize the PC-FF [1], controlling absorption wavelength of SA-grown QDs is also required. For that purpose, we employed an insertion of a strain-reducing-layer (SRL) on QDs. This method is operated by covering grown QDs with an $In_{0.2}Ga_{0.8}As$ layer to reduce strain in the QDs, and it results in a red-shift of the PL emission peak of QDs [12].

We have investigated the effectiveness of SRL for controlling the absorption wavelength of QDs. Figure 13.7 (a) summarizes the PL spectra from strain reduced

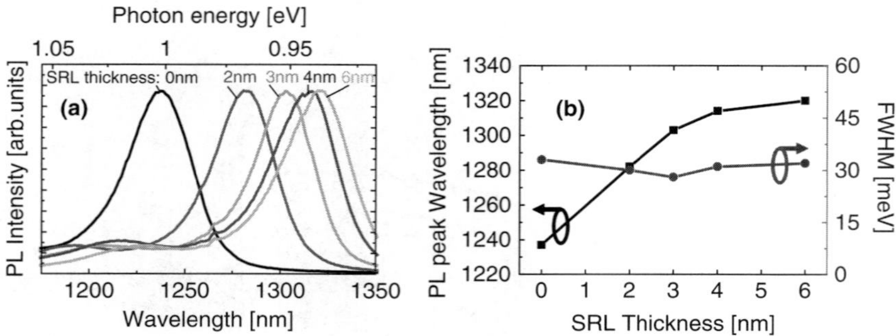

Fig. 13.7 (a) Red-shift of PL spectrum from QDs as a function of SRL thickness. (b) Plotted peak wavelength and FWHM of the PL spectra vs. SRL thickness

QDs with different SRL thicknesses ranging from 0 to 6 nm. The PL spectra exhibits the PL peak wavelength that shifted from 1240 nm to 1320 nm and with the FWHM that was almost maintained at approximately 30 meV, as shown in Fig. 13.7 (b).

This result clearly indicates that the SRL can control the PL wavelength of QDs without degradation of the QD optical quality. Thus, a combination of the MM method and the insertion of the SRL appears to have potential to develop the PC-FF, which requires SAG of QDs with different absorption wavelengths at different areas.

13.3 Advanced Techniques for the Growth of Site-controlled InAs Quantum Dots

In the previous section, we reported on a selective-area growth technique involving self-assembled InAs QDs that uses a MM. The MM method is a simple technique to obtain high-quality InAs QDs in selected areas for PC-SMZ and PC-FF by merely positioning shadow masks on the sample surface. However, the QDs fabricated by the MM method are randomly distributed on the sample surface, since their growth proceeds in the Stranski-Krastanov (SK) mode. In other words, the MM method cannot control the nucleation sites of QDs precisely.

In this section, we report on two advanced techniques for obtaining site-controlled QDs that enable the formation of a required number of QDs at the desired locations. One technique involves GaAs substrates patterned with electron beam (EB) lithography and dry etching [10, 11]. Regular arrays of In(Ga)As QDs with up to 70 nm periodicity have been successfully fabricated by controlling the nucleation sites using artificially prepared nano-hole arrays on the GaAs substrates. The other technique is new and employs an *in situ* AFM probe with a specially designed cantilever, referred to as the Nano-Jet Probe (NJP). The NJP has afforded the capacity to fabricate high-density 2D indium (In) nano-dot arrays on GaAs substrates within a selected area; these arrays can be directly converted into InAs arrays by subsequent annealing with arsenic flux irradiation [8, 9]. Using these methods, we have achieved site-controlled QD formation at desired locations on the sample surface.

13.3.1 Site Control of QDs Using Patterned Substrate

First, we report on the site-control technique using a patterned substrate with nano-hole arrays. Regular arrays of GaInAs QDs with up to 70 nm periodicity have been successfully formed by controlling the nucleation sites on nano-hole arrays prepared artificially on GaAs substrates fabricated by EB lithography and subsequent dry etching.

The patterned GaAs substrates were prepared in the following manner (Fig. 13.8). First, Si-doped 300-nm-thick GaAs layers were grown on Si-doped GaAs (001) substrates to obtain flat surfaces. Next, nano-hole arrays were formed in 50 μm × 50 μm square regions by field-emission-type EB lithography and reactive ion-beam etching (RIBE) with chlorine gas. After the RIBE process, the EB resist was removed with an organic solution, O_2 plasma ashing, and an acid solution. The samples were then loaded into a UHV system connected to an MBE chamber and their surfaces were further cleaned by irradiation with atomic hydrogen for 1h at a substrate temperature of 500°C. Just before MBE growth, the sample surface was observed using an *in situ* scanning tunneling microscope (STM). Figure 13.9 (a) shows an STM image of a regular nano-hole array with a periodicity of 100 nm; here, the array is observed to be free of dust. The diameters of the nano-holes are ∼ 50 nm and their depths are ∼ 50 nm or less.

On the patterned substrate, we fabricated 3D QD arrays. We began by depositing the first InGaAs QD layer (10-monolayer (ML)-thick $In_{0.33}Ga_{0.67}As$) at a low temperature of 400°C with an InGaAs growth rate of 0.3 ML/s; this layer was annealed at 450°C. As shown in Fig. 13.10 (a), the regular nucleation of a QD array with 100 nm periodicity (corresponding to a high density of $1 \times 10^{10}\ cm^{-2}$) was observed

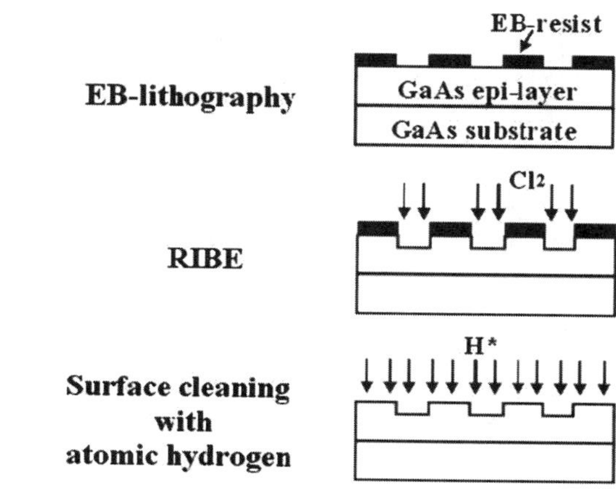

Fig. 13.8 Schematic illustrations of the experimental procedures involved in the fabrication of a GaAs patterned substrate with nano-hole arrays

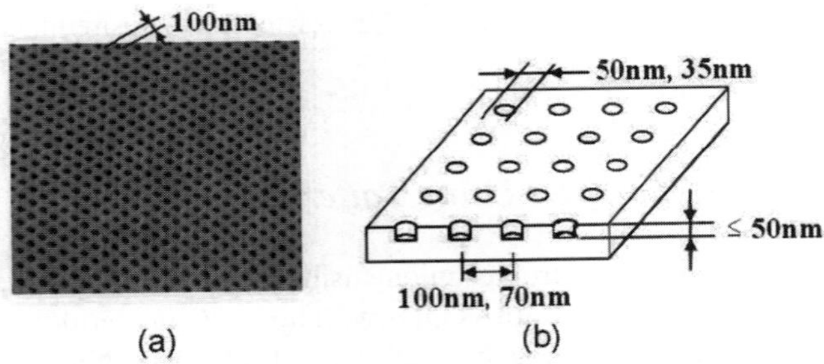

Fig. 13.9 (a) STM images of a patterned GaAs substrate with nano-hole arrays and a periodicity of 100 nm. (b) Schematic illustration of the STM image

after the annealing. In this case, the nucleation probability of the QDs reached ~100% across ~400 sites and the average height of the QDs was ~7 nm. In the case of the sample with a nano-hole array with a 70 nm periodicity (corresponding to a higher density of 2×10^{10} cm^{-2}), the regular nucleation of the QDs was achieved by reducing the InGaAs deposition temperature from 400°C to ~ 300° C. Figure 13.10 (b) shows an STM image of the surface after annealing a 17-ML-thick InGaAs layer, which also exhibits ~100% nucleation probability across ~1,000 sites; in this case, the average height of the QDs is ~8 nm.

On the regular QD arrays with 100 nm periodicity, three other InGaAs QD layers with thicknesses of 10–12 ML were grown under conventional conditions of self-assembled QDs at 480° C in the SK mode. We measured the optical property of the stacked QD array. Figure 13.11 shows a PL spectrum of the fabricated site-controlled QD stacked arrays. The sample structure is shown in the inset of Fig. 13.11. A nano-hole region with an area of 50 μm × 50 μm was excited using a He-Ne laser ($\lambda = 633$ nm) with an optical microscope; here, the size of

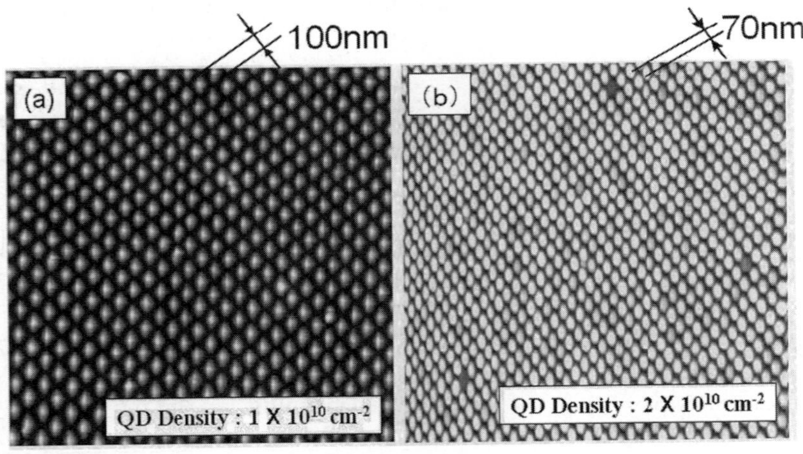

Fig. 13.10 STM images of regular arrays of InGaAs QDs with periodicities of (a) 100 nm and (b) 70 nm

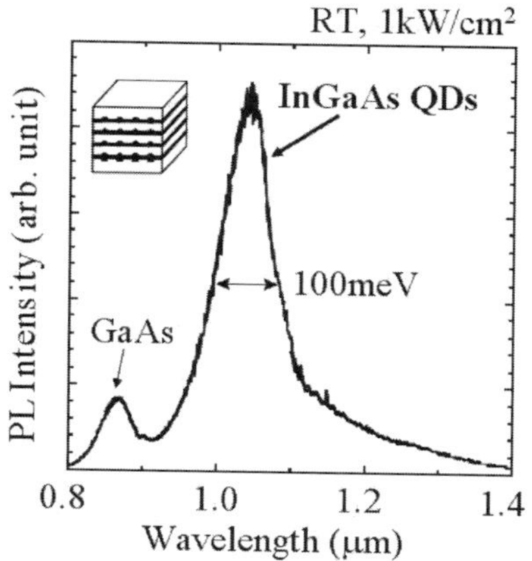

Fig. 13.11 PL spectrum of the stacked QD arrays. The inset shows the sample structure

the focal point on the sample was ~10 μm and the excitation power density was ~1 kWcm^{-2}. The PL results revealed a peak of the InGaAs QD array at 1.04 μm, and another peak at 0.87 μ m, corresponding to the GaAs layer, was also observed. The peak of the InGaAs QD array has a linewidth of ~100 meV. This broad PL linewidth can be explained by the inhomogeneity of the QD layers and/or the superposition of different PL peaks of the three QD layers.

13.3.2 Site Control of QDs Using the Nano-Jet Probe (NJP)

We have developed a new nano-probe-assisted technique which enables the formation of site-controlled InAs QDs. By using a specially designed AFM probe (NJP), we have successfully fabricated 2D arrays of ordered indium (In) nano-dots on a GaAs substrate. These In nano-dots can be directly converted to InAs QD arrays by subsequent irradiation of arsenic flux with annealing.

Figure 13.12 shows a schematic illustration of the NJP and nano-dot formation procedure developed in this study. The nano-dot formation was realized using the NJP, having a hollow pyramidal tip with a sub-micron size aperture on the apex and an In-reservoir tank within the stylus. This cantilever is used for nano-dot fabrication as well as for sensing the atomic force in AFM observations. By applying a voltage pulse between the pyramidal tip and the sample, In clusters were extracted from the reservoir tank within the stylus through the aperture, resulting in the In nano-dot formation.

The developed cantilever was embedded with a lead zirconate titanate (PZT) piezoelectric thin film on the beam, sensing the force during AFM observations as well as actuating. The cantilever for nano-dot formation was prepared as follows. First, we formed a micro aperture of about 500 nm in diameter on the apex of the stylus using focused ion beam etching technique. Next, In was charged to the inside of the hollow stylus by evaporation in a high vacuum from the opposite side. The amount of the charged In is sufficient to form at least one million In nano-dots.

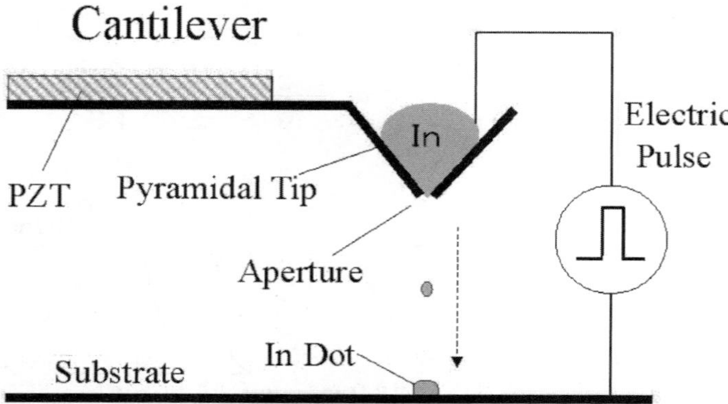

Fig. 13.12 Schematic illustration of the Nano-Jet Probe and the procedures involved in nano-dot formation

Figure 13.13 shows scanning electron microscope (SEM) images of the fabricated cantilever and a long distance optical microscope image of the NJP which is located above the sample surface. In the SEM image of the close-up view as shown in Fig. 13.13, a sub-micron size aperture can be observed at the apex of the stylus.

The experiments to study the In nano-dot formation were performed by using a conventional non-contact UHV-AFM operated at room temperature. Figure 13.14 shows an illustration of the experimental procedures involved. Figure 13.15 shows an AFM image of a 2D In nano-dot array with 200 nm periodicity deposited on a

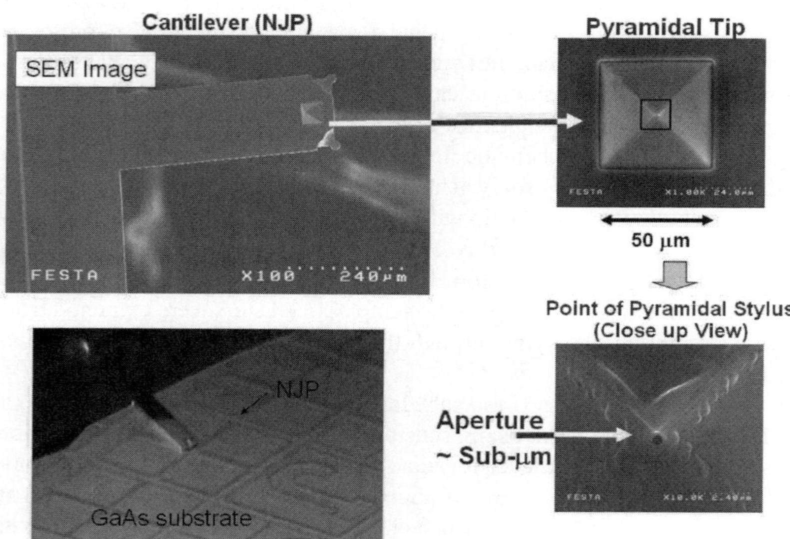

Fig. 13.13 SEM images and long-distance optical microscope image of the NJP

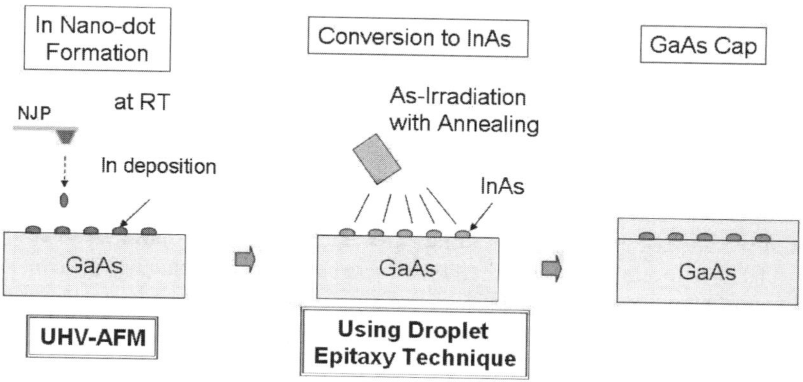

Fig. 13.14 Fabrication process of stacked InAs QDs using In(As) dot arrays formed by using the NJP method

GaAs substrate. Each nano-dot was formed on an MBE-grown GaAs surface by applying a single voltage pulse of 140 V for 10 ms. The value of the applied pulsed voltage for the nano-dot formation varied in the range of 100–140 V, although it was somewhat cantilever-dependent. During the deposition, the feedback loop was switched off.

The success of the In nano-dot formation depends on parameters such as the voltage, the width and shape of the applied electric pulse, and the distance between the tip and the sample during the deposition. Further, the size of the In nano-dots increased with the applied voltage pulse. The mechanism of the In nano-dot formation is thought to be similar to that of electric droplet ejection in inkjet patterning [13]. In other words, the charged In that was melted due the local Joule heating

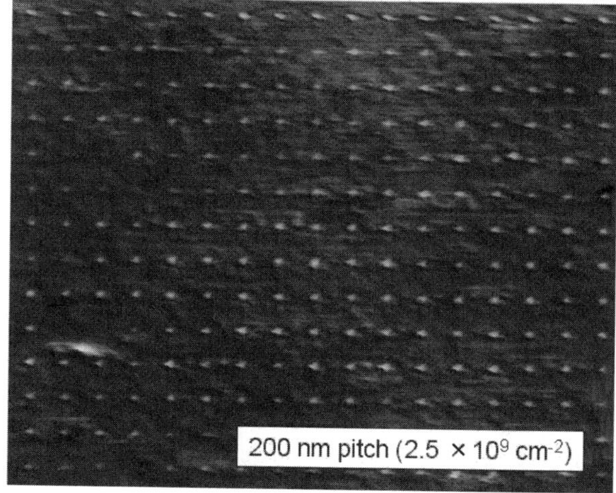

Fig. 13.15 AFM image of a fabricated In nano-dot array

induced by the electric pulse formed a so-called Taylor cone due to the equilibrium between the electrical and surface tension forces [9].

These ordered In nano-dots were directly converted into InAs QD arrays by subsequent irradiation with arsenic flux in the MBE chamber which is connected to the AFM chamber through a UHV tunnel. In other words, by using the droplet epitaxy technique [14], In nano-dots were crystallized into InAs dots. After the In nano-dots were deposited, the sample was transferred to the MBE chamber to crystallize them. The temperature of the sample was gradually raised to 420° C under an arsenic flux of 5×10^{-5} Torr and it was annealed for 40 min. Then it was transferred to the AFM chamber and its surface structure was observed by AFM. The AFM image revealed the formation of InAs nano-dot arrays at the same position where the In nano-dots had been deposited. Figure 13.16 shows the AFM images and schematic illustrations of the nano-dots before and after the conversion. Before the conversion, the In nano-dots appeared cone-shaped as shown in Fig. 13.16 (a). After the conversion, however, they assumed anisotropic shapes as shown in Fig. 13.16 (b). To be precise, the nano-dots were elongated along the [-110] direction after the conversion. Addi-

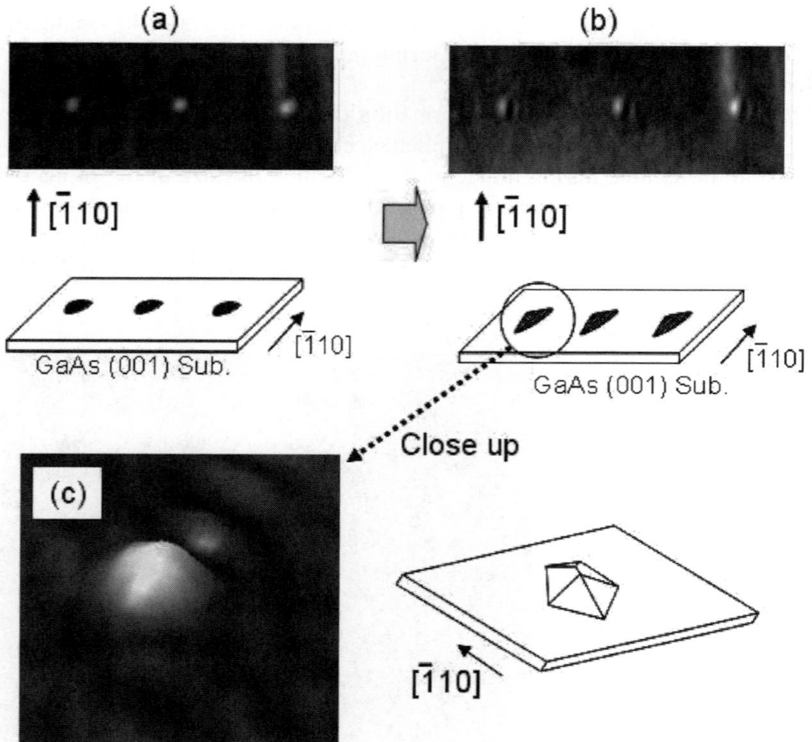

Fig. 13.16 AFM images of the In nano-dots (a) before and (b) after the crystallization process. (c) is a 3D close-up view of the converted InAs dot

tionally, high-index facets were formed on the side-walls of the converted nano-dots (Fig. 13.16 (c)). These features of the converted nano-dots correspond to those of InAs self-assembled QDs. Therefore, based on the observed change in the features of the nano-dots, we concluded that the In nano-dots had, in fact, been converted into InAs dots.

The proposed technology enables the formation of a required number of QDs in desired location with high uniformity and density. Furthermore, the volume of QDs required for optical switching devices can be achieved by first employing this SC-QD formation technique and subsequently stacking the spatially ordered InAs QD arrays thus obtained by using the MBE growth technique [11].

13.4 Summary

We have developed new and advanced techniques for selective-area growth and site-controlled of InAs-QDs with high density and high uniformity. The methods developed in this study provide greater latitude and design flexibility in fabricating optoelectronic and electronic devices with QD structures. Furthermore, they are promising technologies for future high-performance functional devices, such as regular arrays of quantum bits and single photon emitters for quantum computers and quantum communications.

References

1. Asakawa K, Sugimoto Y, Watanabe Y, Ozaki N, Mizutani A, Takata Y, Kitagawa Y, Ishikawa H, Ikeda N, Awazu K, Wang X, Watanabe A, Nakamura S, Ohkouchi S, Inoue K, Kristensen M, Sigmund O, Borel PI and Baets R (2006) Photonic crystal and quantum dot technologies for all-optical switch and logic device. New J. Phys. 8: 208
2. Tajima K (1993) All-optical switch with switch-off time unrestricted by carrier lifetime. Jpn. J. Appl. Phys. 32: L1746–1749
3. Nakamura H, Sugimoto Y, Kanamoto K, Ikeda N, Tanaka Y, Nakamura Y, Ohkouchi S, Watanabe Y, Inoue K, Ishikawa H and Asakawa K (2004) Ultra-fast photonic crystal/quantum dot all-optical switch for future photonic network. Opt. Express 12: 6606–6614
4. Ozaki N, Takata Y, Ohkouchi S, Sugimoto Y, Nakamura Y, Ikeda N and Asakawa K (2007) Selective area growth of InAs quantum dots with a metal mask towards optical integrated circuit devices. J. Cryst. Growth 301–302: 771–775
5. Yoshie T, Scherer A, Hendrickson J, Khitrova G, Gibbs HM, Rupper G, Ell C, Schchekin OB and Deppe DG (2004) Vacuum Rabi splitting with a single quantum dot in a photonic crystal nanocavity. Nature 432: 200–203
6. Waks E, Inoue K, Santori C, Fattal D, Vuckovic J, Solomon GS and Yamamoto Y (2002) Secure communication: Quantum cryptography with a photon turnstile. Nature 420: 762
7. Imamoglu A, Awschalom DD, Burkard G, DiVincenzo DP, Loss D, Sherwin M and Small A (1999) Quantum information processing using quantum dot spins and cavity QED. Phys. Rev. Lett. 83: 4204–4207
8. Ohkouchi S, Nakamura Y, Nakamura H and Asakawa K (2004) Indium nano-dot arrays formed by field-induced deposition with a Nano-Jet Probe for site-controlled InAs/GaAs quantum dots. Thin Solid Films 464–465: 233–236

9. Ohkouchi S, Nakamura Y, Nakamura H and Asakawa K (2005) InAs Nano-Dot array formation using Nano-Jet Probe for photonics applications. Jpn. J. Appl. Phys. 44: 5777–5780
10. Nakamura Y, Ikeda N, Ohkouchi S, Sugimoto Y, Nakamura H and Asakawa K (2004) Regular array of InGaAs quantum dots with 100 nm-periodicity formed on patterned GaAs substrates. Physica. E. 21: 551–554
11. Nakamura Y, Ikeda N, Ohkouchi S, Sugimoto Y, Nakamura H and Asakawa K (2004) Two-dimensional InGaAs Quantum-Dot arrays with periods of 70–100 nm on artificially prepared nanoholes. Jpn. J. Appl. Phys. 43: L362–L364
12. Nishi K, Saito H, Sugou S and Lee JS (1999) A narrow photoluminescence linewidth of 21 meV at 1.35um from strain-reduced InAs quantum dots covered by $In_{0.2}Ga_{0.8}As$ grown on GaAs substrates. Appl. Phys. Lett. 74: 1111–1113
13. Hartman RPA, Brunner DJ, Camelot DMA, Marijnissen JCM and Scarlett B (1999) Electrohydrodynamic atomization in the cone-jet mode physical modeling of the liquid cone and jet. J. Aerosol Sci. 30: 823–849
14. Chikyow T and Koguchi N (1990) MBE growth method for pyramid-shaped GaAs micro crystals on ZnSe(001) surface using Ga droplets. Jpn. J. Appl. Phys. 29: L2093–L2095

Chapter 14
Detailed Analysis of the Shape-dependent Deformation Field in 3D Ge Islands on Si(001)

G. Vastola, R. Gatti, A. Marzegalli, F. Montalenti, and Leo Miglio

14.1 Introduction

More than 15 years already passed since the formation of 3D nanometric-size coherent Ge islands spontaneously formed on Si(001) was first reported [1, 2]. However, a large community of researchers is still trying to shed more light on several aspects of island formation, stability, and evolution. The reason why the Si/Ge system is so popular resides on one side in the hope of building future-generation devices based on Ge quantum dots, and on the other in the fascinating physics emerging from years of experimental measurements and theoretical modeling. Despite being a simple heteroepitaxial system (e.g. with respect to III-V or II-VI compounds), indeed, the Stranski-Krastanow growth mode [3, 4] of Ge on Si(001) has revealed a sequence of well reproducible stages, where islands of qualitatively different shapes are formed on the substrate. Small islands [5], which start to form at early stages of growth after a thin (3–4ML) wetting layer (WL) is formed, are bounded by {105} facets [1] and appear as square-base pyramids [6] (or, elongated huts) of 0.1 height-to-base aspect ratio (AR). After reaching a critical volume, pyramids start to develop steps at their top [7], and finally reach the steeper (AR$\sim$0.2) dome shape [6, 8], characterized by more complex facets, such as {113} and {15 3 23}, and by a more rounded base. Coherent islands with aspect ratio larger than dome have also been observed, and carefully characterized [9, 10, 11] as barns bounded by {111}, {20 4 23}, and {23 4 20} facets, keeping a rather rounded base. A value of $\sim$0.32 was reported for their AR. The general reason why these islands form and tend to reach higher AR values is well understood and many details will be given in this work. The Ge lattice parameter is $\sim$4% larger than the Si one, so that the system naturally looks for elastic-energy relaxation channels. Pseudomorphic growth only allows for tetragonal strain relaxation. After the WL is completed, the formation of islands becomes favored by the balance between the superior volumetric strain relaxation E_V allowed by the formation of new free tilted facets and the extra-surface energy cost E_S [12]. Disregarding differences in surface energies (γ) between alternative facet orientation, it then comes as natural to predict shallow islands for small volumes, evolving towards steeper ones while increasing their dimensions. It is worth noticing that, indeed, refined ab initio calculations [13, 14] have recently pointed

Z. M. Wang, *Self-Assembled Quantum Dots.*

out that, due to the existence of suitable reconstructions, Ge(001), (105), (113), and (111) facets are characterized by extremely close γ values. Following the above line of thoughts, one could expect islands to evolve at late stages of growth into {111} pyramids, reaching ~0.7 AR. The situation, however, is more complex. After a further critical volume is reached, plastic strain relaxation becomes competitive, and islands are reported to loose their coherence due to the insertion of misfit dislocations [15, 16], deeply influencing further evolution [11]. While on flat Si(001) substrates the ultimate coherent shape seems to be the barn one [17], growth on pit-patterned Si(001) allows for a delay in dislocation injection, making it possible to observe {111} pyramids, although heavily truncated (0.37AR) [18].

The above presented explanation is clearly only qualitative and, in fact, several papers appeared in the literature, aimed at quantifying critical volumes, strain fields, and alternative elastic-energy relaxation channels, such as Ge/Si intermixing (of particular relevance at high temperatures, see [19, 20, 21, 22]), or trenches formation [23].

In this work we shall focus on volumetric strain relaxation, presenting a systematic analysis of the influence of the actual 3D island shape in inducing elastic-energy relieve, and in producing distortions in the Si substrate. To this goal, we shall mainly make use of finite-element methods (FEM) calculations, while atomistic simulations will be used to establish FEM predictive power.

14.2 Methodology: Atomistic Simulations vs Continuum Elasticity Theory

Several approaches can be applied to the computation of the elastic field in Ge/Si islands. Continuum models built in the framework of elasticity theory and yielding analytical solutions are surely a valid tool to obtain fast, semi-quantitative estimates. Starting from the Green function associated to a point-like inclusion of Ge on a semi-infinite Si substrate [24], the elastic field due to a macroscopic island is obtained integrating this Green function over the island volume. Based on this starting point, a method generally known as "flat-island approximation" was developed [3] [12] and shown to be valid for low-aspect ratio islands (i.e. {105} pyramids), while expected to generally fail in determining the elastic field in steeper islands. Recently, an improvement of this method has been obtained by taking into explicit account the base-to-top stress relaxation in islands. The improved method has shown to yield a good estimate of the elastic field also for pyramids with AR similar to domes [25].

However, the critical shortcoming is that the analytical solution can be computed only for simple island shapes, and/or in two dimensions. Ge/Si systems, on the other hand, are characterized by a rather broad spectrum of island shapes, some of them being very complex, like domes and barns.

Two approaches particularly suited for treating any geometry are classical molecular dynamics (MD) simulations and calculations based on continuum elasticity theory as solved by Finite Element Methods (FEM).

In MD, the system is built atom by atom. Possible many-body interatomic potentials have been proposed for the Si/Ge system. Here we recall the very popular ones introduced by Stillinger and Weber [26] and by Tersoff [27]. One of the advantages of atomistic simulations is that they can handle atomic-scale features like surface reconstructions, defects [28, 29], or non-homogenous distributions of atomic species [21]. The latter phenomenon, called intermixing, has been widely investigated experimentally [30, 31, 20], and shown to strongly influence the global relaxation inside the island [19].

MD simulations have been shown to yield results consistent with experiments in several investigations dealing with Ge islands on Si(001). For example, a total-energy calculation based on the Tersoff potential was used to demonstrate the critical role of facets reconstruction to determine the stability of {105} pyramids [32]. Atomistic calculations represent also a useful tool for the investigation of the elastic-field induced interaction between islands. When a sequence of Ge/Si layers are grown in a stacked array, islands show a general trend of enhancing the spatial order as the number of Ge/Si layers is increased. While a beautifully simple model based on dipolar fields was shown to capture the main physics of the ordering process [33], MD simulations allowed for a detailed characterization of the strain field generated in Si by buried Ge islands [34, 35]. Interestingly, an increase in lateral ordering was reported also for a single Ge-dot layer on Si(001), and shown to be triggered by partial capping with silicon [36]. Also in this scenario MD simulations proved to be useful, suggesting that the ordering driving force was the elastic interaction between adjacent islands, inducing an energetic unbalance among the facets of the dots, eventually determining an effective lateral motion aimed at increasing the island-island separation [36, 37]. Asymmetric alloying has also been shown to promote lateral motion [38, 39]. In the total-energy island budget, it has been shown that, at least for small islands, the contributions due to island edges are critical to predict the correct size of island nucleation [40]; classical MD simulations have been recently used to directly compute this term [41].

Despite the above reported examples of success in determining key phenomena associated with island-induced strain fields, MD simulations become computationally very demanding when handling systems with more than a few million atoms, so that FEM-based calculations are often applied to treat realistic-scale systems. Among several possible applications, here we would like to recall the application of FEM to the calculation of the elastic energy stored in quantum dots used, in turn, to predict the correct equilibrium island shape [42] and to establish the gain in relaxation characterizing the introduction of a defect in the island [43]. Despite being fast and allowing one to treat complex geometries, FEM calculations do not allow one to easily handle defects and non-homogenous intermixing.

While the strong limitations in the treatable size prevent the use of ab initio calculations in computing volumetric relaxation terms, so that no further reference to such approaches will be found in this work, it is worth emphasizing the key role they play when a quantitative estimate of the surface energies of the exposed surfaces is needed, e.g. for determining the stability of Ge heteroepitaxial islands [13, 14, 40].

14.3 Elastic Energy Relieve and Redistribution in Ge/Si Systems

As already mentioned in the Introduction, the formation of 3D stable island is possible because it allows for a partial relaxation of its lattice with respect to a flat film. Two main mechanisms are responsible of such relaxation as fully described in Ref. [25].

The in-plane compression can be reduced by exploiting all of the free surfaces characterized by an inclination with respect to the interface. In fact, all the forces acting along the free surfaces normal vanish, being the material free to change its effective lattice parameter in that direction. The force components parallel to the surfaces, on the other hand, can be reduced by involving a deformation of the substrate. It is important to notice that while the first mechanism allows for an effective reduction of the elastic energy, the second one acts as a redistribution of the energy between the island and the substrate, lowering the elastic energy in the former and enhancing it in the latter.

In order to give a quantitative estimation of the relative redistribution of the elastic energy due to the two mechanisms, we report in Fig. 14.1 the results of a FEM calculation performed on a {105} pyramid initially compressed as the underlying Ge wetting layer and computing the subsequent relaxation. Panels (a) and (b) show a gray-scale map of the elastic energy density in the island and in the substrate, respectively. The elastic energy turns out to be stored in both the island and the

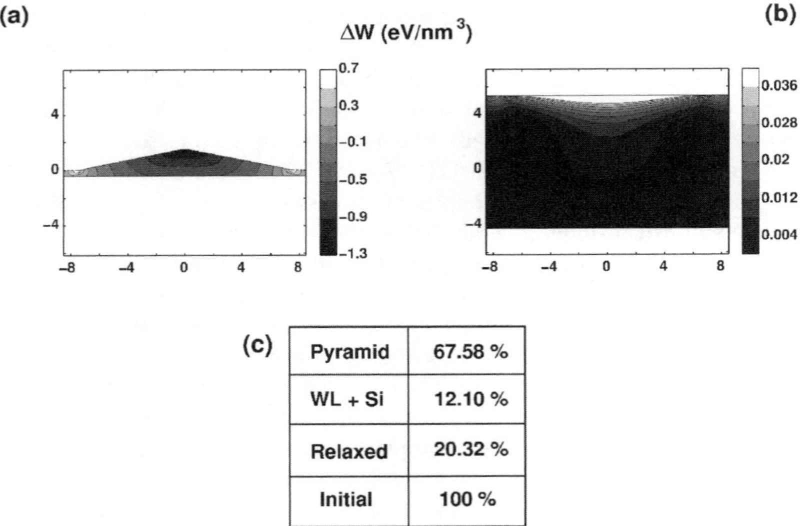

Pyramid	67.58 %
WL + Si	12.10 %
Relaxed	20.32 %
Initial	100 %

Fig. 14.1 Elastic-energy-density plot for a vertical scan crossing through the island apex, for a {105} pyramid. In (a) the elastic energy is referred to the wetting-layer one. Negative values correspond to higher relaxation; (b) elastic energy density stored in the Si substrate; (c) Table summarizing the elastic energy redistribution and relieve in a {105} island. The initial value was computed as the energy stored in a volume equivalent to that of the pyramid but tetragonally strained

Table 14.1 Aspect Ratio, Facet inclination and Elastic energy density for full {105},{113} and {111} Ge pyramids

Crystalline facet	Aspect ratio	Facet inclination (degrees)	W (meV/Ang^3)
{105}	0.1	11.3	0.9578
{113}	0.2357	25.2	0.6170
{111}	0.707	54.7	0.2407

substrate; in particular it can be noticed that the upper part of the island has lower energy density with respect to the initial condition (used as reference) via free surface relaxation and substrate loading, while the substrate has higher elastic energy with respect to the initial condition. In Fig. 14.1(c) it is quantified the correspondent estimation of the redistribution and relieve of the total initial elastic energy. Quantitative results always depend on the chosen geometry. As it is well known in the literature (see, e.g. [42]) and here quantified in Table 14.1, the steeper is the island, the more effective is the volumetric relaxation.

It is important to point out that, beyond surfaces, also the lateral edges joining adjacent island facets act as further centers of deformation relieve. Indeed, in this region each surface atom can exploit enhanced degrees of freedom to locally adjust distances.

Notice that, being the elastic field self-similar, it is the detailed island shape only which determines all of the main characteristics of the local elastic field inside nanometric islands. Several examples will be given in Section 14.4.

14.3.1 General Features of the Elastic Fields in Ge/Si Islands

The island relaxation guarantees a global reduction of the elastic energy, establishing a non uniform elastic field in the island and in the substrate. In the following we will describe such field in its general features and in the island-shape dependent characteristics.

An elastic field can be described by the strain tensor. In a (x,y,z) reference frame, denoting by u the displacement vector of each point in space with respect to its bulk position, the strain tensor ε is defined as

$$\varepsilon_{ij} = \frac{1}{2}\left(\frac{\partial u_i}{\partial x_j} + \frac{\partial u_j}{\partial x_i}\right), \qquad x_i = \text{x,y,z}$$

Such 3×3 tensor is symmetric. When diagonal strain tensor components are positive a local tensile deformation along the corresponding directions is present, while a negative sign means a local compressed lattice.

Figures 14.2, 14.3, 14.4, 14.5, 14.6, 14.7 depict the xx, zz, xz strain tensor components for a vertical cross section inside each island, both for the Ge and the Si substrate lattice obtained via FEM calculations. In addition, a horizontal cross

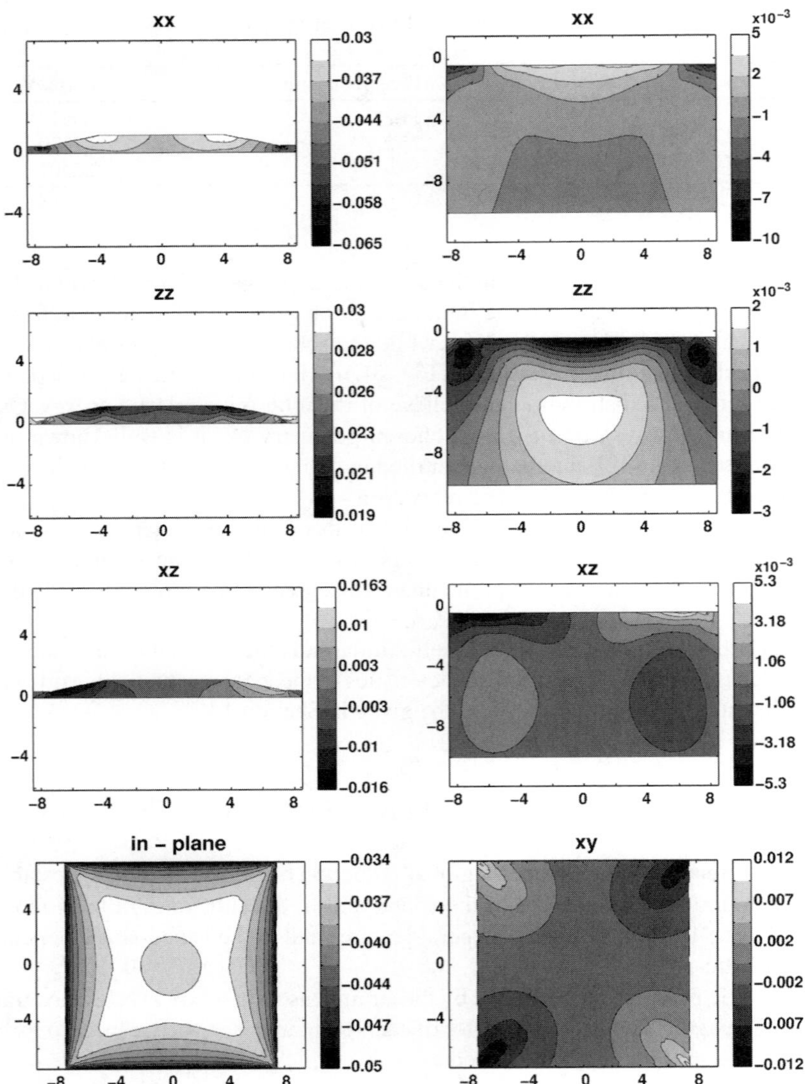

Fig. 14.2 Gray-scale strain tensor mapping of the elastic field of a {105} truncated Ge pyramid with aspect ratio 0.05, on a Si substrate. Maps on the left show the island, the ones on the right the substrate. Here and in the following maps, due to the self-similarity of the elastic field (neglecting the WL, see text), the lengths in x and z are expressed in arbitrary units. The in-plane and the xy map are computed for an horizontal plane passing through the island bottom

cut at the island base is exploited to map the xy and in-plane (i.e. $\frac{1}{2}(\varepsilon_{xx} + \varepsilon_{yy})$) strain components. Each figure is focused upon a shape: truncated {105} pyramid, full {105} pyramid, dome, barn, truncated {111} pyramid, full {111} pyramid, respectively, as summarized in Fig. 14.8.

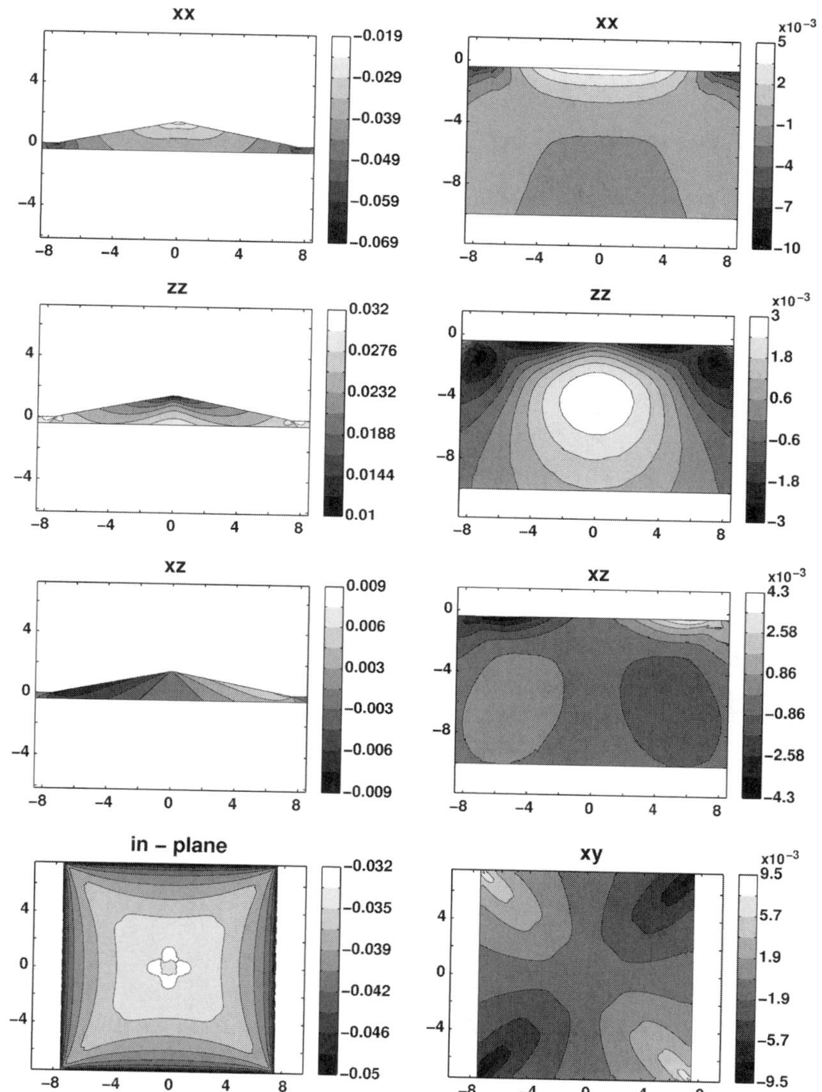

Fig. 14.3 Gray-scale strain tensor mapping of the elastic field for a full {105} pyramid. The aspect ratio is 0.1. Panels ordering is the same described in Fig. 14.2

Let us start to describe the common features of the xx component. Two regions are clearly visible, located at the island corners, where the lattice is highly compressed. This compression is higher than the misfit f, so that this effect cannot be only due to the pseudomorphic arrangement of the Ge lattice on the Si substrate. In fact, this compression is one of the fingerprints of the elastic energy

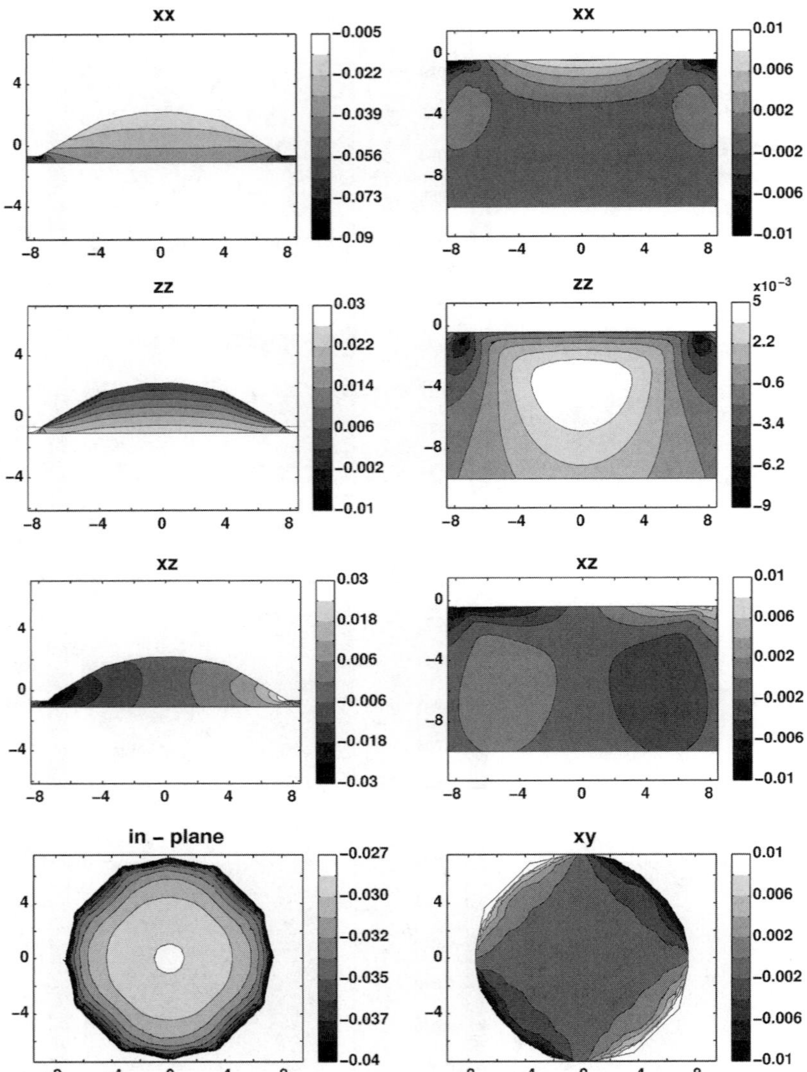

Fig. 14.4 Gray-scale strain tensor mapping of the elastic field for a dome island. The aspect ratio is 0.2. Panels ordering is the same described in Fig. 14.2

redistribution previously mentioned. The wetting layer shows a compression that can raise up to 8%. Underneath the island, on the other hand, it is visible a region where the Si lattice is expanded with respect to its native lattice parameter. This is another effect of the energy redistribution due to the presence of the island. Notice that this deformation is absent in the case of a flat film, so that the elastic energy

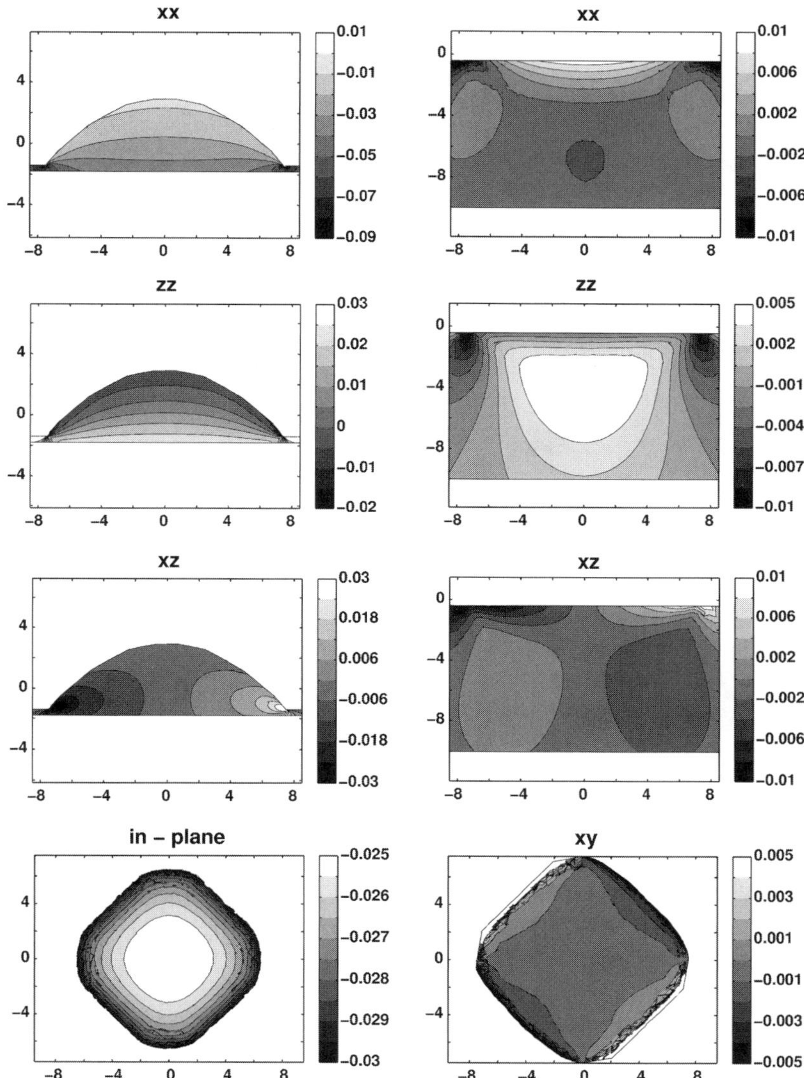

Fig. 14.5 Gray-scale strain tensor mapping of the elastic field for a barn island. The aspect ratio is 0.3. Panels ordering is the same described in Fig. 14.2

does not propagate inside the substrate and is instead confined exactly at the Ge/Si interface.

Inside the island, from bottom to top, the horizontal compression decreases. This is the effect of surface relaxation, which continues to relieve elastic energy as soon as the island top is reached.

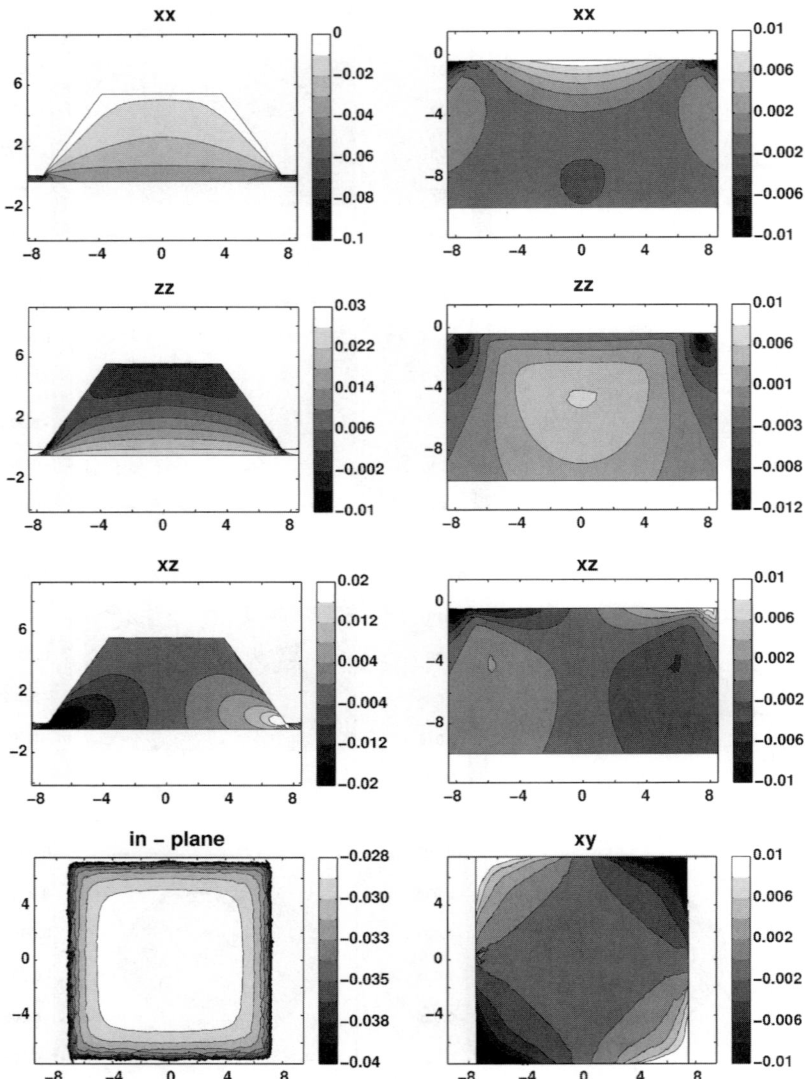

Fig. 14.6 Gray-scale strain tensor mapping of the elastic field for a truncated {111} pyramid. The aspect ratio is 0.37. Panels ordering is the same described in Fig. 14.2

The zz component is directly linked to the xx (or yy) component, through the analytical relation $\varepsilon_{zz} = -\frac{\nu}{1-\nu}\left(\varepsilon_{xx} + \varepsilon_{yy}\right)$ (ν is the Poisson ratio), that is exact for a flat film and approximated for an island. The sign of this formula points out that once the xx and yy components are negative, the zz is positive and vice versa. This means, as can be directly seen from the figures, that the centers of local horizontal compression correspond to a tensile deformation in the z direction. For the

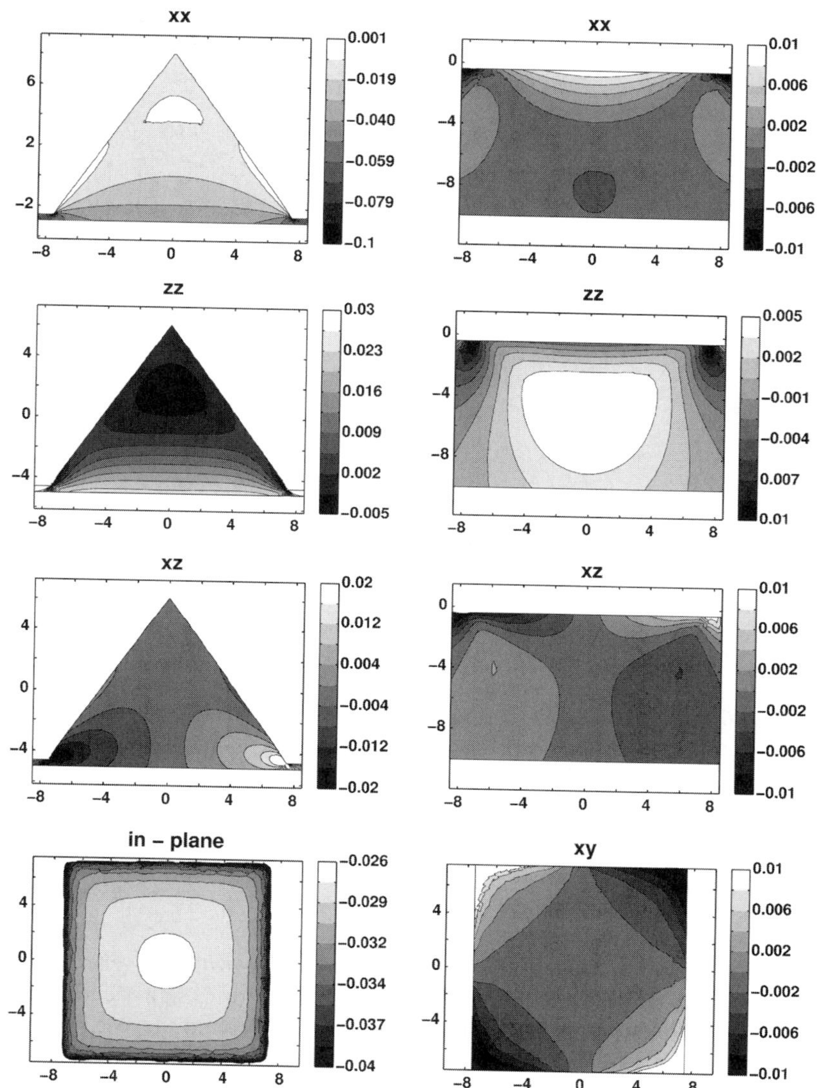

Fig. 14.7 Gray-scale strain tensor mapping of the elastic field for a full {111} pyramid. The aspect ratio is 0.7. Panels ordering is the same described in Fig. 14.2

same reason, the Si lattice, if expanded in the x,y directions, is compressed in the z direction.

The off-diagonal strain tensor components quantify the lattice bending. For instance, the xz component gives the local lattice bending in the (x,z) plane, as well as the xy component gives the torsion in the (x,y) horizontal plane. A careful analysis of these components reveals that the elastic field is symmetric in sign

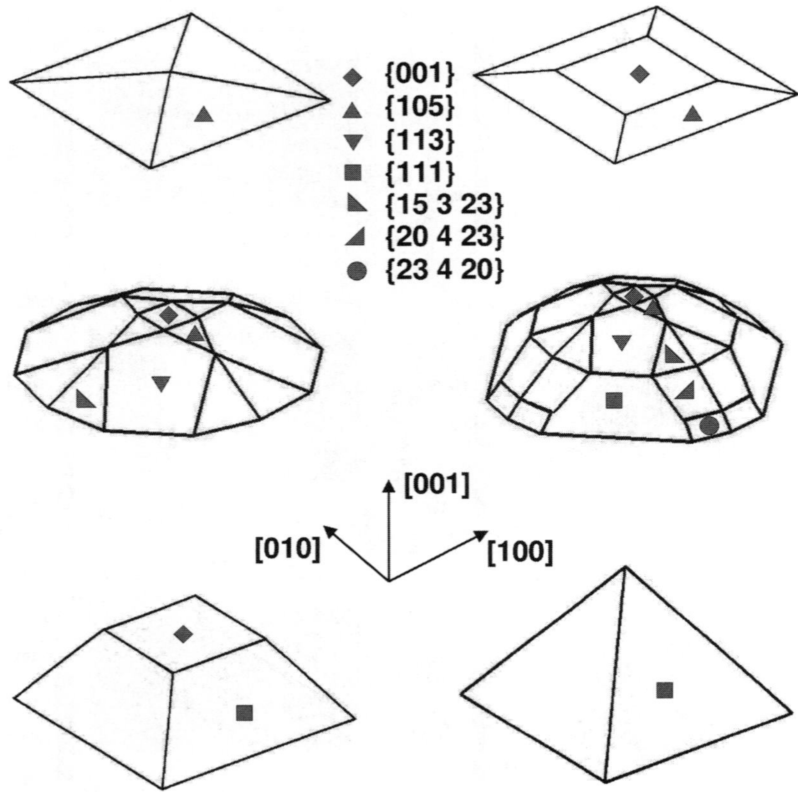

Fig. 14.8 Summary of the geometries used to calculate the elastic field reported in the previous figures. Crystallographic facets are specified for each structure. The two multifaceted structures correspond to a dome (left) and to a barn (right)

with respect to the island center: at every point lattice with a given strain value, there exists another point with the same value and opposite sign meaning that the curvature inside the island follows the symmetry of the island shape.

Finally, the in-plane strain component displays, by definition, symmetry with respect to the island center. It shows how the deformation propagates inside the inner island regions, far from surfaces. It is a common feature that the in-plane elastic field decreases moving from the island boundaries to its center; this is another fingerprint of edges as local sites of high deformation.

14.3.2 Assessing the Reliability of FEM Calculations

The elastic relaxation computed by FEM can be directly compared with atomistic simulations. We built an atomistic {105} pyramid over a 3-ML thick wetting layer covering the Si substrate. As already mentioned, due to the computational cost of the atomistic simulations we can not consider realistic-size island as done in FEM calculations. In this case the island base length was set to 220 Å. Periodic boundary conditions were applied in the horizontal directions. The Ge-atom position lattice was initially imposed to be determined by the Si lattice parameter. The Tersoff interatomic potential was chosen, and a quenching algorithm was applied with a timestep of 2fs in order to obtain the minimum-energy configuration. Convergence was declared when further iterations were observed to change the estimate by less than 10^{-8} eV. Then, the strain tensor was computed at each atomic site and compared with FEM results. To make the comparison closer, we inserted in the FEM code the Ge and Si elastic constants as computed via the Tersoff potential, instead of the experimental ones.

In Fig. 14.9, we provide a direct comparison of ε_{XX} along a line moving from the bottom to island top. The curves mainly overlap, with a slight divergence at the island top where we can see some fluctuations due to atomistic effects.

This accurate test allows us to conclude that FEM calculations are absolutely reliable when applied to Ge/Si nanometric systems. Moreover the FEM capability to perform calculations using the experimental elastic constants represent a further advantage when a quantitative comparison with experiments is necessary.

14.3.3 Wetting Layer and Trenches

In FEM calculations, modeling the wetting layer under the island can require a particularly high number of mesh elements, increasing the computational resources needed, because of its characteristic small thickness. It is thus relevant to analyze to what extent the wetting layer can be neglected, allowing for faster three-dimensional

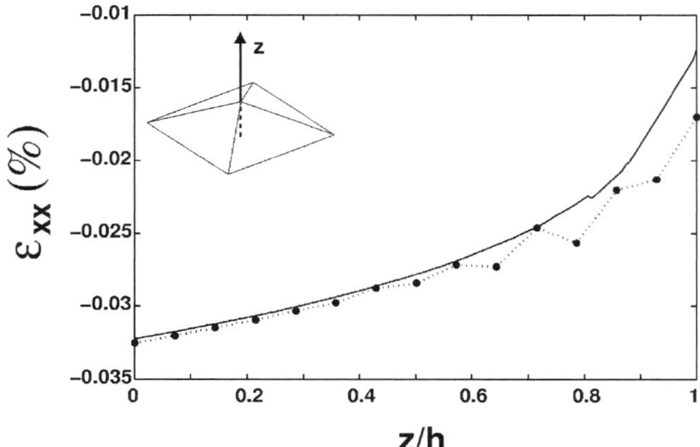

Fig. 14.9 Direct comparison between MD (dashed line) and FEM (solid line) calculations of the xx strain-tensor component along a line from bottom to island apex (see inset)

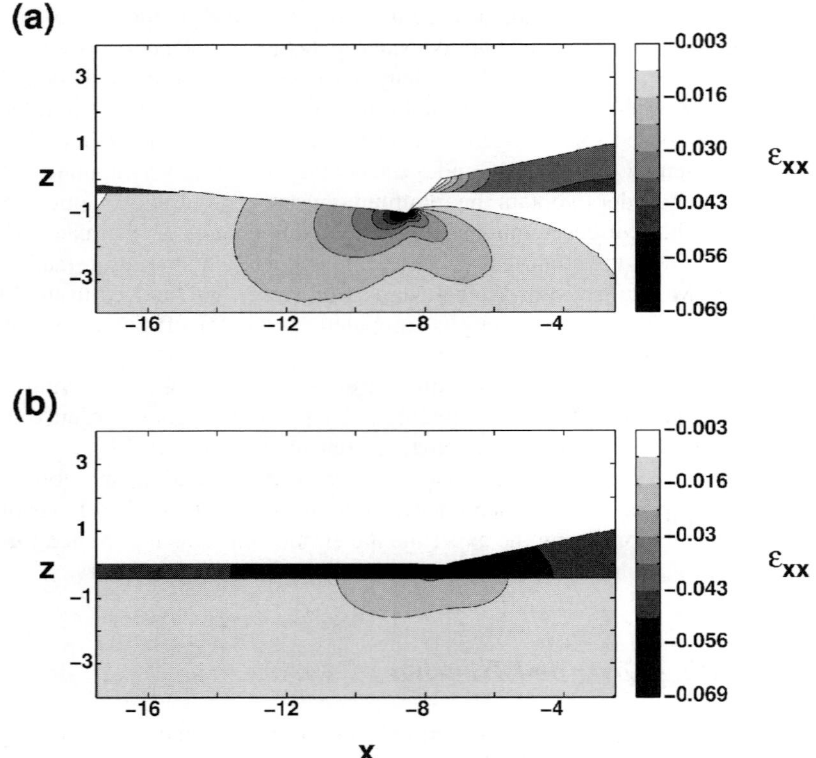

Fig. 14.10 (a) FEM-computed xx strain component of the elastic field surrounding a realistically-modeled trench. The map focuses on the trench and on the island edge which appears on the right. For comparison, the same strain-tensor component in the absence of the trench is reported

FEM simulations. When the wetting layer is absent the elastic loading underneath the island is directly exerted on Si, so that the deformation is expected to be slightly different with respect to the case when the same loading is exerted through a thin Ge film (i.e., the wetting layer). By direct comparison of the computed elastic energy density stored in the same {105} pyramid in the presence or absence of the wetting layer, the difference turns out to be around 2% when the island base is smaller than ~15nm, but it fast becomes negligible for larger dimensions (~0.8% for a base of 30nm; notice that the WL dimensions are fixed, so that calculations including the WL are not self-similar).

Another important effect of the strain field in real Ge/Si system is given by the presence of trenches around the islands (see, e.g. [23]). Trenches originate from removal of material around the island; this removal is energetically favorable because, as we have seen, the elastic field surrounding the island can be very strong, so that the best way to relieve deformation is to directly remove the highly-strained material around the island perimeter. Their depths are usually sufficient to completely dig the wetting layer, reaching the Si substrate [44]. A dedicated two dimensional FEM calculation has been performed and the results are shown in Fig. 14.10, in terms of ε_{XX}. We accurately reproduced trenches as reported in (experimental and theoretical) descriptions of Refs. [23, 44]. The typical compression at the island edge due to the island relaxation (see bottom panel, no trenches are present) is now focused in the Si substrate under the wetting layer (upper panel). The extension of the compressed region inside the island is clearly decreased, meaning a more effective island relaxation.

14.4 Quantitative Analysis of Elastic Fields in Ge/Si Islands

In this section we systematically compare the elastic field characterizing the different islands we have previously introduced. While in Sec. 14.3.1 we analyzed the features of the elastic field that are common among the different shapes, we will tackle here, for every tensor component, the effect of shape in modifying the elastic field. Let us still keep our attention on Figs. 14.2, 14.3, 14.4, 14.5, 14.6, 14.7, using Fig. 14.8 as reference for the sampled geometries.

14.4.1 Diagonal Components: ε_{xx}, ε_{zz}

The mapping of the ε_{xx} component shows that island edges are more compressed as the aspect ratio increases. The maps for the truncated pyramids show that the top edges (convex) are centers of relaxation, with respect to the high compression located at the bottom edges (concave). By increasing aspect ratio, the strain value reached at the island apex decreases, from –3% for a {105} truncated pyramid to almost zero for a truncated {111} pyramid. This trend represents the net effect of surface relaxation, which is more effective in high-aspect ratio islands since they have a surface-to-volume ratio larger than shallow, {105} pyramids. In this trend, the dome and barn shape display the most uniform strain map: we ascribe this finding to the presence of the multiple faceting characterizing these shapes.

The behavior of the ε_{zz} component resembles, as already mentioned, the one of ε_{xx} with opposite sign. It can be seen that the strain field inside {105} pyramids is quite non-uniform, while the rounded shape of domes and barn produces a more uniform strain field.

14.4.2 Off-diagonal Components: ε_{xz}

The trend of this component shows that low aspect ratio islands are bended as a whole. With increasing aspect ratio, two symmetrical lobes located at the island bottom edges become more and more defined, also increased in absolute value.

14.4.3 Horizontal Components: In-plane and ε_{xy}

The in-plane strain component shows that the deformation is always decreasing moving from the island boundary to the center. The truncated and full {105} pyramids display the interesting feature that the elastic field reaches its minimum value in a region surrounding the island center, while is slightly larger exactly in that position. This feature is found to disappear in higher aspect-ratio islands. The shape of the contour lines always resembles the shape of the corresponding boundary: squares in square-base pyramids and circular in domes and barns.

The ε_{xy} component shows a very non-uniform strain distribution, being always larger at the corners than at the island center. Four lobes are present: their shape and size evolves moving towards steeper islands. While they are small and very localized in truncated and full {105} pyramids, these lobes are more spread – and correspond to higher values – in steeper islands. This effect can be ascribed, as expected, to the continuous addition of stressing materials piling up at different heights, which inevitably follows the increase in aspect ratio.

14.4.4 Over-relaxation in {111} Pyramids

Finally, we would like to comment a striking feature of the ε_{xx} map for a full {111} pyramid: it displays (white region in xx component in Fig. 14.7) an over-relaxed region of the lattice close to the island top. This feature has been already theoretically proven in the approximation of a ball-shaped island [45], and experimentally observed in the case of Ge/Si(111) heteroepitaxy [46].

We analyzed quantitatively the behavior of the ε_{xx} strain component from the base of the pyramid to its apex along a vertical line exactly located at the island center, the result being reported in Fig. 14.11. The ε_{xx} increase stops slightly over the half pyramid height, becoming even positive. We here speculate that the presence of the above described over-relaxed region causes actual {111} islands to be

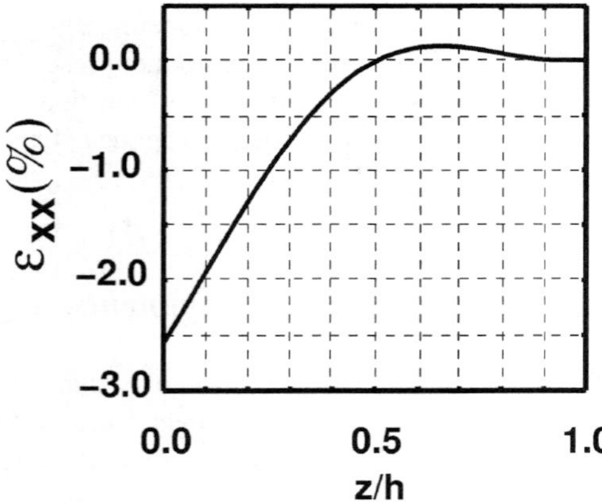

Fig. 14.11 Line plot of the xx strain tensor component for a full {111} pyramid, for a vertical path passing from the bottom to the island apex. The lattice over-expansion is clearly visible

preferentially truncated, as recently reported in Ref. [18]. The truncation allows the material to avoid the formation of the unfavorable top region, while keeping the advantages of the more pronounced effective relaxation typical of high aspect-ratio islands.

14.5 Conclusions

We have here reported a careful, systematic analysis of the elastic-energy relaxation dependence on the actual island morphology. A wide range of aspect ratios and of realistic shapes were examined. The reliability of our FEM-based elasticity-theory calculations in yielding volumetric relaxation was assessed by a close, quantitative comparison with more refined atomistic simulations. The experimentally observed sequence of different shapes vs. volume is fully consistent with our results, including the tendency towards top-flattening in high aspect-ratio islands bounded by {111} facets.

We acknowledge financial support under the European d-DOT-FET STREP project (Contract number: 012150).

References

1. Y.W. Mo, D.E. Savage, B.S. Swartzentruber, Phys. Rev. Lett. **65**, 1020 (1990)
2. D.J. Eaglesham, M. Cerullo, Phys. Rev. Lett. **64**, 1943 (1990)
3. V.A. Shchukin, D. Bimberg, Rev. Mod. Phys. **71**, 1125 (1999)
4. J. Stangl, V. Holy, G. Bauer, Rev. Mod. Phys. **76**, 725 (2004)
5. J. Tersoff *et al.*, Phys. Rev. Lett. **89**, 196104 (2002)
6. G. Medeiros-Ribeiro *et al.*, Science **279**, 353 (1998)
7. F. Montalenti, P. Raiteri, D.B. Migas *et al.*, Phys. Rev. Lett. **93**, 216102 (2004)
8. F.M. Ross *et al.*, Science **286**, 1931 (1999)
9. M. Stoffel *et al.*, Phys. Rev. B **74**, 155326 (2006)
10. P. Sutter, P. Zahl, E. Sutter, Appl. Phys. Lett. **82**, 3454 (2003)
11. G. Capellini, M. De Seta, F. Evangelisti, J. Appl. Phys. **93**, 291 (2003)
12. J. Tersoff, F.K. Le Goues, Phys. Rev. Lett. **72**, 3570 (1994)
13. D.B. Migas, S. Cereda, F. Montalenti *et al.*, Surf. Sci. **556**, 121 (2004)
14. A.A. Stekolnikov, J. Furtmüller, F. Bechstedt, Phys. Rev. B **65**, 115318 (2002)
15. F.K. Le Goues, M.C. Reuter, J. Tersoff, Phys. Rev. Lett. **73**, 300 (1994)
16. T. Merdzhanova, S. Kiravittaya, A. Rastelli *et al.*, Phys. Rev. Lett. **96**, 226103 (2006)
17. A. Marzegalli, V.A. Zinovyev, F. Montalenti, A. Rastelli, M. Stoffel, T. Merdzhanova, O.G. Schmidt, L. Miglio, Submitted
18. Z. Zhong, W. Schwinger, F. Schäffler, G. Bauer, G. Vastola, F. Montalenti, L. Miglio, Phys. Rev. Lett. **98**, 176102 (2007).
19. Y. Tu, J. Tersoff, Phys. Rev. Lett. **98**, 096103 (2007)
20. G. Medeiros-Ribeiro, R. Stanley Williams, Nano. Lett. **72**, 223 (2007)
21. P. Sonnet, P.C. Kelires, Phys. Rev. B **66**, 205307 (2002)
22. M. De Seta, G. Capellini, F. Evangelisti *et al.*, J. Appl. Phys. **92**, 614 (2000)
23. S.A. Chaparro, Y. Zhong, J. Drucker, Appl. Phys. Lett. **76**, 3534 (2000)
24. L.D. Landau, Theory of Elasticity, Pergamon Press (1959)
25. V.A. Zinovyev *et al.*, Surf. Sci. **600**, 4777 (2006)

26. F. Stillinger, T.A. Weber, Phys. Rev. B **31**, 5262 (1985)
27. J. Tersoff, Phys. Rev. B **39**, 5566 (1989)
28. M.G. Fyta, I.N. Ramediakis, P.C. Kelires *et al.*, Phys. Rev. Lett. **96**, 185503 (2006)
29. L. Martinelli, A. Marzegalli, P. Raiteri *et al.*, Appl. Phys. Lett. **84**, 2895 (2004)
30. R. Magalhães-Paniago, G. Medeiros-Ribeiro, A. Malachia *et al.*, Phys. Rev. B **66**, 245312 (2002)
31. G. Katsaros, G. Costantini, M. Stoffel *et al.*, Phys. Rev. B **72**, 195320 (2005)
32. P. Raiteri, D.B. Migas, L. Miglio *et al.*, Phys. Rev. Lett. **88**, 256103 (2002)
33. J. Tersoff, C. Teichert, M.G. Lagally, Phys. Rev. Lett. **76**, 1675 (1996)
34. M.A. Makeev, A. Madhucar, Phys. Rev. Lett. **86**, 5542 (2001)
35. R. Marchetti, F. Montalenti, L. Miglio *et al.*, Appl. Phys. Lett. **87**, 261919 (2005)
36. G. Capellini, M. De Seta, F. Evangelisti, V. Zinovyev, G. Vastola, F. Montalenti, L. Miglio, Phys. Rev. Lett. **96**, 106102 (2006)
37. F. Montalenti, A. Marzegalli, G. Capellini, M. De Seta, Leo Miglio J. Phys. Condens. Matter **19**, 225001 (2007)
38. U. Denker, A. Rastelli, M. Stoffel *et al.*, Phys. Rev. Lett. **94**, 216103 (2005)
39. M. Stoffel, A. Rastelli, S. Kiravittaya *et al.*, Phys. Rev. B **72**, 205411 (2005)
40. O.E. Shklyaev M.J. Beck, M. Asta *et al.*, Phys. Rev. Lett. **94**, 176102 (2005)
41. C.M. Retford, M. Asta, M.J. Miksis *et al.*, Phys. Rev. B **75**, 075311 (2007)
42. H.T. Johnson, L.B. Freund, J. Appl. Phys. **81**, 6081 (1997)
43. K. Tillman, A. Forster, Thin Solid Films **368**, 93 (2000)
44. D.T. Tambe, V.B. Shenoy, Appl. Phys. Lett. **85**, 1586 (2004)
45. B.J. Spencer, J. Tersoff, Phys. Rev. Lett. **79**, 4858 (1997)
46. S.K. Theiss, D.M. Chen, J.A. Golovchenko, Appl. Phys. Lett. **66**, 448 (1994)

Chapter 15
Growth and Characterization of III-Nitride Quantum Dots and their Application to Emitters

Tao Xu and Theodore D. Moustakas

15.1 Introduction

GaN and its alloys with InN and AlN are III-V semiconductor compounds having the wurtzite crystal structure and a direct energy bandgap which is suitable for optoelectronic applications, such as light emitting diodes and laser diodes. The relationship between the direct energy gap and the lattice constant (a) of wurtzite (Al,Ga,In)N is shown in Fig. 15.1. The recent discovery that InN has a bandgap of $\sim$0.7 eV (Lu et al. 2001, Walukiewicz et al. 2004, Monemar et al. 2005) indicates that the Al-Ga-In-N system encompasses a broad spectral region ranging from the deep UV to infra red (200 nm – 1770 nm). Another advantage of the nitride semiconductors over other wide bandgap semiconductor is the strong chemical bond which makes the material very stable and resistant to degradation under conditions of high electric current injection and intense light illumination.

Quantum dots (QDs) can be used as the active region of opto-electronic devices, such as LEDs, lasers, modulators, and detectors. The fabrication methods of QDs can be classified into two major categories: the post-growth method and the self-assemble method. The post-growth lateral patterning of 2D quantum wells, as an indirect method, suffers from insufficient lateral resolution and interface damage caused by the patterning process. Self-organization of QDs on crystal surface is a more promising method to fabricate quantum dots. These techniques require large lattice mismatch between the growing layer and the substrate to achieve elastic relaxation. Specifically, elastic strain relaxation on facet edges and island interaction via the strained substrate are the driving forces for the self-organization of ordered arrays of uniform, strained islands on crystal surfaces. Stranski-Krastanov (SK) growth (with wetting layer, WL) and Volmer-Weber (VW) growth (without WL) in highly lattice-mismatched material systems like InAs-(Al)GaAs, GaSb-GaAs or InAs-Si are used to fabricate self-assembled 3D quantum dots. Both molecular beam epitaxy (MBE) and metal organic vapor-phase epitaxy (MOVPE) are currently used to grow QDs using the above mechanisms.

GaN QDs grown on an AlN buffer follow the classical pattern of the Stranski-Krastanov mode (the lattice mismatch between GaN and AlN is $\sim$2.5%), where the formation of a wetting layer is followed by the formation of GaN QDs (Xu

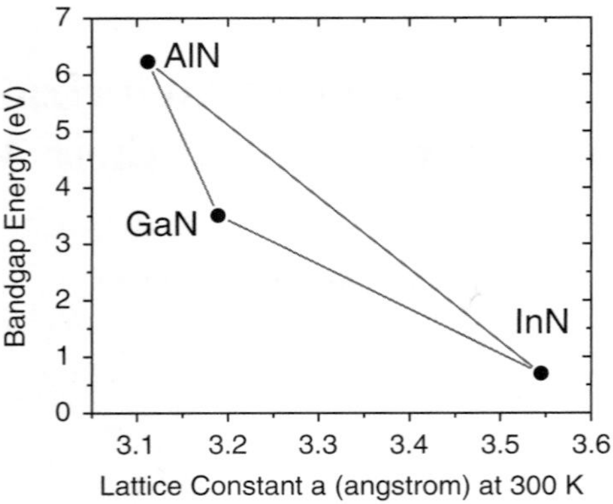

Fig. 15.1 The relationship between bandgap energy and lattice constant (a) for AlN, GaN and InN (Vurgaftman et al. 2001, Lu et al. 2001)

et al. 2005). The growth and characterization of GaN QDs are discussed in Section 15.2. The growth of InN QDs on a GaN buffer follows the pattern of the Volmer-Weber mode, where 3D InN QDs are formed directly on GaN without a wetting layer, a result attributed to the larger lattice mismatch between GaN and InN ($\sim$11%). The growth and characterization of InN dots are reported in Section 15.3 (Zhou et al. 2006). The growth mode of InGaN QDs on GaN is more complicated depending on the InGaN composition. The InGaN QDs grown on GaN buffer and their properties are reported in Section 15.4. The application of III-nitride QDs to LED structures is discussed in detail in Section 15.5.

15.2 GaN QDs

The challenges in the growth of GaN QDs lie in the development of methods to obtain dots with good size uniformity and high spatial density. Recently, there has been significant progress in control of GaN QD growth using Stranski-Krastanov mode by MBE (Adelmann et al. 2004, Gogneau et al. 2003, Brown et al. 2004, Brault et al. 2003). Adelmann et al. (Adelmann et al. 2004) studied GaN QD growth under N-rich conditions at 730 °C. Under these conditions, QDs are formed in the presence of N_2-plasma. The GaN QD growth under Ga-rich conditions was also studied, and was referred to as the "modified SK mode" (Gogneau et al. 2003, Brown et al. 2004). The 2-dimensional (2D) GaN film grown on AlN under Ga-rich conditions is spontaneously rearranged into 3-dimensional (3D) islands when allowing the excess Ga to desorb under vacuum. Furthermore, it has been reported (Adelmann et al. 2004) that the GaN QD density often saturates at about 10^{11} dots/cm^2. Multiple stacks of GaN QDs can be used to increase the spatial density in the vertical direction. Electron microscopy observation showed that wurtzite GaN QDs grown on AlN may be coherent and dislocation-free (Arley et al. 1999, Sarigiannidou et al. 2005).

GaN optoelectronic devices are usually grown by MBE at 750–800 °C. However, for most of the results reported in the literature, the growth temperature for GaN

QDs is 650–750 °C. In order to apply GaN QDs in optoelectronic devices, it is necessary to get a better understanding of the GaN QD growth at higher temperatures. In this section, we discuss the MBE growth of GaN QDs and multiple-stack-QDs (MQDs) using AlN barriers at 770 ° C under slightly Ga-rich conditions using the modified Stranski-Krastanov mode. The morphology and structural properties of the GaN QD and MQD samples were characterized by atomic force microscopy (AFM) and transmission electron microscopy (TEM) methods. Factors that influence the dot density and distribution, such as GaN coverage, number of GaN QD stacks in the superlattice and lateral ordering of the substrates, were investigated.

15.2.1 MBE Growth of GaN QDs

The GaN quantum dot samples were grown in a Varian Gen II MBE system. An EPI RF plasma source was used to activate the molecular nitrogen and sapphire wafers with *c*-plane orientation were used as substrates. Prior to the MBE growth, some of the sapphire substrates were annealed at 1400 °C for 1 h, to reveal the step structure associated with the substrate miscut, and characterized carefully by AFM. The sapphire substrates were nitridated at 870 °C for 1 h, followed by a 300 nm thick AlN film grown under Al-rich conditions at the same temperature. The substrate temperature was then reduced to 770 °C for the GaN MQD growth. The GaN MQD structure consists of GaN QDs obtained from depositing several monolayers (MLs) of GaN, separated with 10–20 nm of AlN barriers. The GaN QDs were grown under Ga-rich conditions using the modified Stranski-Krastanov growth mode. In this mode, several MLs of GaN were deposited under Ga-rich conditions, followed by evaporation of excess Ga in vacuum and subsequent rearrangement of 2D GaN islands into 3D GaN dots. The 2D to 3D transition was confirmed by a transition of the RHEED pattern from streaky to spotty during growth interruption (see Fig. 15.2). The last

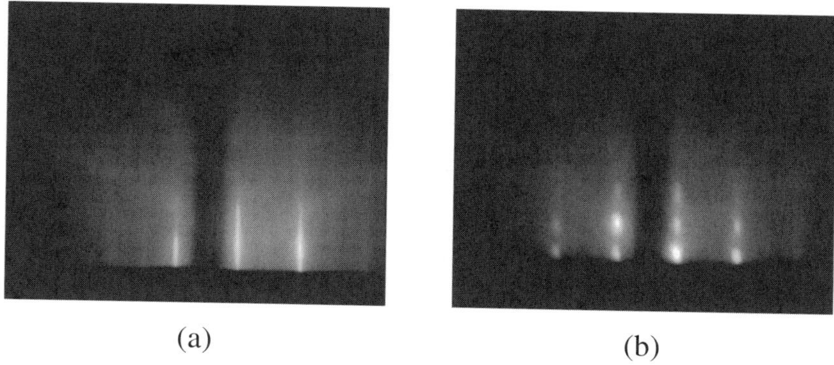

Fig. 15.2 RHEED patterns of (a) AlN template before GaN QD deposition; (b) GaN QDs ripen in vacuum

GaN QD layer was left in vacuum for 1 min before the substrate temperature was ramped down to 200 °C at 20 ° C/min.

15.2.2 Characterization of GaN QDs and MQDs

15.2.2.1 Structure of GaN QDs and MQDs

A GaN MQD structure is shown in Fig. 15.3(a) (Xu et al. 2005). It contains 4 stacks of GaN QDs obtained from deposition of 5 MLs of GaN, separated with 18-nm-AlN barriers. The dot height is around 4 nm and the dot diameter is around 15 nm as shown in the high resolution TEM (HRTEM) picture of Fig. 15.3(b). The QDs in the free surface have a truncated pyramidal shape faceted along {1–103} planes. The surface morphology of this GaN QD SL sample is shown in the AFM height image of Fig. 15.3(c). The dot density on the top surface is 9×10^{10} cm^{-2}. The height of the dots has a Gaussian distribution as described in Fig. 15.3(d). The mean height for the dots is 3.2 nm with FWHM of 1.9 nm.

15.2.2.2 Bimodal Height Distribution of GaN QDs

The GaN coverage (Θ), stated in monolayer units (where 1 ML $= c/2 = 0.259$ nm), is an important parameter directly affecting the evolution of GaN QDs during growth (Adelmann et al. 2004). For a GaN MQD sample of four stacks and GaN coverage of 3 MLs or 4 MLs, the dot height distribution obtained from AFM analysis shows a clear bimodal distribution: Mode 1 has a narrow Gaussian distribution with average height about 1.5–2 nm and FWHM of 0.8 nm. Mode 2 has a broader distribution with average height around 3.5 nm and FWHM of 2 nm. As the GaN coverage increases, Mode 2 becomes the dominant one. When the coverage is higher than 5MLs, the height distribution becomes practically Gaussian, with the average height between 3.1–3.4 nm.

From the plan-view and cross-sectional TEM images of a GaN MQD sample with 30 stacks and GaN coverage of 4 MLs, it was observed that the bigger GaN QDs are located adjacent to the edge-type threading dislocations (TDs) while some smaller dots are located in dislocation-free regions. Thus, the bimodal distribution can be attributed to the interaction of dislocation with the GaN QD growth.

15.2.2.3 Lateral Ordering of GaN QDs

The QD ordering was studied by growing one stack of QDs on the vicinal surfaces of miscut sapphire substrates. Figure 15.4(a) shows a sapphire substrate as acquired from the manufacturer prior to thermal annealing. As seen in this figure the substrate is decorated by a network of scratches due to mechanical polishing. The surface roughness is 0.25 nm. The Fast Fourier Transform (FFT) of the surface does not show any preferred orientation as shown in the inset of Fig. 15.4(a). After the substrate was annealed at 1400 °C for 1 h, the vicinal steps, due to miscut of the

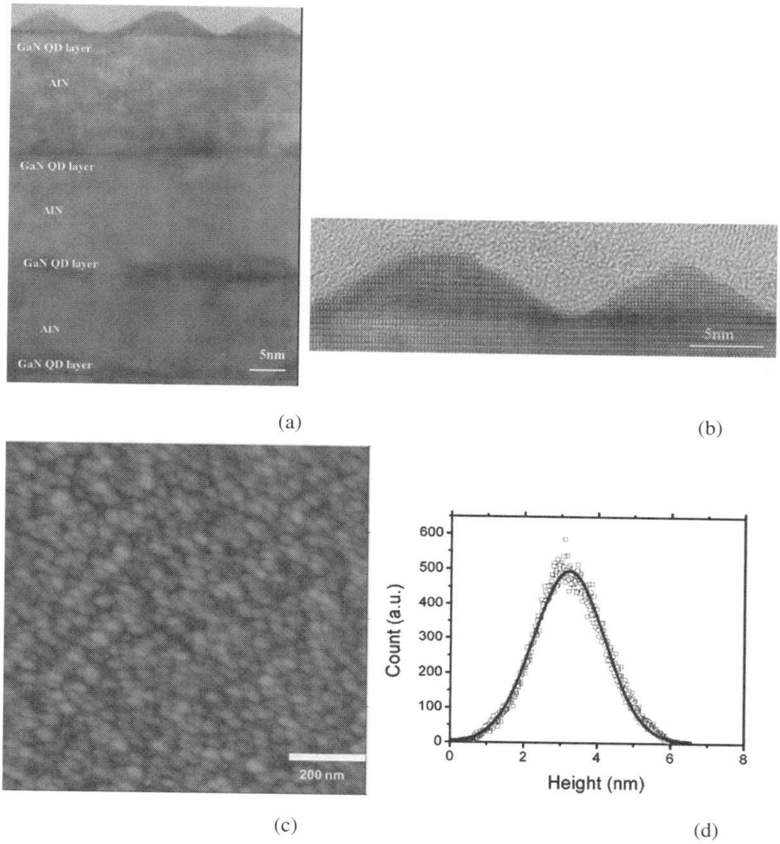

Fig. 15.3 (a) Cross-sectional TEM for a GaN MQD sample of 4 stacks of QDs, grown with GaN coverage of $\Theta = 5$ MLs; (b) HRTEM for the top layer of dots for the same sample; (c) $1\,\mu\text{m} \times 1\,\mu\text{m}$ AFM height image of the same sample (the z scale is 10 nm); (d) height distribution of the dots (Xu et al. 2005)

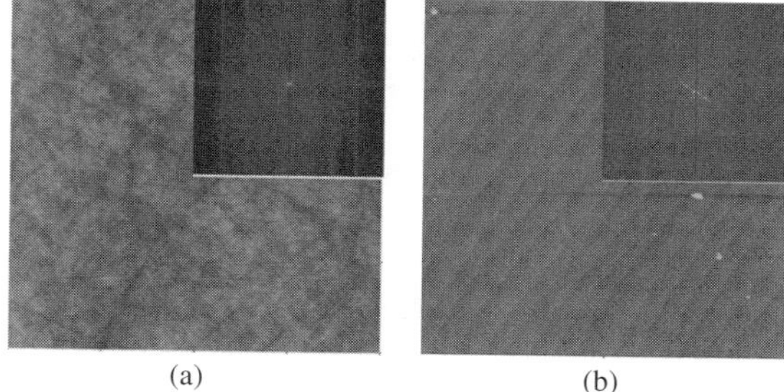

Fig. 15.4 (a) $10\,\mu\text{m} \times 10\,\mu\text{m}$ AFM image of an unannealed sapphire substrate; (b) $10\,\mu\text{m} \times 10\,\mu\text{m}$ AFM image of an annealed sapphire substrate. The insets are their FFT spectra

substrate, are revealed as shown in Fig. 15.4(b). This surface rearrangement resulted in the removal of the scratches and the rms roughness is reduced to 0.083 nm. The FFT of the surface shows the preferential arrangement of steps in the reciprocal space.

One stack of GaN QDs was grown on these vicinal c-plane sapphire substrates after growing 125 nm AlN. The surface morphology and its FFT of 1 stack of GaN QDs grown on unannealed and annealed sapphire substrates are compared in Fig. 15.5. GaN QDs were observed to align along the steps as shown in Fig. 15.5(b). This alignment can be better seen in FFT spectra as shown in the inset of Fig. 15.5(b). The isotropic ring in the inset of Fig. 15.5(a) indicates the random distribution of QDs grown on unannealed sapphire substrates while the tilted ellipse shown in the inset of Fig. 15.5(b) indicates the preferential alignment of QDs along the surface steps grown on annealed sapphire substrates.

15.2.2.4 Dislocation Filtering Effect of GaN QDs

From the cross-sectional TEM image of a GaN MQD sample with 30 stacks and 4 ML GaN coverage, it was observed that the TD lines in the sample are deflected and annihilated as more GaN QD stacks were grown. Based on this observation, GaN QDs were applied as a dislocation filtering layer in the early stage of the GaN cladding layer growth. Two stacks of GaN QDs with GaN coverage of 5 MLs and AlN spacer of 8 nm were grown directly on the AlN buffer layer. Then GaN bulk films were grown on top of the GaN MQDs. At the interface of GaN MQDs and GaN bulk film, the TDs in the GaN film are deviating, probably by the stress field introduced by the GaN QDs. The dislocation lines are forming loops, which leads to reduction of TDs as the thickness of the GaN film increases. The structure of these GaN films was determined by studying the on-axis (0002) and off-axis $(10\bar{1}2)$ x-ray diffraction rocking curves. The FWHMs of $(10\bar{1}2)$ XRD rocking curves of the GaN films with GaN QDs in the buffer are smaller than those without GaN QDs in the buffer, which indicates that GaN QDs in the nucleation layer help reduce the TD density in GaN films.

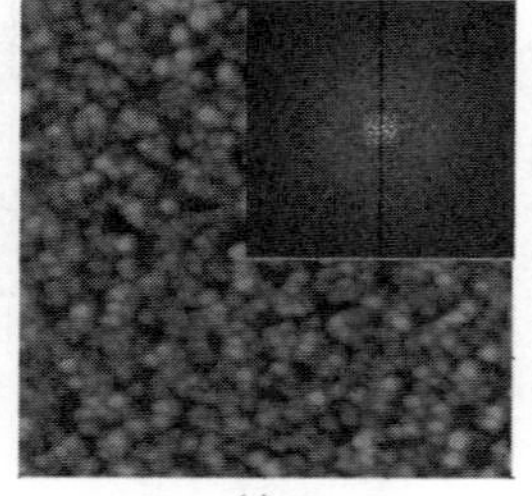

(a)

(b)

Fig. 15.5 (a) AFM image of 1 stack of GaN QDs grown on unannealed sapphire substrate; (b) AFM image of 1 stack of GaN QDs grown on annealed sapphire substrate. The insets are the FFT spectra of the AFM images

15.3 InN QDs

Indium nitride (InN) is the end-member of the isomorphous (In, Ga, Al)N series of wurtzite semiconductors which are of much scientific and technological importance for solid-state lighting and device applications because of the possibility of spanning a wide and continuous range of wavelengths from ultraviolet to infrared (Wu et al. 2002, Wang and Yoshikawa 2004). Recent research has established that InN has a bandgap of $\sim$0.7 eV (Walukiewicz et al. 2004, Monemar et al. 2005), meaning that InN and In_xGa_{1-x} N alloys would be potentially strong candidates for future devices such as temperature-insensitive high-efficiency infrared laser diodes for optical communications and high-efficiency tandem solar-cells. Much attention has been given to the growth of InN thin films using the techniques of MOVPE (Yamamoto et al. 2002, Maleyre et al. 2004), as well as MBE (Nanishi et al. 2003, Dimakis et al. 2005). High substrate temperature ($\sim$550–600°C) is used for the growth of N-polar InN by MBE, whereas lower temperature (<500°C) is used for In-polar growth to avoid InN decomposition (Xu and Yoshikawa 2003).

The initiation of InN growth on intermediary GaN buffer layers is an important and interesting topic. InN has a lattice mismatch of $\sim$11% with GaN so that a layer-to-island transition (Stranski-Krastanov growth mode) is to be anticipated after formation of a very thin wetting layer (<2 InN monolayers). The effects of growth temperature and III/V molar ratio on the size, density and morphology of InN quantum dots (QDs) grown by MOVPE (Briot et al. 2003, Maleyre et al. 2004, Ruffenach et al. 2005, Lozano et al. 2005), and by MBE (Monemar et al. 2005, Cao et al. 2003, Yoshikawa et al. 2005, Kehajias et al. 2005), have recently been investigated. In this section, the growth and characterization of InN QDs on Ga-polar GaN buffer layers by plasma-assisted MBE (PAMBE) is reported. The island/buffer misfit is found to be accommodated primarily by periodic arrays of misfit dislocations, with separations of $\sim$2.8 nm, which is indicative of highly relaxed islands.

15.3.1 PAMBE Growth of InN QDs

The samples were grown in a Varian Gen II MBE system with an EPI RF plasma source to activate molecular nitrogen and standard effusion cells to provide In, Ga and Al atoms. The plasma power was kept at 300 W and the nitrogen flow rate was 1.0 sccm. Sapphire wafers with c-plane orientation were used as substrates. These wafers were first nitridated at 870°C for 2 h, and a 60-nm-thick AlN buffer layer was then deposited at the same temperature. An intermediary 500-nm-thick Ga-polar GaN buffer layer was grown at 770°C under slightly Ga-rich conditions. At the end of the intermediate GaN layer growth, the GaN surface was always nitridated for sufficient time to consume any Ga adlayer until a clear 2 × 2 reconstruction RHEED pattern was observed. Then the N-flux was interrupted and the substrate temperature was dropped to 425 °C for InN QD layer deposition. Before the InN QD deposition, a 3 × 1 RHEED pattern was typically observed for Ga-polar GaN at this low temperature. The relatively low temperature was chosen primarily because of the low decomposition temperature of In-polar InN. For InN deposition, high F_N/F_{In} ratio was used because of the high vapor pressure of nitrogen. The typical equivalent growth rate for InN deposition was ~ 0.05 nm/s. The sample was annealed for 5 min at the growth temperature under N atmosphere before cooling down at the rate of 15 °C/min. The results reported here correspond to a total equivalent coverage of ~9 InN MLs and ~ 18 MLs (where 1 InN ML ~ 0.285 nm).

RHEED patterns with the electron-beam incident along the $[11\bar{2}0]$ azimuth, recorded during and after InN layer growth, had features that were attributable to the GaN buffer layer (streaks) as well as the InN QDs (spotty pattern), as shown in Figs. 15.6 (a) and (b), for the InN QD samples with 9 ML InN coverage and 18 ML InN coverage respectively. Coexistence of both features in the RHEED pattern indicates that the InN did not cover the entire GaN surface, thus suggesting the likelihood that no InN wetting layer was formed. AFM images and cross-sectional transmission electron micrographs, as described in the following section, also confirmed this apparent absence of an InN wetting layer.

15.3.2 Characterization of InN QDs

AFM was extensively used *ex situ* to characterize the InN QD size and morphology. TEM provided valuable microstructural information concerning the morphology of the InN QDs. Samples were prepared for observation by standard mechanical polishing and dimpling, followed by Ar ion-milling with the specimen held at liquid nitrogen temperature to avoid In droplet formation. Observations were made in both

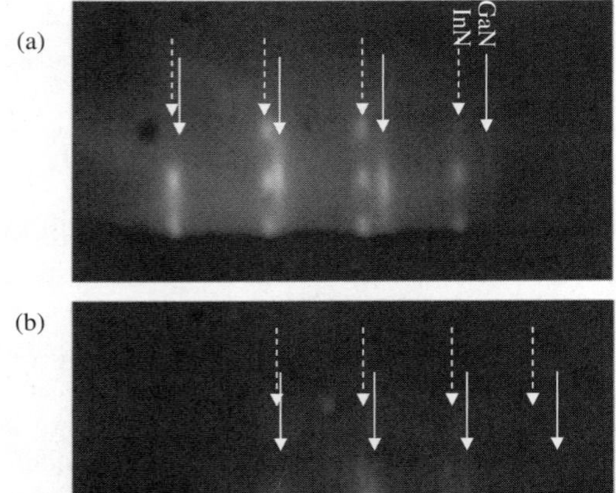

Fig. 15.6 RHEED pattern with the electron-beam incident along the $[11\bar{2}0]$ azimuth taken from InN QD/GaN buffer layer samples during growth showing coexistence of streaks from GaN and spots from InN QDs (a) $\Theta = 9$ MLs; (b) $\Theta = 18$ MLs

plan-view and cross-section geometries with a JEM-4000EX high-resolution electron microscope ($C_S = 1.0$ mm), equipped with a top-entry, double-tilt specimen holder.

15.3.2.1 InN QD Size Distribution

Figures 15.7(a) and (b) are the AFM images in 10 μm × 10 μm area for the InN QDs with InN coverage of 9 MLs and 18 MLs, respectively. Both images show a random distribution of InN QDs. The analysis of the size distribution and density of the InN QDs is summarized in Table 15.1. For the InN QD sample with $\Theta = 9$ MLs, particle analysis of the AFM micrograph (Fig. 15.7(a)) showed that the QD density is $\sim 2 \times 10^9/\text{cm}^2$, whilst the mean length, width, and height dimensions are 157nm, 72 nm, and 14.8 nm, respectively. For the InN QD sample with $\Theta = 18$ MLs, the QD density, mean length, width and height are $1.3 \times 10^9/\text{cm}^2$, 234 nm, 95 nm, 21 nm, respectively. Estimates of the total InN volume per unit substrate area based on these average QD dimensions and surface density gave values close to the nominal as-deposited 9 MLs and 18 MLs, once again confirming the absence of any substantial amounts of InN in regions between the QDs. As the InN coverage increases, the dot mean height and mean diameter both increase whilst the dot density decreases probably due to the dot coalescence at higher coverage.

15.3.2.2 InN QD Microstructure

Standard *g.b* diffraction contrast analysis (Hirsch et al. 1965) revealed that dislocations of either edge- or mixed edge-screw type are the primary threading defects present in the GaN buffer layers. Plan-view observations revealed that the dislocation densities for these relatively thin GaN films is typically in the range of 2 to $5 \times 10^{10}/\text{cm}^2$.

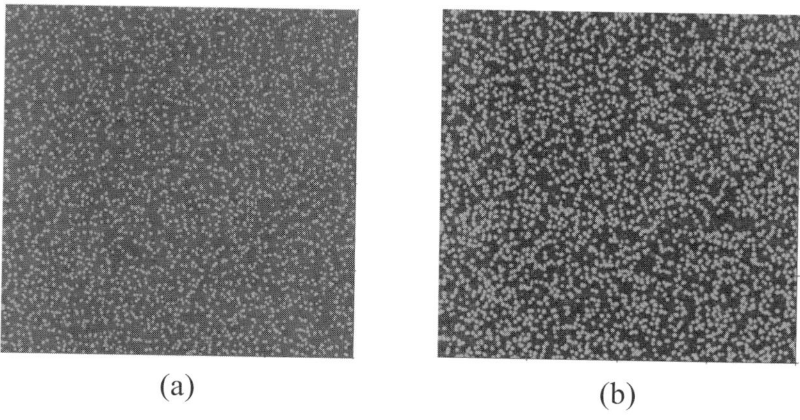

Fig. 15.7 μm × 10 μm AFM images of InN QDs with (a) $\Theta = 9$ MLs and (b) $\Theta = 18$ MLs

Table 15.1 Summary of InN QD density and size distribution for the InN QDs with $\Theta = 9$ MLs and $\Theta = 18$ MLs from the AFM image analysis

-	$\Theta = 9$ MLS	$\Theta = 18$ MLS
Dot Density (cm^{-2})	2×10^9	1.3×10^9
Mean Height (nm)	14.8	20.6
Mean Diameter (nm)	115	163
Mean Length (nm)	157	233
Mean Width (nm)	72	95

Figure 15.8(a) is a low magnification, diffraction-contrast electron micrograph showing the cross sections of several InN QDs, as viewed along a $[11\bar{2}0]$ projection of the GaN buffer layer, and Fig. 15.8(b) is a higher magnification image showing a larger InN island. These electron micrographs are representative of the general finding that almost all of the QDs and islands had nucleated directly above threading dislocations (TDs) present in the GaN buffer layer. Similar association of GaN QDs with threading dislocations in AlN has been reported previously (Rouviere et al. 1999). It was also observed that very few of the QDs showed any visible evidence for faceting, either in cross-sectional electron micrographs or in images obtained by AFM.

Detailed examination at higher magnification confirmed that the vast majority of the InN QDs had the wurtzite crystal structure, and that the QDs also maintained a well-defined epitaxial relationship with the underlying GaN buffer layer. This epitaxy was also apparent from selected–area electron diffraction patterns recorded in the plan-view geometry. Small isolated regions with cubic structure were visible in some QDs, usually at positions where the GaN buffer layer underneath the QDs was not perfectly flat. Periodic arrays of moiré fringes with spacings of ~ 2.8 nm were visible along the InN QD/GaN buffer layer interface, as would be expected for perfect InN/GaN heteroepitaxy. Lattice-fringe imaging confirmed that this periodicity also corresponded to the spacing between the misfit dislocations that would accommodate the lattice mismatch between the two materials. Similar findings relating to

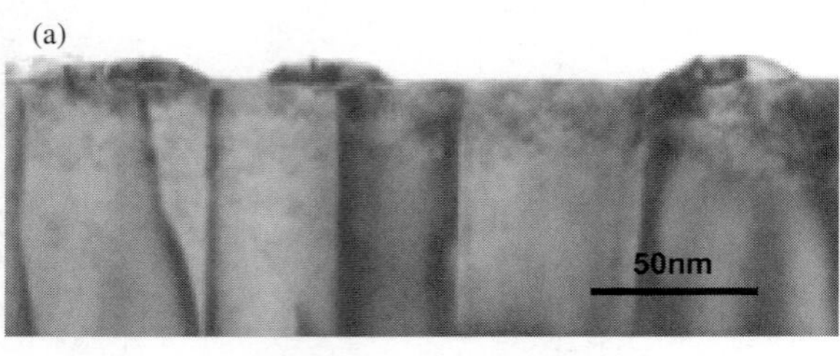

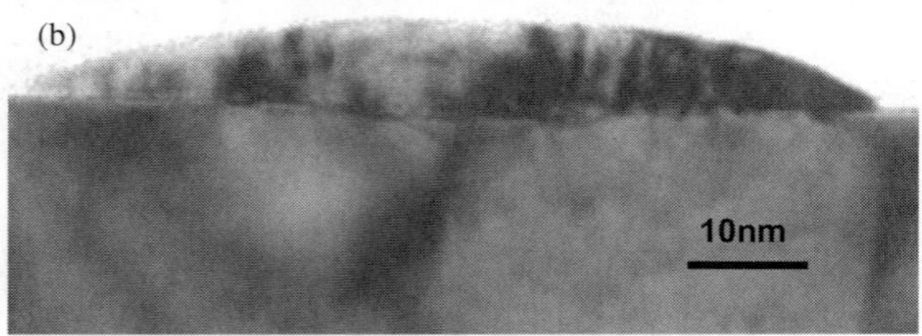

Fig. 15.8 Cross-sectional electron micrographs showing epitaxial InN QDs, grown by MBE at 425 °C. Note contrast from threading dislocations in the GaN buffer layer. (Zhou et al. 2006)

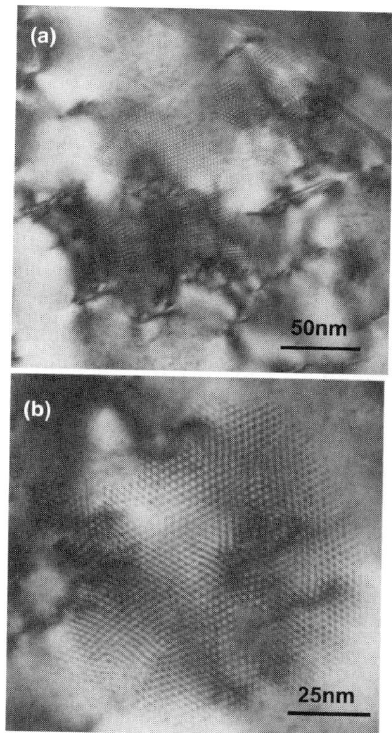

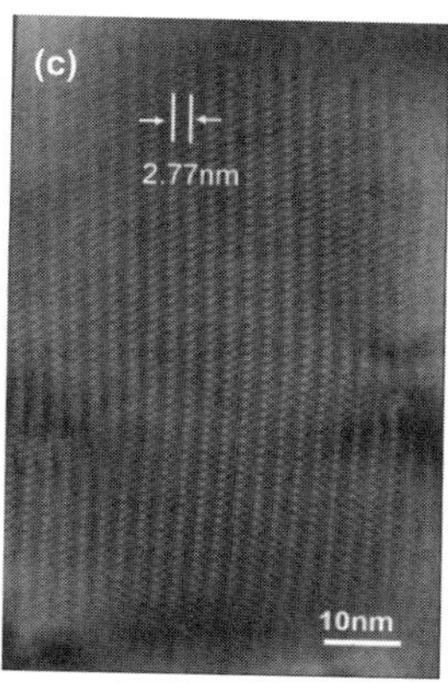

Fig. 15.9 Plan-view electron micrographs of epitaxial InN QDs, grown on GaN buffer layer. Note hexagonal array of moiré fringes with spacings that correspond to the mismatch between the lattice spacings of InN and GaN, as well as disruptions in the fringe periodicity caused by GaN threading dislocations. (Zhou et al. 2006)

the separation of misfit dislocations for the growth of epitaxial InN thin films on GaN (0001) have been recently reported (Kehajias et al. 2005).

Plan-view imaging of the InN QDs provides an alternative perspective on the InN QD/GaN buffer-layer epitaxy. As shown by the examples in Figures 15.9(a) and (b), overlap of the two wurtzite materials in projection gives rise to a primarily hexagonal array of moiré fringes, again with an average spacing of ~ 2.8 nm (see Fig. 15.9(c)), which would correspond to the separation between misfit dislocations at a relaxed InN QD/GaN buffer interface. Note, however, the disruptions in the fringe periodicity that are clearly visible in Fig. 15.9(b), are attributable directly to the effects of threading dislocations penetrating to the surface of the underlying GaN buffer layer.

15.4 InGaN QDs

The lattice mismatch between $In_xGa_{1-x}N$ and GaN depends on the InN mole fraction in the $In_xGa_{1-x}N$ alloys. It is known that when $In_xGa_{1-x}N$ ($x > 18\%$) is grown on GaN, 3D islands are formed automatically to reduce the strain energy (Adelmann et al. 2000). Daudin et al. (Daudin et al. 2000) proposed a quantum dot nucleation mechanism for InGaN. They observed that below an In content of about 18%, the growth is 2D, above 18% In, SK growth mode is observed, the critical thickness being about 2–3 MLs. They found that in the case of InN deposited on GaN, no photoluminescence from the dots was observed. They attributed this phenomenon to the formation of interface dislocations at the border of the InN islands, as evidenced by electron microscopy, i.e. InN exhibits an incoherent SK growth mode on GaN. Hence, as far as InGaN alloys are concerned, an upper limit in the In content is expected which should correspond to the transition between coherent and incoherent SK mode of growth. Downes et al. (Downes et al. 1994) calculated by numerical integration the energy of a strained layer in an infinite isotropic continuous elastic medium with and without a misfit dislocation dipole of pure edge character. The equilibrium critical thickness is the thickness at which these energies are the same. Thus they proposed an expression to predict equilibrium critical thickness. The deposition thickness of InGaN on GaN should be in between two critical thicknesses ($t_1 < t < t_2$): it should be larger than t_1 at which the 2D to 3D transition starts to occur, and it should be smaller than t_2 at which misfit dislocations start to generate in the InGaN film (Daudin et al. 2000). Within this thickness range, coherent and dislocation-free InGaN dots are formed.

There are many reports on the formation of InGaN QDs. Self-assembled InGaN QDs have been grown in the Stranski-Krastanov mode by plasma-assisted MBE (Adelmann et al. 2000, Yamaguchi et al. 2005, Smeeton et al. 2006), NH_3 MBE (Grandjean and Massies 1998, Dalmasso et al. 2000) and MOCVD (Tachibana et al. 1999). A number of growth techniques were studied, including post-growth nitrogen annealing (Oliver et al. 2003), the use of In (Zhang et al. 2002) and Si as anti-surfactants (Hirayama et al. 1999), the employment of low temperature passivation (Chen et al. 2002, Qu et al. 2003) etc. Moreover, there are numerous reports on InGaN QDs formed in InGaN QWs that enhance exciton localization and hence the internal quantum efficiency of LEDs. In this section, the study on self-assembled InGaN QDs grown on GaN by PAMBE is reported.

15.4.1 PAMBE Growth of InGaN QDs and MQDs

All of the InGaN QD and MQD samples were grown in a Varian Gen II system which uses standard effusion cells for Ga, In, Si and Mg. An EPI RF plasma source was used as the nitrogen source, operated at 300 W with a N_2 gas flow rate of 1.0 sccm. Quarters of 2 inch c-plane sapphire wafers coated previously with GaN by MBE were used as substrates. A 100-nm-thick undoped GaN layer was overgrown on the GaN template at 770 °C under Ga-rich condition after which the remaining Ga on the surface was evaporated in vacuum at 770 °C. Then the substrate temperature was ramped down to various lower temperatures between 520 ° C to 620 °C for InGaN QDs growth. For InGaN MQD samples, GaN barrier was grown with the plasma power of 230 W and the nitrogen flow rate of 0.5 sccm, while the InGaN QD layer was grown with the plasma power of 300 W and the nitrogen flow rate of 1.0 sccm.

15.4.2 Characterization of InGaN QDs and MQDs

The surface dot density and height of the InGaN QD samples were studied using AFM. The optical properties of these InGaN QDs were investigated using cathodoluminescence (CL) measurements at 3 kV acceleration voltage. The InGaN MQD samples were characterized by high-resolution θ-2θ x-ray diffraction (HRXRD) scans (Philips Four Circle Diffractometer) using Cu $K_{\alpha 1}$ radiation and CL measurements at 13 kV acceleration voltage.

15.4.2.1 Growth Temperature Effect

The growth temperature effect on the formation of InGaN QDs was studied by growing InGaN thin films at different temperatures and characterizing them using AFM and CL measurements. Figures 15.10 (a)–(c) show the AFM surface height images and the corresponding CL spectra of InGaN dots samples grown at 520 °C, 550 °C and 580 °C with nominal InGaN coverages (Θ) of 12 MLs, 12 MLs and 32 MLs, respectively. In all the three samples, the beam equivalent pressure of In and Ga were kept the same and the InGaN materials were deposited under nitrogen-rich conditions.

For the InGaN QD sample grown at 520 °C with 12 ML InGaN coverage, Fig. 15.10(a) clearly shows dot formation, with an average dot height of 3 nm, an average dot diameter of 30 nm and a dot density of 7×10^{10} cm^{-2}. The CL emission energy of this sample shown in Fig. 15.10(d) has a peak at 2.58 eV with FWHM of 308 meV. The InGaN QD sample grown at 550 °C with also 12 ML

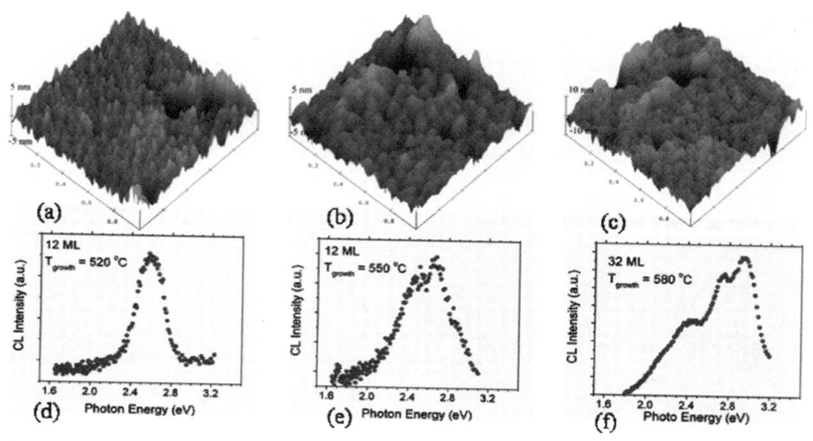

Fig. 15.10 AFM images of InGaN films grown at different temperatures and the corresponding CL spectra: (a) and (d) $T_g = 520\,^\circ C$; (b) and (e) $T_g = 550\,^\circ C$; (c) and (f) $T_g = 580\,^\circ C$

InGaN coverage (see Fig. 15.10(b)) has a smoother surface with a larger average dot diameter of 70 nm, a lower average dot height of 2.3 nm and a lower dot density of 1.5×10^{10} cm^{-2} than the one grown at 520 °C. Its CL spectra, as shown in Fig. 15.10(e), have 2 peaks at 2.65 eV and 2.51 eV, respectively. The sample grown at 580 °C with 32 ML InGaN coverage does not show dot formation as indicated in the AFM image of Fig. 15.10(c). Its CL spectra shown in Fig. 15.10(f) are quite broad with 3 peaks at 2.96 eV, 2.73 eV and 2.49 eV, respectively.

It should be pointed out that although these films were grown using the same In and Ga fluxes, the composition of the InGaN materials grown at different temperatures is different: the film grown at 580 °C has the least InN mole fraction while the film grown at 520 °C has the highest InN mole fraction. The difference in the InGaN composition induces different strain in InGaN with respect to the underlying GaN template. The lattice mismatch and the growth temperature may both contribute to the critical thickness difference of these InGaN films. These results are consistent with Yamaguchi's (Yamaguchi et al. 2005) observations that for higher growth temperatures, the critical layer thickness increases.

15.4.2.2 Structural and Optical Properties of InGaN MQDs

The structural and optical properties of an InGaN/GaN MQD sample containing ten stacks of InGaN QDs, each of which was grown with 12 ML InGaN coverage, are discussed in this section. The on-axis XRD profile around the (0002) Bragg peaks of InGaN and GaN layers is shown in Fig. 15.11(a). The red line is the result of a kinematical simulation of the multiple quantum well structure. The composition and thickness of the InGaN dot layer and GaN barrier can be calculated from the simulation. The InGaN QD composition was determined to be 41% and the InGaN QDs and barrier thicknesses were found to be 4 nm and 6 nm respectively. The CL emission peak of this sample was found to be at 2.25 eV with the FWHM of

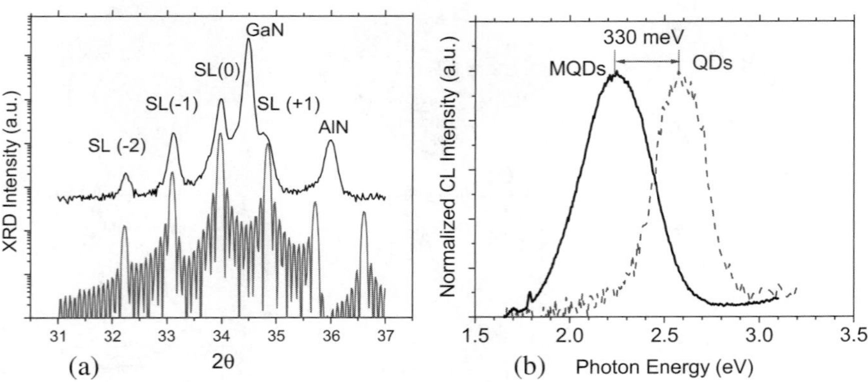

Fig. 15.11 (a) HRXRD θ-2θ scan measured from the 10-period $In_{0.41}Ga_{0.59}N$/GaN MQDs; (b) The CL emission spectra of the same sample has a peak at 2.25 eV with the FWHM of 415 meV

415 meV. Comparing with the CL spectra of the single layer of InGaN QDs grown at 520 °C shown in Fig. 15.11(b), the InGaN MQD sample emission peak is red shifted by 330 meV and broadened by 107 meV. The red shift is probably due to the Quantum Confined Stark effect (Perlin et al. 1998, Bai et al. 2001). The InGaN QDs are sandwiched between GaN barriers in the MQD sample and are highly strained due to the large lattice mismatch. The large piezoelectric field induced by the strain tilts the potential profile and makes the optical transition energy to be red-shifted (QCSE).

15.5 Application of III-nitride QDs in LEDs

Visible light emitting diodes (LEDs) are generally based on InGaN multiple quantum wells (MQWs) (Nakamura et al. 1995, Chichibu et al. 2000, Lester et al. 1998, Nakamura 1996, Laukkanen et al. 2001). The internal quantum efficiency of III-nitride LEDs decreases substantially with the percentage of InN mole fraction in the InGaN/GaN QWs. It is likely that the origin of this reduction of efficiency is related to problems associated with the polarization effects in the QWs, which gives rise to the Quantum Confined Stark effect, or/and problems associated with the crystal and defect structure of the InGaN alloys in the wells. Alloy phenomena, such as phase separation, have been observed both in bulk InGaN films as well as InGaN/GaN MQWs (Singh et al. 1997, Romano et al. 1999). It has been reported that such phenomena give rise to potential fluctuations, which are beneficial to emission from LEDs due to exciton localization (Chichibu et al. 1996). Another alloy phenomenon, which is an alternative mechanism of strain relief, is atomic ordering (Doppalapudi et al. 1999, Misra et al. 1999). While the first observations of ordering in these materials were simple alternating Ga-rich and In-rich (0001) layers in the InGaN, it was reported later that more complex ordering exhibits ordered superlattice structures that are incommensurate with the wurtzite crystal lattice (Wang et al. 2006). Doppalapudi et al. found that ordering occurs in certain crystal domains, while others maintain their random alloy structure (Doppalapudi et al. 1999). Experimental studies (Misra et al. 1999) as well as theoretical studies (Dudiy and Zunger 2004) indicate that the energy gaps of the ordered and random domains are different and form a type II heterostructure. As a result, when electrons and holes are injected either optically or electrically, they become spatially separated and thus unable to recombine efficiently. Therefore inhomogeneous alloy ordering is undesirable for devices such as LEDs or lasers. The lateral dimension of the ordered and random domains was found to be around 100 nm (Doppalapudi et al. 1999). Thus by confining InGaN in nanoscale-structures such as quantum dots, it is possible that inhomogeneous alloy ordering can be eliminated and hence the LED internal quantum efficiency can be enhanced.

InGaN/GaN multi-quantum dot LEDs and lasers grown by MOVPE (Ji et al. 2004, Tachibana et al. 1999) and ammonia-source MBE (Damilano et al. 2000) methods have been reported. However, there are only limited reports on the device application of InGaN QDs grown by plasma-assisted MBE. In this section, the application of III-nitride QDs in visible LED structures is reported. Specifically, GaN QDs were used in the nucleation stage of LED structures as a dislocation filtering layer and InGaN QDs were used in the active region of LED structures to enhance carrier localization and eliminate detrimental alloy phenomena such as partial atomic ordering.

The surface of the as-grown LED epitaxial layers was partially etched until the n-type GaN layer was exposed. A V (15 nm)/Al (85 nm)/V (20 nm)/Au (9 nm) contact was deposited onto the exposed n-GaN layer to serve as the n-electrode (France et al. 2007), and a Ni (5 nm)/Au (20 nm) contact was evaporated onto the p-GaN layer to serve as the p-electrode. The current-voltage (I-V) measurements were performed at room temperature using an HP semiconductor parameter analyzer. The electroluminescence (EL) output intensity was measured by injecting 1 kHz 50% duty cycle pulsed current into the device.

15.5.1 Application of GaN QDs to LEDs

As it is discussed in Section 15.2.2.4, GaN QDs can be used in the nucleation stage of LED structure to reduce dislocation density in the GaN cladding layer. Figure 15.12 shows a schematic of an LED structure. It consists of a 60-nm thick AlN nucleation layer grown at 870 °C, a GaN QD dislocation filtering layer grown at 770 °C, a 3-μm-thick Si-doped GaN cladding layer grown at 770 °C, a 3-period $In_{0.18}Ga_{0.82}N$/GaN MQW active region grown at 620 °C, a 10-nm thick Mg-doped $Al_{0.2}Ga_{0.8}N$ electron blocking layer grown at 620 °C and a 250-nm thick Mg-doped GaN cap layer grown at 770 °C.

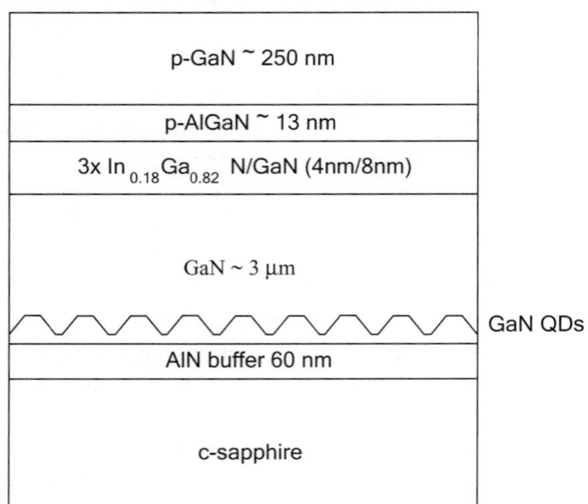

Fig. 15.12 Schematic of a blue LED structure with GaN QDs in the buffer

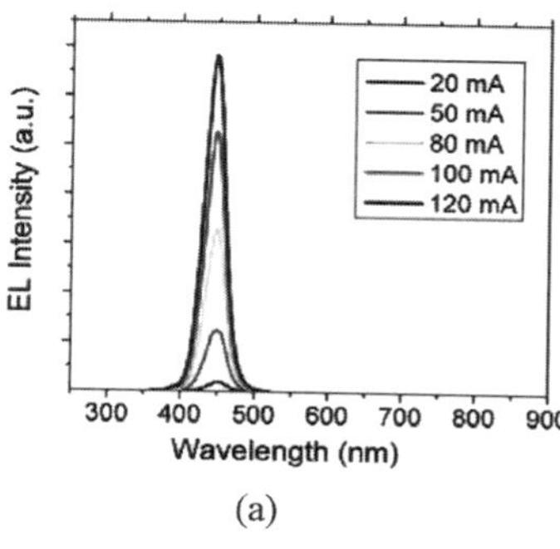

(a)

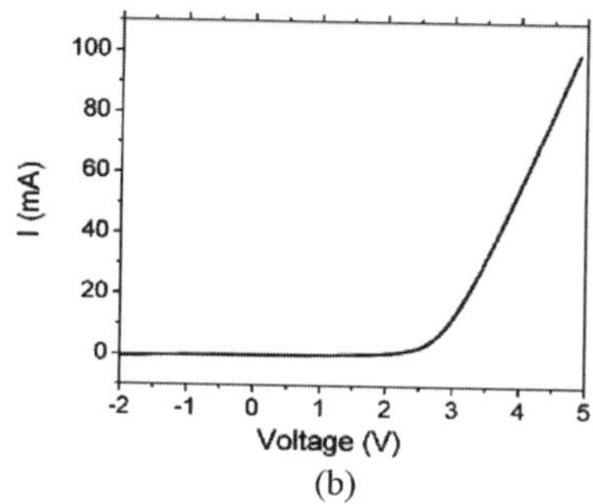

(b)

Fig. 15.13 (a) Electroluminescence spectra of a blue LED under various pulsed injection current; (b) I-V characteristics of a blue LED

Figure 15.13(a) shows representative electroluminescence (EL) spectra of a blue LED having the structure described in Fig. 15.12. The EL peak occurs at 440 nm with FWHM of 30 nm under 20 mA pulsed-current injection. The I-V characteristics of this device are shown in Fig. 15.13(b). It has a threshold voltage of 3.2 V and a series resistance of 23 Ohm.

Another LED structure which does not have GaN QDs in the nucleation layer was also investigated to study the influence of the GaN QD layer on the device performance. It was found that the integrated EL intensity of the blue LED with GaN QDs in the buffer is six times of that of the device without GaN QDs in the buffer at an injection current of 120 mA.

15.5.2 Application of InGaN QDs to LEDs

A schematic of the InGaN/GaN MQD LED structure is given in Fig. 15.14. It consists of a 3-μm thick Si-doped GaN cladding layer with a GaN QD dislocation filtering layer underneath, a 5-period $In_{0.41}Ga_{0.59}N$/GaN MQD active region grown at 520 °C, a 10-nm thick Mg-doped $Al_{0.2}Ga_{0.8}N$ electron blocking layer grown at 800 °C and a 250-nm thick Mg-doped GaN layer grown at 770 °C. Both the green and the red LEDs were made using 5 pairs of InGaN/GaN MQDs with the InGaN composition of 41% and 46%, respectively. From the simulation of XRD θ-2θ scan, the InGaN QD layer height and barrier thickness were estimated to be 4 nm and 6 nm, respectively.

Figure 15.15(a) shows that the green LED emits at 560 nm with FWHM of 87 nm while Fig. 15.15(b) shows that the red LED emits at 640 nm with FWHM of 97 nm under 20 mA injection current (50% duty cycle, 1 kHz). The I-V characteristics of the green InGaN/GaN MQD LED has a turn-on voltage of 2.6 V and a series resistance of 19.2 Ω (Fig. 15.15(c)), while that of the red InGaN/GaN MQD LED has a turn-on voltage of 2.35 V and a series resistance of 26.5 Ω (Fig.15.15(d)).

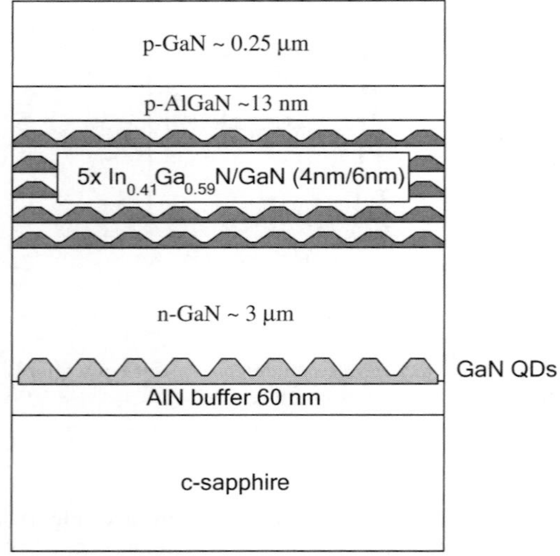

Fig. 15.14 Schematic of the InGaN/GaN MQD LED structure

15.5.3 Discussion

Figure 15.16(a) illustrates the EL peak position as a function of the injection current for the blue, green and red LEDs. Fig. 15.16(b) shows the emission peak wavelength dependence on the injection current for the blue, green and red LEDs. For the blue InGaN/GaN MQW LEDs, the emission photon energy is blue shifted by 25 meV when the injection current increases from 20 mA to 90 mA. For the green InGaN/GaN MQD LEDs, the EL peak is blue shifted by 112 meV as the injection current increases from 26 mA to 70 mA. For the red InGaN/GaN MQD LEDs, the emission energy is blue shifted by 127 meV as the injection current increases from 25 mA to 72 mA. Large blue shift in the EL peak position with increasing injection current were found for an MOVPE grown blue InGaN/GaN MQD LED (Ji et al. 2004). There are two possible reasons for this large EL blue shift with increasing injection current in the MQD LEDs: (A) the band-filling effect (Ding et al. 1997, Narukawa et al. 1997, Chichibu et al. 2000, Tribe et al. 1997); (B) the screening of the piezoelectric field by the high injection current (Perlin et al. 1998, Bai et al. 2001, Bunker et al. 2005). For the blue LEDs based on MQWs, because the In% in the QWs is only 18%, the degree of compositional inhomogeneities is expected to be small and thus the observed blue shift is primarily due to screening of the piezoelectric field at high injection. On the other hand, for the green and red LEDs whose In% in the active region are 41% and 46%, one expect more compositional inhomogeneities. And thus both mechanisms are responsible for the observed blue shift of the spectra.

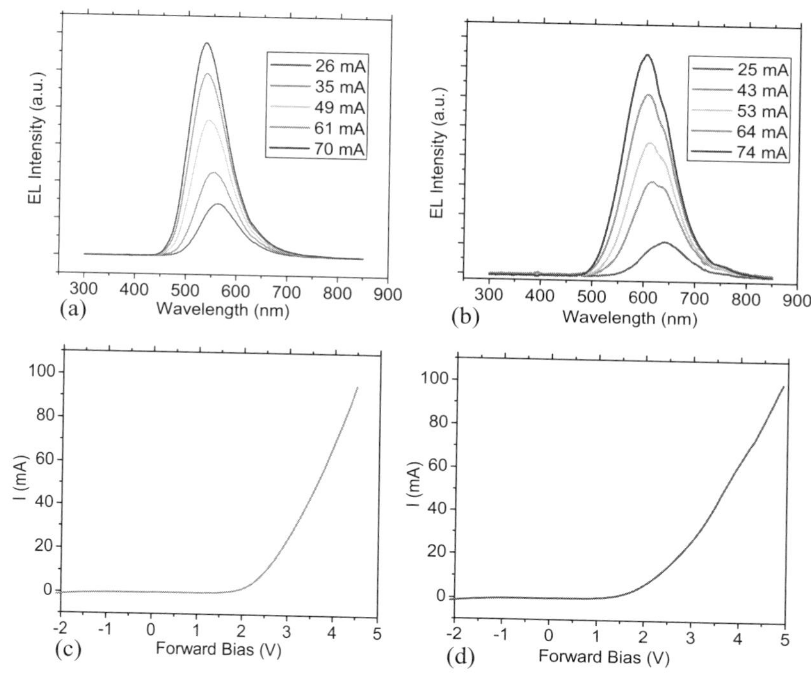

Fig. 15.15 (a) EL spectra of a green InGaN/GaN MQD LED; (b) EL spectra of a red InGaN/GaN MQD LED; (c) I-V characteristics of the green LED showing $V_t = 2.6\,V$, $R_s = 19.2\,\Omega$; (d) I-V characteristics of the red LED showing $V_t = 2.35\,V$, $R_s = 26.5\,\Omega$

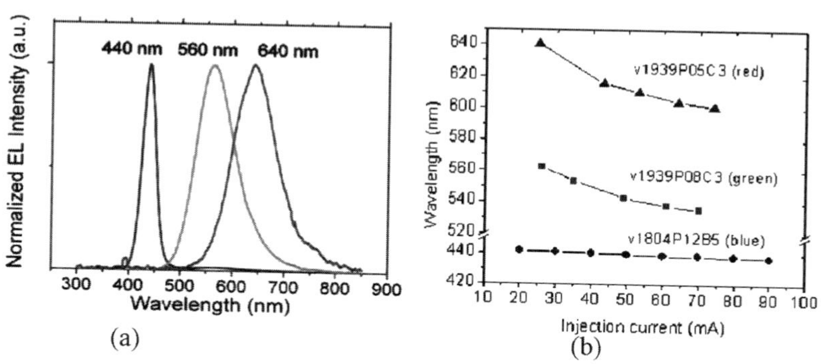

Fig. 15.16 (a) EL spectra of the blue, green and red LEDs under the injection of 50% duty cycle 20 mA pulsed current; (b) EL peak position dependence on the injection current (50% duty cycle) for the blue, green and red LEDs

15.6 Summary

This chapter discusses the growth and characterization of III-nitride QDs and their application to LEDs. GaN QDs were grown on an AlN buffer by PAMBE in the modified Stranski-Krastanov mode. It was found that the bimodal height distribution of GaN QDs can be attributed to the existence of high density of threading dislocations in the buffer: the larger GaN QDs are located adjacent to edge-type dislocations, while the smaller dots are located in the dislocation-free regions. Growth on vicinal surfaces of miscut sapphire substrates shows evidence of QD aligning along the steps. Contrary to GaN QDs grown on AlN, TEM studies show that InN QDs grown on a GaN buffer to be free of a wetting layer and fully relaxed, a result attributed to the larger lattice mismatch between GaN and InN (Volmer-Weber Mode). Misfit dislocations at the InN/GaN interface serve as a strain relief mechanism for the relaxed InN dots. The majority of the InN islands are associated with threading dislocations in the GaN buffer layer having edge-type components. The formation of InGaN QDs on a GaN buffer is more difficult and the mode of growth should depend on the InN mole fraction. AFM studies indicate that InGaN QDs with up to 40% InN mole fraction are formed on a GaN buffer, and their density and size depend strongly on deposition temperature. It was found that the cathodoluminescence emission spectra are significantly red shifted in multiple layers of InGaN/GaN quantum dots, a result attributed to Quantum Confined Stark effect due to the additional stress on QDs by the GaN barriers.

LED structures based on InGaN/GaN MQWs, emitting at 440 nm, were grown with or without GaN QDs in the nucleation layer. TEM studies found that the inclusion of QDs in the nucleation layer acts as a dislocation filtering mechanism leading to superior devices. Green and red LEDs based on InGaN MQDs emitting at 560 nm and 640 nm, respectively, were grown. Such structures were fabricated into 800 μm × 800 μm LED devices using standard photolithography and metallization schemes. The electroluminescence spectra of these devices were investigated as a function of injection current. The electroluminescence spectra of the green and red InGaN MQD LEDs show larger blue-shift with increasing injection current than the blue InGaN/GaN MQW LEDs. Both compositional inhomogeneities and screening of the internal fields due to piezoelectric polarization are considered to be the source of the observed large blue-shift in the EL spectra of the green and red LEDs.

Acknowledgments This work was partially supported by the U.S. Department of Energy under the Solid State Light Initiative.

References

Adelmann C, Daudin B, Oliver RA, Briggs GAD, Rudd RE (2004) Phys. Rev. B 70: 125427

Adelmann C, Simon J, Feuillet G, Pelekanos NT, Daudin B, Fishman G (2000) Appl. Phys. Lett. 76: 1570

Arley M, Rouviere JL, Widmann F, Daudin B, Feuillet G, Mariette H (1999) Appl. Phys. Lett. 74:3287

Bai J, Wang T, Sakai S (2001) J. Appl. Phys. 90: 1740
Brault J, Tanaka S, Sarigiannidou E, Rouviere JL, Daudin B, Feuillet G, Nakagawa H (2003) J. Appl. Phys. 93:3108
Briot O, Maleyre B, Ruffenach S (2003) Appl. Phys. Lett. 83:2919
Brown J, Wu F, Petroff PM, Speck JS (2004) Appl. Phys. Lett. 84: 690
Bunker KL, Garcia R, Russell PE (2005) Appl. Phys. Lett. 86: 082108
Cao YG, Xie MH, Liu Y, Xu SH, Ng YF, Wu HS, Tong SY (2003) Appl. Phys. Lett., 83: 5157
Chen Z, Lu D, Yuan H, Han P, Liu X, Li Y, Wang X, Lu Y, Wang Z (2002) J. Cryst. Growth 235:188
Chichibu S,Azuhata T, Sota T, Nakamura S (1996) Appl. Phys. Lett. 69:4188
Chichibu S. Azuhata T, Sota T, Mukai T, Nakamura S (2000) J. Appl. Phys., 88:5153
Dalmasso S, Damilano B, Grandjean N, Massies J, Leroux M, Reverchon JL, Duboz JY (2000) Thin Solid Films 380:195
Damilano B, Grandjean N, Massies J, Siozade L, Leymarie J (2000) Appl. Phys. Lett. 77:1268
Daudin B, Feuillet G, Mula G, Mariette H, Rouviere JL, Pelekanos N, Fishman G, Adelmann C, Simon J (2000) Diamond and Related Materials 9:506
Dimakis E, Iliopoulos E, Tsagaraki K, Kehajias T, Komninou P, Georgakilas A (2005) J. Appl. Phys. 97:113520
Ding YJ, Reynolds DC, Lee SJ, Khurgin JB, Rabinovich WS, Katzer DS (1997) Appl. Phys. Lett. 71: 2581
Doppalapudi D, Basu SN, Moustakas TD (1999) J. Appl. Phys. 85: 883
Downes JF, Dunstan DJ, Faux DA (1994) Semicond. Sci. Technol. 9: 1265
Dudiy SV, Zunger A (2004) Appl. Phys. Lett. 84:1874
Gogneau N, Jalabert D, Monroy E, Shibata T, Tanaka M, Daudin B (2003) J. Appl. Phys. 94: 2254
Grandjean N, Massies J (1998) Appl. Phys. Lett. 72:1078
Hirayama H, Tanaka S, Aoyagi Y (1999) Microelectronic Engineering 49: 287
Hirsch PB, Howie A, Nicholson RB, Pashley DW, Whelan MJ (1965) *Electron Microscopy of Thin Crystals*, Butterworths, London
Ji LW, Su YK, Chang SJ, Chang CS, Wu LW, Lai WC, Du XL, Chen H (2004) J. of Cryst. Growth 263:114
Kehajias T, Delimitis A, Komninou P, Iliopoulos E, Dimakis E, Georgakilas A, Nouet G (2005) Appl. Phys. Lett. 86:151905
Laukkanen P, Lehkonen S, Uusimaa P, Pessa M, Seppala A, Ahlgren T, Rauhala E (2001) J. Cryst. Growth 230:503
Lester SD, Ludowise MJ, Killeen KP, Perez BH, Miller JN, Rosner SJ (1998) J. Cryst. Growth 189/190: 786
Lozano JG, Sanchez AM, Garcia R, Gonzalez D, Araujo D, Ruffenach S, Briot O (2005) Appl. Phys. Lett. 87:263104
Lu H, Schaff WJ, Hwang J, Wu H, Koley G, Eastman LF (2001) Appl. Phys. Lett. 79:1489
Maleyre B, Briot O, Ruffenach S (2004) J. Cryst. Growth 269:15
Misra M, Korakakis D, Ng HM, Moustakas TD (1999) Appl. Phys. Lett. 74:2203
Monemar B, Paskov PP, Kosic A (2005) Superlattices Microstructures 38:38
Nakamura S (1996) Diamond and Related Materials 5:496
Nakamura S, Senoh M, Iwasa N, Nagahama S (1995) Jpn. J. of Appl. Phys. 34: L797
Nanishi Y, Saito Y, Yamaguchi T, Hori M, Matsuda F, Araki T, Suzuki A, Miyajima T (2003) Phys. Stat. Sol. (a) 200:202
Narukawa Y, Kawakami Y, Funato M, Fujita S, Fujita S, Nakamura S (1997) Appl. Phys. Lett. 70:981
Oliver RA, Briggs GAD, Kappers MJ, Humphreys CJ, Yasin S, Rice JH, Smith JD, Taylor RA (2003) Appl. Phys. Lett. 83:755
Perlin P, Kisielowski C, Iota V, Weinstein BA, Mattos L, Shaprio NA, Kruger J, Weber ER, Yang J (1998) Appl. Phys. Lett. 73:2778
Qu B, Chen Z, Lu D, Han P, Liu X, Wang X, Wang D, Zhu Q, Wang Z (2003) J. Cryst. Growth 252:19

Romano LT, McCluskey MD, Van de Walle CG, Northrup JE, Bour DP, Kneissl M, Suski T, Jun J (1999) Appl. Phys. Lett. 75:3950
Rouviere JL, Simon J, Pelekanos N, Daudin B, Feuillet G (1999) Appl. Phys. Lett. 75: 2632
Ruffenach S, Maleyre B, Briot O, Gil B (2005) Phys. Stat. Sol. (c) 2:826
Sarigiannidou E, Monroy E, Daudin B, Rouviere JL, Andreev AD (2005) Appl. Phys. Lett.87: 203112
Singh R, Doppalapudi D, Moustakas TD, Romano LT (1997) Appl. Phys. Lett. 70:1089
Smeeton TM, Senes M, Smith KL, Hooper SE, Heffernan J (2006) MRS Fall meeting 2006, I7.28
Tachibana K, Someya T, Arakawa Y (1999) Appl. Phys. Lett. 74:383
Tribe WR, Steer MJ, Mowbray DJ, Skolnick MS, Forshaw AN, Roberts JS, Hill G, Pate MA, Whitehouse CR, Williams GM (1997) Appl. Phys. Lett. 70:993
Vurgaftman I, Meyer JR, Ram-Mohan LR (2001) J. Appl. Phys. 89:5815
Walukiewicz W, Li SX, Yu KM, Ager JW, Haller EE, Lu H, Schaff WJ (2004) J. Cryst. Growth 269:119
Wang XQ, Yoshikawa A (2004) Prog. Cryst. Growth and Characterization Mater. 48–49:42
Wang YY, Ozcan AS, Ludwig KF, Bhattacharyya A, Moustakas TD, Zhou L, Smith DJ (2006) Appl Phys. Lett. 88:181915
Wu J, Walukiewicz W, Yu KM, Ager JW, Haller EE, Lu H, Schaff WJ (2002) Appl. Phys. Lett. 80: 4741
Xu K, Yoshikawa A (2003) Appl. Phys. Lett. 83:251
Xu T, Nikiforov AY, France R, Thomidis C, Williams A, Moustakas TD, Zhou L, Smith DJ (2006) *Mat. Res. Soc. Proc.* I. 5.5
Xu T, Williams A, Thomidis C, Zhou L, Smith DJ and Moustakas TD (2005) Mat. Res. Soc. Proc. 831: E2.4
Yamaguchi T, Einfeldt S, Figge S, Kruse C, Roder C, Hommel D (2005) Mat. Res. Soc. Proc. 831:E2.2
Yamamoto A, Tanaka T, Koide K, Hashimoto A (2002) Phys. Stat. Sol. (a) 194:510
Yoshikawa A, Hashimoto N, Kikukawa N, Che SB, Ishitani Y (2005) Appl. Phys. Lett. 86:153115
Zhang J, Hao M, Li P, Chua SJ (2002) Appl. Phys. Lett. 80: 485
Zhou L, Xu T, Smith DJ, Moustakas TD (2006) *Appl Phys. Lett.* 88: 231906

Subject Index

Surface mass transport, 8

Printed in the United States of America